SYSTEMS ANALYSIS

AND DESIGN

An Object-Oriented Approach with UML

ALAN DENNIS

Indiana University

BARBARA HALEY WIXOM

University of Virginia

DAVID TEGARDEN

Virginia Tech

John Wiley & Sons, Inc.

www.wiley.com/college/dennis

CREDITS

ACQUISITIONS EDITOR: Beth Lang Golub
MARKETING MANAGER: Gitti Lindner
SENIOR PRODUCTION EDITOR: Patricia McFadden
SENIOR DESIGNER: Dawn L. Stanley
PRODUCTION MANAGEMENT SERVICES: Hermitage Publishing Services

This book was set in 10.5/12 Times New Roman by Hermitage Publishing Services
and printed and bound by Von Hoffmann Press. The cover was printed by Von Hoffmann Press.

Cover images provided as a courtesy from the following contributors:

- Copyright 1995–2001 ©ComputerJobs.com, Inc. All rights reserved. ComputerJobs.com is a registered trademark of ComputerJobs, Inc.
- Information Systems Hunter, Inc.
- Developers.Net™ is a service of Tapestry.Net™, Inc. Copyright ©1998–2002. All rights reserved.
- ©2001 Critical Mass Systems and www.SystemsAnalyst.com are trademarks of Critical Mass Systems.
- Technojobs.co.uk/Advance Your Technical Career
- Computerwork.com
- Copyright ©Dice.com. Reproduced with permission.
- Techies.com
- Webhire.com is the property of Webhire, Inc.
- Monster.com

This book is printed on acid-free paper. ∞

ISBN 0-471-41387-9

Printed in the United States of America

10 9 8 7 6 5 4 3 2 1

BRIEF CONTENTS

CONTENTS

CHAPTER 5 REQUIREMENTS GATHERING 119

PREFACE

PURPOSE OF THIS BOOK

Systems Analysis and Design (SAD) is an exciting, active field in which analysts continually learn new techniques and approaches to develop systems more effectively and efficiently. However there is a core set of skills that all analysts need to know - no matter what approach or methodology is used. All information systems projects move through the four phases of planning, analysis, design, and implementation; all projects require analysts to gather requirements, model the business needs, and create blueprints for how the system should be built; and all projects require an understanding of organizational behavior concepts like change management and team building. Today, the cost of developing modern software is composed primarily of the cost associated with the developers themselves and not the computers. As such, object-oriented approaches to developing information systems hold much promise in controlling these costs.

Today, the most exciting change to systems analysis and design is the move to object-oriented techniques, which view a system as a collection of self-contained objects that have both data and processes. This change has been accelerated through the creation of the Unified Modeling Language (UML). UML provides a common vocabulary of object-oriented terms and diagramming techniques that is rich enough to model any systems development project from analysis through implementation.

This book captures the dynamic aspects of the field by keeping students focused on doing SAD while presenting the core set of skills that we feel every systems analyst needs to know today and in the future. This book builds on our professional experience as systems analysts and on our experience in teaching SAD in the classroom.

This book will be of particular interest to instructors who have students do a major project as part of their course. Each chapter describes one part of the process, provides clear explanations on how to do it, gives a detailed example, and then has exercises for the students to practice. In this way, students can leave the course with experience that will form a rich foundation for further work as a systems analyst.

OUTSTANDING FEATURES

A Focus on Doing SAD

The goal of this book is to enable students to do SAD — not just read about it, but understand the issues so they can actually analyze and design systems. The book introduces each major technique, explains what it is, explains how to do it, presents an example, and provides opportunities for students to practice before they do it for real in a project. After reading each chapter, the student will be able to perform that step in the system development life cycle (SDLC) process.

Rich Examples of Success and Failure

The book includes a running case about a fictitious company called CD Selections. Each chapter shows how the concepts are applied in situations at CD Selections. Unlike running cases in other books we have tried to focus these examples on planning, managing, and executing the activities described in the chapter, rather than on detailed dialogue between fictious actors. In this way, the running case serves as a template that students can apply to their own work. Each chapter also includes numerous Concepts in Action boxes that describe how real companies succeeded—and failed—in performing the activities in the chapter. Many of these examples are drawn from our own experiences as systems analysts.

Real World Focus

The skills that students learn in a systems analysis and design course should mirror the work that they ultimately will do in real organizations. We have tried to make this book as "real" as possible by building extensively on our experience as professional systems analysts for organizations such as Arthur Andersen, IBM, the U.S. Department of Defense, and the Australian Army. We have also worked with a diverse industry advisory board of IS professionals and consultants in developing the book and have incorporated their stories, feedback, and advice throughout. Many students who use this book will eventually use the skills on the job in a business environment, and we believe they will have a competitive edge in understanding what successful practitioners feel is relevant in the real world.

Project Approach

We have presented the topics in this book in the SDLC order in which an analyst encounters them in a typical project. Although the presentation is necessarily linear (because students have to learn concepts in the way in which they build on each other), we emphasize the iterative, complex nature of SAD as the book unfolds. The presentation of the material should align well with courses that encourage students to work on projects because it presents topics as students need to apply them.

Graphic Organization

The underlying metaphor for the book is doing SAD through a project. We have tried to emphasize this graphically throughout the book so that students better understand how the major elements in the SDLC are related to each other. First, at the start of every major phase of the system development life cycle, we have a graphic illustration that shows the major deliverables that will be developed and added to the "project binder" during that phase. Second, at the start of each chapter, we present a checklist of key tasks or activities that will be performed to produce the deliverables associated with this chapter. These graphic elements — the binder of deliverables tied to each phase and the task checklist tied to each chapter — can help students better understand how the tasks, deliverables, and phases are related and flow from one to another.

Finally, we have highlighted important practical aspects throughout the book by marking boxes and illustrations with a "pushpin." These topics are particularly important in the practical day-to-day life of systems analysts and are the kind of topics that junior analysts should pull out of the book and post on the bulletin board in their office to help them avoid costly mistakes!

ORGANIZATION OF THIS BOOK

This book is organized by the phases of the Systems Development Life Cycle (SDLC). Each chapter has been written to teach students specific tasks that analysts need to accomplish over the course of a project, and the deliverables that will be produced from the tasks. As students complete the book, tasks will be "checked off" and deliverables will be completed and filed in a Project Binder. Along the way, students will be reminded of their progress using roadmaps that indicate where their current task fits into the larger context of SAD.

Chapter 1 introduces the SDLC, object-orientation, UML, and describes the roles and skills needed for a project team. Part 1 contains Chapters 2 and 3, which describe the first phase of the SDLC, the Planning Phase. Chapter 2 presents Project Initiation, with a focus on the System Request and Feasibility Analysis. In Chapter 3, students learn about Project Management, with emphasis on the Workplan, Staffing Plan, Project Charter, and Risk Assessment that are used to help manage and control the project.

Part Two presents techniques needed during the Analysis Phase. In Chapter 4, students create an Analysis Plan after learning a variety of analysis techniques to help with Business Automation, Business Improvement, and Business Process Reengineering. Chapter 5 focuses on requirements for gathering techniques that are used to create Use Cases and Use Case Models (Chapter 6), Structural Models (Chapter 7), and Behavioral Models (Chapter 8).

The Design Phase is covered in Part 3 of the textbook. In Chapter 9, students learn how to evolve the analysis models into design models via the use of factoring, partitions, and layers. They create an Alternative Matrix that compares custom, packaged, and outsourcing alternatives. Chapter 10 focuses on the system architecture design, which includes the Infrastructure Design, Network Model, Hardware/Software Specification, and Security Plan. Chapters 11 and 12 present interface design, and students learn how to create the Interface Structure, Interface Standards, User Interface Template, and User Interface Design. Chapter 13 presents the issues involved in designing persistence for objects. These issues include the different storage formats that can be used for object persistence and how to map an object-oriented design into the chosen storage format. Finally, Chapter 14 concentrates on designing the individual classes and their respective methods through the use of contracts and method specifications.

The Implementation Phase is presented in Chapters 15 and 16. Chapter 15 focuses on system construction, and students learn how to build and test the system. It includes information about the Test Plan and User Documentation. Installation is covered in Chapter 16, and students learn about the Conversion Plan, Change Management Plan, Support Plan, and the Project Assessment.

SUPPLEMENTS

Web Site (www.wiley.com/college/dennis)

The Web site includes a variety of resources that will help in the instruction and learning processes:

- Word and RTF templates for the project deliverables that students can use as starting points for building their own project binders
- PowerPoint slides, prepared by Fred Niederman of St. Louis University, that instructors can tailor to their classroom needs and that students can use to guide their reading and studying activities
- Relevant Internet links

Instructors Resources (www.wiley.com/college/dennis)

Instructors Resources are provided on a password protected section of the website. Resources include:

- A test bank of different kinds of questions ranging from multiple choice and short answer to essay style
- Answers to end-of-chapter questions are provided
- PowerPoint slides as described above
- Short experiential exercises that instructors can use to help students experience and understand key topics
- Short stories have been provided by people working in both corporate and consulting environments for instructors to insert into lectures to make concepts more colorful and real
- Additional mini-cases

Cases in Systems Analysis and Design

A separate CD-based Case Book, edited by Jonathan Trower of Baylor University, provides a set of more than a dozen cases that can be used to supplement the book and provide exercises for students to practice with. The cases are a mixture of shorter cases designed to be used to support one phase within the SDLC, and longer cases than can be used to support an entire semester-long project. The cases are primarily drawn from the U.S. and Canada, but also include a number of international cases. We are always looking for new cases, so if you have a case that might be appropriate please contact us directly (or your local Wiley sales representative).

CASE Software

Visible Systems Corporation's Visible Analyst Student Edition CASE (Computer-Aided Software Engineering) tools can be purchased with the text. Contact your local Wiley sales representative for details, including pricing and ordering information.

Project Management Software

A 120-Day Trial Edition of Microsoft Project 2000 can be purchased with the textbook. Contact your local Wiley sales representative for details.

ACKNOWLEDGMENTS

Our thanks to the many people who contributed to the preparation of this first edition. We are indebted to the staff at John Wiley & Sons for their support, including Beth Lang Golub, Information Systems Editor, Jennifer Battista, Editorial Program Assistant, Gitti Lindner, Marketing Manager, Trish McFadden, Production Editor, Larry Meyer and the staff at Hermitage Publishing Services, Dawn L. Stanley, Designer, and Norman Christensen, Freelance Illustrator.

We would like to thank the following reviewers for their helpful and insightful comments:

Evans Adams	Fort Lewis College
Noushin Ashrafi	University of Massachusetts, Boston
Dirk Baldwin	University of Wisconsin-Parkside
Qing Cao	University of Missouri-Kansas City
Ahmad Ghafarian	North Georgia College & State University
Daniel V. Goulet	University of Wisconsin-Stevens Point
Harvey Hayashi	Loyalist College of Applied Arts and Technology
Jean-Piere Kuilboer	University of Massachusetts, Boston
Daniel Mittleman	DePaul University
Fred Niederman	Saint Louis University
H. Robert Pajkowski	DeVry Institute of Technology, Scarborough, Ontario
June S. Park	University of Iowa
Tom Pettay	DeVry Institute of Technology, Columbus, Ohio
Neil Ramiller	Portland State University
Eliot Rich	University at Albany, State University of New York
Carl Scott	University of Houston
Keng Siau	University of Nebraska-Lincoln
Jonathan Trower	Baylor University
Anna Wachholz	Sheridan College
Randy S. Weinberg	Carnegie Mellon University
Eli J. Weissman	DeVry Institute of Technology, Long Island City, NY
Heinz Roland Weistroffer	Virginia Commonwealth University
Amy Wilson	DeVry Institute of Technology, Decatur, GA
Vincent C. Yen	Wright State University

We would like to thank these practitioners for helping us add a real world component to this project:

Matthew Anderson	*Andersen Consulting*
James Chapman	*Bell Atlantic*
Jeff Elgin	*Capital One*
Graig Galef	*American Management Systems*
Jane Griffin	*Arthur Andersen & Co.*
Mary Haffey	*Arthur Anderson & Co.*
Rixey Jones	*Home Depot*
Brice Marsh	*Computer Sciences Corporation*
Darrell Piatt	*First USA Bank*
David Price	*Connect Knowledge*
Troy Venis	*CheckFree Corporation*
Gillen Young	*Ernst & Young*

We would also like to acknowledge Brian Henderson-Sellers (University of Technology, Sydney) and David E. Monarchi (University of Colorado, Boulder) for discussions about object-orientation over the last decade and especially Steven D. Sheetz (Virginia Tech) for discussions on specific topics covered, and not covered, in this book.

Thanks also to our families and friends for their patience and support along the way, especially Linda Tegarden, and Chris and Nathan Tegarden; Christopher Wixom; and Eileen Dennis, and Alec Dennis. .

Alan Dennis
ardennis@indiana.edu

Barb Wixom
bwixom@mindspring.com

David Tegarden
david.tegarden@vt.edu

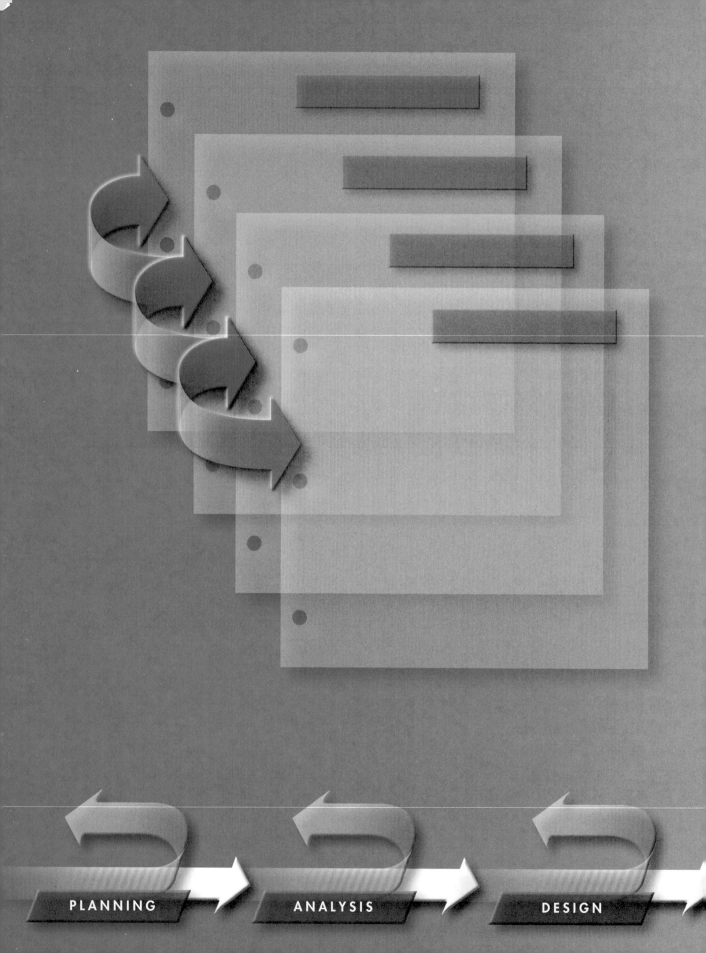

PLANNING

ANALYSIS

DESIGN

CHAPTER 1

INTRODUCTION

THIS chapter introduces the systems development life cycle, the fundamental four-phase model (planning, analysis, design, and implementation) that is common to all information system development projects. It also describes the evolution of system development methodologies and introduces the basic characteristics of object-oriented systems as well as the Unified Modeling Language. Finally, the chapter closes with a discussion of the roles and skills necessary within the project team.

OBJECTIVES

- Understand the fundamental systems development life cycle and its four phases.
- Understand the evolution of systems development methodologies.
- Understand the basic characteristics of object-oriented systems.
- Understand the benefits of using the Unified Modeling Language in object-oriented systems analysis and design.
- Be familiar with the different roles of the project team.

CHAPTER OUTLINE

IMPLEMENTATION

INTRODUCTION

The systems development life cycle (SDLC) is the process of understanding how an information system (IS) can support business needs, designing the system, building it, and delivering it to users. If you have taken a programming class or have programmed on your own, this process probably sounds pretty simple. Unfortunately, however, this is not the case. More than half of all IS development projects fail; either they are canceled before the system is completed, the system is never used after it is installed, or it fails to provide the expected benefits. Of the half that don't fail, most are delivered to the users significantly late, end up costing far more than planned, and have fewer features than originally planned.

Most of us would like to think that these problems only occur to "other" people or "other" organizations, but they affect most companies. Even Microsoft has a history of failures and overdue projects (e.g., Windows 1.0, Windows 95).[1]

Although we would like to promote this book as a "silver bullet" that will keep you from IS failures, we readily admit that a silver bullet guaranteeing IS development success simply does not exist. Instead, this book will provide you with several fundamental concepts and many practical techniques that you can use to improve the probability of success.

The key person in the SDLC is the systems analyst who analyzes the business situation, identifies opportunities for improvements, and designs an information system to implement them. Being a systems analyst is one of the most interesting, exciting, and challenging jobs around. As a systems analyst, you will work with a variety of people and learn how they conduct business. Specifically, you will work with a team of systems analysts, programmers, and others on a common mission. You will feel the satisfaction of seeing systems that you designed and developed make a significant business impact, while knowing that you contributed your unique skills to make that happen.

This book will introduce you to the fundamental skills you need to be a systems analyst. It is a pragmatic book that discusses best practices in systems development; it does not present a general survey of systems development that exposes you to everything about the area. By definition, systems analysts *do things* and challenge the current way that organizations work. To get the most out of this book, you will need to actively apply the ideas and concepts in the examples, the "Your Turn" exercises that are presented throughout, and, ideally, on your own systems development project. This book will guide you through all the steps for delivering a successful information system. Also, we will illustrate how one organization (which we call CD Selections) applies the steps in one project (developing a Web-based CD sales system). By the time you finish the book, you won't be an expert analyst, but you will be ready to start building systems for real.

In this chapter, we first introduce the basic systems development life cycle that IS projects follow. This life cycle is common to all projects, although the focus and approach to each phase of the life cycle may differ. In the next section, we describe the evolution of systems development methodologies. Next we introduce the basic characteristics of object-oriented systems and the Unified Modeling Language.

---

[1] For more information on the problem, see Capers Jones, *Assessment and Control of Software Project Risks,* Englewood Cliffs, NJ: Yourdon Press, 1994; Julia King, "IS Reins in Runaway Projects," *Computerworld,* February 24, 1997.

CONCEPTS   **1-A WHOLESALER SHELVES PROJECT**

IN ACTION

Nash Finch Co., a $4 billion Minnesota-based supermarket and food wholesaler operator, shelved a major 100-person project to improve its retail information systems. Nash Finch had purchased SAP's R/3 retail system with the intention of improving operations, providing better data analysis, and preventing Y2K problems, but the R/3 system required so much additional programming to make it meet Nash Finch's needs that it became clear that the system would not be ready by January 2000. The company suspended the project after spending $50 million and was unable to estimate the additional cost needed to make its existing systems Y2K compliant.

Source: "Big Retail SAP Project Put on Ice," *Computerworld,* November 2, 1998.

QUESTION
How might Nash Finch have avoided this failure?

Finally, one of the most challenging aspects of systems development—the depth and breadth of skills that are required of systems analysts—is discussed. Today, most organizations use project teams that contain members with unique but complementary skills. This chapter closes with a discussion of the key roles played by members of the systems development team.

THE SYSTEMS DEVELOPMENT LIFE CYCLE

In many ways, building an information system is similar to building a house. First, the house (or the information system) starts with a basic idea. Second, this idea is transformed into a simple drawing that is shown to the customer and refined (often through several drawings, each improving on the other) until the customer agrees that the picture depicts what he or she wants. Third, a set of blueprints is designed that present much more detailed information about the house (e.g., the type of water faucets and the location of the telephone jacks). Finally, the house is built following the blueprints—and often with some changes and decisions made by the customer as the house is erected.

The *systems development life cycle* (SDLC) has a similar set of four fundamental phases: planning, analysis, design, and implementation (see Figure 1-1). Different projects may emphasize different parts of the SDLC or approach the SDLC phases in different ways, but all projects have elements of these four phases. Each *phase* is itself composed of a series of *steps,* which rely on *techniques* that produce *deliverables* (specific documents and files that provide understanding about the project).

For example, when students apply for admission to a university, they must all go through several phases: information gathering, applying, and accepting. Each of these phases has steps; information gathering, for example, includes steps like searching for schools, requesting information, and reading brochures. Students then use techniques (e.g., Internet searching) that can be applied to steps (e.g., requesting information) to create deliverables (e.g., evaluations of different aspects of universities).

Figure 1-1 suggests that the SDLC phases and steps proceed in a logical path from start to finish. This is indeed the case in some projects, but in many other

Phase	Chapter	Step	Sample Techniques	Deliverable
Planning (Why build the system?)	2	Identifying business value	System request	System request
		Analyze feasibility	Technical feasibility Economic feasibility Organizational feasibility	Feasibility study
	3	Develop work plan	Task identification Time estimation	Work plan
	3	Staff the project	Creating a staffing plan Creating a project charter	Staffing plan Project charter
	3	Control and direct project	Refine estimates Track tasks Coordinate project Manage scope Mitigate risk	GANTT chart PERT/CPM CASE tool Standards list Project binder(s) Risk assessment
Analysis (Who, what, when, where will the system be?)	4	Analysis	Problem analysis Benchmarking Reengineering	Analysis plan
	5	Information gathering	Interviews Questionnaires	Information
	6	Use case modeling	Use cases Use case model	Functional model
	7	Structural modeling	CRC cards Class diagram Object diagram	Structural models
	8	Behavioral modeling	Sequence diagram Collaboration diagram Statechart diagram	Dynamic models
Design (How will the system work?)	9	System design	Custom development Package development Outsourcing	Design Strategy
	10	Network architecture design	Hardware design Network design	Architecture design Infrastructure design
	11, 12	Interface design	Interface structure chart Input design Output design	Interface design
	13	Database and file design	Selecting a data storage format Optimizing data storage	Data storage design
	14	Object design	Program structure chart Program specifications	Program design
Implementation (System delivery)	15	Construction	Programming Testing	Test plan Programs Documentation
	16	Installation	Direct conversion Parallel conversion Phased conversion	Conversion plan Training plan
		Support	Support strategy Post-Implementation Review	Support plan

FIGURE 1-1
System Development Life Cycle Phases

projects, the project teams move through the steps consecutively, incrementally, iteratively, or in yet other patterns. In this section, we describe the phases and steps, as well
as some of the techniques, that are used to accomplish the steps at a very high level.
We should emphasize that not all organizations follow the SDLC in exactly the way
described here. As we shall shortly see, there are many variations on the overall SDLC.

For now, there are two important points you must understand about the SDLC.
First, you should get a general sense of the phases and steps that IS projects move
through and some of the techniques that produce certain deliverables. Second, it is
important to understand that the SDLC is a process of *gradual refinement*. The
deliverables produced in the analysis phase provide a general idea of the shape of
the new system. These deliverables are used as input to the design phase, which
then refines them to produce a set of deliverables that describe in much more
detailed terms exactly how the system will be built. These deliverables in turn are
used in the implementation phase to produce the actual system. Each phase refines
and elaborates on the work done previously.

Planning

The *planning phase* is the fundamental process of understanding *why* an information system should be built and determining how the project team will go about
building it. The first step is called Project Initiation, during which the system's business value to the organization is identified—how will it lower costs or increase
profits? Most ideas for new systems come from outside the IS area (from the marketing department, accounting department, etc.) in the form of a *system request*. A
system request presents a brief summary of a business need and explains how a system that supports the need will create business value. The IS department works
together with the person or department that generated the request (called the *project sponsor*) to conduct a *feasibility analysis* that examines the idea's technical feasibility (Can we build it?), the economic feasibility (Will it provide business
value?), and the organizational feasibility (If we build it, will it be used?).

The system request and feasibility analysis are presented to an information
systems *approval committee* (sometimes called a steering committee), which
decides whether the project should be undertaken. If the committee approves the
project, then the second step of Project Initiation occurs—*project management*.
During project management, the *project manager* creates a *work plan*, staffs the
project, and puts techniques in place to help control and direct the project through
the entire SDLC. The deliverable for project management is a *project plan* that
describes how the project team will go about developing the system. This plan
includes all of the project management deliverables.

Analysis

The *analysis phase* answers the questions of *who* will use the system, *what* the system will do, and *where* and *when* it will be used. During this phase, the project team
investigates any current system(s), identifies improvement opportunities, and develops a concept for the new system (see Figure 1-1).

Analysis begins with development of an *analysis strategy* that guides the project team's efforts. Such a strategy usually includes an analysis of the current system
(called the *As-Is system*) and its problems, and an examination of ways to design a
new system (called the *To-Be system*). The next step is *information gathering* (e.g.,

through interviews or questionnaires). The analysis of this information—in conjunction with input from the project sponsor and many other people—leads to the development of a concept for a new system. The system concept is then used as a basis to develop a set of business *analysis models* that describe how the business will operate if the new system were developed. The set of models typically includes those that represent the data and processes necessary to support the underlying business process.

The analysis, system concept, and models are combined into a document called the *system proposal,* which is presented to the project sponsor and other key decision makers (e.g., members of the approval committee) who decide whether the project should continue to move forward (see Figure 1-1).

The system proposal is the initial deliverable that describes what the new system should look like. Because it is really the first step in the design of the new system, some experts argue that it is inappropriate to use the term "analysis" for this phase; some argue that a better name would be "analysis and initial design." However, most organizations continue using the name "analysis" for this phase, so we will use it in this book as well. However, it is important to remember that the deliverable from the analysis phase is both an analysis, and a high-level initial design for the new system.

Design

The *design phase* decides *how* the system will operate, in terms of the hardware, software, and network infrastructure; the user interface, forms, and reports; and the specific programs, databases, and files that will be needed. Although most of the strategic decisions about the system were made in the development of the system concept during the analysis phase, the steps in the design phase determine exactly how the system will operate.

The first step in the design phase is to develop the *design strategy:* whether the system will be developed by the company's own programmers, whether it will be outsourced to another firm (usually a consulting firm), or whether the company will buy an existing software package. This leads to the development of the basic *architecture design* for the system that describes the hardware, software, and network infrastructure that will be used. In most cases, the system will add or change the infrastructure that already exists in the organization. The *interface design* specifies how the users will move through the system (e.g., navigation methods such as menus and on-screen buttons) and the forms and reports that the system will use. Next, the *database and file specifications* that define exactly what data will be stored where are developed. Finally, the analyst team develops the *program design,* which defines the programs that need to be written and exactly what each program will do.

This collection of deliverables (architecture design, interface design, database and file specifications, and program design) is the *system specification* that is handed to the programming team for implementation. At the end of the design phase, the feasibility analysis and project plan are reexamined and revised, and another decision is made by the project sponsor and approval committee about whether to terminate the project or continue (see Figure 1-1).

Implementation

The final phase in the SDLC is the *implementation phase,* during which the system is actually built (or purchased, in the case of a packaged software design). This phase usually gets the most attention because for most systems it is the longest and most expensive single part of the development process.

The first step in implementation is system *construction,* during which the system is built and tested to ensure it performs as designed. Testing is one of the most critical steps in implementation because the cost of bugs can be immense; most organizations spend more time and attention on testing than on writing the programs in the first place. Once the system has passed a series of tests, it is installed. *Conversion* is the process by which the old system is turned off and the new one is turned on. It may include a direct cutover approach (in which the new system immediately replaces the old system), a parallel conversion approach (in which both the old and new systems are operated for a month or two until it is clear that there are no bugs in the new system), or a phased conversion strategy (in which the new system is installed in one part of the organization as an initial trial and then gradually installed in others). One of the most important aspects of conversion is the development of a *training plan* to teach users how to use the new system and help manage the changes caused by the new system.

Once the system has been installed, the analyst team establishes a *support plan* for the system. This plan usually includes a formal or informal post-implementation review, as well as a systematic way to identify the major and minor changes needed for the system.

THE EVOLUTION OF SYSTEMS DEVELOPMENT

A *methodology* is a formalized approach to implementing the SDLC (i.e., it is a list of steps and deliverables). There are many different systems development methodologies, and each one is unique based on the order and focus it places on each SDLC phase. Some methodologies are formal standards used by government agencies, whereas others have been developed by consulting firms to sell to clients. Many organizations have their own internal methodologies that have been honed over the years, and they explain exactly how each phase of the SDLC is to be performed in that company.

Methodologies have evolved over time as we have learned more about how to develop information systems.[2] In the early days of computing, the need for formal and well-planned life cycle methodologies was not well understood. Programmers tended to move directly from a very simple planning phase right into the construction step of the implementation phase; in other words, they moved directly from a very fuzzy, not well-thought-out system request into writing code, which is the same approach that you sometimes use when writing programs for a programming class. This approach can work for small programs that require only one programmer, but if the requirements are complex or unclear you may miss important aspects of the problem and have to start all over again, throwing away part of the program (and the time and effort spent writing it). This approach also makes teamwork difficult because members have little idea about what needs to be accomplished and how to work together to produce a final product.

---

[2] A good reference for comparing systems development methodologies is Steve McConnell, *Rapid Development* (Redmond, WA: Microsoft Press, 1996).

Structured Design

The first category of systems development methodologies is called *structured design.* These methodologies became dominant in the 1980s, replacing the previous ad hoc, undisciplined approach. Structured design methodologies adopt a formal step-by-step approach to the SDLC that moves logically from one phase to the next. The original structured design methodology (that is still used today) is *waterfall development.* In waterfall development-based methodologies, the analysts and users proceed in sequence from one phase to the next (see Figure 1-2). The key deliverables for each phase are typically produced on paper (often hundreds of pages in length) and are presented to the project sponsor for approval as the project moves from phase to phase. Once the sponsor approves the work that was conducted for a phase, the phase ends and the next one begins. These methodologies are referred to as waterfall development because it moves forward from phase to phase in the same manner as a waterfall. Although it is possible to go backwards in the SDLC (e.g., from design back to analysis), it is extremely difficult (imagine yourself as a salmon trying to swim "upstream" against a waterfall as shown in Figure 1-2).

Structured design also introduced the use of formal modeling or diagramming techniques to describe the basic business processes and the data that support them. Traditional structured design uses one set of diagrams to represent the processes and a separate set of diagrams to represent data. Because two sets of diagrams are used, the systems analyst must decide which set to develop first and use as the core of the system: process model diagrams or data model diagrams. There is much debate over which should come first, the processes or the data, because both are important to the system. As a result, several different structured design methodologies have evolved that follow the basic steps of the waterfall model but use different modeling approaches at different times. Those that attempt to emphasize process model diagrams as the core of the system are called *process-centered*

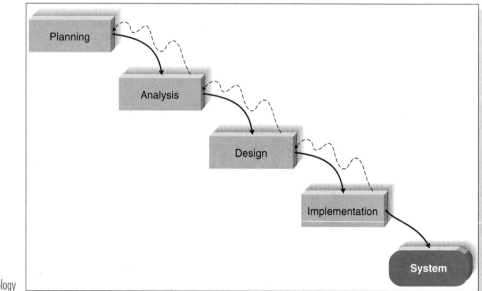

FIGURE 1-2
The Waterfall Development Methodology

methodologies, whereas those that emphasize data model diagrams as the core of the system concept are called *data-centered methodologies.*[3]

The key advantage of the structured design waterfall approach is that it identifies system requirements long before programming begins, and it minimizes changes to the requirements as the project proceeds. It has two key disadvantages: the design must be completely specified on paper before programming begins, and a long time elapses between completion of the system proposal in the analysis phase and delivery of the system (usually many months or years). A paper document is often a poor communication mechanism, so important requirements can be overlooked in the hundreds of pages of documentation. Users rarely are prepared for their introduction to the new system, which occurs long after the initial idea for the system was introduced. If the project team misses important requirements, expensive post-implementation programming may be needed. (Imagine yourself trying to design a car on paper; how likely would you be to remember interior lights that come on when the doors open, or to specify the right number of valves on the engine?).

A system may also require significant rework because the business environment has changed from the time that the analysis phase occurred. When changes do occur, it means going back to the initial phases and following the change through each of the subsequent phases in turn.

Rapid Application Development (RAD)

Rapid Application Development-based methodologies are a newer class of systems development methodologies that emerged in the 1990s. RAD-based methodologies attempt to address both weaknesses of structured design methodologies by adjusting the SDLC phases to get some part of the system developed quickly and into the hands of the users. In this way, the users can better understand the system and suggest revisions that will bring the system closer to what is needed.[4]

Most RAD-based methodologies recommend that analysts use special techniques and computer tools to speed up the analysis, design, and implementation phases, such as CASE tools (see Chapter 3), JAD sessions (see Chapter 5), fourth-generation/visual programming languages that simplify and speed up programming (e.g., Visual Basic), and code generators that automatically produce programs from design specifications. It is the combination of the changed SDLC phases and the use of these tools and techniques that improves the speed and quality of systems development. However, there is one possible subtle problem with RAD-based methodologies: user expectation management. Because of the use of the tools and techniques that can improve the speed and quality of systems development, user expectations of what is possible may dramatically change. As the user better understands the information technology, the systems requirements tend to expand. This was never a real problem when using paper-based models. Today, however, the actual strength of RAD-based methodologies can be a double-edged sword. There

[3] The classic modern process-centered methodology is that by Edward Yourdon, *Modern Structured Analysis* (Englewood Cliffs, NJ: Yourdon Press, 1989). An example of a data-centered methodology is information engineering by James Martin, *Information Engineering*, Vols. 1–3 (Englewood Cliffs, NJ: Prentice Hall, 1989). One widely accepted standardized methodology that can be either process-oriented or data-oriented depending on how it is used is IDEF; see FIPS 183, *Integration Definition for Function Modeling,* Federal Information Processing Standards Publications, U.S. Department of Commerce, 1993.

[4] One of the best RAD books is that by McConnell, *Rapid Development.*

are process-centered, data-centered, and object-oriented methodologies that follow the basic approaches of the three RAD categories described below.

Phased Development   A *phased development*-based methodology breaks the overall system into a series of *versions* that are developed sequentially. The analysis phase identifies the overall system concept, and the project team, users, and system sponsor then categorize the requirements into a series of versions. The most important and fundamental requirements are bundled into the first version of the system. The analysis phase then leads into design and implementation, but only with the set of requirements identified for version 1 (see Figure 1-3).

Once version 1 is implemented, work begins on version 2. Additional analysis is performed based on the previously identified requirements and combined with

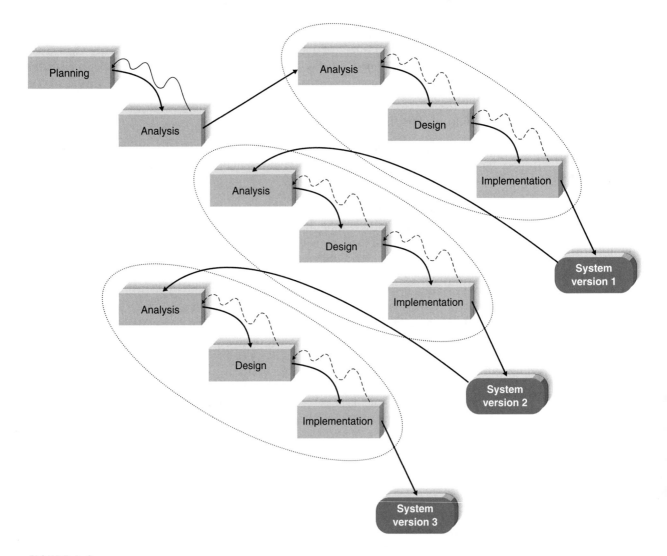

FIGURE 1-3
The Phased Development Methodology

new ideas and issues that arose from the users' experience with version 1. Version 2 then is designed and implemented, and work immediately begins on the next version. This process continues until the system is complete or is no longer in use.

Phased development-based methodologies have the advantage of quickly getting a useful system into the user's hands. Although the system does not perform all the functions the users need at first, it does begin to provide business value sooner than if the system were delivered after completion, as is the case with the waterfall methodology. Similarly, because users begin to work with the system sooner, they are more likely to identify important additional requirements sooner than with structured design situations.

The major drawback of phased development is that users begin to work with systems that are intentionally incomplete. It is critical to identify the most important and useful features and to include them in the first version, while managing users' expectations along the way.

Prototyping   A *prototyping*-based methodology performs the analysis, design, and implementation phases concurrently, and all three phases are performed repeatedly in a cycle until the system is completed. With these methodologies, the basics of analysis and design are performed, and work immediately begins on a *system prototype,* a "quick-and-dirty" program that provides a minimal amount of features. The first prototype is usually the first part of the system that the user will utilize. This is shown to the users and project sponsor who provide comments, which are used to re-analyze, re-design, and re-implement a second prototype that provides a few more features. This process continues in a cycle until the analysts, users, and sponsor agree that the prototype provides enough functionality to be installed and used in the organization. After the prototype (now called the "system") is installed, refinement occurs until it is accepted as the new system (see Figure 1-4).

The key advantage of a prototyping-based methodology is that it *very* quickly provides a system for the users to interact with, even if it is not ready for widespread organizational use at first. Prototyping reassures the users that the project team is working on the system (there are no long delays in which the users see little progress), and it helps to more quickly refine real requirements. Rather than attempting to understand a system specification on paper, the users can interact with the prototype to get a better understanding of what it can and cannot do.

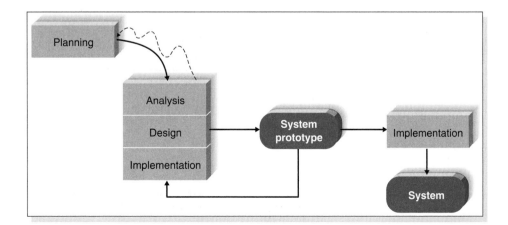

FIGURE 1-4
The Prototyping Methodology

The major problem associated with prototyping is that its fast-paced system releases challenge attempts to conduct careful, methodical analysis. Often the prototype undergoes such significant changes that many initial design decisions become poor ones. This can cause problems in the development of complex systems because fundamental issues and problems are not recognized until well into the development process. Imagine building a car and discovering late in the prototyping process that you have to take the whole engine out to change the oil (because no one thought about the need to change the oil until after it had been driven 10,000 miles).

Throw-Away Prototyping *Throw-away prototyping*-based methodologies are similar to prototyping-based methodologies in that they include the development of prototypes; however, throw-away prototypes are done at a different point in the SDLC. These prototypes are used for a very different purpose than the ones previously discussed, and they have a very different appearance (see Figure 1-5).[5]

The throw-away prototyping-based methodologies have a relatively thorough analysis phase that is used to gather information and to develop ideas for the system concept. However, many of the features suggested by the users may not be well understood, and challenging technical issues may need to be solved. Each of these issues is examined by analyzing, designing, and building a *design prototype*. A design prototype is not a working system; rather, it is a product that represents a part of the system that needs additional refinement, and it only contains enough detail to enable users to understand the issues under consideration. For example,

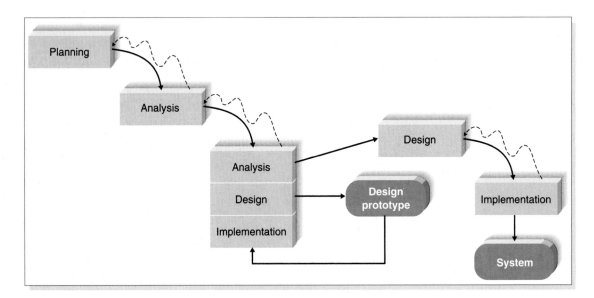

FIGURE 1-5
The Throw-away Prototyping Methodology

[5] Our description of the throw-away prototyping methodology is a modified version of the spiral development methodology developed by Barry Boehm, "A Spiral Model of Software Development and Enhancement," *Computer* (May 1988): 61–72.

suppose users are not completely clear on how an order entry system should work. The analyst team might build a series of HTML pages viewed using a Web browser to help the users visualize such a system. In this case, a series of mock-up screens *appear* to be a system, but they really do nothing. Or suppose that the project team needs to develop a sophisticated graphics program in Java. The team could write a portion of the program with sample data to ensure that they could do a full-blown program successfully.

A system that is developed using this type of methodology probably relies on several design prototypes during the analysis and design phases. Each of the prototypes is used to minimize the risk associated with the system by confirming that important issues are understood before the real system is built. Once the issues are resolved, the project moves into design and implementation. At this point, the design prototypes are thrown away, which is an important difference between these methodologies and prototyping methodologies, in which the prototypes evolve into the final system.

Throw-away prototyping-based methodologies balance the benefits of well-thought-out analysis and design phases with the advantages of using prototypes to refine key issues before a system is built. It may take longer to deliver the final system as compared to prototyping-based methodologies (since the prototypes do not become the final system), but this type of methodology usually produces more stable and reliable systems.

Object-Oriented Analysis and Design

As RAD-based methodologies have become more popular, we have come to realize that they are not as effective as they should be. These techniques, adopted from structured design methodologies, are either process centric or data centric. That is, they focus the decomposition of problems into a set of processes (process centric) or a set of entities (data centric). However, processes and data are so closely related that it is difficult to pick one or the other as the primary focus. Based on this lack of congruence with the real world, new *object-oriented methodologies* emerged that use the RAD-based sequence of SDLC phases but attempt to balance the emphasis between process and data by focusing the decomposition of problems on objects that contain both data and processes.

In the early days, object-oriented systems were implemented in different ways by different developers. Each developer had his or her own methodology and notation, making it difficult for analysts in different organizations to speak a common language, for companies to develop object-oriented tools, and for universities to teach generally accepted object-oriented concepts. Object-oriented approaches were used for some projects, but most organizations did not adopt them for widespread use.

In 1995, Rational Software brought together three leaders of the object-oriented world to create a single approach to object-oriented development. Grady Booch, Ivar Jacobson, and James Rumbaugh worked with others to create a standard set of diagramming techniques known as the *Unified Modeling Language,* or UML. In November 1997, the *Object Management Group* (OMG) formally accepted UML as the standard diagramming notation for all object-oriented development. It was at this point that the modern object-oriented systems development era began.

According to the creators of UML, Grady Booch, Ivar Jacobson, and James Rumbaugh,[6] any modern object-oriented approach to developing information systems must be (1) use-case driven, (2) architecture centric, and (3) iterative and incremental.

Use-Case Driven   *Use-case driven* means that *use cases* are the primary modeling tool for defining the behavior of the system. A use case describes how the user interacts with the system to perform some activity, such as placing an order, making a reservation, or searching for information. The use cases are used to identify and communicate the business requirements for the system to the programmers who must write the system. All other details of the system are derived from the system's use cases.

Use cases are inherently simple because they focus on only one activity at a time. In contrast, the process model diagrams used by traditional structured and RAD methodologies are far more complex because they require the system analyst and user to develop models of the entire system at a time rather than just one activity. With traditional methodologies, each business activity is decomposed into a set of subprocesses, which are in turn decomposed into further subprocesses, and so on. This goes on until no further process decomposition makes sense, and it often requires dozens of pages of interlocking diagrams that often can only be understood by examining several pages simultaneously. By developing use cases that focus on only one activity at once, developing models becomes simpler.[7]

Architecture Centric   Any modern approach to systems analysis and design should be *architecture centric*. This term means that the underlying software architecture of the evolving system specification drives the specification, construction, and documentation of the system. Modern object-oriented systems analysis and design approaches should support at least three separate, but interrelated, architectural views of a system: functional, static, and dynamic.

The *functional view* describes the external behavior of the system from the perspective of the user. Use cases and use-case diagrams are used to depict the functional view. The *static view* describes the structure of the system in terms of attributes, methods, classes, relationships, and messages. CRC (class-responsibility-collaboration) cards, class diagrams, and object diagrams are used to portray the static view of an evolving object-oriented information system. The *dynamic view* describes the internal behavior of the system in terms of messages passed between objects and state changes within an object. The dynamic view is represented in UML by sequence diagrams, collaboration diagrams, and statecharts. Each of these views, and their related UML notation, is described in greater detail in Chapters 6, 7, and 8.

---

[6] Grady Booch, Ivar Jacobson, and James Rumbaugh, *The Unified Modeling Language User Guide* (Reading, MA: Addison-Wesley, 1999).

[7] For those of you who have experience with traditional structured analysis and design, this will be one of the most unusual aspects of object-oriented analysis and design using UML. Structured approaches emphasize the decomposition of the complete business process into subprocesses and sub-subprocesses. Object-oriented approaches stress focusing on just one use case activity at a time and distributing that single use case over a set of communicating and collaborating objects. Therefore, use-case modeling may initially seem unsettling or counterintuitive, but in the long run this single focus does make analysis and design simpler.

Iterative and Incremental   Modern object-oriented systems analysis and design approaches using UML emphasize *iterative* and *incremental* development that undergoes continuous testing and refinement throughout the life of the project. Each iteration of the system brings the system closer and closer to the real needs of the users. UML does not dictate any formal approach to developing information systems, but its iterative nature is best suited to RAD-based approaches such as phased development (see Figure 1-3).

The same UML diagramming techniques are used throughout all phases of the SDLC (although some diagrams are more important in some phases). The UML diagrams are rather conceptual and abstract at the start of the SDLC. As the developers move through the SLDC, the diagrams gradually become more detailed. In other words, the diagrams move from documenting the requirements to laying out the design. Overall, the consistent notation, integration among the diagramming techniques, and application of the diagrams across all phases of the SDLC make UML a powerful language for analysts and developers.

BASIC CHARACTERISTICS OF OBJECT-ORIENTED SYSTEMS

The ideas behind today's object-oriented systems analysis and design approaches can be traced back to the object-oriented programming languages Simula, created in the 1960s, and Smalltalk, created in the early 1970s. However, until the mid-1980s, object-oriented approaches were not feasible due to the lack of computational power of mainframe computers of that era. Today, with the increase in processor power and the decrease in processor cost, object-oriented approaches are practical. Nonetheless, for object-oriented approaches to be adopted, new terminology must be learned. The following sections overview the key characteristics of object-oriented systems.

Classes, Objects, Methods, and Messages

The basic building block of the system is the *object,* which contains both data and processes. A *class* is the template we use to define objects; a class explains all the data and processes that each object of that type contains. If we were building an appointment system for a doctor's office, classes might include doctor, patient, and appointment. The specific patients like Jim Maloney, Mary Wilson, and Theresa Marks are considered *instances,* or objects, of the patient class.

Each object has *attributes* that describe data about the object, such as a patient's name, birth date, address, and phone number. The *state* of an object is defined by the value of its attributes and its relationships with other objects at a particular point in time. For example, a patient might have a state of "new" or "current" or "former." Each object also has *methods,* that specify what processes the object can perform, much like a function or procedure in a programming language such as Visual Basic or C. For example, a patient object would likely contain data attributes such as names, address, and birth date and process methods such as insert new patient and delete patient (see Figure 1-6).

In order to get an object to perform a method (e.g., to delete itself), a *message* is sent to the object. A message is essentially a function or procedure call from one object to another object. For example, if a patient is new to the doctor's office, the

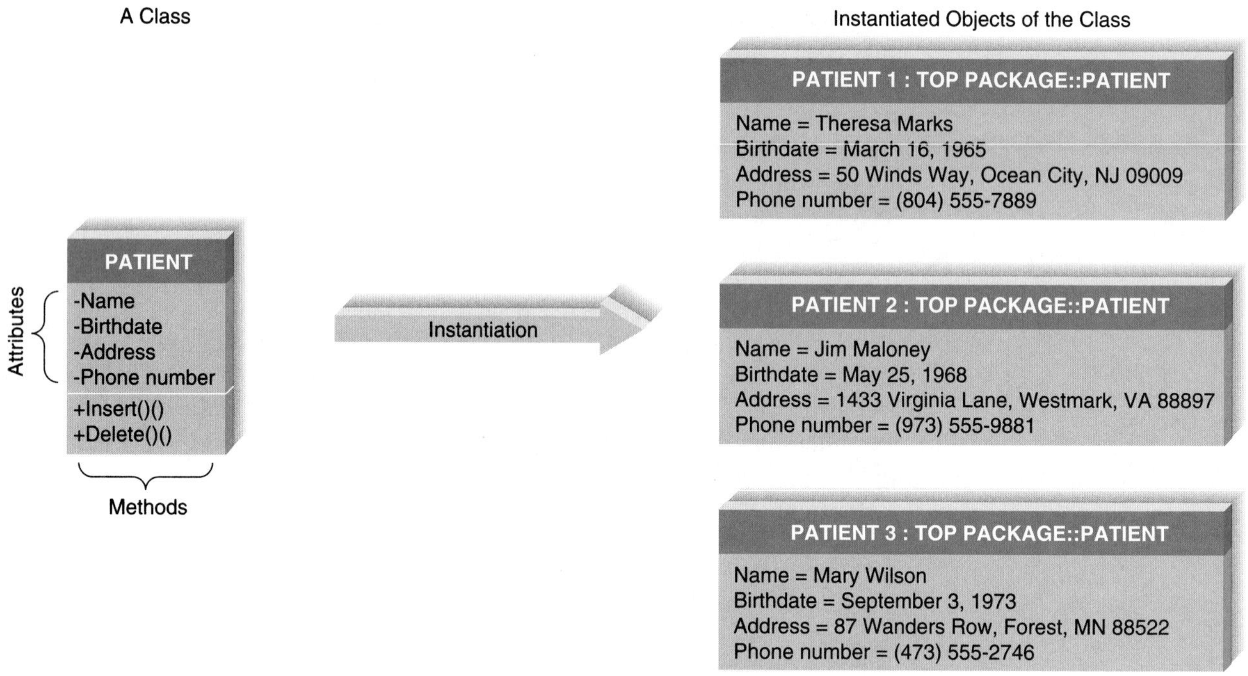

A Class

Instantiated Objects of the Class

PATIENT 1 : TOP PACKAGE::PATIENT

Name = Theresa Marks
Birthdate = March 16, 1965
Address = 50 Winds Way, Ocean City, NJ 09009
Phone number = (804) 555-7889

PATIENT

Attributes {
-Name
-Birthdate
-Address
-Phone number

+Insert()()
+Delete()()
}

Methods

Instantiation

PATIENT 2 : TOP PACKAGE::PATIENT

Name = Jim Maloney
Birthdate = May 25, 1968
Address = 1433 Virginia Lane, Westmark, VA 88897
Phone number = (973) 555-9881

PATIENT 3 : TOP PACKAGE::PATIENT

Name = Mary Wilson
Birthdate = September 3, 1973
Address = 87 Wanders Row, Forest, MN 88522
Phone number = (473) 555-2746

FIGURE 1-6
A Class and Its Objects

system will send an insert message to the object. The patient object will receive a message (instruction) and do what it needs to do to go about inserting the new patient into the system (execute its method).

Encapsulation and Information Hiding

The ideas of *encapsulation* and *information hiding* are interrelated in object-oriented systems. Encapsulation is a mechanism that combines the processes and data into a single object. The principle of information hiding suggests that only the information required to use an object should be available outside the object. Exactly how the object stores data or performs methods is not relevant; we really do not care as long as the object functions as it should. In short, objects are treated like black boxes.

The fact that we can use an object by sending a message that calls methods is the key to reusability because it shields the internal workings of the object from changes in the outside system and it keeps the system from being affected when changes are made to an object. All an object should need to know is the set of methods that other objects can perform and what messages need to be sent to trigger them. If we change the data in a patient object or the way it executes its methods, we just need to change the object; we do not need to change the other objects that send it messages.

Inheritance

Inheritance means that classes can reuse attributes and methods that have been defined in other classes. In many cases, there are similarities among different classes in an information system. A medical system, for example, might contain information about patients and doctors. Much of the data attributes and methods we need for both patients and doctors may be quite similar (e.g., name, address, phone number, insert, delete).

Although we could define each class separately, it might be simpler to define one general class called person that contains the data and methods needed by both patients and doctors, and then have these classes inherit the properties of the person class. When using inheritance, we arrange classes in a hierarchy in which the general classes (called *superclasses*) are at the top and the *subclasses,* or specific classes, are at the bottom. In Figure 1-7, person is a superclass to the classes doctor and patient. Doctor, in turn, is a superclass to general practitioner and specialist. Notice how a class (e.g., doctor) can concurrently serve as a superclass and subclass.

Subclasses *inherit* the appropriate attributes and methods from the superclasses above them. That is, each subclass contains attributes and methods from its parent superclass. For example, Figure 1-8 shows that both doctor and patient are subclasses of person and therefore will inherit the attributes and methods of the person class. Inheritance makes it simpler to define classes. Instead of repeating the attributes and methods in the doctor and patient classes separately, the attributes and methods that are common to both are placed in the person class and inherited

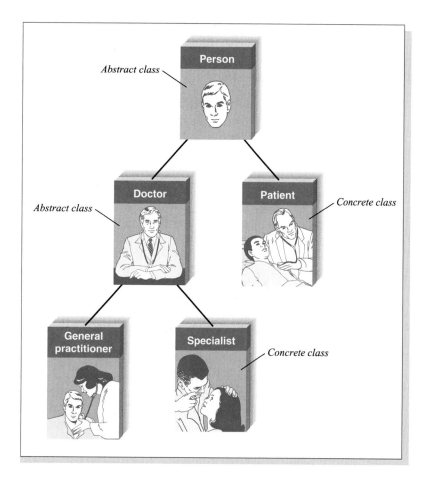

FIGURE 1-7
Class Hierarchy

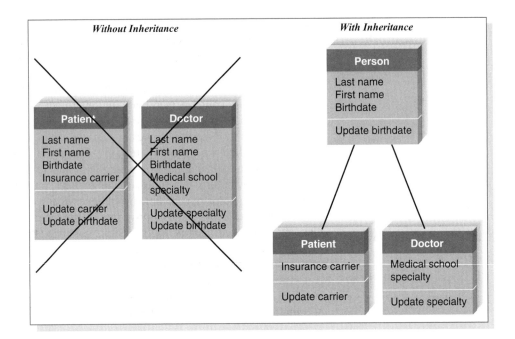

FIGURE 1-8
Inheritance

by those classes below it. Notice how much more efficient hierarchies of object classes are than the same objects without a hierarchy in Figure 1-8.

Polymorphism

Polymorphism means that the same message can be interpreted differently by different classes of objects. For example, inserting a patient means something different from inserting an appointment. As such, different pieces of information need to be collected and stored. Luckily, we do not have to be concerned with *how* something is done when using objects. We can simply send a message to an object, and that object will be responsible for interpreting the message appropriately. For example, if we sent the message "draw yourself" to a square object, a circle object, and a triangle object, the results would be very different, even though the message would be the same.

OBJECT-ORIENTED ANALYSIS AND DESIGN USING UML

UML is an object-oriented modeling language used to describe information systems. It provides a common vocabulary of object-oriented terms and a set of diagramming techniques that are rich enough to model any systems development project from analysis through implementation. Although UML defines a large set of diagramming techniques, this book focuses on a smaller set of the most commonly used techniques.[8] However, it should be stressed that UML is nothing more than a notation. UML does not dictate any specific approach to developing information

---

[8] For a complete description of all diagramming techniques, see www.rational.com/uml.

systems. Even though it is unlikely, it is possible to develop an information system using a traditional approach, such as structured systems development or information engineering, and to document the analysis and design using the UML.

So far we have described several major concepts that permeate the object-oriented approach in general and UML in particular, but you may be wondering how these concepts affect the performance of a project team. The answer is simple. Concepts such as:

> use-case driven and use cases,
>
> architecture centric and functional, static, and dynamic views,
>
> iterative and incremental development,
>
> classes, objects, methods, and messages,
>
> encapsulation and information hiding,
>
> inheritance, and
>
> polymorphism

Concept	What Concept Supports	What Concept Leads To
Classes, objects, methods, and messages	A more realistic way for people think about their business Highly cohesive units that contain both data and processes	Better communication between user and analyst Reusable objects Benefits from having a highly cohesive system
Encapsulation and information hiding	Loosely coupled units	Reusable objects Fewer ripple effects from changes within an object or in the system itself Benefits from having a loosely coupled system
Inheritance	Allows us to use classes as standard templates from which other classes can be built	Less redundancy Faster creation of new classes Standards and consistency within and across development efforts Ease in supporting exceptions
Polymorphism	Minimal messaging that is interpreted by objects themselves	Simpler programming of events Ease in replacing or changing objects in a system Fewer ripple effects from changes within an object or in the system itself
Use-case driven and use cases	Allows users and analysts to focus on how a user will interact with the system to perform a single activity	Better understanding and gathering of user needs Better communication between user and analyst
Architecture centric and functional, static, and dynamic views	Viewing the evolving system from multiple points of view	Better understanding and modeling of user needs More complete depiction of information system
Iterative and incremental development	Continuous testing and refinement of the evolving system	Meeting real needs of users Higher quality systems

FIGURE 1-9
Benefits of the Object Approach

enable analysts to break a complex system into smaller, more manageable components, work on the components individually, and more easily piece the components back together to form a system. This modularity makes system development easier to grasp, easier to share among members of a project team, and easier to communicate to users who are needed to provide requirements and confirm how well the system meets the requirements throughout the SDLC.

By modularizing system development, the project team is actually creating reusable pieces that can be plugged into other systems efforts or can be used as starting points for other projects. Ultimately, this can save time because new projects don't have to start completely from scratch, and learning curves aren't as steep.

Finally, many people argue that "object-think" is a much more realistic way to think about the real world. Users typically do not think in terms of data or process; instead, they see their business as a collection of logical units that contain both—so communicating in terms of objects improves the interaction between the user and the analyst or developer. Figure 1-9 summarizes the major concepts of the object approach and how each contributes to the benefits.

PROJECT TEAM ROLES AND SKILLS

As should be clear from the various phases and steps performed during the SDLC, the project team needs a variety of skills. The systems analyst is first and foremost a *change agent* who identifies ways to improve an organization, builds an information system to support them, and trains and motivates others to use the system. Leading a successful organizational change effort is one of the most difficult jobs that someone can do because understanding what to change, how to change it, and convincing others of the need for change requires a wide range of skills. Analysts must understand technology and how to apply it to business situations. They must be able to communicate effectively one-on-one with users and business managers (who often have little experience with technology) and with programmers (who often have more technical expertise than the analyst). They must be able to give presentations to large and small groups, write reports, and manage people. Analysts require the ability to adapt to unexpected events and to accept and manage the pressure and risks associated with unclear situations. Finally, analysts must deal fairly, honestly, and ethically with other project team members, managers, and system users. Analysts often deal with confidential information or information that if shared with others could cause harm (e.g., dissent among employees). It is important to maintain confidence and trust with all people.

In addition to these general skills, analysts require many specific skills. In the early days of systems development, most organizations expected one person, the analyst, to have all of the specific skills needed to conduct a systems development project. Some small organizations still expect one person to perform many roles, but because organizations and technology have become more complex, most large organizations now build project teams that contain several individuals with clearly defined roles. Different organizations divide the responsibilities differently, but Figure 1-10 presents one commonly used set of project team roles. Most IS project teams include many other individuals, such as the *programmers,* who actually write the programs that make up the system, and *technical writers,* who prepare the help screens and other documentation (e.g., users manuals, systems manuals).

Role	Responsibilities
Business analyst	Analyzing the key business aspects of the system
	Identifying how the system will provide business value
	Designing the new business processes and policies
Systems analyst	Identifying how technology can improve business processes
	Designing the new business processes
	Designing the information system
	Ensuring that the system conforms to information systems standards
Infrastructure analyst	Ensuring the system conforms to infrastructure standards
	Identifying infrastructure changes needed to support the system
Change management analyst	Developing and executing a change management plan
	Developing and executing a user training plan
Project manager	Managing the team of analysts, programmers, technical writers, and other specialists
	Developing and monitoring the project plan
	Assigning resources
	Serving as the primary point of contact for the project

FIGURE 1-10
Project Team Roles

Business Analyst

The *business analyst* focuses on the business issues surrounding the system. These issues include identifying the business value that the system will create, developing ideas and suggestions for how the business processes can be improved, and designing the new processes and policies in conjunction with the systems analyst. This individual will likely have business experience and some type of professional training (e.g., the business analyst for accounting systems will likely be a CPA [in the United States] or a CA [in Canada]). He or she represents the interests of the project sponsor and the ultimate users of the system. The business analyst assists in the planning and design phases but is most active in the analysis phase.

Systems Analyst

The *systems analyst* focuses on the IS issues surrounding the system. This person develops ideas and suggestions for how information technology can improve business processes, formulates the new business processes with help from the business analyst, designs the new information system, and ensures that all IS standards are maintained. The systems analyst likely will have significant training and experience in analysis and design, programming, and even areas of the business. He or she represents the interests of the IS department and works intensively through the project, but perhaps less so during the implementation phase.

Infrastructure Analyst

The *infrastructure analyst* focuses on the technical issues involved in how the system will interact with the organization's technical infrastructure (e.g., hardware, software, networks, and databases). The infrastructure analyst tasks include ensuring that the new information system conforms to organizational standards and identifying infrastructure changes needed to support the system. This individual will likely have significant training and experience in networking, database administration, and various hardware and software products. He or she represents the interests of the organization and IS group that will ultimately have to operate and support the new system once it has been installed. The infrastructure analyst works throughout the project, but perhaps less so during planning and analysis phases.

Change Management Analyst

The *change management analyst* focuses on the people and management issues surrounding the system installation. The roles of this person include ensuring that adequate documentation and support are available to users, providing user training on the new system, and developing strategies to overcome resistance to change. This individual likely will have significant training and experience in organizational behavior in general and change management in particular. He or she represents the interests of the project sponsor and users for whom the system is being designed. The change management analyst works most actively during the implementation phase but begins laying the groundwork for change during the analysis and design phases.

Project Manager

The *project manager* is responsible for ensuring that the project is completed on time and within budget and that the system delivers all benefits that were intended by the project sponsor. The role of the project manager includes managing the team members, developing the project plan, assigning resources, and being the primary point of contact when people outside the team have questions about the project. This individual likely will have significant experience in project management and likely has worked for many years as a systems analyst beforehand. He or she represents the interests of the IS department and the project sponsor. The project manager works intensely during all phases of the project.

YOUR TURN   **1-1 BEING AN ANALYST**

Suppose you decide to become an analyst after you graduate. What type of analyst would you most prefer to be? What type of courses should you take before you graduate? What type of summer job or internship should you seek?

QUESTION:
Develop a short plan that describes how you will prepare for your career as an analyst.

SUMMARY

Today, the most exciting change in systems analysis and design is the move to object-oriented techniques, which view a system as a collection of self-contained objects that have both data and processes. However, the ideas underlying object-oriented techniques are simply ideas that have evolved from traditional approaches to systems development or old ideas that have now become practical due to the cost-performance ratios of modern computer hardware in comparison to the ratios of the past. Today, the cost of developing modern software is composed primarily of the cost associated with the developers themselves and not the computers. As such, object-oriented approaches to developing information systems hold out much promise in controlling these costs.

The Systems Development Life Cycle

All systems development projects follow essentially the same fundamental process called the systems development life cycle (SDLC). SLDC starts with a planning phase in which the project team identifies the business value of the system, conducts a feasibility analysis, and plans the project. The second phase is the analysis phase, in which the team develops an analysis strategy, gathers information, and builds a set of analysis models. In the next phase, the design phase, the team develops the physical design, architecture design, interface design, database and file specifications, and program design. In the final phase, implementation, the system is built, installed, and maintained.

The Evolution of Systems Development Methodologies

Systems development methodologies are formalized approaches to implementing an SDLC that have evolved over the decades. Structured design methodologies emphasize decomposition of a problem by either focusing on process decomposition (process-centric methodologies) or data decomposition (data-centric methodologies). They produce a solid, well-thought-out system, but can overlook requirements because users must specify them early in the design process before seeing the actual system. RAD-based methodologies attempt to speed up development and make it easier for users to specify requirements by having parts of the system developed sooner either by producing different versions or by using prototypes through the use of CASE tools and fourth-generation/visual programming languages. However, RAD-based methodologies still tend to be either process centric or data centric.

Object-oriented systems analysis and design methodologies are most closely associated with a phased development RAD-based methodology where the time spent in each phase is very short. They focus on object decomposition where objects contain both processes and data. Object-oriented systems analysis and design methodologies should use a use-case driven, architecture-centric, and iterative and incremental information systems development approach. They should support three different views of the evolving system: functional, static, and dynamic.

Basic Characteristics of Object-Oriented Systems

An object is a person, place, or thing about which we want to capture information. Each object has attributes that describe information about it and behaviors that are described by methods that specify what an object can do. Objects are grouped into classes, which are collections of objects that share common attributes and methods.

The classes can be arranged in a hierarchical fashion in which low-level classes, or subclasses, inherit attributes and methods from superclasses above them to reduce the redundancy in development. Objects communicate with each other by sending messages, which trigger methods. Polymorphism allows a message to be interpreted differently by different kinds of objects. This form of communication allows us to treat objects like black boxes and to ignore the inner workings of the objects. Encapsulation and information hiding allow an object to conceal its inner processes and data from the other objects.

Object-Oriented Analysis and Design Using UML

Object-oriented analysis and design using UML allows the analyst to decompose complex problems into smaller, more manageable components using a commonly accepted set of notations. UML is a standard set of diagramming techniques that provides a graphical representation that is rich enough to model any systems development project from analysis through implementation. Typically, today most object-oriented analysis and design approaches use the UML to depict an evolving system. Finally, many people believe that users do not think in terms of data or processes but instead in terms of a collection of collaborating objects. As such, object-oriented analysis and design using UML allows the analyst to interact with the user employing objects from the user's environment instead of a set of separate processes and data.

Project Team Roles and Skills

The project team needs a variety of skills. All analysts need to have general skills, such as change management, ethics, communications, and technical. However, different kinds of analysts require specific skills in addition to these. Business analysts usually have business skills that help them understand the business issues surrounding the system, whereas systems analysts also have significant experience in analysis and design and programming. The infrastructure analyst focuses on technical issues surrounding how the system will interact with the organization's technical infrastructure, and the change management analyst focuses on people and management issues surrounding the system installation. In addition to analysts, project teams will include a project manager, programmers, technical writers, and other specialists.

KEY TERMS

Analysis model	Construction	Gradual refinement
Analysis phase	Conversion	Implementation phase
Analysis strategy	Database and file specification	Incremental
Approval committee	Data-centered methodology	Information gathering
Architecture centric	Deliverable	Information hiding
Architecture design	Design phase	Infrastructure analyst
As-Is system	Design prototype	Inherit
Attribute	Design strategy	Inheritance
Business analyst	Dynamic view	Instance
Change agent	Encapsulation	Interface design
Change management analyst	Feasibility analysis	Iterative
Class	Functional view	Message

Method
Methodology
Object
Object Management Group (OMG)
Object-oriented methodology
Phase
Phased development
Planning phase
Polymorphism
Process-centered methodology
Program design
Programmer
Project management
Project manager
Project plan

Project sponsor
Prototyping
Rapid Application Development
    (RAD)
State
Static view
Step
Structured design
Subclass
Superclass
Support plan
Systems development life cycle
    (SDLC)
System proposal
System prototype

System request
System specification
Systems analyst
Technical writer
Technique
Throw-away prototyping
To-Be system
Training plan
Unified Modeling Language
    (UML)
Use Case
Use-case driven
Version
Waterfall development
Work plan

QUESTIONS

1. Compare and contrast phases, steps, techniques, and deliverables.
2. Describe the major phases in the systems development life cycle (SDLC).
3. Describe the principal steps in the planning phase. What are the major deliverables?
4. Describe the principal steps in the analysis phase. What are the major deliverables?
5. Describe the principal steps in the design phase. What are the major deliverables?
6. Describe the principal steps in the implementation phase. What are the major deliverables?
7. What are the roles of a project sponsor and the approval committee?
8. What does "gradual refinement" mean in the context of SDLC?
9. Compare and contrast process-centered methodologies, data-centered methodologies, and object-oriented methodologies.
10. Compare and contrast structured design-based methodologies in general to RAD-based methodologies in general.

11. Describe the major elements and issues associated with waterfall development.
12. Describe the major elements and issues associated with phased development.
13. Describe the major elements and issues associated with prototyping.
14. Describe the major elements and issues associated with throw-away prototyping.
15. Describe what it means for an object-oriented analysis and design methodology to be use-case driven, architecture centric, and iterative and incremental.
16. What are the basic characteristics of object-oriented systems?
17. What are the benefits of using an object-oriented analysis and design methodology that uses UML?
18. What are the major roles of a project team?
19. Compare and contrast the roles of the systems analyst, business analyst, and infrastructure analyst.

EXERCISES

A. Suppose you are a project manager using a waterfall development-based methodology on a large and complex project. Your manager has just read the latest article in *Computerworld* that advocates replacing this methodology with prototyping and comes to

your office requesting you to switch. What do you say?

B. The basic types of methodologies discussed in this chapter can be combined and integrated to form new hybrid methodologies. Suppose you were to combine throw-away prototyping with the use of waterfall development. What would the methodology look like? Draw a picture (similar to those in Figures 1-2 through 1-5). How would this new methodology compare to the others?

C. Suppose you were an analyst working for a small company to develop an accounting system. What type of methodology would you use? Why?

D. Suppose you were an analyst developing a new executive information system intended to provide key strategic information from existing corporate databases to senior executives to help in their decision making. What type of methodology would you use? Why?

E. Suppose you were an analyst developing a new information system to automate the sales transactions and manage inventory for each retail store in a large chain. The system would be installed at each store and exchange data with a mainframe computer at the company's head office. What type of methodology would you use? Why?

F. Look in the classified section of your local newspaper. What kinds of job opportunities are available for people who want analyst positions? Compare and contrast the skills that the ads ask for to the skills that we presented in this chapter.

G. Think about your ideal analyst position. Write a newspaper ad to hire someone for that position. What requirements would the job have? What skills and experience would be required? How would applicants be able to demonstrate that they had the appropriate skills and experience?

MINICASES

1. Barbara Singleton, manager of western regional sales at the WAMAP Company, requested that the IS department develop a sales force management and tracking system that would enable her to better monitor the performance of her sales staff. Unfortunately, due to the massive backlog of work facing the IS department, her request was given a low priority. After six months of inaction by the IS department, Barbara decided to take matters into her own hands. Based on the advice of friends, Barbara purchased a PC and simple database software, and constructed a sales force management and tracking system on her own.

Although Barbara's system has been "completed" for about six weeks, it still has many features that do not work correctly, and some functions are full of errors. Barbara's assistant is so mistrustful of the system that she has secretly gone back to using her old paper-based system, since it is much more reliable.

Over dinner one evening, Barbara complained to a systems analyst friend, "I don't know what went wrong with this project. It seemed pretty simple to me. Those IS guys wanted me to follow this elaborate set of steps and tasks, but I didn't think all that really applied to a PC-based system. I just thought I could build this system and tweak it around until I got what I wanted without all the fuss and bother of the methodology the IS guys were pushing. I mean, doesn't that just apply to their big, expensive systems?"

Assuming you are Barbara's systems analyst friend, how would you respond to her complaint?

2. Marcus Weber, IS project manager at ICAN Mutual Insurance Co., is reviewing the staffing arrangements for his next major project, the development of an expert system-based underwriters assistant. This new system will involve a whole new way for the underwriters to perform their tasks. The underwriters assistant system will function as sort of an underwriting supervisor, reviewing key elements of each application, checking for consistency in the underwriter's decisions, and ensuring that no critical factors have been overlooked. The goal of the new system is to improve the quality of the underwriters' decisions and to improve underwriter productivity. It is expected that the new system will substantially change the way the underwriting staff do their jobs.

Marcus is dismayed to learn that due to budget constraints, he must choose between one of two available staff members. Barry Filmore has had considerable experience and training in individual and organizational behavior. Barry has worked on several other projects in which the end users had to make significant adjustments to the new system, and Barry seems to have a knack for anticipating problems and smoothing the transition to a new work environment. Marcus had hoped to have Barry's involvement in this project.

Marcus's other potential staff member is Kim Danville. Prior to joining ICAN Mutual, Kim had considerable work experience with the expert system technologies that ICAN has chosen for this expert system project. Marcus was counting on Kim to help integrate the new expert system technology into ICAN's systems environment, and also to provide on-the-job training and insights to the other developers on this team.

Given that Marcus's budget will only permit him to add Barry or Kim to this project team, but not both, what choice do you recommend for him? Justify your answer.

PART ONE
PLANNING PHASE

PROJECT BINDER

The Planning Phase is the fundamental process of understanding why an information system should be built, and determining how the project team will build it. The first step is called Project Initiation. The project team creates the System Request and Feasibility Analysis that describe the system's business value to the organization--how will the system lower costs or increase profits? The second step is Project Management. The project manager creates a Workplan, Staffing Plan, Project Charter, and Risk Assessment to manage and control the project. The deliverables from both steps are presented to the project sponsor and approval committee at the end of the Planning Phase. They decide whether it is advisable to proceed with the system.

Project Initiation   CHAPTER 2

Project Management   CHAPTER 3

System Request

Feasibility Analysis

Work Plan

Staffing Plan

Charter and Standards

Risk Assessement

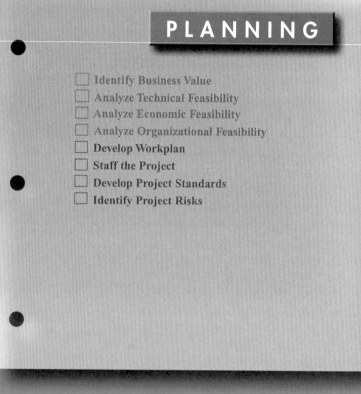

PLANNING

- ☐ Identify Business Value
- ☐ Analyze Technical Feasibility
- ☐ Analyze Economic Feasibility
- ☐ Analyze Organizational Feasibility
- ☐ **Develop Workplan**
- ☐ **Staff the Project**
- ☐ **Develop Project Standards**
- ☐ **Identify Project Risks**

T A S K C H E C K L I S T

PLANNING   ANALYSIS   DESIGN

CHAPTER 2
PROJECT INITIATION

THIS chapter describes project initiation, the point at which an organization creates and assesses the original goals and expectations for a new system. The first step in the process is to identify the business value for the system by developing a system request that provides basic information about the proposed system. The next step is to perform a feasibility analysis to determine the technical, economic, and organizational feasibility of the system. If all is appropriate, the system is approved and the development project begins.

OBJECTIVES

- Understand the importance of linking the information system to business objectives.
- Be able to create a system request.
- Understand how to assess technical, economic, and organizational feasibility.
- Be able to perform a feasibility analysis.

CHAPTER OUTLINE

IMPLEMENTATION

INTRODUCTION

The first step in any new development project is for someone—a manager, staff member, sales representative, or systems analyst—to see an opportunity to improve the business. New systems usually originate from some business need or opportunity. Many ideas for new systems or improvements to existing ones arise from the application of a new technology, but an understanding of technology is usually secondary to a solid understanding of the business and its objectives.

This may sound like common sense, but unfortunately, many projects are started without a clear understanding of how the system will improve the business. The information system (IS) field is filled with thousands of buzzwords, fads, and trends (e.g., Java, the Web, data warehousing). The promise of these innovations can appear so attractive that organizations begin projects even if they are not sure what value they offer, because they believe that the technologies are somehow important in their own right. However, as we discussed in Chapter 1, half of all IS projects are abandoned before completion. Most times, problems can be traced back to the very beginning of the systems development life cycle (SDLC), where too little attention was given to the identifying business value and understanding the risks associated with the project.

Does this mean that technical people should not recommend new systems projects? Absolutely not. In fact, the ideal situation is for both information technology (IT) people (i.e., the experts in systems) and businesspeople (i.e., the experts in business) to work closely to find ways for technology to support business needs. In this way, organizations can leverage the exciting technologies that are available while ensuring that projects are based on real business objectives, such as increasing sales, improving customer service, and decreasing operating expenses. Ultimately, information systems need to affect the organization's bottom line (in a positive way!).

Project initiation begins when someone (or some group) in the organization (called the *project sponsor*) identifies the business value that can be gained from using IT. The proposed project is described very briefly using a technique called the *system request,* which is submitted to an *approval committee* for consideration. This approval committee could be an IS steering committee of senior executives that meets regularly to make IS decisions, a senior executive who has control of organizational resources, or any other decision-making body that governs the use of business investments. The approval committee reviews the system request and makes an initial determination, on the basis of the information provided, of whether to investigate the proposal. If so, the next step is the feasibility analysis.

The *feasibility analysis* plays an important role in deciding whether to proceed with an IS development project. It examines the technical, economic, and organizational pros and cons of developing the system and gives the organization a slightly more detailed picture of the advantages of investing in the system as well as of any obstacles that could arise. In most cases, the project sponsor works together with an analyst (or analyst team) to develop the feasibility analysis for the approval committee.

Once the feasibility analysis has been completed, the system request is revised and submitted, along with the final feasibility study, back to the approval committee. The board then decides whether to approve the project, decline it, or table it until additional information is available.

IDENTIFYING BUSINESS VALUE

All systems have to address business needs or they likely will fail; therefore, most organizations have a process for ensuring that business value has been identified before a development project can begin. Every organization has its own way of initiating a system, but most start with a technique called a *system request*. A system request documents the business reasons for building the system and the value that the system is expected to provide. When the project is initiated, business users complete this form as part of the formal system approval process within the organization.

System Request

Most system requests include four elements: project sponsor, business need, functionality, and expected value (Figure 2-1).

The *project sponsor* is the person who has an interest in seeing the system succeed, someone who will work throughout the SDLC to make sure that the project is moving in the right direction from the perspective of the business. Usually

```
                                                              [Date]

                          SYSTEM REQUEST


Project Name:

Project Sponsor:
Name:
Department:
Phone:                          E-mail:

Business Need:


Functionality:


Expected Value:
Tangible:

Intangible:


Special Issues or Constraints:
```

FIGURE 2-1
System Request

2-1 IDENTIFY TANGIBLE AND INTANGIBLE VALUE

Virginia Power, a Richmond-based utility that serves more than 2 million customers across an area the size of South Carolina, overhauled some of its core processes and technology. The goal was to improve customer service and cut operations costs by developing a new workflow and geographic information system. When the project is finished, service engineers who used to sift through thousands of paper maps will be able to pinpoint the locations of electricity poles with computerized searches. The project will help the utility improve management of all its facilities, records, maps, scheduling, and human resources. That, in turn, will help Virginia Power increase employee productivity, improve customer response times, and reduce the costs of operating crews.

Source: Julia King "Utility hopes IT overhaul recharges customer service," *Computerworld*, November 10, 1997.

QUESTIONS:

1. What kinds of things does Virginia Power do that requires it to know power pole locations? How often does it do these things? Who benefits if Virginia Power can locate power poles faster?
2. On the basis of your answers to question 1, describe three tangible benefits that Virginia Power can receive from its new computer system. How can these be quantified?
3. On the basis of your answers to question 1, describe three intangible benefits that Virginia Power can receive from its new computer system. How can these be quantified?

this is the person who initiated the proposed project and who will serve as the primary point of contact for the system on the business side.

The *business need* describes why the information system should be built and explains to the approval committee why the organization should fund the project. The business need should be clear and concise but at this point is not likely to be completely defined.

The *functionality* of the system (what the information system will do) needs to be explained at a high level so that the approval committee and ultimately the project team understands what the business unit expects from the final product. Functionality indicates what features and capabilities the information system will have to include, such as the ability to collect customer orders over the Web or enable organization employees to access statistics about vendor performance.

The *expected value* to be gained from the system should include both tangible and intangible value. *Tangible value* can be quantified and measured (e.g., 2% reduction in operating costs). An *intangible value* results from an intuitive belief that the system provides important but hard-to-measure benefits to the organization (e.g., improved customer service). This portion of the form is most useful when it includes values that can be measured after the system is in place so that the actual results can be compared to the original goals in a post-system audit.

Special issues or constraints also are included as a catchall for other information that should be considered in assessing the project. For example, the project may need to be completed by a specific deadline. Project teams need to be aware of any special circumstances that could affect the outcome of the system.

Once the system request is completed, it is sent to the *approval committee*, which could be an information system's steering committee that meets regularly to make information systems decisions, a senior executive who has control of organizational resources, or any other decision-making body that governs the use of business investments. Steering committees are quite common because they include senior executives representing departments or divisions across the organization. They assess the system requests from an organization-wide perspective, trying to prioritize them based on the needs of the company and ensuring that projects are integrated into the technical environment that exists and is envisioned by management.

Applying the Concepts at CD Selections

Throughout this book, we will apply the concepts in each chapter to a fictitious company called CD Selections. For example, in this section, we will illustrate the creation of a system request. CD Selections is a chain of 50 music stores located in California, with headquarters in Los Angeles. Annual sales last year were $50 million, and they have been growing at about 3% to 5% per year for the past few years.

Background Margaret Mooney, vice president of marketing, has recently become both excited by and concerned with the rise of Internet sites selling CDs. The Internet has the potential to enable CD Selections to reach beyond its regional focus and significantly increase sales by moving firmly into the North American market (or even the global market). However, the Internet also poses a significant threat, as Web competitors such as CDnow and Amazon.com continue to take sales from CD Selections' traditional stores.

CD Selections currently has a simple Web page that provides information about the company and about each of its stores (e.g., map, operating hours, relative size). The page was developed by an Internet consulting firm and is hosted by a prominent local Internet service provider (ISP) in Los Angeles. Neither CD Selections nor the ISP has much experience with Internet-based electronic commerce.

System Request At CD Selections, new IS projects are reviewed and approved by an IS steering committee that meets monthly. For Margaret, the first step was to prepare a system request for the steering committee.

Figure 2-2 shows the system request she prepared. The sponsor is Margaret, and the business need is "to increase sales by reaching a new market: Internet customers." Notice that the need does not focus on the technology, such as the need "to upgrade our Web page." The focus is on the business aspects: increasing sales.

The functionality is described at a very high level of detail. In this case, Margaret's vision for the functionality includes the ability to complete a sale over the Web. Specifically, customers should be able to search for products over the Internet, place an order for a product or products on-line, receive immediate credit approval, receive a confirmation for orders that are placed, and enjoy timely delivery of the products.

The expected value describes how this functionality will affect the business. Margaret found estimating intangible business value to be fairly straightforward in this case. The Internet is a hot area, so she expects the Internet to improve customer recognition and satisfaction. Estimating tangible value is more difficult. She expects that Internet ordering will increase sales, but by how much? Margaret started by trying to find some estimates of Internet-based CD sales—where else but on the Web? She started searching the Web sites of the leading business magazines

November 15, 1999

SYSTEM REQUEST

Project Name: Internet sales

Project sponsor: Margaret Mooney, Vice President of Marketing
Department: Marketing
Phone: 555-3242
E-mail: mmooney@cdselections.com

Business need: To increase CD sales by reaching a new market: Internet cus-
                       tomers.

Functionality:
Using the Web, customers should be able to complete a purchase of our products. The
initial focus should be on CDs, but we should also consider the other products that we
sell (e.g., tapes, posters, books, accessories). Customers should be able to:
- Search through a list of products we sell.
- Place an order for one or more products.
- Receive immediate credit approval.
- Receive confirmation that their order has been placed and an expected delivery date.
- Receive their order in a timely manner.

Expected value:
Tangible:
- A $3 million to $5 million increase in annual sales after the business has been operat-
  ing for 2 to 3 years.
Intangible:
- Improved customer satisfaction.
- Improved recognition of the CD Selections brand name that may increase traffic in
  our traditional stores.

Special issues or constraints:
- The Marketing Department views this as a strategic system. According to Jupiter
  Communications LLC, Internet music industry revenues were $23 million in 1998,
  and they are predicted to reach $2 billion to $3 billion in sales by 2002 (or about
  20% to 25% of total industry sales). The leading Internet CD retailer in 1998,
  CDnow, had revenues of about $8 million. Today, the top two retailers, CDnow and
  Amazon.com, have combined sales of $30 million. Without Internet sales capability,
  we are unable to compete in this market.
- The system should be in place for the holiday shopping season next year.

FIGURE 2-2
System Request for CD Selections

such as *The Economist, Business Week, and CIO Magazine.*[1] She quickly learned
that according to Jupiter Communications LLC, Internet music industry revenues
were $23 million in 1998 (compared with $12 billion for all music retailing) and

---

[1] Okay, we know a vice president of marketing probably wouldn't immediately search *CIO Magazine*'s Web
site, (www.cio.com), but it is one of the best sources of information for IS professionals, particularly for Web-
based businesses. We started our search for this information on that Web site and found the exact information
Margaret needed in less than 2 minutes (no kidding!).
See http://www.cio.com/archive/webbusiness/120197_music.html

Think about your own university or college and choose an idea that could improve student satisfaction with the course enrollment process. Currently, can students enroll for classes from anywhere? How long does it take? Are directions simple to follow? Is on-line help available?

Next, think about how technology can help support your idea. Would you need completely new technology? Can the current system be changed?

QUESTION:

Create a system request that you could give to the administration that explains the sponsor, business need, functionality, and potential value of the project. Include any constraints or issues that should be considered.

that the Internet is predicted to account for $2 billion to $3 billion in sales by 2002 (or about 20% to 25% of total industry sales). The leading Internet retailer in 1998, CDnow, had revenues of about $8 million. During 1999, Amazon.com pulled ahead of CDnow. Margaret had expected that the Internet was a potentially important source of revenue, but she had never expected that industry analysts had predicted such a phenomenal growth rate—more than 200% per year compared to a 3% growth in sales through traditional music stores.

Estimating how much revenue CD Selections should anticipate from a new Internet music business was not simple. One approach would be to try to use some of CD Selections' standard models for predicting sales of new stores. Retail stores average about $1 million in sales per year (after they have been open 1 or 2 years), depending on such location factors as city population, average incomes, proximity to universities, and so on. However, these models really don't lend themselves to the Internet—population in the local area is not a useful concept. Margaret thought the Internet business might end up generating revenue similar to that for a half dozen traditional stores. This would suggest revenues of $6 million, give or take $1 million or $2 million, after the Internet business had been operating for a few years.

Another approach would be to base CD Selections' Internet revenues on some percentage of CDnow's revenues and assume some annual growth rate to predict future sales. For example, if CD Selections earns only 10% of CDnow's revenues in the first year ($800,000) and experiences a 100% growth rate for the next 2 years (somewhat conservative given the predicted industry growth rate of over 200%), this would give a third-year revenue of $3.2 million. Assuming a 200% growth rate, this would mean sales of about $7 million.

A third approach to estimating revenues would be to assume that CD Selections' Internet revenues should ultimately be about 20% of its revenues from traditional stores, because the Internet is expected to account for 20% of total industry sales. This would suggest predicted revenue of $10 million per year (20% of $50 million).

Although there was considerable variation among the three approaches, they suggested a range of values from $3 million to $10 million. Margaret is conservative, so she decided to estimate an increase in sales of only $3 to 5 million. See Figure 2-2 for the completed system request.

FEASIBILITY ANALYSIS

Once the need for the system and its basic functionality have been defined, it is time to create a more detailed business case to better understand the opportunities and limitations associated with the proposed project. *Feasibility analysis* guides the organization in determining whether to proceed with a project. Feasibility analysis also identifies the important risks associated with the project that must be addressed if the project is approved. As with the system request, each organization has its own process and format for the feasibility analysis, but most include three techniques: technical feasibility, economic feasibility, and organizational feasibility. Some firms also conduct a formal legal feasibility analysis that includes both an analysis of civil law to assess potential liabilities and the possibility of lawsuits, and an analysis of criminal law to assess conformance with laws. Assessing criminal law is particularly important in an international context where laws may be less well understood. For example, it is against the law in France to operate a Web server that provides information in languages other than French, unless the information is also available in French. The results of these techniques are combined into a *feasibility study* deliverable that is given to the approval committee at the end of project initiation (Figure 2-3).

Although we will discuss feasibility analysis now within the context of project initiation, most project teams will revise their feasibility study throughout the SDLC and revisit its contents at various checkpoints during the project. If at any point the project's risks and limitations outweigh its benefits, the project team may decide to cancel the project or make necessary improvements.

Technical Feasibility

The first technique in the feasibility analysis is to assess the *technical feasibility* of the project, the extent to which the system can be successfully designed, developed,

---

Technical feasibility: can we build it?
- Familiarity with application: less familiarity generates more risk.
- Familiarity with technology: less familiarity generates more risk.
- Project size: large projects have more risk.

Economic feasibility: should we build it?
- Development costs
- Annual operating costs
- Annual benefits (cost savings and revenues)
- Intangible costs and benefits

Organizational feasibility: if we build it, will they come?
- Project champion(s)
- Senior management
- Users
- Other stakeholders

FIGURE 2-3

Feasibility Analysis Assessment Factors

and installed by the IS group. Technical feasibility analysis is in essence a *technical risk analysis*[2] that strives to answer the question "*Can* we build it?"[3]

Many risks that endanger the successful completion of the project. Most important is the users' *and* analysts' *familiarity with the application.* When analysts are unfamiliar with the business application area, they have a greater chance of misunderstanding the users or missing opportunities for improvement. The risks increase dramatically when the users themselves are less familiar with an application, such as with the development of a system to support a new business innovation (e.g., Microsoft's starting up a new Internet dating service). In general, the development of new systems is riskier than extensions to an existing system because existing systems tend to be better understood.

Familiarity with the technology is another important source of technical risk. When a system will use technology that has not been used before *within the organization,* there is a greater chance that problems will occur and delays will be incurred because of the need to learn how to use the technology. Risk increases dramatically when the technology itself is new (e.g., a new Java development tool kit).

Finally, *project size* is an important consideration, whether measured as the number of people on the development team, the length of time it will take to complete the project, or the number of distinct features in the system. Larger projects present more risk, both because they are more complicated to manage and because there is a greater chance that some important system requirements will be overlooked or misunderstood. The extent to which the project is highly integrated with other systems (which is typical of large systems) can cause problems because complexity is increased when many systems must work together.

The assessment of a project's technical feasibility is not cut and dried, because in many cases, some interpretation of the underlying conditions is needed (e.g., how large does a project need to grow before it becomes less feasible?). The best approach is to compare the project under consideration with prior projects undertaken by the organization.

Economic Feasibility

The second element of a feasibility analysis is to perform an *economic feasibility* analysis (also called a *cost–benefit analysis*) that identifies the financial costs and benefits associated with the project. This attempts to answer the question "*Should* we build the system?" Economic feasibility is determined by identifying costs and benefits associated with the system, assigning values to them, and then calculating the cash flow and return on investment for the project.

Identify Costs and Benefits

The first task when developing an economic feasibility is to identify the kinds of costs and benefits the system will have and list them along the left-hand column of a spreadsheet, grouped with benefits first and costs second (Figure 2-4).

---

[2] One of the best books on risk analysis, as it pertains to information systems development, is Robert N. Charette, *Software Engineering Risk Analysis and Management* (New York: McGraw-Hill, 1989).

[3] We use the words *build it* in the broadest sense. Organizations can also choose to buy a commercial software package and install it, in which case the question might be, "Can we select the right package and successfully install it?"

Economic feasibility includes an assessment of financial impacts in four categories: (1) development costs, (2) operational costs, (3) tangible benefits, and (4) intangible costs and benefits. *Development costs* are those tangible expenses that are incurred during the construction of the system, such as salaries for the project team, hardware and software expenses, consultant fees, training, and office space and equipment. Development costs are usually thought of as one-time costs. *Operational costs* are those tangible costs that are required to operate the system, such as salaries for operations staff, software licensing fees, equipment upgrades, and communications charges. Operational costs are usually thought of as variable costs. Figure 2-5 lists examples of development and operational costs that can be used to determine economic feasibility.

Revenues and cost savings are those *tangible benefits* that the system enables the organization to collect or tangible expenses that the system enables the organization to avoid. Revenues often include increased sales, reductions in staff, and reductions in inventory. Of course, a project also can affect the organization's bottom line by reaping *intangible benefits* or incurring *intangible costs*. Intangible costs and benefits are more difficult to incorporate into the economic feasibility

Benefits							
Increased sales							
Improved customer service							
Reduced inventory costs							
Total benefits							
Development costs							
2 servers @ $125,000							
Printer							
Software licenses							
Server software							
Development labor							
Total development costs							
Operational costs							
Hardware							
Software							
Operational labor							
Total operational costs							
Total costs							
Net benefits							

FIGURE 2-4

Identify Costs and Benefits

Development Costs	Operational Costs
Development team salaries	Software upgrades
Consultant fees	Software licensing fees
Development training	Hardware repairs
Hardware and software	Hardware upgrades
Vendor installation	Operational team salaries
Office space and equipment	Communications charges
Data conversion costs	User training

FIGURE 2-5
Example Costs for Economic Feasibility

because they are based on intuition and belief rather than "hard numbers." Nonetheless, they should be listed along with the tangible economic feasibility items.

Assign Values to Costs and Benefits   Once the types of costs and benefits have been identified, you will need to assign specific dollar values to them. This may seem impossible—how can someone quantify costs and benefits that haven't happened yet? How can those predictions be realistic? Although this task is very difficult, you have to do the best you can to come up with reasonable numbers for all of the costs and benefits. Only then can the approval committee make an educated decision about whether to move ahead with the project.

The best strategy for estimating costs and benefits is to rely on the people who have the best understanding of them. For example, costs and benefits that are related to the technology or the project itself can be provided by the company's IS group or external consultants, and business users can develop the numbers associated with the business (e.g., sales projections, order levels). You also can consider past projects, industry reports, and vendor information, although these approaches probably will be a bit less accurate. Likely, all of the estimates will be revised as the project proceeds.

What about the *intangible* costs and benefits? Sometimes it is acceptable to list intangible benefits, such as improved customer service, without assigning a dollar value, whereas other times estimates have to made regarding how much an intangible benefit is "worth." We suggest that you quantify intangible costs or benefits if at all possible. If you do not, how will you know if they have been realized? Suppose that a system is supposed to improve customer service. This is intangible, but let's assume that the greater customer service will decrease the number of customer complaints by 10% each year over 3 years and that $200,000 is spent on phone charges and phone operators who handle complaint calls. Suddenly we have some very tangible numbers with which to set goals and measure the original intangible benefit.

Figure 2-6 shows the economic feasibility spreadsheet with values assigned to the various costs and benefits. Notice that the customer service intangible benefit has been quantified on the basis of fewer customer complaint phone calls. The intangible benefit of being able to offer services that competitors currently offer was not quantified, but it was listed so that the approval committee will consider the benefit when assessing the system's economic feasibility.

Benefits[a]	
Increased sales	500,000
Improved customer service[b]	70,000
Reduced inventory costs	68,000
Total benefits	**638,000**
Development costs	
2 servers @ $125,000	250,000
Printer	100,000
Software licenses	34,825
Server software	10,945
Development labor	1,236,525
Total development costs	**1,632,295**
Operational costs	
Hardware	54,000
Software	20,000
Operational labor	111,788
Total operational costs	**185,788**
Total costs	**1,818,083**

[a] An important yet intangible benefit will be the ability to offer services that our competitors currently offer.

[b] Customer service numbers have been based on reduced costs for customer complaint phone calls.

FIGURE 2-6
Assign Values to Costs and Benefits

Determine Cash Flow   A formal cost–benefit analysis usually contains costs and benefits over a selected number of years (usually 3 to 5 years) to show cash flow over time (Figure 2-7). When this *cash flow method* is used, the years are listed across the top of the spreadsheet to represent the time period for analysis and numeric values are entered in the appropriate cells within the spreadsheet's body. Sometimes fixed amounts are entered into the columns. For example, Figure 2-7 lists the same amount for customer service and inventory costs for all 5 years. Usually, amounts are augmented by some rate of growth to adjust for inflation or business improvements, as shown by the 6% increase that is added to the sales numbers in the sample spreadsheet. Finally, totals are added to determine what the overall benefits will be, and the higher the overall total, the more feasible the solution is in terms of the economic feasibility.

Determine Return on Investment   There are several problems with the cash flow method because it does not show the "bang for the buck" that the organization is getting on the investment, and it does not consider the time value of money (i.e., $1

	2000 ($)	2001 ($)	2002 ($)	2003 ($)	2004 ($)	Total ($)
Benefits[a]						
Increased sales	500,000	530,000	561,800	595,508	631,238	2,818,546
Improved customer service[b]	70,000	70,000	70,000	70,000	70,000	350,000
Reduced inventory costs	68,000	68,000	68,000	68,000	68,000	340,000
Total benefits	638,000	668,000	699,800	733,508	769,238	3,508,546
Development costs						
2 servers @ $125,000	250,000	0	0	0	0	250,000
Printer	100,000	0	0	0	0	100,000
Software licenses	34,825	0	0	0	0	34,825
Server software	10,945	0	0	0	0	10,945
Development labor	1,236,525	0	0	0	0	1,236,525
Total development costs	1,632,295	0	0	0	0	1,632,295
Operational costs						
Hardware	54,000	81,261	81,261	81,261	81,261	379,044
Software	20,000	20,000	20,000	20,000	20,000	100,000
Operational labor	111,788	116,260	120,910	125,746	130,776	605,480
Total operational costs	185,788	217,521	222,171	227,007	232,037	1,084,524
Total costs	1,818,083	217,521	222,171	227,007	232,037	2,716,819
Total	(1,180,083)	450,479	477,629	506,501	537,201	791,728

[a] An important yet intangible benefit will be the ability to offer services that our competitors currently offer.

[b] Customer service numbers have been based on reduced costs for customer complaint phone calls.

FIGURE 2-7
Determine Cash Flow

today is *not* worth $1 tomorrow). Therefore, some project teams need to add calculations to the spreadsheet to provide the approval committee with a more accurate picture of the project's worth.

The *return on investment* (ROI) is a calculation that is listed somewhere on the spreadsheet that measures the amount of money an organization receives in return for the money it spends—a high ROI results when benefits far outweigh costs. One drawback of ROI is that it considers only the end points of the investment, not the cash flow in between, so it should not be used as the sole indicator of a project's worth (Figure 2-8).

The ROI figure in Figure 2-8 does not account for the present value of future money. For instance, consider Figure 2-9, which shows the worth of $1 given different numbers of years and different interest rates. If you have a friend who owes you $1 today but instead of returning it now gives you a dollar 3 years from now, you've been had! Given a 10% increase in value, you'll be receiving the equivalent of $0.75 in today's terms.

	2000 ($)	2001 ($)	2002 ($)	2003 ($)	2004 ($)	Total ($)
Benefits[a]						
Increased sales	500,000	530,000	561,800	595,508	631,238	2,818,546
Improved customer service[b]	70,000	70,000	70,000	70,000	70,000	350,000
Reduced inventory costs	68,000	68,000	68,000	68,000	68,000	340,000
Total benefits	**638,000**	**668,000**	**699,800**	**733,508**	**769,238**	**3,508,546**
Development costs						
2 servers @ $125,000	250,000	0	0	0	0	250,000
Printer	100,000	0	0	0	0	100,000
Software licenses	34,825	0	0	0	0	34,825
Server software	10,945	0	0	0	0	10,945
Development labor	1,236,525	0	0	0	0	1,236,525
Total development costs	**1,632,295**	**0**	**0**	**0**	**0**	**1,632,295**
Operational costs						
Hardware	54,000	81,261	81,261	81,261	81,261	379,044
Software	20,000	20,000	20,000	20,000	20,000	100,000
Operational labor	111,788	116,260	120,910	125,746	130,776	605,480
Total operational costs	**185,788**	**217,521**	**222,171**	**227,007**	**232,037**	**1,084,524**
Total costs	**1,818,083**	**217,521**	**222,171**	**227,007**	**232,037**	**2,716,819**
Total	**(1,180,083)**	**450,479**	**477,629**	**506,501**	**537,201**	**791,728**

[a] An important yet intangible benefit will be the ability to offer services that our competitors currently offer.

[b] Customer service numbers have been based on reduced costs for customer complaint phone calls.

Return on investment (ROI): 29.14%

ROI = Total/Total Costs

FIGURE 2-8
Determine Return on Investment

Number of Years	6%	10%	15%
1	.943	.909	.870
2	.890	.826	.756
3	.840	.751	.572
4	.792	.683	.497

FIGURE 2-9
The Value of a Dollar

CONCEPTS   **2-A   RETURN ON INVESTMENT**

IN ACTION

In 1996, International Data Corp. conducted a study of 62 companies that had implemented data warehousing. They found that the implementations generated an average 3-year return on investment (ROI) of 401%. The interesting part is that although 45 companies reported results between 3% and 1,838%, the range varied from –1,857% to as high as 16,000%!

QUESTIONS:
1. What kinds of reasons would you give for the varying degrees of ROI?
2. How would you interpret these results if you were a manager thinking about investing in data warehousing for your company?

Net present value (NPV) is used to calculate the present value of future cash flows with the investment outlay required to implement the project. It can be calculated in many different ways, some of which are extremely complex. See Figure 2-10 for a basic calculation that can be applied to your total figure to get a more realistic value.

Organizational Feasibility

The final technique used for feasibility analysis is to assess the *organizational feasibility* of the system, how well the system ultimately will be accepted by its users and incorporated into the ongoing operations of the organization. Many organizational factors can have an impact on the project, and seasoned developers know that organizational feasibility can be the most difficult feasibility dimension to assess. In essence, an organizational feasibility analysis attempts to answer the question "If we build it, will they come?"

One way to assess organizational feasibility is to conduct a *stakeholder analysis*.[4] A *stakeholder* is a person, group, or organization that can affect (or will be affected by) a new system. In general, the most important stakeholders in the introduction of a new system are the project champion, system users, and organizational management (Figure 2-11), but systems sometimes affect other stakeholders as well. For example, the IS department can be a stakeholder of a system because IS jobs or roles may be changed significantly after its implementation. One key stakeholder outside of the champion, users, and management in Microsoft's project that embedded Internet Explorer as a standard part of Windows was the U.S. Department of Justice.

The *champion* is a high-level non-IS executive who is usually but not always the project sponsor who initiated the system request. The champion supports the project by providing time, resources (e.g., money), and political support within the organization by communicating the importance of the system to other organizational decision makers. More than one champion is preferable because if the champion leaves the organization, the support could leave as well.

[4] A good book on stakeholder analysis that presents a series of stakeholder analysis techniques is that by R. O. Mason and I. I. Mittroff, *Challenging Strategic Planning Assumptions: Theory, Cases, and Techniques* (New York: John Wiley & Sons, 1981).

	2000 ($)	2001 ($)	2002 ($)	2003 ($)	2004 ($)	Total ($)
Benefits[a]						
Increased sales	500,000	530,000	561,800	595,508	631,238	2,818,546
Improved customer service[b]	70,000	70,000	70,000	70,000	70,000	350,000
Reduced inventory costs	68,000	68,000	68,000	68,000	68,000	340,000
Total benefits	638,000	668,000	699,800	733,508	769,238	3,508,546
Development costs						
2 servers @ $125,000	250,000	0	0	0	0	250,000
Printer	100,000	0	0	0	0	100,000
Software licenses	34,825	0	0	0	0	34,825
Server software	10,945	0	0	0	0	10,945
Development labor	1,236,525	0	0	0	0	1,236,525
Total development costs	1,632,295	0	0	0	0	1,632,295
Operational costs						
Hardware	54,000	81,261	81,261	81,261	81,261	379,044
Software	20,000	20,000	20,000	20,000	20,000	100,000
Operational labor	111,788	116,260	120,910	125,746	130,776	605,480
Total operational costs	185,788	217,521	222,171	227,007	232,037	1,084,524
Total costs	1,818,083	217,521	222,171	227,007	232,037	2,716,819
Total	(1,180,083)	450,479	477,629	506,501	537,201	791,728

[a] An important yet intangible benefit will be the ability to offer services that our competitors currently offer.

[b] Customer service numbers have been based on reduced costs for customer complaint phone calls.

Return on investment (ROI): 29.14%

Net present value (NPV): 682,951

$NPV = Total/(1 + interest\ rate)^n$

where n is the number of years into the future

interest rate = .03

$n = 5$

FIGURE 2-10
Determine Net Present Value

Although champions provide day-to-day support for the system, *organizational management* also needs to support the project. Such management support conveys to the rest of the organization the belief that the system will make a valuable contribution and that necessary resources will be made available. Ideally, management should encourage people in the organization to use the system and to accept the many changes that the system will likely create.

A third important set of stakeholders is the *system users* who ultimately will use the system once it has been installed in the organization. Too often, the project team meets with users at the beginning of a project and then disappears until after the system is created. In this situation, rarely does the final product meet the expectations and needs of those who are supposed to use it, because

Stakeholders	Roles	Techniques for Improvement
Champion(s)	Initiating the project Promoting the project Allocating time to the project Providing resources	Making a presentation about the objectives of the project and the proposed benefits to those executives who will benefit directly from the system Creating a prototype of the system to demonstrate its potential value
Organizational management	Knowing about the project	Making a presentation to management about the objectives of the project and the proposed benefits
	Budgeting enough money for the project Encouraging users to accept and use the system	Marketing the benefits of the system using memos and organizational newsletters Encouraging the champion(s) to talk about the project with peers
System users	Making decisions that influence the project Performing hands-on activities for the project Ultimately determining whether the project is successful by using or not using the system	Assigning users official roles on the project team Assigning users specific tasks to perform with clear deadlines Asking for feedback from users regularly (e.g., at weekly meetings)

FIGURE 2-11
Some Important Stakeholders for Organizational Feasibility

needs change and users become more savvy as the project progresses. User participation should be promoted throughout the development process to make sure that the final system will be accepted and used by getting users actively involved in the development of the system (e.g., performing tasks, providing feedback, and making decisions).

One other area of organizational feasibility that should be considered is that of politics. It is always possible that the creation of a new information system causes a change in the power structure of a firm. As developers of an information system, we should be aware of what Machiavelli stated about 500 years ago:

> There is nothing more difficult to plan, more doubtful of success, nor more dangerous to manage than the creation of new order of things. . . . Whenever his enemies have occasion to attack the innovator they do so with the passion of partisans, while the others defend him sluggishly so that the innovator and his party alike are vulnerable.

As an information system's developer, you will be a creator of a new order of things. As such, you should be aware of the changes, both direct and indirect, that the new information system will put into place. For example, many of the enterprise resource planning systems such as SAP force organizational changes to take place.

The final feasibility study helps organizations make wiser investments regarding ISs because it forces project teams to consider technical, economic, and organizational factors that can impact their projects. It protects IS professionals from criticism by keeping the business units educated about decisions and positioned as the leaders in the decision-making process. Remember—the feasibility study should be revised several times during the project at points when the project team makes critical decisions about the system (e.g., before the design begins). It can be used to support and explain the critical choices that are made throughout the SDLC.

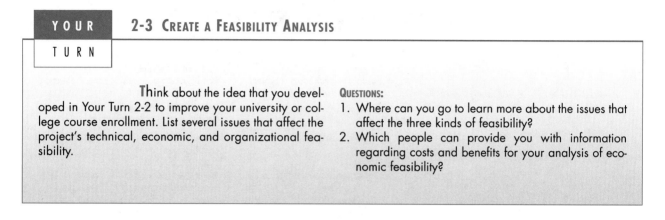

Applying the Concepts at CD Selections

The steering committee met and placed the Internet sales project high on its list of projects. A senior systems analyst, Alec Adams, was assigned to conduct a feasibility analysis because of his familiarity with CD Selections' sales and distribution systems. He also was an avid user of the Web and had been offering suggestions for the improvement of CD Selections' Web site.

Alec and Margaret worked closely together over the next few weeks on the feasibility analysis. Figure 2-12 is the summary page of the feasibility analysis; the report itself would provide additional detail and supporting documentation and would probably run for 5 to 10 pages.

As shown in Figure 2-12, the project is a high-risk project for CD Selections. They have little experience with the application or the technology. One solution may be to hire a consultant with experience to support the project.

The economic feasibility analysis has refined the initial assumptions Margaret made for the system request. Figure 2-13 shows the summary spreadsheet that led to the conclusions on the feasibility analysis. Development costs are expected to be about $250,000. This is a very rough estimate, as Alec has had to make some assumptions about the amount of time it will take to design and program the system. These estimates will be revised after a detailed project plan has been developed and as the project proceeds.[5] Traditionally, operating costs include the costs of the computer operations. In this case, CD Selections has had to include the costs of business staff because they are creating a new business unit, resulting in a total of $1,472,633. Margaret and Alec have decided to use a conservative estimate for revenues ($5 million), although they note the potential for higher revenues. This shows that the project can still add significant business value, even if the underlying assumptions prove to be overly optimistic. The spreadsheet was projected over 3 years, and both the ROI and NPV were included.

The organizational feasibility is presented in the bottom of Figure 2-12. There is a strong champion, well placed in the organization to support the project. The project originated in the business or functional side of the company, not the IS department, and Margaret has carefully built up support for the project among the senior management team. This is an unusual system in that the ultimate end users are the consumers exter-

---

[5] Some of the salary information may seem high to you. Most companies use a full-cost model for estimating salary cost in which all benefits (e.g., health insurance, retirement, payroll taxes) are included in salaries when estimating costs.

FEASIBILITY ANALYSIS

Technical feasibility (risky)

Familiarity with application (low):
- The Marketing Department has no experience with Internet-based marketing and sales.
- The Information Systems Department has very little experience with the application area, aside from traditional sales and distribution systems.

Familiarity with technology (low):
- The Information Systems Department has never developed an Internet application of this scope before, although there is some limited experience from the corporate Intranet.
- Development tools and products for commercial Web application development are available in the marketplace, but we have little experience with them.
- The system will run off of our inventory database system, which is solid and stable.

Project size (moderate):
- We estimate that the project is moderate in size.
- With some effort, we can design the system to be fairly independent from existing systems to reduce complexity.

Economic feasibility (excellent)
Please see attached spreadsheet for details.

Tangible costs and benefits:
- 197% ROI over a 3-year period.
- Total benefits after 3 years equals $3,109,073 (adjusted for present value).

Intangible costs and benefits:
- Improved customer satisfaction.
- Improved recognition of the CD Selections brand name that may increase traffic in our traditional stores.
- Possible decreased sales in traditional stores as customers purchase on the Internet. However, if we are not on the Internet, those sales may go to our competitors who are on the Internet.

Organizational feasibility (excellent):

Project champion:
- Vice president of Marketing.

Senior management:
- There is strong support of the project within the senior management team.

Users:
- The ultimate end users are the consumers external to CD Selections who have shown significant and increasing interest in the use of the Internet for retail sales.

Other stakeholders:
- Management of our traditional stores may see Internet sales as competition, so we will need to carefully manage the project to ensure we can avoid problems associated with "channel conflict."

Additional comments:
- The Marketing Department views this as a strategic system. Without an Internet sales capability we are unable to compete in a large and rapidly growing market.
- Industry experts predict that by 2002 20% to 25% of industry sales will be from the Internet. Using these assumptions, revenue could increase to $10 million, although this is very optimistic.
- We should consider hiring a consultant with expertise in Internet marketing to assist in the project.
- We will also need to hire new staff to operate the business, from both the technical and business operations aspects.
- We need to conduct some market research with potential customers as the project proceeds.

FIGURE 2-12
Feasibility Analysis for CD Selections

	Year 1 ($)	Year 2 ($)	Year 3 ($)	Total ($)
Benefits[a]				
Increased sales	0	1,000,000	4,000,000	5,000,000
Improved customer service[b]	0	60,000	60,000	120,000
Total benefits	0	1,060,000	4,060,000	5,120,000
Development costs				
Labor				
Analysis and design	42,000	0	0	42,000
Programming	120,000	0	0	120,000
Web design	21,000	0	0	21,000
External consultant	25,000	0	0	25,000
Training	5,000	0	0	5,000
Hardware	25,000	0	0	25,000
Software	10,000	0	0	10,000
Vendor costs	0	0	0	0
Office space and equipment	2,000	0	0	2,000
Data conversion	0	0	0	0
Total development costs	250,000	0	0	250,000
Operational costs				
Software upgrades	1,000	1,000	1,000	3,000
Software licenses	3,000	1,000	1,000	5,000
Hardware repairs	1,000	1,000	1,000	3,000
Hardware upgrades	3,000	3,000	3,000	9,000
Labor (technical)				
Webmaster	85,000	87,550	90,177	262,727
Network technician	60,000	61,800	63,654	185,454
Computer operations	50,000	51,500	53,045	154,545
User training	2,000	1,000	1,000	4,000
Communications charges	20,000	20,000	20,000	60,000
Labor (business)				
Business manager	60,000	61,800	63,654	185,454
Assistant manager	50,000	51,500	53,045	154,545
Three staff	90,000	92,700	95,481	278,181
Office expense	30,000	30,900	31,827	92,727
Marketing Expenses	25,000	25,000	25,000	75,000
Total operational costs	480,000	489,750	502,883	1,472,633
Total costs	730,000	489,750	502,883	1,722,633
Net benefits	(730,000)	570,250	3,557,118	3,397,368
ROI	197%			
NPV	3,109,073			

[a] Additional benefit includes greater brand-name recognition for CD Selections.

[b] Customer service will reduce costs associated with customer complaints.

ROI = return on investment; NPV = net present value.

FIGURE 2-13
Economic Feasibility Analysis for
CD Selections

nal to CD Selections. Margaret and Alec have not done any specific market research to see how well potential customers will react to the CD Selections system, so this is a potential risk. However, general industry trends are probably a reasonable predictor. As the project proceeds, they should probably do some formal market research.

One potential additional stakeholder in the project is the management team responsible for the operations of the traditional stores, and the store managers. They may view the Internet system as stealing sales from their stores. Margaret and Alec need to manage this carefully as the project proceeds. It needs to be made clear that Internet stores will exist and that unless CD Selections has an Internet store it will lose the sales to competitors.

SUMMARY

Project Initiation

Project initiation is the point at which an organization creates and assesses the original goals and expectations for a new system. The first step in the process is to identify the business value for the system by developing a system request that provides basic information about the proposed system. The next step is for the analysts to perform a feasibility analysis to determine the technical, economic, and organizational feasibility of the system. If the analysis findings are appropriate, the system is approved and the development project begins.

System Request

The business value for an information system is identified and then described using a system request. This form contains the project's sponsor, business need, desired functionality, and expected value of the information system, along with any other issues or constraints that are important to the project. The document is submitted to an approval committee which determines whether the project would be a wise investment of the organization's time and resources.

Feasibility Analysis

A feasibility analysis is then used to provide more detail about the consequences of the investment in the proposed system, and it includes technical, economic, and organizational feasibilities. The technical feasibility focuses on whether the system *can* be built by examining the risks associated with the users' and analysts' familiarity with the application, familiarity with the technology, and project size. The economic feasibility addresses whether the system *should* be built. It includes a cost–benefit analysis of development costs, operational costs, tangible benefits, and intangible costs and benefits. Finally, the organizational feasibility assesses how well the system will be accepted by its users and incorporated into the ongoing operations of the organization. A stakeholder analysis of the project champion, system users, and senior management can be used to assess this feasibility dimension.

KEY TERMS

Approval committee	Cost–benefit analysis	Familiarity with the application
Business need	Development cost	Familiarity with the technology
Cash flow method	Economic feasibility	Feasibility analysis
Champion	Expected value	Feasibility study

Functionality
Intangible benefits
Intangible costs
Intangible value
Net present value (NPV)
Operational cost
Organizational feasibility

Organizational management
Project initiation
Project size
Project sponsor
Return on investment (ROI)
Special issues
Stakeholder

Stakeholder analysis
System request
System users
Tangible benefits
Tangible value
Technical feasibility
Technical risk analysis

QUESTIONS

1. Give three examples of business reasons for a system to be built.
2. What is the purpose of an approval committee?
3. Why should the system request be created by a businessperson as opposed to an information system (IS) professional?
4. What is the difference between intangible value and tangible value? Give three examples of each.
5. What are the purposes of the system request and the feasibility analysis? How are they used in the project approval process?
6. Describe two special issues that may be important to list on a system request.
7. Describe the three techniques for feasibility analysis.
8. What factors are used to determine project size?
9. Describe a risky project in terms of technical feasibility. Describe a project that would *not* be considered risky.
10. What are the four steps for assessing economic feasibility?
11. List two intangible benefits. Describe how these benefits can be quantified.

12. List two tangible benefits and two operational costs for a system. How would you determine the values that should be assigned to each item?
13. Explain the net present value (NPV) and return on investment (ROI) for a cost–benefit analysis. Why would these calculations be used?
14. What is stakeholder analysis? Discuss three stakeholders that would be relevant for most projects.
15. Which of the three feasibility analysis techniques is subject to the most errors (i.e., is the one experienced analysts are most likely to be concerned about making mistakes with)? Why?
16. Some companies believe that intangible benefits should not be considered in deciding whether to proceed with a system. Other companies do not bother to perform extensive ROI calculations, but rather rely most on the intangible benefits and quick calculations. What reasons do you think the companies might have for being so different (think about the characteristics of the different companies)?
17. What do you think are three common mistakes made by novice analysts in conducting a feasibility analysis?

EXERCISES

A. Locate a news article in an information technology (IT) trade magazine (e.g., *Computerworld*) about an organization that is implementing a new computer system. Describe the tangible and intangible value that the organization likely will realize from the new system.
B. Car dealers have realized how profitable it can be to sell automobiles using the Web. Pretend you work for a local car dealership that is part of a large chain such as CarMax. Create a system request you

might use to develop a Web-based sales system. Remember to list special issues that are relevant to the project.
C. Suppose that you are interested in buying yourself a new computer. Create a cost–benefit analysis that illustrates the return on investment (ROI) that you would receive from making this purchase. Computer-related Web sites (e.g., those for Dell and for Compaq) should have real tangible costs that you can include in your analysis. Project your numbers

out to include a 3-year period and provide the net present value (NPV) of the final total.

D. Consider the Amazon.com Web site. The management of the company decided to extend its Web-based system to include products other than books (e.g., wine, specialty gifts). How would you have assessed the feasibility of this venture when the idea first came up? How risky would you have considered the project that implemented this idea? Why?

E. Interview someone who works in a large organization and ask him or her to describe the approval process that exists for approving new development projects. What does this person think about the process? What are the problems? What are the benefits?

F. Reread Your Turn 2.1 ("Identify Tangible and Intangible Value"). Create a list of the stakeholders that should be considered in a stakeholder analysis of this project.

MINICASES

1. The Amberssen Specialty Company is a chain of 12 retail stores that sell a variety of imported gift items, gourmet chocolates, cheeses, and wines in the Toronto area. Amberssen has an IS staff of three people who have created a simple but effective information system of networked point-of-sale registers at the stores, and a centralized accounting system at the company headquarters. Harry Hilman, the head of Amberssen's IS group, has just received the following memo from Bill Amberssen, sales director (and son of Amberssen's founder).

> Harry—It's time Amberssen Specialty launched itself on the Internet. Many of our competitors are already there, selling to customers without the expense of a retail storefront, and we should be there too. I project that we could double or triple our annual revenues by selling our products on the Internet. I'd like to have this ready by Thanksgiving, in time for the prime holiday gift-shopping season. Bill

After pondering this memo for several days, Harry scheduled a meeting with Bill so that he could clarify Bill's vision of this venture. Using the standard content of a system request as your guide, prepare a list of questions that Harry needs to have answered about this project.

2. The Decker Company maintains a fleet of 10 service trucks and crews that provide a variety of plumbing, heating, and cooling repair services to residential customers. Currently, it takes on average about 6 hours before a service team responds to a service request. Each truck and crew averages 12 service calls per week, and the average revenue earned per service call is $150. Each truck is in service 50 weeks per year. Due to the difficulty in scheduling and routing, there is considerable slack time for each truck and crew during a typical week.

In an effort to more efficiently schedule the trucks and crews and improve their productivity, Decker management is evaluating the purchase of a prewritten routing and scheduling software package. The benefits of the system will include reduced response time to service requests and more productive service teams, but management is having trouble quantifying these benefits.

One approach is to make an estimate of how much service response time will decrease with the new system, which then can be used to project the increase in the number of service calls made each week. For example, if the system permits the average service response time to fall to 4 hours, management believes that each truck will be able to make 16 service calls per week on average—an increase of 4 calls per week. With each truck making 4 additional calls per week and the average revenue per call at $150, the revenue increase per truck per week is $600 (4* $150). With 10 trucks in service 50 weeks per year, the average annual revenue increase will be $300,000 ($600 * 10 * 50).

Decker Company management is unsure whether the new system will enable response time to fall to 4 hours on average, or will be some other number. Therefore, management has developed the following range of outcomes that may be possible outcomes of the new system, along with probability estimates of each outcome occurring.

New Response Time	# Calls/Truck/Week	Likelihood
2 hours	20	20%
3 hours	18	30%
4 hours	16	50%

Given these figures, prepare a spreadsheet model that computes the expected value of the annual revenues to be produced by this new system.

- ☑ **Identify Business Value**
- ☑ **Analyze Technical Feasibility**
- ☑ **Analyze Economic Feasibility**
- ☑ **Analyze Organizational Feasibility**
- ☐ Develop Workplan
- ☐ Staff the Project
- ☐ Develop Project Standards
- ☐ Identify Project Risks

TASK CHECKLIST

PLANNING   ANALYSIS   DESIGN

CHAPTER 3
PROJECT MANAGEMENT

THIS chapter describes the important steps of project management, the second part of the planning phase. At this time, the project manager creates a work plan, which includes the project's tasks, and estimates how long it will take to perform them. Next, he or she staffs the project and then puts several life-cycle activities in place to help control and direct the project throughout the systems development life cycle. These steps produce a project plan that includes a variety of project management deliverables.

OBJECTIVES

- Be able to create a project work plan.
- Become familiar with estimation.
- Understand why project teams use timeboxing.
- Become familiar with how to staff a project.
- Understand how computer aided software engineering, standards, and documentation improve the efficiency of a project.
- Understand how to reduce risk on a project.

CHAPTER OUTLINE

IMPLEMENTATION

INTRODUCTION

Think about major projects that occur in people's lives, such as throwing a big party like a wedding or graduation celebration. Months are spent in advance identifying and performing all of the tasks that need to get done, like sending out invitations and selecting a menu, and time and money are carefully allocated among them. Along the way, decisions are recorded, problems are addressed, and changes are made. The increasing popularity of the party planner, a person whose sole job is to coordinate a party, suggests how tough this job can be. In the end, the success of any party has a lot to do with the effort that went into planning along the way. System development projects can be much more complicated than the projects we encounter in our personal lives—usually, more people are involved (e.g., the organization), the costs are higher, and more tasks need to be completed. Therefore, you should not be surprised to learn that "party planners" exist for information system projects—they are called project managers.

Project management is the process of planning and controlling the development of a system within a specified time frame at a minimum cost with the right functionality.[1] A *project manager* has the primary responsibility for managing the hundreds of tasks and roles that need to be carefully coordinated. Nowadays, project management is an actual profession, and analysts spend years working on projects prior to tackling the management of them. In a 1999 *ComputerWorld* survey, more than half of 103 companies polled said they now offer formal project management training for information technology (IT) project teams. There also is a variety of project management software available like Microsoft Project and Symantic's Timeline (two popular personal computer packages) that support project management activities.

Although training and software are available to help project managers, unreasonable demands set by project sponsors and business managers can make project management very difficult. Too often, the approach of the holiday season, the chance at winning a proposal with a low bid, or a funding opportunity pressures project managers to promise systems long before they are able to deliver them. These overly optimistic timetables are thought to be one of the biggest problems that projects face; instead of pushing a project forward faster, they result in delays.

Thus, a critical success factor for project management is to start with a realistic assessment of the work that needs to be accomplished and then manage the project according to that assessment. This can be achieved by *carefully* following the three steps that are presented in this chapter: creating the work plan, staffing the project, and controlling and directing the project. The project manager ultimately creates a *project plan,* which includes all of the deliverables that result from these steps.

CREATING THE WORK PLAN

The first step in project management is to create a *work plan,* a dynamic schedule that records and keeps track of all of the tasks that need to be accomplished over

---

[1] A good book on project management is by Jack R. Meredith and Samuel J. Mantel, *Project Management: A Managerial Approach* (New York: John Wiley & Sons, 1995). Also the Project Management Institute (www.pmi.org) and the Information Systems Special Interest Group of the Project Management Institute (www.pmi-issig.org) have valuable resources on project management in information systems.

3-A HASTE SLOWS MICROSOFT

In 1984, Microsoft planned for the development of Microsoft Word to take one year. At the time, this was two months less than the most optimistic estimated deadline for a project of its size. In reality, it took Microsoft five years to complete Word. Ultimately, the overly aggressive schedule for Word *slowed* its development for a number of reasons. The project experienced high turnover due to developer burn-out from unreasonable pressure and work hours. Code was "finalized" prematurely, and the software spent much longer in "stabilization" (i.e., fixing bugs) than was originally expected (i.e., 12 months versus 3 months). And, the aggressive scheduling resulted in poor planning (the delivery date consistently was off by more than 60% for the first four years of the project.)

Source: Marco Iansiti, *Microsoft Corporation: Office Business Unit,* Harvard Business School Case Study 9-691-033, revised May 31, 1994, Boston: Harvard Business School, 1994.

QUESTION:
Suppose you take over as project manager in 1986 after the previous project manager has been fired. Word is now one year overdue. Describe three things you would do on your first day on the job to improve the project.

the life of the project. The work plan lists each task, along with important information about it, such as when it needs to be completed, the person assigned to do the work, and any deliverables that will result (Figure 3-1). The level of detail and the amount of information captured by the work plan depend on the needs of the project (and the detail usually increases as the project progresses). Usually, the work plan is the main component of the project management software that we mentioned earlier.

To create a work plan, you will need to perform two steps: identify the tasks that need to be accomplished and estimate the time that it will take to complete them. If the deadline needs to be shortened after the work plan is completed, a technique called timeboxing can be used to deliver important parts of the system to its users much faster.

Work Plan Information	Example
Name of the task	Perform economic feasibility
Start date	Jan 5, 2001
Completion date	Jan 19, 2001
Person assigned to the task	Project sponsor; Mary Smith
Deliverable(s)	Cost–benefit analysis
Completion status	Open
Priority	High
Resources that are needed	Spreadsheet software
Estimated time	16 hours
Actual time	14.5 hours

FIGURE 3-1
Kinds of Information Found on a Work Plan

Identifying Tasks

The overall objectives for the system should be listed on the system request, and when the system request is made the project manager must identify all of the tasks that need to be accomplished, from the beginning through the end of the project, to meet those objectives. One way to do this is by using a structured, top-down approach whereby the high-level tasks are defined first and then these are broken down into the subtasks that make up each. For example, a project manager can first list the four main phases of the systems development life cycle (SDLC)—planning, analysis, design, and implementation—and then list the steps that occur during each phase. The steps can be broken down until the desired level of detail is achieved.

Figure 3-2 shows that the planning phase includes project initiation and project management, and project initiation in turn involves identifying business value and performing a feasibility analysis. As each step is broken down, details like expected deliverables and hours required to complete the task are included. The number of items required in a breakdown of each step depends on the complexity and size of the project. The larger the project, the more important it becomes to define tasks at the very lowest level of detail so that important steps are not overlooked.

Another approach for building a work plan is to use a standard list of tasks, or a *methodology,* as a starting point. As we stated in Chapter 1, a methodology is a formalized approach to implementing the SDLC (i.e., it is a list of steps and deliverables). Although all of a methodology may not be needed for your current project, you can select the parts of the methodology that do apply and add them to the work plan. If an existing methodology or approach is not available in house, methodologies can be purchased from consultants or vendors, or books like this textbook can serve as guidance. Using an existing methodology is the most popular way to create a work plan because most organizations have a methodology that they use for projects.

Time Estimation

Once the tasks are identified, the project manager must then estimate the amount of time and effort that will be needed to complete the project. *Estimation*[2] is the process of assigning projected values for time and effort, and it can be performed manually or with the help of an estimation software package like Costar and Construx—there are over 50 available on the market. The estimates developed at the start of a project are usually based on a range of possible values (e.g., the design phase will take 3 to 4 months) and gradually become more specific as the project moves forward (e.g., the design phase will be completed on March 22).

The numbers used to calculate these estimates can come from several sources. Many methodoligies include formal estimating procedures. They also can be provided with the methodology that is used, taken from projects with similar tasks and technologies, or provided by experienced developers. Generally speaking, the numbers should be conservative. A good practice is to keep track of the actual time and effort values during the SDLC so that numbers can be refined along the way and the next project can benefit from real data. One of the greatest strengths of systems consulting firms is the past experience that they offer to

---

[2] A good book for further reading on software estimation is that by Capers Jones—*Estimating Software Costs* (New York: McGraw-Hill, 1989).

Phases
Planning
Analysis
Design
Implementation

Phases with High-Level Steps
Planning
Project initiation
System request
Feasibility analysis
Project management
Create work plan
Staff the project
Control and direct the project
Analysis
Design
Implementation

Work Plan	Deliverables	Estimated Hours	Actual Hours	Assigned To
Planning				
Project initiation				
Identify business value	System request	8		Sponsor
Feasibility analysis	Feasibility analysis	8		Sponsor
Perform technical feasibility		12		Sponsor
Perform economic feasibility	Cost–benefit analysis	12		Sponsor
Perform organizational feasibility	Stakeholder analysis	8		Sponsor
Project management				
Create work plan	Work plan	12		Project mgr
Staff the project	Staffing plan	8		Project mgr
Control and direct the project		4/week		Project mgr
Set up CASE tool		8		Project mgr
Prepare documentation		4		Project mgr
Determine standards	Standards list	4		Project mgr
Analysis				
Design				
Implementation				

CASE = Computer-aided software engineering; mgr = manager.

FIGURE 3-2
Top-down Approach to Identifying Project Tasks

a project; they have estimates and methodologies that have been developed and honed over time and applied to hundreds of projects.

Whether manual or automated, estimation is difficult because it involves making *tradeoffs* among its three components: the size of the system (in terms of what

CONCEPTS   **3-B TRADEOFFS**

IN ACTION

I was once on a project to develop a system that should have taken a year to build. Instead, the business need demanded that the system be ready within 5 months—impossible!

On the first day of the project, the project manager drew a triangle on a white board to illustrate some trade-offs that he expected to occur over the course of the project. The corners of the triangle were labeled *Quality*, *Time*, and *Money*. The manager explained, "We have too little time. We have an unlimited budget. We will not be measured by the bells and whistles that this system contains. So over the next several weeks, I want you as developers to keep this triangle in mind and do everything it takes to meet this 5-month deadline."

At the end of the five months, the project was delivered on time; however, the project was incredibly over budget, and the final product was "thrown away" after it was used because it was unfit for regular usage. Remarkably, the business users felt that the project was very successful, and they believed that the tradeoffs that were made were worthwhile. *Barbara Wixon*

QUESTIONS:
1. What are the risks in stressing only one corner of the triangle?
2. How would you have managed this project? Can you think of another approach that might have been more effective?

it does), the time taken, and the cost. If a project is estimated to take more time than is available, then the only solutions are to decrease the size of the system (by eliminating some of its functions) or to increase costs by adding more people or having them work overtime. Often, a project manager will have to work with the project sponsor to change the goals of the project, such as developing a system with less functionality or extending the deadline for the final system, so that the project has reasonable goals that can be met.

There are two basic ways to estimate the time required to build a system. The simplest method uses the amount of time spent in the planning phase to predict the time required for the entire project. The idea is that a simple project will require little planning and a complex project will require more planning, so using the amount of time spent in the planning phase is a reasonable way to estimate overall project time requirements.

With this approach, you take the time spent in (or estimated for) the planning phase and use industry standard percentages (or percentages from the organization's own experiences) to calculate estimates for the other SDLC phases. Industry standards suggest that a "typical" business application system spends 15% of its effort in the planning phase, 20% in the analysis phase, 35% in the design phase, and 30% in the implementation phase. This would suggest that if a project takes 4 months in the planning phase, then the rest of the project likely will take a total of 22.66 person-months ($4 \div .15 = 22.66$). These same industry percentages are then used to estimate the amount of time in each phase (Figure 3-3). The obvious limitation of this approach is that it can be difficult to take into account the specifics of your individual project, which may be simpler or more difficult than the "typical" project.

The second approach to estimation uses a more complex—and, it is hoped, more reliable—three-step process (Figure 3-4). First, the project manager estimates the size of the project in terms of number of lines of code required (on the basis of

	Planning	Analysis	Design	Implementation
Typical industry standards for business applications	15%	20%	35%	30%
Estimates based on actual figures for first stages of SDLC	Actual: 4 person-months	Estimated: 5.33 person-months	Estimated: 9.33 person-months	Estimated: 8 person-months

SDLC = systems development life cycle.

FIGURE 3-3
Estimating Project Time Schedules Using the Planning Phase Time

some rough assumptions about the system). This size estimate is then converted into the amount of effort required to develop the system in terms of the number of person-months. The estimated effort is then converted into an estimated schedule time in terms of the number of months from start to finish.

Estimate Size   One way to estimate the size of a project uses function points, a concept developed in 1979 by Allen Albrecht of IBM.[3] A *function point* is a measure of program size that is based on the number and complexity of inputs, outputs, queries, files, and program interfaces.

First, the number and complexity of each component of the system are estimated and recorded on a worksheet (Figure 3-5) and used to calculate the *total unadjusted function points* (TUFP). Each input, output, or query is one major element of the system, such as a data entry screen, a report, or a database search. The complexity of each of these elements is a subjective assessment made by the

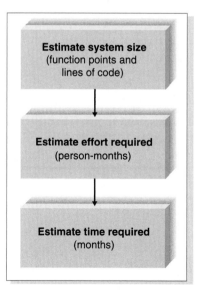

FIGURE 3-4
Estimating Project Time Schedules Using a More Complex Approach

[3] Albrecht's original article is out of print, but much additional research has been done and can be found at www.ifpug.org.

Description	Complexity			Total
	Low	**Medium**	**High**	
Inputs	__ × 3	__ × 4	__ × 6	__
Outputs	__ × 4	__ × 5	__ × 7	__
Queries	__ × 3	__ × 4	__ × 6	__
Files	__ × 7	__ × 10	__ × 15	__
Program interfaces	__ × 5	__ × 7	__ × 10	__

Total unadjusted function points (TUFP): _____

(0 = no effect on processing complexity; 5 = great effect on processing complexity)

	0–5
Data communications	__
Heavy use configuration	__
Transaction rate	__
End-user efficiency	__
Complex processing	__
Installation ease	__
Multiple sites	__
Performance	__
Distributed functions	__
On-line data entry	__
On-line update	__
Reusability	__
Operational ease	__
Extensibility	__

Project complexity (PC): _____

Adjusted project complexity (PCA) = 0.65 + (0.01 · _____)

FIGURE 3-5
Function Point Estimation Worksheet

Total adjusted function points (TAFP): _____ · _____ = [_____]

project manager. Complexity depends on the system as well as on the familiarity of the project team with the business area and the technology that will be used to implement the project. A project that might be very complex for a team with little experience might have little complexity for a team with lots of experience.

At this point, the exact nature of the system has not been determined, so it is impossible to know exactly how many inputs, outputs, and so forth will be in the system. It is up to the project manager to make a reasonable guess. Later in the project, once more is known about the system, the project manager will revise the estimates using this better knowledge to produce more accurate estimates.

Because a variety of factors can impact the complexity of the project, 14 factors, such as end-user efficiency, reusability, and data communications are assessed in terms of their effect on the project's complexity (see Figure 3-5). These assessments are totaled and placed into a formula to calculate an *adjusted project complexity* (PCA) score. Finally, the TUPF value is multiplied by the PCA value to determine the ultimate size of the project in terms of *total adjusted function points* (TAFP).

Sometimes a shortcut is used to determine the complexity of the project. Instead of calculating the complexity for the 14 factors listed in Figure 3-5, project managers choose to assign PCA a value that ranges from .65 for very simple systems to 1.00 for "normal" systems to as much as 1.35 for complex systems. For example, a very simple system that has 200 unadjusted function points would have a size of 130 adjusted function points ($200 \times .65 = 130$). However, if the system with 200 unadjusted function points were very complex, its function point size would be 270 ($200 \times 1.35 = 270$).

Once you have the estimated number of function points, then you need to convert the number of function points into the lines of code that will be required. The number of lines of code depends on the programming language you choose to use. Figure 3-6 presents a very rough conversion guide for some popular languages. The actual number of lines of code depends on the complexity of the system.

For example, suppose you had a system with 100 function points. If you were to develop the system in COBOL, it would typically require around 11,000 lines of code to write it. Conversely, if you were to use Visual Basic, it would typically take 3,000 lines of code. If you could develop the system using a package such as Excel or Access, it would take between 1,500 and 4,000 lines of code. There is a great

Language	Approximate Number of Lines of Code per Function Point
C	130
COBOL	110
Java	55
C++	50
Turbo Pascal	50
Visual Basic	30
PowerBuilder	15
HTML	15
Packages (e.g., Access, Excel)	10–40

Source: Capers Jones, Software Productivity Research, http://www.spr.com
HTML = hypertext mark-up language.

FIGURE 3-6
Converting from Function Points to Lines of Code

Imagine that job hunting has been going so well that you need to develop a system to support your effort. The system should allow you to input information about the companies with which you interview, the interviews and office visits that you have scheduled, and the offers that you receive. It should be able to produce reports, such as a company contact list, an interview schedule, and an office visit schedule, as well as produce thank-you letters to be brought into a word processor to customize. You also need the system to answer queries, such as the number of interviews by city and your average offer amount.

QUESTION:

Given these requirements, determine the number of function points associated with this program.

range for packages because different packages enable you to do different things, and not all systems can be built using certain packages. Sometimes you end up writing lots of extra code to do some simple function because the package does not have the capabilities you need.

There is also a very important message from the data in this figure. Since a direct relationship exists between lines of code and the amount of effort and time required to develop a system, the choice of development language has a significant impact on the time and cost of projects.

Estimate Effort   Once an understanding is reached about the size of the project, the next step is to estimate the effort that is required to tackle its tasks. Effort is a function of the size combined with *production rates* (how much work someone can complete in a given time). Much research has been done on software production rates. One of the most popular algorithms, the COCOMO[4] model, was designed by Barry W. Boehm to convert a lines-of-code estimate into a person-month estimate. There are different versions of the COCOMO model that depend on the complexity of the software, the size of the system, the experience of the developers, and the type of software you are developing (e.g., business application software such as the registration system at your university; commercial software such as Word; or system software such as Windows).

For small to moderate-size business software projects (i.e., 100,000 lines of code and 10 or fewer programmers), the model is quite simple:

$$effort\ (in\ person\text{-}months) = 1.4 \times thousands\ of\ lines\ of\ code$$

For example, let's suppose that we were going to develop a business software system requiring 10,000 lines of code. This project would typically take 14 person-months to complete. But once again, this may change depending on the system's complexity, size, and so on.

---

[4] The original COCOMO model is presented by Barry W. Boehm in *Software Engineering Economics*, (Englewood Cliffs, NJ: Prentice-Hall, 1981). Since then, much additional research has been done. For the latest updates, see http://sunset.usc.edu/COCOMOII/cocomo.html

Refer to the function point worksheets that you completed for Your Turn 3.1 and Figure 3-5.

Suppose that you are considering building the system in Visual Basic, PowerBuilder, or hypertext mark-up language (HTML).

QUESTION:
1. What is the total effort required to develop the system for each language?
2. Select the language that you will use to develop the system. Why did you choose that language?

Estimate Schedule Time   Once the effort is understood, the optimal schedule for the project can be estimated. Historical data or estimation software can be used as aids, or one rule of thumb is to determine schedule using the following equation:

$$\text{schedule time (months)} = 3.0 \times \text{person-months}^{1/3}$$

This equation is widely used, although the specific numbers vary (e.g., some estimators may use 3.5 or 2.5 instead of 3.0). The equation suggests that a project that has an effort of 40 person-months should be scheduled to take a little more than 10 months to complete. It is important to note that this estimate is for the analysis, design, and implementation phases; it does not include the planning phase.

The average number of staff needed for the project also can be determined at this point. It is the number of total person-months of effort divided by the optimal schedule. So to complete a 40 person-month project in 10 months, a team should have an average of four full-time staff members, although this may change over the project schedule as different specialists enter and leave the team (e.g., business analysts, programmers, technical writers).

Many times, the temptation is to assign more staff to a project to shorten the project's length, but this is not a wise move. Adding staff resources does not translate into increased productivity; staff size and productivity share a disproportionate relationship, mainly because a large number of staff members is more difficult to coordinate. The more a team grows, the more difficult it becomes to manage. Imagine how easy it is to work on a two-person project team: the team members share a single line of communication. But adding two people increases the number of communication lines to six, and greater increases lead to more dramatic gains in communication complexity. Figure 3-7 illustrates the impact of adding team members to a project team.

One way to reduce efficiency losses on teams is to understand the complexity that is created in numbers and to build in a reporting structure that tempers its effects. The rule of thumb is to keep team sizes under 8 to 10 people; therefore, if more people are needed, create subteams. In this way, the project manager can keep the communication effective within small teams, which in turn communicate to a contact at a higher level in the project.

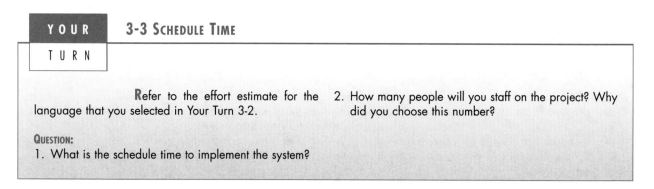

| YOUR | **3-3 SCHEDULE TIME** |
| TURN | |

Refer to the effort estimate for the
language that you selected in Your Turn 3-2.

QUESTION:
1. What is the schedule time to implement the system?

2. How many people will you staff on the project? Why
did you choose this number?

Timeboxing

Up until now, we have described projects that are *task oriented*. In other words, we
have described projects that have a schedule that is driven by the tasks that need to
be accomplished, so the greater number of tasks and requirements, the longer the

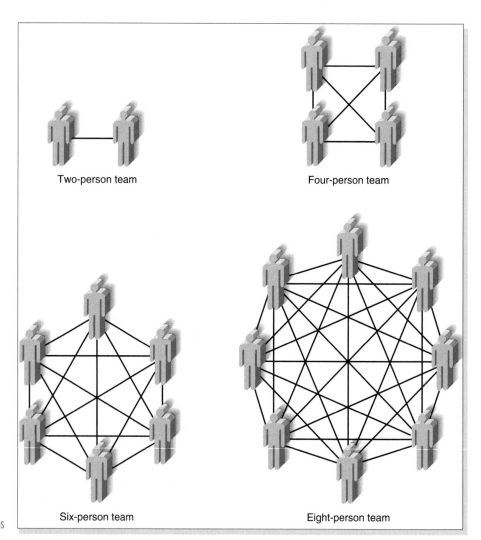

FIGURE 3-7
Increasing Complexity with Larger Teams

project will take. Some companies have little patience for development projects that take a long time, and these companies take a *time-oriented* approach that places meeting a deadline above delivering functionality.

Think about your use of word processing software. For 80% of the time, you probably use only 20% of the features, such as the spelling checker, boldfacing, and cutting and pasting. Other features, such as document merging and creation of mailing labels, may be nice to have, but they are not a part of your day-to-day needs. The same goes for other software applications; most users rely on only a small subset of their capabilities. Ironically, most developers agree that typically 75% of a system can be provided relatively quickly, with the remaining 25% of the functionality demanding most of the time.

To resolve this incongruency, a technique called *timeboxing* has become quite popular, especially when using rapid application development (RAD) methodologies. This technique sets a fixed deadline for a project and delivers the system by that deadline no matter what, even if functionality needs to be reduced. Timeboxing ensures that project teams don't get hung up on the final "finishing touches" that can drag out indefinitely, and it satisfies the business by providing a product within a relatively fast time frame. Typically, in object-oriented development, the time frame for each timebox varies from 1 to 2 weeks to 1 to 2 months depending on the size and complexity of the project.

There are several steps to implement timeboxing on a project (Figure 3-8). First, set the date of delivery for the proposed goals. The deadline should not be impossible to meet, so it is best to let the project team determine a realistic due date. Next, build the core of the system to be delivered; you will find that timeboxing helps create a sense of urgency and helps keep the focus on the most important features. Because the schedule is absolutely fixed, functionality that cannot be completed needs to be postponed. It helps if the team prioritizes a list of features beforehand to keep track of what functionality the users absolutely need. Quality cannot be compromised, regardless of other constraints, so it is important that the time allocated to activities is not shortened unless the requirements are changed (e.g., don't reduce the time allocated to testing without reducing features). At the end of the time period, a high-quality system is delivered; future iterations will likely be needed, to make changes and enhancements, and the timeboxing approach can be used once again.

Applying the Concepts at CD Selections

Alec Adams was very excited about managing the Internet sales system project at CD Selections, but he realized that his project team would have very little time to deliver at least some parts of the system because the company wanted to conduct sales over the Internet during the holiday season. Therefore, he decided that

---

1. Set the date for system delivery.
2. Prioritize the functionality that needs to be included in the system.
3. Build the core of the system (the functionality ranked as most important).
4. Postpone functionality that cannot be provided within the time frame.
5. Deliver the system with core functionality.
6. Repeat steps 3 through 5, to add refinements and enhancements.

FIGURE 3-8
Steps for Timeboxing

CONCEPTS   **3-C TIMEBOXING**

IN ACTION

DuPont was one of the first companies to use timeboxing. Timeboxing originated in the company's fibers division, which was moving to a highly automated manufacturing environment. It was necessary to create complex application software quickly, and DuPont recognized that it is better to get a basic version of the system working, learn from the experience of operating with it, and then design an enhanced version than it is to wait for a comprehensive system at a later date.

DuPont's experience implies that:

- The first version must be built quickly.
- The application must be built so that it can be changed and added to quickly.

DuPont stresses that the timebox methodology works well for the company and is highly practical. It has resulted in automation being introduced more rapidly and effectively. DuPont quotes large cost savings from the methodology, and variations of timebox techniques have since been used in many other corporations.

Source: James Martin, "Within the Timebox, Development Deadlines Really Work," *PC Week,* March 12, 1990.

QUESTIONS:
1. Why do you think DuPont saves money by using this technique?
2. Are there situations in which timeboxing would not be appropriate?

the project should follow a RAD phased development-based methodology, combined with the timeboxing technique. In this way, he could be sure that some version of the product would be in the hands of the users within several months, even if the completed system would be delivered at a later date.

As project manager, Alec's first job was to identify the tasks that would be needed to complete the system. He used a RAD phased development-based methodology that CD Selections had in house, and he borrowed its high-level phases (e.g., analysis) and the major tasks associated with them (e.g., gathering information, analyzing information, creating the system concept). These were recorded in a work plan using the Microsoft Project. Alec expected to define the steps in much more detail at the beginning of each phase (Figure 3-9).

Next came estimation—one of Alec's least favorite jobs because of how tough it is to do at the very beginning of the project. But he knew that the users would expect at least general ranges for a product delivery date. He began by attempting to estimate the number of inputs, outputs, queries, files, and program interfaces in the new system. For the Web part of the system to be used by customers, he could think of four main queries (searching by artist, by CD title, by song title, and by current sales promotions), two input screens (selecting a CD to buy and entering credit card and other order information), three output screens (the home page with general information, information about CDs, information about the customer's order), two files (CD information, customer orders), and four program interfaces (one to the credit card clearance center, and three to other CD Selections systems: inventory, shipping, and accounting). For the part of the system to be used by CD Selections staff (to maintain the marketing materials), he identified three additional inputs, three outputs, four queries, one file, and one program interface. He believed all of these to be of medium complexity. He entered these numbers in a worksheet (Figure 3-10).

Step	Deliverables	Estimated Hours	Actual Hours	Assigned To
Planning phase				
Analysis phase				
Examine current system	Behavioral models Structural models			
Identify improvements	Ideas for system			
Develop concept for new system	Behavioral models Structural models			
Design phase				
Develop factored models	Factored models			
Design system architecture	System architecture			
Design infrastructure	Infrastructure design			
Design interface	Windows navigation diagram Interface standards Interface design Use scenarios Real use cases			
Design data storage	Data storage design			
Program design	Restructured models Class design Contracts Method design			
Implementation phase				
Construction	Test plan Documentation Complete system			
Implementation	Implementation plan Change management plan Training plan Implemented system			

FIGURE 3-9
Work Plan for CD Selections

Rather than attempt to assess the complexity of the system in detail, Alec chose to use a value of 1.20 for PCA. He reasoned that the system was of medium complexity, but his staff had had no prior experience with the Web, so it would be somewhat complex for them. This produced a TAFP of about 176.

Converting function points into lines of code was challenging. The project would use a combination of C (for most programs) and HTML for the Web screens. Alec decided to assume that about 75% of the function points would be C and 25% would be HTML, which produced a total of about 17,700 lines of code [$(.75 \times 176 \times 130) + (.25 \times 176 \times 15)$].

Using the COCOMO formula, he found that this translated into about 25 person-months of effort (1.4×17.7). This in turn suggested a schedule time of about 9 months ($3.0 \times 25^{1/3}$). After much consideration, Alec decided to pad the estimate by 10% (by adding 1 extra month). On the basis of the estimates, it appeared that about 3 people would be needed to deliver the system by the holidays (25 person-months over 10 months of calendar time means 2.5 people, rounded up to 3).

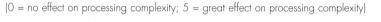

Description	Complexity			Total
	Low	**Medium**	**High**	
Inputs	___ × 3	_5_ × 4	___ × 6	_20_
Outputs	___ × 4	_6_ × 5	___ × 7	_30_
Queries	___ × 3	_8_ × 4	___ × 6	_32_
Files	___ × 7	_3_ × 10	___ × 15	_30_
Program interfaces	___ × 5	_5_ × 7	___ × 10	_35_

Total unadjusted function points (TUFP): _____141_____

(0 = no effect on processing complexity; 5 = great effect on processing complexity)

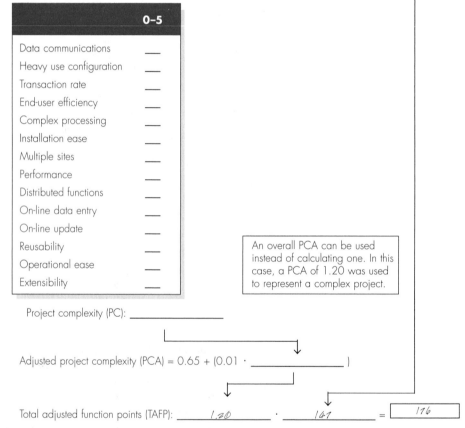

	0–5
Data communications	___
Heavy use configuration	___
Transaction rate	___
End-user efficiency	___
Complex processing	___
Installation ease	___
Multiple sites	___
Performance	___
Distributed functions	___
On-line data entry	___
On-line update	___
Reusability	___
Operational ease	___
Extensibility	___

An overall PCA can be used instead of calculating one. In this case, a PCA of 1.20 was used to represent a complex project.

Project complexity (PC): _____

Adjusted project complexity (PCA) = 0.65 + (0.01 · _____)

Total adjusted function points (TAFP): ____1.20____ · ____141____ = [176]

FIGURE 3-10
Function Point Estimation for CD
Selections

STAFFING THE PROJECT

Staffing means much more than determining how many people should be assigned to the project. It involves matching people's skills with the needs of the project, motivating them to meet the project's objectives, and minimizing the conflict that will occur over time. The deliverable for this part of project management is a *staffing plan,* which describes the kinds of people who will work on the project and the overall reporting structure, and the *project charter,* which describes the project's objectives and rules.

After the estimates are complete and the project manager understands how many people are needed, he or she creates a staffing plan that lists the roles that are required for the project and the proposed reporting structure. Typically, a project will have one project manager who oversees the overall progress of the development effort, with the core of the team comprising the various types of analysts described in Chapter 1. A *functional lead* usually is assigned to manage a group of analysts, and a *technical lead* oversees the progress of a group of programmers and more technical staff members.

There are many structures for project teams; Figure 3-11 illustrates one possible configuration of a project team. After the roles are defined and the structure is in place, the project manager needs to think about which people can fill each role. Often, one person fills more than one role on a project team.

When you make assignments, remember that people have *technical skills* and *interpersonal skills,* and both are important on a project. Technical skills are useful when working with technical tasks (e.g., programming in Java) and in trying to understand the various roles that technology plays in the particular project (e.g., how a Web server should be configured on the basis of a projected number of hits from customers).

Interpersonal skills, on the other hand, include interpersonal and communication abilities that are used when dealing with business users, senior management executives, and other members of the project team. They are particularly critical when performing the requirements-gathering activities and when addressing organizational feasibility issues. Each project will require unique technical and interpersonal skills. For example, a Web-based project may require Internet experience or Java programming knowledge, or a highly controversial project may need analysts who are particularly adept at managing political or volatile situations.

When the skills of the project team do not match what is actually required, the project manager has several options to improve the situation. First, outside help—such as a consultant or vendor—can be hired to train team members and start them off on the right foot. Training classes are usually available for both technical and interpersonal instruction, if time is available. Mentoring may also be an option; a

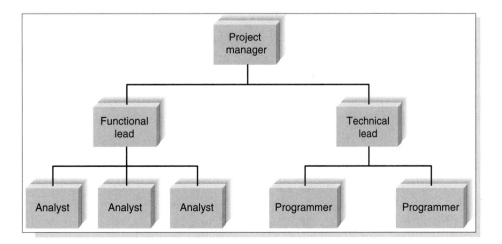

FIGURE 3-11
Possible Reporting Structure

3-4 STAFFING PLAN

Now it is time to staff the project that was described in Your Turn 3-1. On the basis of the number of people that you estimated would be needed, (from Your Turn 3-3) select classmates who will work with you on your project.

QUESTIONS:

1. What roles will be needed to develop the project? List them and write short descriptions for each of these roles, almost as if you had to advertise the positions in a newspaper.

2. Which roles will each classmate perform? Will some people perform multiple roles?

3. What will the reporting structure be for the project?

project team member can be sent to work on another similar project so that he or she can return with skills to apply to the current job.

Motivation

Assigning people to tasks isn't enough; project managers need to motivate the people to make the project a success. Motivation has been found to be the number one influence on people's performance,[5] but determining how to motivate the team can be quite difficult. You may think that good project managers motivate their staff by rewarding them with money and bonuses, but most project managers agree that this is the *last* thing that should be done. The more often you reward team members with money, the more they expect it—and most times monetary motivation won't work.

Assuming that team members are paid a fair salary, technical employees on project teams are much more motivated by recognition, achievement, the work itself, responsibility, advancement, and the chance to learn new skills.[6] If you feel like you need to give some kind of reward for motivational purposes, try a pizza or free dinner, or even a kind letter or award. They often have much more effective results. Figure 3-12 lists some other motivational don'ts that you should avoid to ensure that motivation on the project is as high as possible.

Handling Conflict

The third component of staffing is organizing the project to minimize conflict among group members. *Group cohesiveness* (the attraction that members feel to the group and to other members) contributes more to productivity than do project members' individual capabilities or experiences.[7] Clearly defining the roles on

---

[5] Barry W. Boehm, *Software Engineering Economics* (Englewood Cliffs, NJ: Prentice-Hall, 1981). One of the best books on managing project teams is that by Tom DeMarco and Timothy Lister, *Peopleware: Productive Projects and Teams,* (New York: Dorset House, 1987).

[6] F. H. Hertzberg, "One More Time: How Do You Motivate Employees?" *Harvard Business Review* (January–February 1968).

[7] B. Lakhanpal, "Understanding the Factors Influencing the Performance of Software Development Groups: An Exploratory Group-Level Analysis," *Information and Software Technology* 35, no. 8 (1993):468–473.

Don'ts	Reasons
Assign unrealistic deadlines	Few people will work hard if they realize that a deadline is impossible to meet.
Ignore good efforts	People will work harder if they feel like their work is appreciated. Often, all it takes is public praise for a job well done.
Create a low-quality product	Few people can be proud of working on a project that is of low quality.
Give everyone on the project a raise	If everyone is given the same reward, then high-quality people will believe that mediocrity is rewarded—and they will resent it.
Make an important decision without the team's input	Buy-in is very important. If the project manager needs to make a decision that greatly affects the members of her team, she should involve them in the decision-making process.
Maintain poor working conditions	A project team needs a good working environment or motivation will go down the tubes. This includes lighting, desk space, technology, privacy from interruptions, and reference resources.

Source: Adapted from Steve McConnell, *Rapid Development*, (Redmond, WA: Microsoft Press, 1996).

FIGURE 3-12
Motivational Don'ts

the project and holding team members accountable for their tasks is a good way to begin mitigating potential conflict on a project. Some project managers develop a *project charter* that lists the project's norms and ground rules. For example, the charter may describe when the project team should be at work, when staff meetings will be held, how the group will communicate with each other, and the procedures for updating the work plan as tasks are completed. Figure 3-13 lists additional techniques that can be used at the start of a project to keep conflict to a minimum.

Applying the Concepts at CD Selections

Alec next turned to the task of how to staff his project with the three people from his earlier estimates. First, he created a list of the various roles that he needed to fill. He thought he would need several analysts to work with the analysis and

- Clearly define plans for the project.
- Make sure the team understands how the project is important to the organization.
- Develop detailed operating procedures and communicate these to the team members.
- Develop a project charter.
- Develop schedule commitments ahead of time.
- Forecast other priorities and their possible impact on project.

Source: H. J. Thamhain and D. L. Wilemon, "Conflict Management in Project Life Cycles," *Sloan Management Review* (Spring 1975).

FIGURE 3-13
Conflict Avoidance Strategies

YOUR TURN 3-5 PROJECT CHARTER

Get together with several of your classmates and pretend that you are all staffed on the project described in Your Turn 3-1. Discuss what would most motivate each of you to perform well on the project. List three potential sources of conflict that could surface as you work together.

QUESTION:
Develop a project charter that lists five rules that all team members will need to follow. How might these rules help prevent potential team conflict?

design of the system as well as an infrastructure analyst to manage the integration of the Internet sales system with CD Selections' existing technical environment. Alec also needed people who had good programmer skills and who could be responsible for ultimately implementing the system. Ian, Anne, and K.C. are three analysts with strong technical and interpersonal skills (although Ian is less balanced, having greater technical than interpersonal abilities), and Alec believed that he would be able to pull all three onto this project. He wasn't certain if they had experience with the actual Web technology that would be used on the project, but he decided to rely on vendor training or an external consultant to build those skills later when they were needed. Because the project was so small, Alec envisioned all of the team members reporting to him because he would be serving as the project's manager.

Alec created a staffing plan that captured this information, and he included a special incentive structure in the plan (Figure 3-14). Meeting the holiday deadline was very important to the project's success, so he decided to offer a day off to the

Role	Description	Assigned To
Project manager	Oversees the project to ensure that it meets its objectives in time and within budget	Alec
Infrastructure analyst	Ensures the system conforms to infrastructure standards at CD Selections; ensures that the CD Selections infrastructure can support the new system	Ian
Systems analyst	Designs the information system—with a focus on interfaces with the distribution system	Ian
Systems analyst	Designs the information system—with a focus on the process models and interface design	K.C.
Systems analyst	Designs the information system—with a focus on the data models and system performance	Anne
Programmer	Codes system	K.C.
Programmer	Codes system	Anne

Reporting structure: All project team members will report to Alec.

Special incentives: If the deadline for the project is met, all team members who contributed to this goal will receive a free day off, to be taken over the holiday season.

FIGURE 3-14
Staffing Plan for the Internet Sales System

team members who contributed to meeting that date. He hoped that this incentive would motivate the team to work very hard. Alec also planned to budget money for pizza and sodas for times when the team worked long hours.

Before he left for the day, Alec drafted a project charter, to be fine-tuned after the team got together for its *kickoff meeting* (i.e., the first time the project team gets together). The charter listed several norms that Alec wanted to put in place from the start to eliminate any misunderstanding or problems that could come up otherwise (Figure 3-15).

CONTROLLING AND DIRECTING THE PROJECT

The final step of project management—controlling and directing the project until the final product is delivered to the project sponsor and users—will continue throughout the entire project. This step includes a variety of activities, including refining original project estimates, tracking tasks, encouraging efficient development practices, managing scope, and mitigating risk. All these activities occur over the course of the entire SDLC, but it is at this point in the project when the project manager needs to put them in place. Ultimately, these activities ensure that the project stays on track and that the chance of failure is kept at a minimum. The rest of this section will describe each of these activities in more detail.

Refining Estimates

The estimates that were produced during the estimation stage of project management will need to be refined as the project progresses. This means not that estimates were poorly done at the start of the project but that it is virtually impossible to develop an exact assessment of the project's schedule before the analysis and design phases are conducted. A project manager should expect to be satisfied with broad ranges of estimates that become more and more specific as the project's product becomes better defined.

In many respects, estimating what an information system (IS) development project will cost, how long it will take, and what the final system will actually do follows a *hurricane model.* When storms and hurricanes first appear in the Atlantic or Pacific, forecasters watch their behavior and, on the basis of minimal information about them (but armed with lots of data on previous storms), attempt to predict

> **Project objective:** The Internet sales system project team will create a working Web-based system to sell CDs to CD Selections' customers in time for the holiday season.
>
> The Internet sales system team members will
>
> 1. Attend a staff meeting each Friday at 2 P.M. to report on the status of assigned tasks.
> 2. Update the work plan with actual data each Friday by 5 P.M.
> 3. Discuss all problems with Alec as soon as they are detected.
> 4. Agree to support each other when help is needed, especially for tasks that could hold back the progress of the project.
> 5. Post important changes to the project on the team bulletin board as they are made.

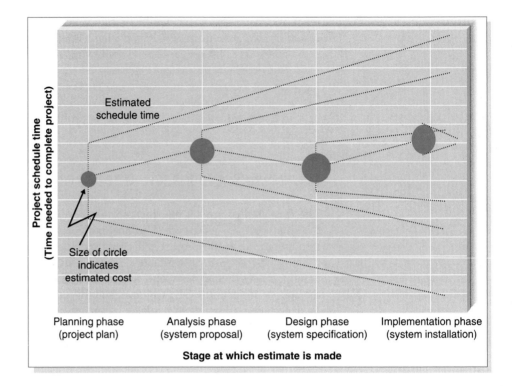

FIGURE 3-16
Hurricane Model

when and where the storms will hit and what damage they will do when they arrive. As storms move closer to North America, forecasters refine their tracks and develop better predictions about where and when they are most likely to hit and their force when they do. The predictions become more and more accurate as the storms approach a coast, until they finally arrive.

In the planning phase when a system is first requested, the project sponsor and project manager attempt to predict how long the SDLC will take, how much it will cost, and what it will ultimately do when it is delivered (i.e., its *functionality*). However, the estimates are based on very little knowledge of the system. After the feasibility analysis, more information is gained and the estimates for time, cost, and functionality become more accurate and more precise. As the system moves into the analysis phase, more information is gathered and the system concept is developed, and the estimates become even more accurate and precise. As the system moves closer to completion, the accuracy and precision increase until the final system is delivered (Figure 3-16).

According to one of the leading experts in software development,[8] a *well-done* project plan (prepared at the end of the planning phase) has a 100% margin of error for project cost and a 25% margin of error for schedule time. In other words, if a carefully done project plan estimates that a project will cost $100,000 and will take 20 weeks, the project will actually cost between $0 and $200,000 and take between 15 and 25 weeks. Figure 3-17 presents typical margins of error for other stages in the project. It is important to note that these margins of error

[8] Barry W. Boehm and colleagues, "Cost Models for Future Software Life Cycle Processes: COCOMO 2.0," in J. D. Arthur and S. M. Henry (eds.), *Annals of Software Engineering: Special Volume on Software Process and Product Measurement* (Amsterdam: J. C. Baltzer AG Science Publishers, 1995).

| Phase | Deliverable | Typical Margins of Error for Well-Done Estimates | |
		Cost (%)	Schedule Time (%)
Planning phase	System request	400	60
	Project plan	100	25
Analysis phase	System proposal	50	15
Design phase	System specifications	25	10

Source: Barry W. Boehm and colleagues, "Cost Models for Future Software Life Cycle Processes: COCOMO 2.0," in J. D. Arthur and S. M. Henry, eds. *Annals of Software Engineering Special Volume on Software Process and Product Measurement* (Amsterdam: J. C. Baltzer AG Science Publishers, 1995).

FIGURE 3-17
Margins of Error in Cost and Time Estimates

apply only to *well-done* plans; a plan developed without much care has a much greater margin of error.

What happens if you overshoot an estimate (e.g., the analysis phase ends up lasting 2 weeks longer than expected)? There are a number of ways to adjust future estimates. If the project team finishes a step ahead of schedule, most project managers shift the deadlines sooner by the same amount but do not adjust the promised completion date. The challenge, however, occurs when the project team is late in meeting a scheduled date. Three possible responses to missed schedule dates are presented in Figure 3-18. We recommend that if an estimate proves too optimistic

Assumptions	Actions	Level of Risk
If you assume the rest of the project is simpler than the part that was late and is also simpler than believed when the original schedule estimates were made, you can make up lost time	Do not change schedule.	High risk
If you assume the rest of the project is simpler than the part that was late and is no more complex than the original estimate assumed, you can't make up the lost time, but you will not lose time on the rest of the project.	Increase the entire schedule by the total amount of time that you are behind (e.g., if you missed the scheduled date by 2 weeks, move the rest of the schedule dates to 2 weeks later). If you included padded time at the end of the project in the original schedule, you may not have to change the promised system delivery date; you'll just use up the padded time.	Moderate risk
If you assume that the rest of the project is as complex as the part that was late (your original estimates too optimistic), then all the scheduled dates in the future underestimate the real time required by the same percentage as the part that was late.	Increase the entire schedule by the percentage of weeks that you are behind (e.g., if you are 2 weeks late on part of the project that was supposed to take 8 weeks, you need to increase all remaining time estimates by 25%). If this moves the new delivery date beyond what is acceptable to the project sponsor, the scope of the project must be reduced.	Low risk

FIGURE 3-18
Possible Actions When a Schedule Date Is Missed

early in the project, do not expect to make up for lost time—very few projects end up doing this. Instead, change your future estimates to include an increase similar to the one that was experienced. For example, if the first phase was completed 10% over schedule, increase the rest of your estimates by 10%.

Tracking Tasks

The most effective work plans are updated with actual numbers on a frequent basis and shared with everyone on the team. The better a project manager tracks the tasks of the project, the better he or she can make staffing decisions, predict deadlines, calculate accurate costs, and understand how well the project is progressing. The work plan is invaluable to the project manager because it provides a mechanism for tracking the progress of the project team over time. The information is kept as current as possible, and the work plan contents are scrutinized regularly for potential issues or problems (e.g., delays, a lack of resources).

Tasks are very interrelated (e.g., sometimes one task must be completed before another task can begin), and it sometimes is easier to understand how tasks overlap and are interrelated by looking at the work plan graphically. In fact, most software packages let you toggle back and forth to different views of a project. One popular graphic depiction of the work plan is a *Gantt chart,* which shows time versus activities. The Gantt chart, named for its inventor Henry Laurence Gantt, lists project tasks along a *y*-axis and time along an *x*-axis and uses shaded and unshaded boxes to illustrate tasks that are completed and still need to be addressed (Figure 3-19).

Another popular approach, especially on larger projects, involves the Program Evaluation and Review Technique (PERT) chart and the *Critical Path Method (CPM).* PERT is a network analysis technique that can be used when the individual task time estimates are fairly uncertain. Instead of simply providing a point estimate for the duration estimate, PERT uses three time estimates: optimistic, most likely, and pessimistic. It then combines the three estimates into a single weighted average estimate using the following formula:

PERT weighted average =

$$\frac{\text{optimistic estimate} + (4 * \text{most likely estimate}) + \text{pessimistic estimate}}{6}$$

The PERT chart is drawn as a node and arc type of graph that shows the time estimates in the nodes and the task dependencies on the arcs (see Figure 3-20). The CPM simply allows identification of the critical path in the network. The critical path is the set of interrelated tasks that must not fall behind schedule. If any of the tasks on the critical path takes longer than expected, the entire project will fall behind. CPM can be used with or without PERT (see Figure 3-20).

Coordinating Project Activities

Generally speaking, project management relies on good coordination, and three techniques are commonly used to help coordinate activities on a project: computer-aided software engineering (CASE), standards, and documentation.

Computer-Aided Software Engineering Tools *Computer-aided software engineering (CASE)* is a category of software that automates all or part of the development

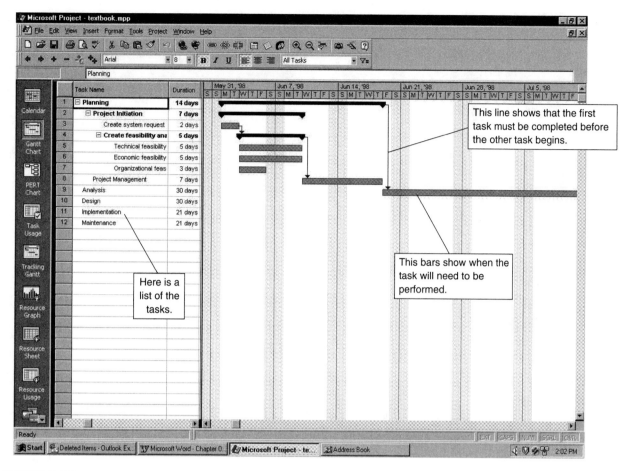

FIGURE 3-19
A Gantt Chart

process. Some CASE software packages are used primarily during the analysis phase to create integrated diagrams of the system and to store information regarding the system components (often called *upper CASE*), whereas others are design-phase tools that create the diagrams and then generate code for database tables system functionality (often called *lower CASE*). *Integrated CASE,* or *I-CASE,* contains functionality found in both upper CASE and lower CASE tools in that it supports tasks that happen throughout the SDLC. CASE comes in a wide assortment of flavors in terms of complexity and functionality, and there are many good programs available in the marketplace, such as the Visible Analyst Workbench, Oracle Designer/2000, Rational Rose, and the Logic Works suite.

Using CASE has numerous benefits. With CASE tools, tasks are much faster to complete and alter, development information is centralized, and information is illustrated through diagrams, which typically are easier to understand. Potentially, CASE can reduce maintenance costs, improve software quality, and enforce discipline. Some project teams even use CASE to assess the magnitude of changes to the project.

Of course, like anything else, CASE should not be considered a silver bullet for project development. The advanced CASE tools are complex applications that require significant training and experience to achieve real benefits. Often, CASE

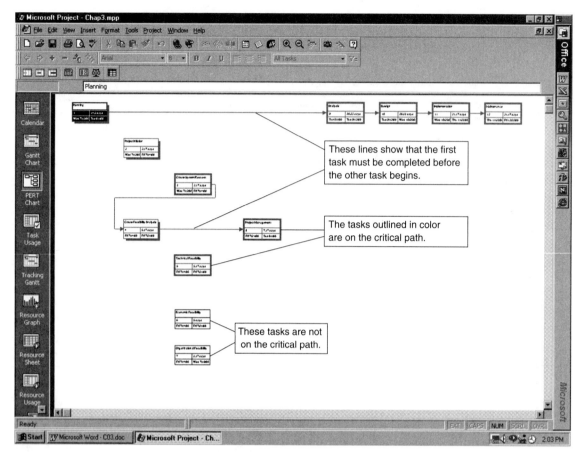

FIGURE 3-20
PERT Chart in Microsoft Project

serves only as a glorified diagramming tool that supports the practices described for behavioral and structural modelling. However, our experience has shown that CASE is a helpful way to support the communication and sharing of project diagrams and technical specifications—as long as it is used by trained developers who have applied CASE on past projects.

The central component of any CASE tool is the *CASE repository,* otherwise known as the information repository or data dictionary. The CASE repository stores the diagrams and other project information, such as screen and report designs, and it keeps track of how the diagrams fit together. For example, most CASE tools will warn you if you place a field on a screen design that doesn't exist in your data model. As the project evolves, project team members perform their tasks using CASE. As you read through the textbook, we will indicate when and how the CASE tool can be used so that you can see how CASE supports the project tasks.

Standards   Members of a project team need to work together, and most project management software and CASE tools provide access privileges to everyone working on the system. When people work together, however, things can get pretty confusing. To make matters worse, people sometimes get reassigned in the middle of a project. It is important that their project knowledge does not leave with them and that their replacements can get up to speed quickly.

One way to make certain that everyone is on the same page by performing tasks in the same way and following the same procedures is to create *standards* that the project team must follow. Standards can range from formal rules for naming files to forms that must be completed when goals are reached to programming guidelines. See Figure 3-21 for some examples of the types of standards that a

Types of Standards	Examples
Documentation standards	The date and project name should appear as a header on all documentation.
	All margins should be set to 1 inch.
	All deliverables should be added to the project binder and recorded in its table of contents.
Coding standards	All modules of code should include a header that lists the programmer, last date of update, and a short description of the purpose of the code.
	Indentation should be used to indicate loops, if-then-else statements, and case statements.
	On average, every program should include one line of comments for every five lines of code.
Procedural standards	Record actual task progress in the work plan every Monday morning by 10 A.M.
	Report to project update meeting on Fridays at 3:30 P.M.
	All changes to a requirements document must be approved by the project manager.
Specification requirement standards	Name of program to be created
	Description of the program's purpose
	Special calculations that need to be computed
	Business rules that must be incorporated into the program.
	Pseudocode
	Due date
User interface design standards	Labels will appear in boldface text, left-justified and followed by a colon.
	The tab order of the screen will move from top left to bottom right.
	Accelerator keys will be provided for all updatable fields.

FIGURE 3-21
A Sampling of Project Standards

project may create. When a team forms standards and then follows them, the project can be completed faster because task coordination becomes less complex.

Standards work best when they are created at the beginning of each major phase of the project and well communicated to the entire project team. As the team moves forward, new standards are added when necessary. Some standards (e.g., file naming conventions, status reporting) will be applied for the entire SDLC, whereas others (e.g., programming guidelines) will be appropriate for certain tasks.

Documentation   A final technique that project teams put in place during the planning phase is good *documentation,* which includes detailed information about the tasks of the SDLC. Often, the documentation is stored in *project binder(s)* that contain all the deliverables and all the internal communication that takes place—the history of the project.

A poor project management practice is waiting until the last minute to create documentation, and this typically leads to an undocumented system that no one understands. In fact, many problems that companies had updating their systems to handle the year 2000 were the result of the lack of documentation. Good project teams learn to document the system's history as it evolves while the details are still fresh in their memory.

The first step to setting up your documentation is to get some binders and include dividers with which to separate content according to the major phases of the project. An additional divider should contain internal communication, such as the minutes from status meetings, written standards, letters to and from the business users, and a dictionary of relevant business terms. Then, as the project moves forward, place the deliverables from each task into the project binder with descriptions so that someone outside of the project will be able to understand it, and keep a table of contents up to date with the content that is added. Documentation takes time up front, but it is a good investment that will pay off in the long run.

Managing Scope

In the beginning of the chapter, we mentioned the problem of projects that run over schedule. By following the steps for good project management, you may assume that your project will be safe from scheduling problems. Unfortunately, the most common reason for schedule and cost overruns surfaces after the project is underway—*scope creep.*

Scope creep occurs when new requirements are added to the project after the original project scope was defined and "frozen." As the First Law of Software Engineering states, "No matter where you are in the system life cycle, the system will change, and the desire to change it will persist throughout the lifecycle."[9] It can happen for many reasons: users may suddenly understand the potential of the new system and realize new functionality that would be useful; developers may discover interesting capabilities to which they become very attached; a senior manager may decide to let this system support a new strategy that was developed at a recent board meeting.

Unfortunately, after the project begins, it becomes increasingly difficult to address changing requirements. The ramifications of change become more extensive, the focus is removed from original goals, and there is at least some impact on

[9] E.H. Bershoff, V.D. Henderson, and S.G. Siegel, *Software Configuration Management* (Englewood Cliffs, NJ: Prentice Hall, 1980).

CONCEPTS

3-D POOR NAMING STANDARDS

IN ACTION

I once started on a small project (four people) in which the original members of the project team had not set up any standards for naming electronic files. Two weeks into the project, I was asked to write a piece of code that would be referenced by other files that had already been written. When I finished my piece, I had to go back to the other files and make changes to reflect my new work.

The only problem was that the lead programmer decided to name the files using his initials (e.g., GG1.prg, GG2.prg, GG3.prg)—and there were over 200 files! I spent 2 days opening every one of those files because there was no way to tell what their contents were.

Needless to say, from then on, the team created a code for file names that provided basic information regarding the file's contents and they kept a log that recorded the file name, its purpose, the date of last update, and programmer for every file on the project. *Barbara Wixon*

QUESTION:
Think about a program that you have written in the past. Would another programmer be able to make changes to it easily? Why or why not?

cost and schedule. Therefore, the project manager plays a critical role in managing this change to keep scope creep to a minimum.

The keys are to identify the requirements as well as possible in the beginning of the project and to apply analysis techniques effectively. For example, if needs are fuzzy at the project's onset, a combination of intensive meetings with the users and prototyping could be used so that users "experience" the requirements and better visualize how the system could support their needs. In fact, the use of meetings and prototyping has been found to reduce scope creep to less than 5% on a typical project.

Of course, some requirements may be missed no matter what precautions you take, but several practices can be helpful to control additions to the task list. First, the project manager should allow only absolutely necessary requirements to be added after the project begins. Even at that point, members of the project team should carefully assess the ramifications of the addition and present the assessment back to the users. For example, it may require two more person-months of work to create a newly defined report, which would throw off the entire project deadline by several weeks. Any change that is implemented should be carefully tracked so that an audit trail exists to measure the change's impact.

Sometimes changes cannot be incorporated into the present system even though they truly would be beneficial. In this case, these additions to scope should be recorded as future enhancements to the system. The project manager can offer to provide functionality in future releases of the system, thus getting around telling someone no.

Managing Risk

One final facet of project management is *risk management,*[10] the process of assessing and addressing the risks that are associated with developing a project. Risks can

[10] One of the best books on risk managment, as it pertains to information systems development, is Robert N. Charette, *Software Engineering Risk Analysis and Management* (New York: McGraw-Hill, 1989).

CONCEPTS 3-E THE REAL NAMES OF THE SYSTEMS DEVELOPMENT LIFE CYCLE (SDLC) PHASES

IN ACTION

Dawn Adams, Senior Manager with Asymetrix Consulting, has renamed the SDLC phases:

1. Pudding (Planning)
2. Silly Putty (Analysis)
3. Concrete (Design)
4. Touch-this-and-you're-dead-sucker (Implementation)

Adams also uses icons, such as a skull and crossbones for the implementation phase. The funny labels lend a new depth of interest to a set of abstract concepts. But her names have had another benefit. "I had one participant who adopted the names wholeheartedly," she says, "including my icons. He posted an icon on his office door for the duration of each of the phases, and he found it much easier to deal with requests for changes from the client, who could see the increasing difficulty of the changes right there on the door."

Source: Learning Technology Shorttakes 1(2), Wednesday, August 26, 1998.

QUESTION:
What would you do if your project sponsor demanded that an important change be made during the "touch-this-and-you're-dead-sucker" phase?

be caused by many things: weak personnel, scope creep, poor design, and overly optimistic estimates. The project team must be aware of potential risks so that problems can be avoided or controlled well ahead of time.

Typically, project teams create a *risk assessment,* or a document that tracks potential risks along with an evaluation of the likelihood of the risk and its potential impact on the project (Figure 3-22). A paragraph or two is also included that explains potential ways that the risk can be addressed. There are many options: risk could be publicized, avoided, or even eliminated by dealing with its root cause. For example, imagine that a project team plans to use new technology but its members have identified a risk in the fact that its members do not have the right technical skills. They believe that tasks may take much longer to perform because of a high learning curve. One plan of attack could be to eliminate the root cause of the risk—the lack of technical experience by team members—by finding time and resources that are needed to provide proper training to the team.

Most project managers keep abreast of potential risks, even prioritizing them according to their magnitude and importance. Over time, the list of risks will change as some items are removed and others surface. The best project managers, however, work hard to keep risks from having an impact on the schedule and costs associated with the project.

Applying the Concepts at CD Selections

Alec wanted the Internet sales system project to be well coordinated, and he immediately put several practices in place to support his responsibilities. First, he acquired the CASE tool used at CD Selections and set up the product so that it could be used for the analysis-phase tasks. The team members would likely start creating diagrams and defining components of the system fairly early on. He pulled out some standards that he uses on all development projects and made a note to review them with his project team at the kickoff meeting for the system. He also

---

RISK ASSESSMENT

RISK #1:	The development of this system likely will be slowed considerably because project team members have not programmed in Java prior to this project.
Likelihood of risk:	High probability of risk
Potential impact on the project:	This risk likely will increase the time to complete programming tasks by 50%.

Ways to address this risk:
It is very important that time and resources are allocated to up-front training in Java for the programmers who are used for this project. Adequate training will reduce the initial learning curve for Java when programming begins. Additionally, outside Java expertise should be brought in for at least some part of the early programming tasks. This person should be used to provide experiential knowledge to the project team so that JAVA-related issues (of which novice Java programmers would be unaware) are overcome.

RISK #2: ...

FIGURE 3-22
Sample Risk Assessment

had his assistant set up binders for the project deliverables that would start rolling in. Already he was able to include the system request, the feasibility analysis, and the project plan, which included the initial work plan, staffing plan, project charter, standards list, and risk assessment.

SUMMARY

Project Management

Project management is the second major component of the planning phase of the systems development life cycle (SDLC), and it includes three steps: creating the work plan, staffing the project, and controlling and directing the project. Project management is important in ensuring that a system is delivered on time, within budget, and with the desired functionality.

Creating the Work Plan

The first step for project management occurs when a project manager creates a work plan that lists the tasks that need to be completed to meet the project's objectives and information about those tasks. After the tasks are identified using a top-down approach or an existing methodology, the project manager estimates the amount of time and effort that will be needed to complete the project. First, the size is estimated by relying on past experiences or industry standards or by calculating

PRACTICAL **3-1 AVOIDING CLASSIC PLANNING MISTAKES**

TIP

As Seattle University's David Umphress has pointed out, watching most organizations develop systems is like watching reruns of *Gilligan's Island*. At the beginning of each episode, someone comes up with a cockamamie scheme to get off the island that seems to work for a while, but something goes wrong and the castaways find themselves right back where they started—stuck on the island. Similarly, most companies start new projects with grand ideas that seem to work, only to make a classic mistake and deliver the project behind schedule, over budget, or both. Here we summarize four classic mistakes in the planning and project management aspects of the project and discuss how to avoid them:

1. **Overly optimistic schedule:** Wishful thinking can lead to an overly optimistic schedule that causes analysis and design to be cut short (missing key requirements) and puts intense pressure on the programmers, who produce poor code (full of bugs).

   Solution: Don't inflate time estimates; instead, explicitly schedule slack time at the *end* of each phase to account for the variability in estimates, using the margins of error from Figure 3-17.

2. **Failing to monitor the schedule:** If the team does not regularly report progress, no one knows if the project is on schedule.

Solution: Require team members to *honestly* report progress (or the lack or progress) every week. There is no penalty for reporting a lack of progress, but there are immediate sanctions for a misleading report.

3. **Failing to update the schedule:** When a part of the schedule falls behind (e.g., information gathering uses all of the slack in item 1 above plus 2 weeks), a project team often thinks it can make up the time later by working faster. It can't. This is an early warning that the entire schedule is too optimistic.

   Solution: *Immediately* revise the schedule and inform the project sponsor of the new end date or use timeboxing to reduce functionality or move it into future versions.

4. **Adding people to a late project:** When a project misses a schedule, the temptation is to add more people to speed it up. This makes the project take longer because it increases coordination problems and requires staff to take time to explain what has already been done.

   Solution: Revise the schedule, use timeboxing, throw away bug-filled code, and add people only to work on an isolated part of the project.

Source: Adapted from Steve McConnell, *Rapid Development,* (Redmond, WA: Microsoft Press, 1996), pp. 29–50.

the function points, a measure of program size based on the number and complexity of inputs, outputs, queries, files, and program interfaces. Next, the project manager calculates the effort for the project, which is a function of size and production rates. Algorithms like the COCOMO model can be used to determine the effort value. Third, the optimal schedule for the project is estimated along with the number of staff persons that should be assigned to the project. If the final schedule will not deliver the system in a timely fashion, timeboxing can be used. Timeboxing sets a fixed deadline for a project and delivers the system by that deadline no matter what, even if functionality must be reduced.

Staffing the Project

Staffing involves assigning project roles to team members, developing a reporting structure for the team, and matching people's skills with the needs of the project. Information from these tasks is placed in the staffing plan. Staffing also includes motivating the team to meet the project's objectives and minimizing conflict among team members. Both motivation and cohesiveness have been found to greatly influence performance of team members in project situations. Team members are motivated most by such nonmonetary things as recognition, achievement, and the work itself. Conflict can be minimized by clearly defining the roles on a project and holding team members accountable for their tasks. Some managers create a project charter that lists the project's norms and ground rules.

Controlling and Directing the Project

The final step of project management includes controlling and directing the project, which includes refining original estimates, tracking tasks, coordinating project activities, managing scope, and mitigating risk. The accuracy and precision of project estimates start out low but increase as the system moves closer to completion, until the final system is delivered. Tasks are closely tracked by using the work plan and special views of the work plan, such as Gantt charts. Project management relies on good coordination, and three techniques are available to help coordinate activities on a project: computer-aided software engineering (CASE), standards, and documentation. The most common reason for schedule and cost overruns is scope creep, the addition of extra requirements to a predefined set of tasks, so the project manager should allow only the most important requirement additions and should postpone others until after the project is delivered. A risk assessment is used to mitigate risk because it identifies potential risks and evaluates the likelihood of risk and its potential impact on the project.

KEY TERMS

Adjusted project complexity (PCA)
Computer-aided software
   engineering (CASE)
CASE repository
Critical Path Method (CPM)
Documentation
Estimation
Function point
Functional lead
Functionality
Gantt chart
Group cohesiveness
Hurricane model
Integrated CASE (I-CASE)
Interpersonal skills

Kickoff meeting
Lower CASE
Methodology
Program Evaluation and Review
   Technique (PERT) chart
Production rate
Project binder
Project charter
Project management
Project manager
Project plan
Risk assessment
Risk management
Scope creep
Staffing plan

Standards
Task-oriented project
Technical lead
Technical skills
Timeboxing
Time-oriented project
Total adjusted function
   points (TAFP)
Total unadjusted function
   points (TUFP)
Tradeoffs
Upper CASE
Work plan

QUESTIONS

1. Why do many projects end up having unreasonable deadlines? How should a project manager react to unreasonable demands?
2. What are the three steps to project management?
3. Name two ways to identify the tasks that need to be accomplished over the course of a project.
4. What is the difference between a methodology and a work plan? How are the two terms related?
5. What are the tradeoffs that you must manage during estimation?
6. What are two basic ways to estimate the size of a project?
7. Describe the three steps of estimation.
8. Name five factors that affect the complexity of a project.
9. What is a function point and how is it used?

10. What is the formula for calculating the effort for a project?
11. What is timeboxing and why is it used?
12. Describe the differences between a technical lead and a functional lead. How are they similar?
13. Describe three technical skills and three interpersonal skills that would be very important to have on any project.
14. What are the best ways to motivate a team? What are the worst ways?
15. List three techniques to reduce conflict.
16. Describe the hurricane model.
17. Create a list of potential risks that could affect the outcome of a project.
18. How does a Gantt chart help a project manager with the role of allocating tasks?
19. Describe three techniques that are available to help coordinate the activities on a project.
20. What is the difference between upper CASE (computer-aided software engineering) and lower CASE?
21. Describe three types of standards and provide examples of each.
22. What belongs in the project binder? How is the project binder organized?
23. Some companies hire consulting firms to develop the initial project plans and manage the project, but use their own analysts and programmers to develop the system. Why do you think some companies do this?
24. What parts of the project plan are most likely to change between the initial project plan and the final project plan produced near the end of the project? Why?
25. What do you think are three common mistakes made by novice analysts in developing a project plan?

EXERCISES

A. Visit the *Computerworld* Web resource center at http://www.computerworld.com/res/index.html. It includes links to pages on all types of information system topics, including project management. Examine some of the links for project management to better understand a variety of Internet sites that contain information related to this chapter.

B. Select a specific project management topic like computer-aided software engineering (CASE), project management software, or timeboxing and search for information on that topic using the Web. The universal resource locator (URL) listed in question A or any search engine (e.g., Yahoo!, AltaVista, Excite, Info-Seek) can provide a starting point for your efforts.

C. Pretend that the career services office at your university wants to develop a system that collects student résumés and makes them available to students and recruiters over the Web. Students should be able to input their résumé information into a standard résumé template. The information then is presented in a résumé format, and it also is placed in a database that can be queried using an on-line search form. You have been placed in charge of the project. As a first step, create a work plan that lists the tasks that will need to be completed to meet the project's objectives.

D. Refer to the situation in question C. You have been told that recruiting season begins a month from today and that the new system must be used. How would you approach this situation? Describe what you can do as the project manager to make sure that your team does not burn out from unreasonable deadlines and commitments.

E. Consider the system described in question C. Develop a plan for estimating the project. How long do you think it would take for you and three other students to complete the project? Provide support for the schedule that you propose.

F. Consider the application that is used at your school to register for classes. Complete a function point worksheet to determine the size of such an application. You will need to make some assumptions about the application's interfaces and the various factors that affect its complexity.

G. Create a sketch of a Gantt chart and describe why and how it is used.

H. Read Your Turn 3-1 near the beginning of this chapter. Create a risk assessment that lists the potential risks associated with performing the project, along with ways to address the risks.

I. Pretend that your instructor has asked you and two friends to create a Web page to describe the course to potential students and provide current class information (e.g., syllabus, assignments, readings) to current students. You have been assigned the role of leader, so you will need to coordinate your activities and those of your classmates until the project is completed. Describe how you would apply the project management techniques that you have learned in this chapter in this situation. Include descriptions of how you would create a work plan, staff the project, and coordinate all activities—yours and those of your classmates.

J. Select two project management software packages and research them using the Web or trade magazines. Describe the features of the two packages. If you were a project manager, which one would you use to help support your job? Why?

K. Select two estimation software packages and research them using the Web or trade magazines. Describe the features of the two packages. If you were a project manager, which one would you use to help support your job? Why?

L. In 1997, Oxford Health Plans had a computer problem that caused the company to overestimate revenue and underestimate medical costs. Problems were caused by the migration of its claims processing system from the Pick operating system to a UNIX-based system that uses Oracle database software and hardware from Pyramid Technology. As a result, Oxford's stock price plummeted, and fixing the system became the number-one priority for the company. Pretend that you have been placed in charge of managing the repair of the claims processing system. Obviously, the project team will not be in good spirits. How will you motivate team members to meet the project's objectives?

M. Suppose that you are in charge of the project that is described in question C, and the project will be staffed by members of your class. Do your classmates have all of the right skills to implement such a project? If not, how will you go about making sure that the proper skills are available to get the job done?

MINICASES

1. Emily Pemberton is an IS project manager facing a difficult situation. Emily works for the First Trust Bank, which has recently acquired the City National Bank. Prior to the acquisition, First Trust and City National were bitter rivals, fiercely competing for market share in the region. Following the acrimonious takeover, numerous staff were laid off in many banking areas, including IS. Key individuals were retained from both banks' IS areas, however, and were assigned to a new consolidated IS department. Emily has been made project manager for the first significant IS project since the takeover, and she faces the task of integrating staffers from both banks on her team. The project they are undertaking will be highly visible within the organization, and the time frame for the project is somewhat demanding. Emily believes that the team can meet the project goals successfully, but success will require that the team become cohesive quickly and that potential conflicts are avoided. What strategies do you suggest that Emily implement in order to help assure a successfully functioning project team?

2. Tom, Jan, and Julie are IS majors at Great State University. These students have been assigned a class project by one of their professors, requiring them to develop a new Web-based system to collect and update information on the IS-program's alumni. This system will be used by the IS graduates to enter job and address information as they graduate, and then make changes to that information as they change jobs and/or addresses. Their professor also has a number of queries that she is interested in being able to implement. Based on their preliminary discussions with their professor, the students have developed this list of system elements:

Inputs: 1 low complexity, 2 medium complexity, 1 high complexity
Outputs: 4 medium complexity
Queries: 1 low complexity, 4 medium complexity, 2 high complexity
Files: 3 medium complexity
Program Interfaces: 2 medium complexity

Calculate Total Unadjusted Function Points for this project.

PART TWO
ANALYSIS PHASE

The Analysis Phase answers the questions of *who* will use the system, *what* the system will do, and *where* and *when* it will be used. During this phase, the project team will learn about the system. The team then produces the Summary, Use Case Model (Use Case Descriptions and Diagram), Structural Model (CRC Cards and Class and Object Diagrams), and Behavioral Models (Sequence, Collaboration, and Statechart Diagrams) that describe the As-Is System. Next, the project team lists improvement opportunities and creates the System Concept, Use Case Model, Structural Model, and Behavioral Models for the To-Be System. All of the deliverables are combined into a System Proposal, which is presented to the project sponsor and other key decision makers (e.g., members of the approval committee) who decide whether the project should continue to move forward.

Systems Analysis — CHAPTER 4

Requirements Gathering — CHAPTER 5

Use Case Modeling — CHAPTER 6

Structural Modeling — CHAPTER 7

Behavioral Modeling — CHAPTER 8

PROJECT BINDER

Analysis Plan

As-Is System Summary

Improvement Opportunities

To-Be System Concept

As-Is System Use Case Model

To-Be System Use Case Model

As-Is System Structural Model

To-Be System Structural Model

As-Is System Behavioral Model

To-Be System Behavioral Model

PLANNING

ANALYSIS

☐ **Develop Analysis Plan**
☐ **Understand As-Is System**
☐ **Identify Improvement Opportunities**
☐ **Develop the To-Be System Concept**
☐ **Develop Use Cases**
☐ **Develop Structural Model**
☐ **Develop Behavioral Model**

TASK CHECKLIST

PLANNING → ANALYSIS → DESIGN →

CHAPTER 4
SYSTEMS ANALYSIS

THE basic process of systems analysis is divided into three steps: understanding the As-Is (current) system, identifying improvements, and developing a concept for the To-Be (new) system. This chapter provides an overview of the analysis process and describes three fundamentally different strategies for analysis: (1) business process automation, which seeks to automate but not to make major changes to the underlying business processes; (2) business process improvement, which seeks to make moderate changes to the business processes; and (3) business process reengineering, which seeks to radically change how the organization runs its business.

OBJECTIVES

- Understand the basic process of systems analysis.
- Understand how to use business process automation techniques.
- Be familiar with business process improvement techniques.
- Be familiar with business process reengineering techniques.
- Understand when to use each analysis strategy.

CHAPTER OUTLINE

IMPLEMENTATION

INTRODUCTION

Systems analysis (sometimes called *requirements analysis*) is the process of gathering information about the current system (called the *As-Is system*), which may or may not be computerized, identifying its strengths and problems, and analyzing them to produce a concept for the new system (called the *To-Be system*). The goal of the analysis phase is to truly understand the requirements for the new system and develop a system concept that addresses them—or decide that a new system is not needed.

The basic process of *analysis* is divided into three steps: understanding the As-Is system, identifying improvements, and developing the concept for the To-Be system. The three steps are tightly coupled and are often iterative, which means the analyst repeatedly flips back and forth between gathering information about the As-Is system, analyzing it, and drawing implications for the To-Be system. The analysis phase ends with a basic plan for the To-Be system that usually is described via behavioral and structural models.

To understand how to do systems analysis, you must understand how to gather information about the As-Is system, how to analyze the information you gather, and how to put it together into a form that documents how the To-Be system will operate. Each of these activities is separate and distinct, but they are tightly connected—how can you gather information unless you know what information you need to do the analysis or what information you need to document the system?

In this chapter, we discuss the basic analysis process in terms of these three basic steps and present three different strategies for conducting the analysis. This chapter also describes several techniques for analyzing information that will help you understand what information you need to gather about the current system. This chapter will help you develop an *analysis plan*—a plan for conducting the analysis stage. However, you won't be ready to actually do the analysis until you have read Chapter 5, which describes various techniques for requirements elicitation (e.g., interviews, group meetings, survey questionnaires). Likewise, you will need to read about behavioral and structural modelling before you will understand what information to gather for building these models.

The line between systems analysis and systems design is very blurry because the To-Be behavioral models and structural models created in the analysis phase are the first step in the design of the new system. In fact, in object-oriented systems development, system design is nothing more than elaborating the behavioral and structural models developed during systems analysis. Many of the major design decisions for the To-Be system are made in the development of these models in the analysis phase. In fact, a better name for the analysis phase would be "analysis and logical design," but because this is a rather long name and because most organizations simply call this phase analysis, we will, too. Nonetheless, it is important to remember that the deliverables from the analysis phase are really the first step in the design of the new system.

THE ANALYSIS PROCESS

Analyzing information system requirements is both a business task and an information technology task. In the early days of computers, there was a presumption that the systems analysts, as experts with computer systems, were in the best position to define how a computer system should operate. Many systems failed because they did not adequately address the true business needs of the users. Gradually, the presumption changed so that the users, as the business experts, were seen as being in the best position to define how a computer system should operate. However, many systems failed to deliver performance benefits because users simply automated an existing inefficient system and failed to incorporate the new opportunities offered by computer technology.

The ideal approach is to balance the expertise of both users and analysts because successful systems build on the business expertise of the users and the systems expertise of the analysts. Perhaps the best analogy is building a house or an apartment. We have all lived in a house or apartment, and most of us have some understanding of what we would like to see in one. However, if we were asked to design one from scratch, it would be a challenge because we lack the design skills and technical engineering skills needed. Likewise, an architect acting alone would probably miss some of the unique requirements that each of us would really want.

The systems development life cycle (SDLC) is the process by which the organization moves from the As-Is system to the To-Be system (Figure 4-1). The initiation phase developed some general ideas for the To-Be system and defined the project's scope. The product of the analysis phase is a system concept for the To-Be system. This concept is then refined in the design phase into a detailed design and is then built and delivered in the implementation phase.

The analysis phase is divided into three steps: understanding the As-Is system, identifying improvements, and developing the concept for the To-Be system. The way in which these steps are performed and the amount of time and effort spent

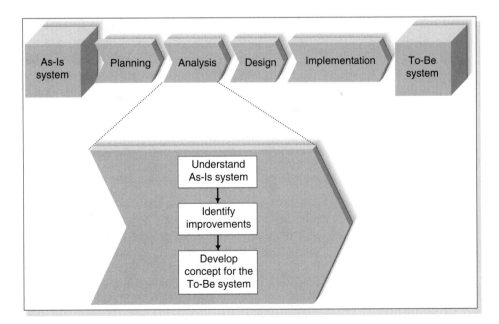

FIGURE 4-1
The Systems Development Life Cycle Process

in each step depends on the analysis strategy adopted by the project team. Nonetheless, each strategy strives to understand the basic business process and implement changes to make it better. A business process is simply a set of activities that are performed to achieve some goal. For example, a typical store would have a business process for ordering products, a business process for selling to customers, a business process for accepting customer returns, and so on. Each of these business processes could in turn be considered one part of the overall business process of running the store. When designing information systems, it is important to understand the scope of the business process under consideration (e.g., whether it is the entire store operations, or just one component).

There are three fundamental analysis strategies: *automating, improving,* and *reengineering* the business processes. *Business process automation* (BPA) means leaving the basic way in which the organization operates unchanged, using computer technology to do some of the work. BPA can make the organization more efficient but has the least impact on the business. *Business process improvement (BPI)* means making moderate changes to the way in which the organization operates to take advantage of new opportunities offered by technology or to copy what competitors are doing. BPI can improve efficiency (i.e., doing things right) and improve effectiveness (i.e., doing the right things). *Business process reengineering (BPR)* means changing the fundamental way in which the organization operates— "obliterating" the current way of doing business and making major changes to take advantage of new ideas and new technology. No one approach is necessarily better than the others. BPA is the most appropriate in some cases, BPI is best in others, and BPR brings about the most improvements in still other cases. BPR is the most challenging because it requires a complete redesign of business before the information systems to support it can be designed. It has the greatest potential to significantly improve profits and reduce expenses, but it can also be the riskiest— most reengineering projects fail to deliver the benefits they promise. Regardless of which analysis strategy is used, the project still flows through the three fundamental steps: understanding the As-Is system, identifying improvements, and developing the To-Be system concept.

Historically, object-oriented analysis and design has most commonly been used on BPR projects, such as those that replace traditional business processes with e-commerce processes. Therefore, most of the effort in object-oriented projects has focused on the To-Be business processes, with little effort spent on modeling the As-Is processes. This may change as the use of object-oriented techniques becomes more common and they are used on a wider array of projects. For this reason, we include a discussion of BPA and BPI approaches, but for the moment, BPR approaches are more common.

Understanding the As-Is System

In most cases, the system under development will replace an existing system. Thus, the first step is to study this system and understand its strengths and weaknesses. This usually requires the project team to apply the requirements elicitation techniques that we will present in Chapter 5.

If the As-Is system is a computerized system, then the team can review the analysis and design documents from the previous systems development project that built the system (if they exist). However, today, since most existing systems are not object-oriented, the project team will most likely have to either convert old analysis and design documents to the equivalent object-oriented ones to develop detailed behavioral and structural models that describe the current system.

In many cases, the tendency is to jump to conclusions about what the new system should be like. There is a temptation to focus only on what the users want in the new system (rather than understanding the As-Is system, whether computerized or not), but this can be dangerous. Without an understanding of the As-Is system, it is hard to understand users' requirements. Although users often know what they would *like* in a new system, it is sometimes not what they really *need*. Similarly, analysts sometimes start a project with a predetermined system concept and fail to listen to users.

Identifying Improvement Opportunities

Once the project team understands the current system, it then identifies ways to improve that system. Again, various information-gathering techniques are used to understand what improvements should be made.

Identifying improvement opportunities requires technology skills and significant business expertise in the functional area being examined. Technology skills are needed because without them, it is impossible to build the needed systems, and because information technology offers new ideas to solve problems and exploit opportunities. Business skills are needed because technology itself cannot make a business successful; only when technology is harnessed to address real business needs can the organization gain value.

An analyst or external consultant with a lot of experience—or the users themselves under the guidance of an experienced analyst or consultant—will work on identifying improvements. Although users will seldom have experience in the analysis of business processes, they are the experts in the processes themselves.

Developing the To-Be System Concept

Once information about the system has been gathered and improvements are identified, the analysts and users then develop the components for the To-Be system. The system concept starts as a fuzzy set of possible improvement ideas that are gradually worked and reworked into a viable concept for the To-Be system. Once the system concept is reasonably well understood, a set of behvioral and structural models are created. These models require very detailed information, and often the analysts must again interview managers and users to gather more information.

Analysis ends with a *system proposal* for the new system that presents an overview of one alternative (sometimes more alternatives). The proposal presents a vision for the new system and outlines its basic design (Figure 4-2). The proposal may present only one recommended concept or may present several alternatives. Each alternative will present an outline of the new system, usually with a set of behavioral and structural models (see Chapters 6, 7, and 8). The analysts will develop a revised work plan for each alternative (Chapter 3) and will again examine the expected costs and benefits and present a more detailed version of the feasibility analysis done when the project began (Chapter 2).

This proposal will be presented to the approval committee, which will decide if the project is to continue, and, if so, which alternative will be used. This is usually done at a system *walk-through,* a meeting at which the concept for the new system is presented in moderate detail to the users, managers, and key decision makers,

1. Table of contents
2. **Executive summary:** A summary of all the essential information in the proposal so a busy executive can read it quickly and decide what parts of the plan to read in more depth.
3. **System request:** The original system request form that initiated the project, with revisions as needed after the analysis phase (see Chapter 2).
4. **Work plan:** The original work plan, revised after completion of the analysis phase (see Chapter 3).
5. **Analysis strategy:** A summary of the activities performed during the analysis phase, such as the analyses performed (see this chapter) and how information was gathered (e.g., who was interviewed, what questionnaires were used; see Chapter 5).
6. **Recommended system:** A summary of the concept for the recommended system, as well as the key facts justifying the decision. Often, a discussion of alternatives considered is also included.
7. **Feasibility analysis:** A revised feasibility analysis using the information from the analysis phase (see Chapter 2).
8. **Behavioral and structural models:** A set of behavioral and structural models and descriptions for the To-Be system (see Chapters 6, 7, and 8). This often also includes behavioral and structural models of the current As-Is system that will be replaced.
9. **Appendices:** These contain additional material relevant to the proposal, often used to support the recommended system. This might include results of a questionnaire survey, interviews, or focus groups with customers; industry reports and statistics; potential design issues to be addressed; and possible hardware and software considerations.

FIGURE 4-2

Outline of a Typical System Proposal

so that all clearly understand it, can identify needed improvements, and are able to decide whether the project should continue. At this point, the project transitions from analysis into the design phase.

BUSINESS PROCESS AUTOMATION

BPA leaves the existing business processes essentially the same but puts in place a new system that makes those processes more efficient. A new system automates existing processes when it replaces existing manual processes with computerized ones, when it adds computer support to processes, or when it improves existing computer systems (but does not change the way in which the business operates).

Business processes are the way in which an organization operates. For example, when the university library loans you a book, it follows a basic business process that probably includes scanning your identification card, scanning the book identification code, recording that you are borrowing it, and demagnetizing the book so an alarm won't go off when you leave the library with the book.

In the case of the library, BPA could be accomplished by installing a new system that runs much faster than the existing computer system or that simplifies the way in which the library staff interact with the system (e.g., reduces the number of keystrokes needed to check out books). Automating processes usually provides improved *efficiency* to users, by speeding up or simplifying their work. Automating processes has the least impact on their jobs because the fundamental business processes—the work that the users do—remains essentially unchanged, although users may now use new tools.

Understanding the As-Is System

BPA devotes considerable time and effort to understanding the As-Is system (whether it is manual or computerized) because the To-Be system will continue to support the As-Is business processes. The new system will do essentially the same things as the old system but in new and better ways. The project team usually conducts extensive information gathering through interviews with users and managers, observation of the system in operation, and analysis of the current system's documentation. (Chapter 5 discusses requirements elicitation techniques in detail.) The team usually builds very detailed behavioral and structural models that document what activities are performed, what information is needed by them, and the type and format of data used. Figure 4-3 illustrates the BPA analysis strategy.

Identifying Improvement Opportunities

As the project team is gathering information about the As-Is system, it also begins to identify improvement opportunities. With BPA, most of the improvement opportunities come from problems in the current system. There are two general techniques commonly used to identify improvements: problem analysis and root cause analysis.

Problem Analysis   The most straightforward BPA analysis technique (and probably the most commonly used one) is *problem analysis*. Problem analysis involves ask-

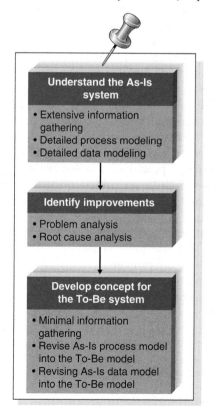

FIGURE 4-3
The Business Process Automation
Analysis Strategy

ing the users and managers to identify problems with the As-Is system and how to solve them in the To-Be system. Most users have a very good idea of the changes they would like to see, and most will be quite vocal about suggesting them. Most changes tend to solve problems rather than capitalize on opportunities, but the latter is possible, too. Improvements from problem analysis tend to be small and incremental (e.g., provide more space in which to type the customer's address).

This type of improvement often is very effective at improving the user's efficiency or the ease of use of the To-Be system compared with the As-Is system. However, they often provide only minor improvements in business value: the new system is better than the old, but it may be hard to identify any significant monetary benefits from the new system.

Root Cause Analysis   The ideas produced by problem analysis tend to be *solutions* to problems. All solutions make assumptions about the nature of the problem, assumptions that may or may not be valid. In our experience, users (and most people in general) tend to quickly jump to solutions without fully considering the nature of the problem. Sometimes the solutions are appropriate, but many times they address a *symptom* of the problem, not the true problem or *root cause* itself.[1]

For example, suppose you notice that a lightbulb is burned out above your front door. You buy a new bulb, get out a ladder, and replace the bulb. A month later, you see that the bulb in the same fixture is burned out, so you buy a new bulb, haul out the ladder, and replace the bulb again. This repeats itself several times. At this point, you have two choices. You could buy a large package of lightbulbs and a fancy lightbulb changer on a long pole so you don't need the haul the ladder out each time (thus saving a lot of trips to the store for new bulbs and a lot of effort in working with the ladder), or you could repair the light fixture that is causing the bulb to burn out in the first place. Buying the bulb changer is treating the symptom (the burned-out bulb), whereas repairing the fixture is treating the root cause.

In the business world, the challenge lies in identifying the root cause—few problems are as simple as the lightbulb problem. The solutions that users propose (or systems analysts think of) may address either symptoms or root causes, but without a careful analysis, it is difficult to tell—and finding out that you've just spent a million dollars on a new lightbulb changer is a horrible feeling!

Root cause analysis therefore focuses on problems, not solutions. The analyst starts by having the users generate a list of problems with the As-Is system and then prioritize the problems in order of importance. Then, the users and/or the analysts generate all the possible root causes for the problems, starting with the most important. Each possible root cause, starting with the most likely or easiest to check, is investigated until the true root cause or causes are identified. If any possible root causes are identified for several problems, those should be investigated first because there is a good chance they are the real root causes influencing the symptom problems.

In our lightbulb example, there are several possible root causes. A tree diagram sometimes helps with the analysis. As Figure 4-4 shows, there are many possible root causes, so that buying a new fixture may or may not address the true root

---

[1] Two good books that discuss the difficulties in finding the root causes to problems are E. M. Goldratt and J. Cox, *The Goal,* (Croton-on-Hudson, NY: North River Press, 1986); and E. M. Goldratt, *The Haystack Syndrome,* (Croton-on-Hudson, NY: North River Press, 1990).

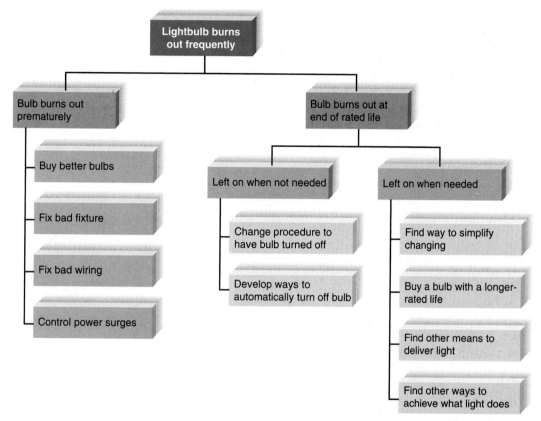

FIGURE 4-4
Root Cause Analysis for the Example of the Burned-Out Lightbulb

cause. In fact, buying a lightbulb changer may actually address the root cause. The key point in root cause analysis is to always challenge the obvious.

Developing the To-Be System Concept

With BPA, the To-Be system is usually quite similar to the As-Is system in terms of the general structure of the business processes. The To-Be models are usually very close to the As-Is models. The principal differences are the way in which information and data are provided to the processes (e.g., from a computer system rather than manual files) and the way in which the processes use them (e.g., manipulated on the screen rather than on paper). Many of these differences are important, but since they don't introduce major changes to the business processes themselves, the models are often very similar, except for certain sections of the process that may exhibit a few alterations. These changes may, of course, require significant design changes, but the changes in screen design and so forth are documented later in the SDLC during the design phase, not during analysis.

Since the changes to the process and data models tend to be slight, the analysts can, in many cases, simply copy the As-Is models and make changes to them to reflect the new system. In general, little additional information gathering is needed.

CONCEPTS **4-A WHEN OPTIMAL ISN'T**

IN ACTION

In 1984, I developed an information system to help schedule production orders in paper mills. Paper is made in huge rolls that are 6 to 10 feet wide. Customer orders (e.g., newspaper, computer paper) are cut from these large rolls. The goal of production scheduling is to decide which orders to combine on one roll to reduce the amount of paper wasted because no matter how you place the orders on the rolls, you almost always end up throwing away the last few inches—customer orders seldom add up to exactly the same width as the roll itself. (Imagine trying to cut several rolls of toilet paper from a paper towel roll.)

The scheduling system was designed to run on an IBM personal computer, which in those days was not powerful enough to use advanced mathematical techniques, such as linear programming, to calculate which orders to combine on which roll. Instead, we designed the system to use the rules of thumb that the production schedulers themselves had developed over many years. After several months of fine-tuning, the system worked very well. The schedulers were happy and the amount of waste decreased.

By 1986, personal computers had increased in power, so we revised the system to use the advanced linear programming techniques to provide better schedules and reduce the waste even more. In our tests, the new system reduced waste by a few percentage points over the original system, so it was installed. However, after several months, there were big problems; the schedulers were unhappy and the amount of waste actually grew.

It turned out that although the new system really did produce better schedules with less waste, the real problem was *not* figuring how to place the orders on the rolls to reduce waste. Instead, the real problem was finding orders for the next week that could be produced with the current batch. The old system combined orders in a way in which the schedulers found it easy to find "matching" future orders. The new system made odd combinations with which the schedulers had a hard time working. The old system was eventually reinstalled. *Alan Dennis*

QUESTION:

What are the problems with the new system? Do you think the project team could have avoided them? Why or why not?

BUSINESS PROCESS IMPROVEMENT

The goal of BPI is to improve the business processes by introducing some moderate changes that are generally incremental or evolutionary in nature.[2] The To-Be system implements these changes and creates business value not only by making the users more efficient but also by changing what they do to make them more effective.

In the library example discussed earlier, a BPI approach might involve attempting to develop new ways to enable the user to check out books that are more effective. For example, the user might be able to search the library catalog from his or her office on campus and then electronically request that the library check out the book and send it to him or her via campus mail. The books are still checked out, but now the user is saved the time and effort of traveling to the library and finding the book on the shelves. In this case, the use of the computer has changed the fundamental business process (or more exactly, added a new process).

---

[2] One good book regarding BPI is that by T. H. Davenport, *Process Innovation,* (Boston: Harvard Business School Press, 1993).

4-1 IBM CREDIT (PART 1)

This is a multi-part example, so you should complete this part before you read the subsequent parts. IBM Credit was a wholly owned subsidiary of IBM responsible for financing mainframe computers sold by IBM. While some customers bought mainframes outright, or obtained financing from other sources, financing computers provided significant additional profit to IBM.

When an IBM sales representative made a sale, he or she would immediately call IBM Credit to obtain a financing quote. The call was received by a credit officer, who would record the information on a request form. The form would then be sent to the credit department to check the customer's credit status. This information would be recorded on the form, which was then sent to the business practices department who would write a contract (sometimes reflecting changes requested by the customer). The form and the contract would then go to the pricing department, which used the credit information to establish an interest rate and recorded it on the form. The form and contract was then sent to the clerical group, where an administrator would prepare a cover letter quoting the interest rate and send the letter and contract via Federal Express to the customer.

The problem at IBM Credit was a major one. Getting a financing quote took anywhere for four to eight days (six days on average), giving the customer time to rethink the order or find financing elsewhere. While the quote was being prepared, sales representatives would often call to find out where the quote was in the process, so they could tell the customer when to expect it. However, no one at IBM Credit could answer the question because the paper forms could be in any department and it was impossible to locate one without physically walking through the departments and going through the piles of forms on everyone's desk.

QUESTION:
How could this situation be improved?

Source: Michael Hammer and James Champy, *Reengineering the Corporation* (New York: HarperCollins 1993).

Understanding the As-Is System

BPI devotes considerable time and effort to understanding the As-Is system (whether it is manual or computerized) because the To-Be system will continue to support most of the As-Is business processes. The new system will do many of the same things as the old system, although some processes will be quite different. As with BPA, the project team conducts extensive information gathering through interviews with users and managers, observation of the system in operation, and analysis of the current system's documentation. The team also usually builds very detailed behavioral and structural models (Figure 4-5).

Identifying Improvement Opportunities

One of the elements that distinguishes BPI from BPA is the real focus on improvements to the *business,* not just to the computer systems that support it. The users and managers actively seek out new business ideas and business opportunities that can be implemented through the To-Be system. Although ideas for improvements can come from the same problem analysis or root cause analysis techniques used for BPA, more powerful techniques are often needed to encourage managers to think more creatively about their business. In this section, we discuss four techniques that can be used. Two focus inside the organization on the business process models developed for the As-Is system (duration analysis and activity-based costing), and the other two focus outside the organization to bring in new ideas developed elsewhere (formal and informal benchmarking). In general, no project would

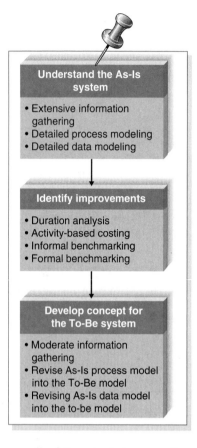

FIGURE 4-5
The Business Process Improvement
Analysis Strategy

use all of these techniques; rather, most use a subset selected by the project sponsor and analysts.

Duration Analysis *Duration analysis* requires a detailed examination of the amount of time it takes to process inputs in the current As-Is business process. The analysts start by determining the total amount of time it takes, on average, to perform the one set of business processes for a typical input. They then examine each of the basic steps (or subprocesses) in the business process and determine the total amount of time it takes on average, to complete that one step for a typical input. The times for the basic steps are then totaled and compared to the total for the overall process. When there is a significant difference between the two—and in our experience the total time often can be 10 or even 100 times longer than the sum of the parts—this indicates that this part of the process is badly in need of a major overhaul. This analysis is repeated for each subprocess to identify the areas in greatest need of improvement.

For example, suppose that the analysts are working on a home mortgage system and discover that, on average, it takes 30 days for the bank to approve a mortgage. They then look at each of the basic steps in the process (e.g., data entry, credit check, title search, appraisal) and find that the total amount of time actually spent on each mortgage is about 8 hours. This is a strong indication that the overall process is badly broken because it takes 30 days to perform 1 day's work.

These problems likely occur because the process is badly fragmented. Many different people must perform different activities before the process finishes. In the mortgage example, the application probably sits on many people's desks for long periods of time before it is processed. Processes in which many different people work on small parts of the inputs are prime candidates for *process integration* or *parallelization.* Process integration involves changing the fundamental process so that fewer people work on the input, which often requires changing the processes and retraining staff to perform a wider range of duties. Process parallelization involves changing the process so that all the individual steps are performed at the same time. For example, in the mortgage application example, there is probably no reason that the credit check cannot be performed at the same time as the appraisal and title check.

In both of these cases, the users, managers, and project sponsor decide to change the business processes first, and the computer system is designed to support the new processes. The decision to change the processes cannot be made by the analysts because first they require an understanding of the business; understanding how to implement the new business processes in the As-Is system comes second.

Activity-Based Costing   *Activity-based costing* is a similar analysis that examines the cost of each major process or step in a business process rather than the time taken.[3] The analysts simply identify the costs associated with each of the basic functional steps or processes, identify the most costly processes, and focus their improvement efforts on them.

Assigning costs is conceptually simple. You just examine the direct cost of labor and materials for each input. Materials costs are easily assigned in a manufacturing process, whereas labor costs are usually calculated on the basis of the amount of time spent on the input and the hourly cost of the staff. However, as you may recall from a managerial accounting course, indirect costs, such as rent and depreciation, can also be included in activity costs.

Informal Benchmarking   *Benchmarking* refers to studying how other organizations perform a business process so you can learn how your organization can do it better. Benchmarking helps the organization by bringing in ideas that employees may never have considered but that have the potential to add value. Benchmarking is often used for process improvement and reengineering.

Informal benchmarking is fairly common for "customer-facing" business processes (i.e., those processes that interact with the customer). With informal benchmarking, the managers and analysts think about other organizations—or visit them—as customers to see how the business process is performed. In many cases, the business studied may be a known leader in the industry or simply a related firm. For example, suppose the team is developing a Web site for a car dealer. The project sponsor, key managers, and key team members would likely visit the Web sites of competitors, as well as those of others in the car industry (e.g., manufacturers, accessories suppliers) and those in other industries that have won awards for their Web sites.

Formal Benchmarking   *Formal benchmarking* is the most thorough and costly of the benchmarking strategies. With this approach, the organization establishes a formal

---

[3] Many books have been written on activity-based costing. Useful ones include K. B. Burk and D. W. Webster, *Activity Based Costing* (Fairfax, VA: American Management Systems, 1994); and Douglas T. Hicks, *Activity-Based Costing: Making It Work for Small and Mid-Sized Companies,* 2nd ed., (New York: John Wiley, 1998). The two books by E. M. Goldratt mentioned in footnote 1 (*The Goal* and *The Haystack Syndrome*) also offer unique insights into costing.

4-2 IBM CREDIT (PART 2)

This is a multi-part example, so you should complete the previous part of this example before you read this. IBM Credit examined the process (see Your Turn 4-1), and changed it so that each credit request was logged into a log book by the credit officer when it was received. After each step (i.e., credit, business practices, pricing, clerical) the forms were returned to the credit officer who recorded the request's status in the log book before taking it to the next department. In this way, sales representatives could call the credit office and quickly learn the status of each application. However, average process times went six days to nine days. Some improvement!

QUESTION:
How could this situation be improved?

relationship with one or more organizations, usually in different industries. The organizations send teams of analysts and managers from each business process to meet with their counterparts in the other organizations. The teams thoroughly review each others' business processes and computer systems, candidly discussing what works and what does not. Often confidential cost information is shared to help pinpoint successes and failures.

Developing the To-Be System Concept

With BPI, the To-Be system is usually similar to the As-Is system in general structure, with some business processes being very different. The To-Be behavioral and structural models are usually very close to the As-Is models in many areas, with the exceptions of the (often few) processes that have been changed. These changes may, of course, require significant rethinking and rework.

In many cases, the analysts can simply copy the As-Is models and make changes to them to reflect the new system. In general, a moderate amount of information gathering is needed for those parts of the business that have changed. Once the processes to be changed have been identified, the project team often embarks on another round of interviews and surveys with the managers and users in those areas.

BUSINESS PROCESS REENGINEERING

BPR is "the fundamental rethinking and radical redesign of business processes to achieve dramatic improvements in critical, contemporary measures of performance, such as cost, quality, service, and speed."[4] BPR is intuitively appealing (who wouldn't want dramatic improvements?) but is very time-consuming and risky because it requires radical change. The goal is to throw away everything about the current business process and redesign it starting with a blank piece of paper.

---

[4] This quote is from the most commonly used book on BPR: Michael Hammer and James Champy, *Reengineering the Corporation,* (New York: HarperCollins, 1993), p. 32. Reengineering can benefit from creative thinking, such as that promoted by Edward De Bono, *Serious Creativity,* (New York: HarperCollins, 1992); and Roger Von Oech, *A Whack on the Side of the Head,* (New York: Warner Books, 1983).

In the library example discussed earlier, a BPR approach might eliminate the need for users to check out books and magazines altogether. Instead, users would read the material on the Web. In this case, the use of the computer has radically changed the fundamental business process. After all, the user doesn't really care about getting the physical book or magazine; the user really wants the information it contains.

Understanding the As-Is System

Because the goal of BPR is to eliminate current business processes, it devotes little time and effort to understanding the As-Is system. The project team ensures it has a basic understanding of the essence of the As-Is system and usually builds only very superficial behavioral models that capture the general idea of the business process. Structural models are seldom done (Figure 4-6).

Identifying Improvement Opportunities

The key focus of BPR is radical improvements, which are not simple to identify. The BPA and BPI techniques discussed above usually produce incremental improvement ideas, and although these techniques are sometimes used in BPR projects, the incremental improvement ideas that they produce (which offer real but minor value) are usually not that useful. BPR requires completely rethinking

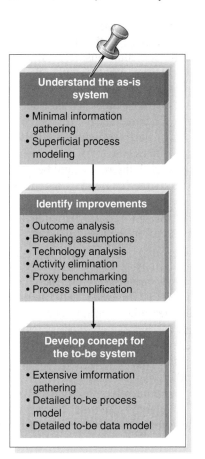

FIGURE 4-6
The Business Process Reengineering
Analysis Strategy

the business, so more powerful techniques are needed to encourage managers to think more creatively about their business. This section describes six different BPR techniques that we have used. Some of the techniques are drawn from the work of Mike Hammer (the father of BPR), and some are drawn from our own work. Once again, we should emphasize that no project would use all of these techniques but rather would use a subset selected by the project sponsor and analysts.

Outcome Analysis   *Outcome analysis* focuses on understanding the fundamental outcomes that provide value to customers. Although these outcomes sound as though they should be obvious, they often aren't. For example, suppose you are an insurance company and one of your customers has just had a car accident. What is the fundamental outcome from the customer's perspective? Traditionally, insurance companies have answered this question by assuming the customer wants to receive the insurance payment quickly. To the customer, however, the payment is only a means to the real outcome: a repaired car. The insurance company might benefit by extending its view of the business process past the traditional boundaries to include not just paying for repairs but also performing the repairs or contracting with an authorized body shop to do repairs.

With this approach, the system analysts encourage the managers and project sponsor to pretend they are customers and think carefully about what the organization's products and services enable the customers to do—and what they *could* enable the customer to do.

Breaking Assumptions   One of the most powerful BPR techniques that we have used is *breaking assumptions*. With this technique, the managers and analysts identify several dozen fundamental business rules about the business process being reengineered that are assumed to be valid and appropriate. Then the group systematically generates ideas about how to break each and every rule and identifies how the business would benefit by breaking the rule.

For example, a fundamental rule for a banking process might be, "Don't cash a check if the customer has insufficient funds in his or her account" (i.e., an NSF [nonsufficient funds] check). At first glance, this is clearly the "right" way to do business. One way to break the rule would be to offer customers the ability to have NSF checks honored by the bank via an automatic cash withdrawal on their credit card. Customers win because they don't have to deal with NSF checks. The bank wins because it, too, doesn't have to deal with an NSF check and because it gets to charge high interest rates on the resulting credit card loans.

Technology Analysis   Many of the major changes in business in the 1990s were due to new technologies. *Technology analysis* therefore starts by having the analysts and managers develop a list of important and interesting technologies. Then the group systematically identifies how each and every technology could be applied to the business process and identifies how the business would benefit.

For example, one useful technology might be the Internet. Saturn, the car manufacturer, took this idea and developed a private Internet-like network to link to its suppliers. Rather than ordering parts for its cars, Saturn makes its production schedule available electronically to its suppliers, who ship the parts Saturn needs so that they arrive at the plant just in time. This saves Saturn significant costs because it eliminates the need for people to monitor the production schedule and issue purchase orders.

PRACTICAL   **4-1 AVOIDING CLASSIC ANALYSIS MISTAKES**

TIP

In Chapter 3, we discussed four classic planning and project management mistakes and how to avoid them. Here, we summarize four classic mistakes in the analysis phase and discuss how to avoid them:

1. **Reduction of Analysis Time:** If time is short, there is a temptation to reduce the time spent in "unproductive" activities such as analysis so that the team can jump into "productive" programming. This results in missing important requirements that have to be added later at a much higher time cost (usually at least 10 times longer).
   Solution: If time pressure is intense, use rapid application development (RAD) techniques and timeboxing to eliminate functionality or move it into future versions.

2. **Requirements gold-plating:** Managers and users sometimes add features that are not really necessary or overspecify features that can significantly increase time and cost (e.g., on-screen boxes should have rounded corners).
   Solution: If a requirement has a high development cost, the person who identified it should be informed of the estimated cost and asked to verify its importance. If there is a lower-cost alternative, it should be explicitly presented and considered.

3. **Developer gold-plating:** Analysts and programmers sometimes add features that are not really necessary but are "cool." Even "small" features, however, have considerable unseen expense in testing, documentation, and training.
   Solution: If an analyst or programmer wants to add a requirement, users should confirm that it provides clear value to *them,* not the developer.

4. **Lack of user involvement:** The number-one reason for project failure is the lack of user involvement—it leads to missed requirements and greatly reduces organizational feasibility (see Chapter 2).
   Solution: Work with the project sponsor to interest and motivate user involvement. If the users are not interested, this suggests the project should be terminated before it wastes additional resources.

Source: Adapted from Steve McConnell, *Rapid Development,* (Redmond, WA: Microsoft Press, 1996).

Activity Elimination   *Activity elimination* is exactly what it sounds like. The analysts and managers work together to identify how the organization could eliminate each and every activity in the business process, how the function could operate without it, and what effects are likely to occur. Initially, managers are reluctant to conclude that processes can be eliminated, but this is a "force-fit" exercise in that they are not permitted to suggest that an activity cannot be eliminated. In some cases, the results are silly; nonetheless, participants must address each and every activity in the business process.

For example, in the home mortgage approval process discussed earlier, the managers and analysts would start by eliminating the first activity, entering the data into the mortgage company's computer. This leads to two immediately obvious possible changes: (1) eliminate the use of a computer system (which probably is not a good idea, but it might be for a small firm) and (2) make someone else do the data entry (e.g., the customer over the Web). They would then eliminate the next activity, the credit check. Silly, right? After all, making sure the applicant has good credit is critical in issuing a loan. Not really. The real answer depends on how many times the credit check identifies bad applications. If all or almost all applicants have good credit and are seldom turned down by a credit check, then the cost of the credit check may not be worth the cost of the few bad loans it prevents. Eliminating it may actually result in lower costs, even considering the cost of bad loans.

Proxy Benchmarking   *Proxy benchmarking* is similar to informal benchmarking, except the target of the benchmarking is a different industry. With this approach,

4-3 IBM Credit (Part 3)

This is a multi-part example, so you should complete parts 1 and 2 before you read this. IBM Credit reexamined its process (see Your Turn 4.1 and 4.2), and decided to automate it. The manual log book was replaced by a computer system so that each department could record an application's status as they completed it and sent it to the next department. The credit officers could quickly see the status of each application in the system when the sales representatives called.

This improved things slightly, but not enough to reduce time much below the original six-day average. IBM credit then applied some sophisticated management science queuing theory analysis to balance workloads and staff across the different departments so that none would be overloaded. They also introduced performance standards for each department (e.g., the pricing decision had to be completed within one day after that department received an application).

However, process times got worse, even though each department was achieving almost 100 % compliance on its performance goals. After some investigation, managers found that when people got busy, they conveniently found errors that forced them to return credit requests to the previous department for correction, thereby removing it from their measurements.

Question:
How could this situation be improved?

the managers and analysts first develop a list of industries that have similar structures to the business processes under consideration. Then they attempt to apply the ideas and techniques from those industries to the process under consideration.

For example, if we were considering the hotel industry, we would look for industries that also have time-dependent inventories that vanish if not used (e.g., airlines, newspapers, rock concerts). This might lead us to consider establishing differential pricing based on the number of unsold rooms, establishing different rates for multiple bookings, or trying to establish a fan club. In our experience, it is helpful to include a few industries that are radically different to see if they spark any ideas (e.g., the army, day care, a religious organization, professional sports).

Process Simplification   Many business processes have become extremely complex over the years because they have been designed to deal with many special cases that occur only rarely. *Process simplification* eliminates complexity from the normal day-to-day operations and isolates it where it belongs—in the special cases. With this approach, a simple process is designed to handle the 80% to 90% percent of inputs that are dealt with on a day-to-day basis. A separate process (or set of processes) is designed to handle complex inputs, thus enabling simple inputs to flow quickly through the process without being delayed by complex inputs.

Developing the To-Be System Concept

With BPR, the To-Be system is usually very different from the As-Is system. Developing the To-Be concept usually requires extensive information gathering after the fundamental changes have been decided on. Analysts often interview managers and users in extensive detail to produce the very detailed behavioral and structural models needed for the design phase.

DEVELOPING AN ANALYSIS PLAN

An *analysis plan* is the plan for activities that the project team will conduct during the analysis phase. The analysis plan outlines what activities will be performed to understand the As-Is system, to identify improvements, and to develop the To-Be system. Each of the analysis strategies and analysis techniques discussed in this chapter has its own strengths and weaknesses. No one strategy or technique is inherently better than the others, and in practice most projects use a combination of techniques and sometimes even a combination of strategies.

The selection of the analysis strategy is determined by the project sponsor and is based on the goal of the project. It is a business decision, not a decision made by the project team, although the project team often plays an important role in providing advice to the project sponsor. It is also important to note that many projects use a mix of strategies—for example, using BPA on much of the business process but using BPI on key parts. Figure 4-7 presents a summary of the strengths and weaknesses of each strategy.

Potential Business Value

The *potential business value* from the project is greatly affected by the analysis strategy chosen. Although BPA has the potential to improve the business, most of the benefits from BPA are tactical and small in nature. Since BPA does not seek to change the business processes, it can only improve their efficiency. BPI usually offers moderate potential benefits, depending on the scope of the project, because it seeks to change the business in some ways. It can increase both efficiency and effectiveness. BPR almost always brings large potential benefits because it seeks to radically improve the nature of the business.

Project Cost

Project cost is always an important factor. In general, BPA requires the least cost because it has the narrowest focus and seeks to make the least changes. BPI can be moderately expensive, depending on the scope of the project. BPR is almost always very expensive, both because of the amount of time required of senior managers and the amount of redesign in both business processes and computer systems it implies.

Breadth of Analysis

Breadth of analysis refers to the extent that the analysis looks throughout the entire business function and even beyond into the adjacent functions and customer and

	Business Process Automation	Business Process Improvement	Business Process Reengineering
Potential business value	Low–moderate	Moderate	High
Project cost	Low	Low–moderate	High
Breadth of analysis	Narrow	Narrow–moderate	Very broad
Risk	Low–moderate	Low–moderate	Very high

FIGURE 4-7
Characteristics of Analysis Strategies

supplier business processes. BPR takes a broad perspective, often across several major business processes. BPI takes a much narrower scope, usually only within one part of a business process. BPA is very narrow in focus.

Risk

One final issue is *risk* of failure, whether due to the inability to design and build a system or due to the system's not providing business value. BPA and BPI have low to moderate risk because the To-Be system is fairly well defined and understood and its potential impact on the business can be assessed before it is implemented. BPR projects, on the other hand, are less predictable, as are the systems they create. BPR is extremely risky and not something to be undertaken unless the organization and its senior leadership are committed to making significant changes. Mike Hammer, the father of BPR, estimates that 70% of BPR projects fail.

APPLYING CONCEPTS AT CD SELECTIONS

Alec Adams, the senior systems analyst, must develop an analysis plan for the Internet sales system for CD Selections. This is an unusual system because it is a new business area. CD Selections does not have an existing system (computerized or not) for Internet sales or any other type of direct sales to consumers; their entire business runs through traditional retail stores. CD Selections does, however, have a series of As-Is systems that may need to interact with the new Internet sales systems, such as the inventory system, the accounting system, and the retail systems used at each of CD Selections' stores.

Alec began by talking with Margaret Mooney, the vice president of marketing, about the project goals. Margaret urged Alec to "be creative." Since this was an entirely new system and a new direction for CD Selections, she wanted to be as innovative as possible in considering ideas for the system. Although Internet sales were new to CD Selections, there are many current Internet sales systems offered by other firms. From Alec's perspective, this suggested the project would mix both BPI (based on the Internet systems of other firms) and BPR (based on CD Selections' own systems).

Understanding the As-Is System

This project was unusual from most projects in that it was a "greenfield" project—a project with no As-Is systems. However, because the Internet sales system would have to interact with CD Selections' existing inventory and accounting systems, Alec decided that the project team should review those to understand their key processes and data. The first step in the analysis plan was therefore to review the current systems documentation, including behavioral and structural models (Figure 4-8).

Step	Who and How
Understanding the As-Is system	
Review documentation	Analysts will review the behavioral and structural models for the inventory, accounting, in-store retail systems, and any other system with which the Internet sales system will interact.
Informal benchmarking	Analysts will review the Web sites for Internet-based CD sellers.
	Analysts will review the Web sites for other leading retail sites on the Internet.
Identifying improvements	
Problem analysis	Analysts will work with marketing staff and customers to identify a basic set of features for the Web site.
	Analysts will work with the current retail staff to identify problems with the in-store retail systems.
Technology analysis	Analysts will develop a list of innovative Internet technologies.
	Analysts, marketing staff, and possibly customers will work together to develop possible applications for CD Selections.
Outcome analysis	Analysts will work with marketing staff and customers to discuss why people buy CDs and attempt to identify other means of delivering the same value.
Developing a To-Be system concept	
Develop behavioral and structural models	Analysts will develop behavioral and structural models for the To-Be system.

FIGURE 4-8
Analysis Plan for CD Selections

Identifying Improvement Opportunities

As is typical in the design of new systems, Alec started by focusing on the part of the system to be used by the primary users. In this case, the primary users are CD Selections' customers, not its own staff (another unusual situation), who will use the Web site to buy CDs. What should CD Selections' Web site look like, and how should customers use it to buy CDs? CD Selections had no existing system to use as a starting point, but there are dozens of Web-based "stores" selling CDs and other similar merchandise. Alec decided the first step should be for the analyst team to do some informal benchmarking of other sites to identify key features for their system (Figure 4-8).

Alec also thought it would be useful to ask customers and the marketing staff for some simple basic requirements for the Web site using a problem analysis technique—in other words, asking them, on the basis of their experience with Web sites selling products, what the Web site should do and should not do. Alec thought there might be some useful lessons from CD Selections' experiences with its own in-store retail systems, so Alec decided to involve the staff from some of CD Selections' stores (see Figure 4-8).

Since the Web offers many new and innovative technologies, Alec decided that a technology analysis would be appropriate, especially since CD Selections was just starting and was not committed to any existing technology. He also thought that the marketing staff—and perhaps even customers—should be involved in brainstorming about how these technologies could be used to create value for CD Selections and its customers (see Figure 4-8).

Finally, since the goal was to identify innovative and radical new ideas, he decided an outcome analysis with customers and perhaps marketing staff could be useful (see Figure 4-8).

Developing the To-Be System Concept

Once these activities were complete, the analysts, senior managers, and project sponsor would work together to identify the key ideas for the Internet sales system and integrate them into a concept. The project team would then develop a set of behavioral and structural models for the new system.

The next step was to take this basic analysis plan and refine it by defining exactly how these steps would be accomplished—that is, adding the information-gathering techniques that described who would gather the required information and how would they do it (via interviews, group meetings, survey questionnaires, etc.). Techniques for information gathering are the subject of the next chapter.

SUMMARY

The Analysis Process

Systems analysis is both a business task and an information technology task because its goal is to create value for the business. The first step in analysis is to understand how the As-Is business processes and information systems operate. Next, ideas for improvements are gathered from users, managers, and other key people. From this information and analysis emerges a concept for a new system, which is documented in a systems proposal that includes a set of behavioral and structural models. The proposal is presented to the approval committee at a walk-through, after which the project is abandoned or moves into the design phase. There are three fundamentally different strategies for the analysis phase (business process automation [BPA], business process improvement [BPI], and business process reengineering [BPR]), although some projects combine elements from two of the strategies.

Business Process Automation

The goal of BPA is to leave the business processes intact but to apply computer technology to make them more efficient. With BPA, analysts spend considerable time gathering information about the As-Is system and develop detailed models for it. This information and models are analyzed using problem analysis and root cause analysis (which takes a deeper look at the situation by thoroughly examining all possible causes for problems). The To-Be models are usually very similar to the As-Is models. BPA usually provides only minor to moderate improvements to the business but is least costly and least risky.

Business Process Improvement

The goal of BPI is to make some minor to moderate changes to the business processes to make them more efficient and effective. With BPI, analysts spend considerable time gathering information about the As-Is system and develop detailed models for it. These are analyzed using problem analysis and root cause analysis or more detailed techniques, such as duration analysis or activity-based costing that examine the time required or costs associated with specific processes. Either formal benchmarking or informal benchmarking, both of which attempt to bring in ideas from other organizations, is also used. The To-Be models are usually somewhat similar to the As-Is models. BPA usually provides moderate improvements to the business, at moderate cost and risk.

Business Process Reengineering

The goal of BPR is to do a fundamental rethinking and radical redesign of business processes. With BPR, analysts usually do not gather much information about the As-Is system and do not develop detailed models for it. More radical and creative techniques are used for identifying improvements, such as outcome analysis, breaking assumptions, technology analysis, activity elimination, proxy benchmarking, and process simplification. Once the improvements have been selected, the team spends considerable time developing the To-Be system concept and its models. BPR's goal is to provide large improvements to the business, but such projects are costly and 70% fail.

KEY TERMS

Activity-based costing
Activity elimination
Analysis
Analysis plan
As-Is system
Automating

Benchmarking
Breadth of analysis
Breaking assumptions
Business process automation (BPA)
Business process improvement
   (BPI)

Business process reengineering
   (BPR)
Duration analysis
Formal benchmarking
Improving
Informal benchmarking

Outcome analysis	Project cost	Root cause analysis
Potential business value	Proxy benchmarking	Solution symptoms
Problem analysis	Reengineering	System proposal
Process integration or	Requirements analysis	Technology analysis
parallelization	Risk	To-Be system
Process simplification	Root cause	Walk-through

QUESTIONS

1. Explain the difference between an As-Is system and a To-Be system.
2. What are the basic steps in the analysis process?
3. Why are business skills important in the analysis phase?
4. What are the principal components of a system proposal?
5. Compare and contrast the business goals of business process automation (BPA), business process improvement (BPI), and business process reengineering (BPR).
6. Compare and contrast problem analysis and root cause analysis. Under what conditions would you use problem analysis? Under what conditions would you use root cause analysis?
7. Compare and contrast informal benchmarking, proxy benchmarking, and formal benchmarking.
8. Compare and contrast duration analysis and activity-based costing.
9. Compare and contrast root cause analysis, breaking assumptions, and outcome analysis.
10. Compare and contrast technology analysis with root cause analysis.
11. Under what conditions would formal benchmarking be important?
12. Assuming time and money were not important concerns, would BPR projects benefit from additional time spent understanding the As-Is system?
13. What are the key factors in selecting an appropriate analysis strategy?
14. Which do you think is most common today BPA, BPI, or BPR? Why?
15. What are the biggest challenges to success in BPA? BPI? BPR?
16. What do you think are three common mistakes novice analysts make in conducting the analysis process?

EXERCISES

A. Perform a root cause analysis for why you might get a low grade in a course.
B. Describe in very general terms how the current As-Is business process works for taking courses at your university. Explain how this process might operate after BPA, BPI, and BPR.
C. Suppose your university is having a dramatic increase in enrollment and is having difficulty finding enough seats in courses for students. Perform a breaking-assumptions analysis to identify new ways to help students complete their studies and graduate.
D. Suppose you are the analyst charged with developing a new system to make students' grades available electronically instead of by mail. Develop an analysis plan.
E. Suppose you are the analyst charged with developing a new retail sales system for the university bookstore. Develop an analysis plan.
F. Suppose you are the analyst charged with developing a new system to help senior managers make better strategic decisions. Develop an analysis plan.
G. Suppose you are the analyst charged with developing a new system to move the airline reservation system used by thousands of travel agents that uses an old and hard-to-learn keyword interface and requires special hardware and software to a Web-based system that will run on personal computers. Develop an analysis plan.

MINICASES

1. The State Firefighters' Association has a membership of 15,000. The purpose of the organization is to provide some financial support to the families of deceased member firefighters and to organize a conference each year bringing together firefighters from all over the state. Annually, members are billed dues and calls. "Calls" are additional funds required to take care of payments made to the families of deceased members. The bookkeeping work for the association is handled by the elected treasurer, Bob Smith, although it is widely known that his wife, Laura, does all of the work. Bob runs unopposed each year at the election, since no one wants to take over the tedious and time-consuming job of tracking memberships. Bob is paid a stipend of $8,000 per year, but his wife spends well over 20 hours per week on the job. The organization, however, is not happy with their performance.

   A computer system is used to track the billing and receipt of funds. This system was developed in 1984 by a Computer Science student and his father. The system is a DOS-based system written using dBase 3. The most immediate problem facing the treasurer and his wife is the fact that the system is not year 2000 compliant. All the calculations for determining dues, calls, and futures (amount paid ahead) are date based. Other problems are apparent in the system as well. One query in particular takes 17 hours to run. Over the years, they have just avoided running this query, although the information in it would be quite useful. Questions from members concerning their statements cannot be easily answered. Usually Bob or Laura just jot down the inquiry and return a call with the answer. Sometimes it takes 3 to 5 hours to find the information needed to answer the question. Often, they have to perform calculations manually since the system was not programmed to handle certain types of queries. When member information is entered into the system, each field is presented one at a time. This makes it very difficult to return to a field and correct a value that was entered. Sometimes a new member is entered but disappears from the records. The report of membership used in the conference materials does not alphabetize members by city. Only cities are listed in the correct order.

   What requirements analysis strategy or strategies would you recommend for this situation? Explain your answer.

2. Brian Callahan, IS project manager, is just about ready to depart for an urgent meeting called by Joe Campbell, manager of manufacturing operations. A major BPI project sponsored by Joe recently cleared the approval hurdle, and Brian helped bring the project through Project Initiation. Now that the approval committee has given the go-ahead, Brian has been working on the project's analysis plan.

   One evening, while playing golf with a friend who works in the manufacturing operations department, Brian learned that Joe wants to push the project's time frame up from Brian's original estimate of 13 months. Brian's friend overheard Joe say, "I can't see why that IS project team needs to spend all that time 'analyzing' things. They've got two weeks scheduled just to look at the existing system! That seems like a real waste. I want that team to get going on building my system."

   Because Brian has a little inside knowledge about Joe's agenda for this meeting, he has been considering how to handle Joe. What do you suggest Brian tell Joe?

ANALYSIS

✓ **Develop Analysis Plan**
☐ Understand As-Is System
☐ Identify Improvement Opportunities
☐ Develop the To-Be System Concept
☐ Develop Use Cases
☐ Develop Structural Model
☐ Develop Behavioral Model

TASK CHECKLIST

PLANNING   ANALYSIS   DESIGN

CHAPTER 5
REQUIREMENTS GATHERING

THE basic process of systems analysis is divided into three steps: understanding the As-Is system, identifying improvements, and developing a concept for the To-Be system. To do any of these steps, however, the analyst must first gather enough information to clearly understand the current business processes and the needs for the new system. This chapter describes five common information-gathering techniques and discusses when to use each.

OBJECTIVES

- Understand how to use interviews to gather requirements.
- Understand how to use joint application design sessions to gather requirements.
- Be familiar with questionnaires.
- Be familiar with document analysis.
- Be familiar with observation.
- Understand when to use each technique.

CHAPTER OUTLINE

IMPLEMENTATION

INTRODUCTION

As we explained in Chapter 4, the basic process of systems analysis is divided into three steps—understanding the As-Is system, identifying improvements, and developing a concept for the To-Be system—and the analysis plan describes what activities will be performed to conduct these three steps. To implement the plan, however, analysts need to gather the right information for each of the activities. The project team members will not be able to automate, improve, or reengineer the business effectively if they have not gotten input from the appropriate people or if they have not uncovered the whole story. Many studies have attributed unmet user needs and runaway projects to a poor understanding of the system requirements early on in the project's life cycle.

When gathering requirements, analysts are like detectives (and business users sometimes are like elusive suspects). They know that there is a problem to be solved and therefore they must look for clues that uncover the solution. Unfortunately, the clues are not always obvious (and often missed), so analysts need to notice details, talk with witnesses, and follow leads just as Sherlock Holmes would have done. The best analysts will thoroughly gather requirements using a variety of information-gathering techniques and make sure that the current business processes and the needs for the new system are well understood before moving into design. You don't want to discover later that you have key requirements wrong—surprises like this late in the systems development life cycle (SDLC) can cause all kinds of problems.

The first challenge when gathering requirements is how to select the "right people." Good analysts let the objectives and needs of the analysis tasks drive the selection of the people who participate during analysis. Think about each task and the information that it requires, and then list the people who can provide that information. For example, if input is needed to understand the As-Is system for a business process automation (BPA) task, then a low-level employee (a data-entry clerk) who works directly with the current system may be a good person to include. By contrast, the analysts may decide to focus on customers and high-level senior managers for business process reengineering (BPR).

Just as important, people are selected to participate in requirements gathering for political reasons. The requirements-gathering process is used for building political support for the project and establishing trust and rapport between the project team building the system and the users who ultimately will choose to use or not use the system. Involving someone in the process implies that the project team views that person as an important resource and values his or her opinions. You *must* include all of the key *stakeholders* (the people who can affect the system or who will be affected by the system) in the requirements-gathering process, such as managers, employees, staff members, and even some customers and suppliers. If you do not involve a key person, that individual may feel slighted, which can cause problems during implementation (e.g., the person who demands, "How could they have developed the system without my input?!").

The second challenge of requirements gathering is choosing the way(s) in which information is collected. There are many techniques for gathering requirements, varying from asking people questions to watching them work. In this chapter, we focus on the five most commonly used techniques: interviews, joint application design (JAD) sessions (a special type of group meeting), document analysis, observation, and questionnaires. Each technique has its own strengths and weaknesses, many of which are complementary, so most projects use a combination of techniques—probably most often interviews, JAD sessions, and document analysis.[1]

INTERVIEWS

The *interview* is the most commonly used requirements-gathering technique. After all, it is natural—usually, if you need to know something, you ask someone. In general, interviews are conducted one-on-one (one interviewer and one interviewee), but sometimes, owing to time constraints, several people are interviewed at the same time. There are five basic steps to the interview process: selecting interviewees, designing interview questions, preparing for the interview, conducting the interview, and follow-up.[2]

Selecting Interviewees

The first step in interviewing is to create an *interview schedule* that lists all the people who will be interviewed, when, and for what purpose (Figure 5-1). The schedule

Name	Position	Purpose of Interview	Meeting
Andria McLellan	Director, Accounting	Strategic vision for new accounting system	Mon., March 1, 1999, 8:00–10:00 A.M.
Jennifer Smalec	Manager, Accounts Receivable	Current problems with accounts receivable process; future goals	Mon., March 1, 1999, 2:00–3:15 P.M.
Mark Goodin	Manager, Accounts Payable	Current problems with accounts payable process; future goals	Mon., March 1, 1999, 4:00–5:15 P.M.
Anne Asher	Supervisor, Data Entry	Accounts receivable and payable processes	Wed., March 3, 1999, 10:00–11:00 A.M.
Fernando Merce	Data Entry Clerk	Accounts receivable and payable processes	Wed., March 3, 1999, 1:00–3:00 P.M.

FIGURE 5-1
Sample Interview Schedule

---

[1] Some excellenet books that address the importance of gathering requirements and various techniques include Alan M. Davis, *Software Requirements: Objects, Functions, & States, Revision* (Englewood Cliffs, NJ: Prenctice Hall, 1993); Gerald Kotonya and Ian Sommerville, *Requirements Engineering* (Chichester, England: John Wiley & Sons, 1998); and Dean Leffingwell and Don Widrig, *Managing Software Requirements: A Unified Approach* (Reading, MA: Addison-Wesley, 2000).

[2] A good book on interviewing is that by Brian James, *The Systems Analysis Interview,* (Manchester: NCC Blackwell, 1989).

CONCEPTS 5-A SELECTING THE WRONG PEOPLE

IN ACTION

In 1990, I led a consulting team on a major development project for the U.S. Army. The goal was to replace eight existing systems used on virtually every army base across the United States. The As-Is process and data models for these systems had been built, and my team's job was to identify improvement opportunities and develop To-Be process models for each of the eight systems.

For the first system, we selected a group of midlevel managers (captains and majors) recommended by their commanders as being the experts in the system under construction. These individuals were the first- and second-line managers of the business function. The individuals were expert at managing the process but did not know the exact details of how the process worked. The resulting To-Be process model was very general and nonspecific. *Alan Dennis*

QUESTION:
Suppose you were in charge of the project. What would your interview schedule for the remaining seven projects look like?

can be an informal list that is used to help set up meeting times or a formal list that is incorporated into the work plan. The people who appear on the interview schedule are selected on the basis of the analyst's information needs. Typically, the analyst examines the analysis plan and identifies information that is needed to support the plan. Then the analyst talks with the sponsor, some business users, and other members of the project team about who in the organization can best provide the information. These people are listed on the interview schedule in the order in which they should be interviewed.

People at different levels of the organization will have different perspectives on the system, so it is important to include both managers and staff who actually perform the business process on the schedule to gain both high-level and low-level perspectives on an issue. Also, the levels of interview subjects that you need may change over time. For example, at the start of the project, the analyst has a limited understanding of the As-Is business process. It is common to begin by interviewing one or two senior managers to get a strategic view and then move to midlevel managers who can provide broad, overarching information about the business process and the expected role of the system being developed. Once the analyst has a good understanding of the big picture, lower-level managers and staff members can fill in the exact details of how the process works. Like most other things about systems analysis, this is an iterative process—starting with senior managers, moving to midlevel managers, then to staff members, back to midlevel managers, and so on, depending on what information is needed along the way. It is quite common for the list of interviewees to grow, often by 50% to 75%. As you interview people, you likely will identify topics on which more information is needed and additional people who can provide the information.

Designing Questions

There are three types of interview questions: closed-ended questions, open-ended questions, and probes. *Closed-ended questions* are those that require a specific answer. You can think of them as being similar to multiple-choice or arithmetic

Types of Questions	Examples
Closed-ended questions	How many telephone orders are received per day?
	How do customers place orders?
	What additional information would you like the new system to provide?
Open-ended questions	What do you think about the current system?
	What are some of the problems you face on a daily basis?
	How do you decide what types of marketing campaigns to run?
Probing questions	Why?
	Can you give me an example?
	Can you explain that in a bit more detail?

FIGURE 5-2
Three Types of Questions

questions on an exam (Figure 5–2). Closed-ended questions are used when the analyst is looking for specific, precise information (e.g., how many credit card requests are received per day). In general, the more precise the question, the better. For example, rather than asking "Do you handle a lot of requests?" it is better to ask "How many requests do you process per day?"

Closed-ended questions enable analysts to control the interview and obtain the information they need. However, these types of questions don't uncover *why* the answer is the way it is, nor do they uncover information that the interviewer does not think to ask ahead of time.

Open-ended questions are those that leave room for elaboration on the part of the interviewee. They are similar in many ways to essay questions that you might find on an exam (see Figure 5–2 for examples). Open-ended questions are designed to gather rich information and give the interviewee more control over the information that is revealed during the interview. Sometimes the information that the interviewee chooses to discuss uncovers information that is just as important as the answer (e.g., if the interviewee talks only about other departments when asked for problems, it may suggest that he or she is reluctant to admit his or her own department's problems).

The third type of question is the *probing question*. Probing questions follow up on what has just been discussed in order to learn more, and they are often used when the interviewer is unclear about an interviewee's answer. They encourage the interviewee to expand on or to confirm information from a previous response, and they are a signal that the interviewer is listening and interested in the topic under discussion. Many beginning analysts are reluctant to use probing questions because they are afraid that the interviewee might be offended at being challenged or because they believe it shows that they don't understand what the interviewee has said. When asked politely, probing questions can be a powerful tool in information gathering.

In general, you should not ask questions about information that is readily available from other sources. For example, rather than asking what information is used to perform to a task, it is simpler to show the interviewee a form or report (see "Document Analysis" later in this chapter) and ask what information on it is used. This helps focus the interviewee on the task and saves time because the interviewee does not need to describe the information in detail—just to point it out on the form or report.

None of these types of questions is any better than another, and usually a combination of questions is used during an interview. At the initial stage of an information system (IS) development project, the As-Is process can be unclear, so the interview process begins with *unstructured interviews,* interviews that seek a broad and roughly defined set of information. In this case, the interviewer has a general sense of the information needed but few close-ended questions to ask. These are the most challenging interviews to conduct because they require the interviewer to ask open-ended questions and probe for important information on the fly.

As the project progresses, the analyst comes to understand the business process much better, and he or she needs very specific information about how business processes are performed (e.g., exactly how a customer credit card is approved). At this time, the analyst conducts *structured interviews,* in which specific sets of questions are developed prior to the interviews. There usually are more close-ended questions in a structured interview than in the unstructured approach.

No matter what kind of interview is being conducted, interview questions must be organized into a logical sequence so that the interview flows well. For example, when the analyst is trying to gather information about the current business process, it can be useful to move in logical order through the process or from the most important issues to the least important.

There are two fundamental approaches to organizing the interview questions: top-down or bottom-up (Figure 5-3). With the *top-down approach,* the interviewer starts with broad, general issues and gradually works toward more specific ones. With the *bottom-up approach,* the interviewer starts with very specific questions and moves to broad questions. In practice, analysts mix the two approaches, starting with broad general issues, moving to specific questions, and then back to general issues.

The top-down approach is an appropriate strategy for most interviews (it is certainly the most common approach). The top-down approach enables the interviewee to become accustomed to the topic before he or she needs to provide specifics. It also enables the interviewer to understand the issues before moving to the details because the interviewer may not have sufficient information at the start of the interview to ask very specific questions. Perhaps most importantly, the top-down approach enables the interviewee to raise a set of big-picture issues before

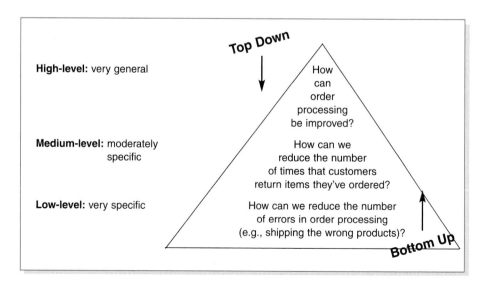

FIGURE 5-3
Top-Down and Bottom-Up Questioning Strategies

becoming enmeshed in details, so the interviewer is less likely to miss important issues.

One case in which the bottom-up strategy may be preferred is when the analyst already has gathered a lot of information about issues (whether regarding the As-Is system, improvement ideas, or the To-Be system) and just needs to fill in some holes with details. The bottom-up strategy may be appropriate if lower-level staff members feel threatened or are unable to answer high-level questions. For example, "How can we improve customer service?" may be too broad a question for a customer-service clerk, whereas a specific question is readily answerable (e.g., "How can we speed up customer returns?"). In any event, all interviews should begin with noncontroversial questions first, then gradually move into more contentious issues after the interviewer has developed some rapport with the interviewee.

Preparing for the Interview

It is important to prepare for the interview in the same way that you would prepare to give a presentation. You should have a general interview plan that lists the questions that you will ask in the appropriate order, that anticipates possible answers and how you will follow up with them, and that identifies segues between related topics. Confirm the areas in which the interviewee has knowledge so you do not ask questions that he or she cannot answer. Review the topic areas, the questions, and the interview plan, and clearly decide which have the greatest priority in case you run out of time.

In general, structured interviews with closed-ended questions take more time to prepare than unstructured interviews, so some beginning analysts prefer unstructured interviews, thinking that they can wing it. This is very dangerous and often counterproductive because (1) any information not gathered in the first interview would require follow-up efforts and (2) most users don't like to be repeatedly interviewed on the same issues.

Be sure to prepare the interviewee as well. When you schedule the interview, inform the interviewee of the reason for the interview and the areas you will be discussing far enough in advance so that he or she has time to think about the issues and organize his or her thoughts. This is particularly important when you are an outsider to the organization and when you are interviewing lower-level employees, who often are not asked for their opinions and who may be uncertain about why you are interviewing them.

Conducting the Interview

When you start the interview, the first goal is to build rapport with the interviewee so that he or she trusts you and is willing to tell you the whole truth, not just give the answers that he or she thinks you want. You should appear to be professional and an unbiased, independent seeker of information. The interview should start with an explanation of why you are there and why you have chosen to interview the person and then move into your planned interview questions.

It is critical to carefully record all the information that the interviewee provides. In our experience, the best approach is to take careful notes—write down *everything* the interviewee says, even if it does not appear immediately relevant. Don't be afraid to ask the person to slow down or to pause while you write, because this is a clear indication that the interviewee's information is important to you. One

PRACTICAL **5-1 DEVELOPING INTERPERSONAL SKILLS**

TIP

Interpersonal skills are those skills that enable you to develop rapport with others, and they are very important for interviewing. They help you to communicate with others effectively. Some people develop good interpersonal skills at an early age; they simply seem to know how to communicate and interact with others. Other people are less "lucky" and need to work hard to develop their skills. Interpersonal skills, like most skills, can be learned. Here are some tips:

- **Don't worry, be happy.** Happy people radiate confidence and project their feelings on others. Try interviewing someone while smiling and then interviewing someone else while frowning and see what happens!
- **Pay attention.** Pay attention to what the other person is saying (which is harder than you might think). See how many times you catch yourself with your mind on something other than the conversation at hand.
- **Summarize key points.** At the end of each major theme or idea that someone explains, you should repeat the key points back to the speaker (e.g., "Let me make sure I

understand. The key issues are . . ."). This demonstrates that you consider the information important—and also forces you to pay attention (you can't repeat what you didn't listen to).

- **Be succinct.** When you speak, be succinct. The goal in interviewing (and in much of life) is to learn information, not impress others. The more you speak, the less time you give to others.
- **Be honest.** Answer all questions truthfully, and if you don't know the answer, say so.
- **Watch body language (yours and theirs).** The way a person sits or stands conveys much information. In general, a person who is interested in what you are saying sits or leans forward, makes eye contact, and often touches his or her face. A person leaning away from you or with an arm over the back of a chair is disinterested. Crossed arms indicate defensiveness or uncertainty, while "steepling" (sitting with hands raised in front of the body with fingertips touching) indicates a feeling of superiority.

potentially controversial issue is whether to tape-record the interview. Recording ensures that you do not miss important points, but it can be intimidating for the interviewee. Most organizations have policies or generally accepted practices about the recording of interviews. If you are an external consultant, make sure you understand organizational policies about taping before you start an interview. If you don't know the policy, don't ask to tape the interview. A better approach is to have a second person with you whose sole job is to take detailed notes.

As the interview progresses, it is important that you understand the issues that are discussed. If you do not understand something, be sure to ask. Don't be afraid to ask "dumb" questions because the only thing worse than *appearing* dumb is to *be* dumb by not understanding something. If you don't understand something during the interview, you certainly won't understand it afterward. Try to recognize and define jargon and be sure to clarify jargon you do not understand. One good strategy to increase your understanding during an interview is to periodically summarize the key points that the interviewee is communicating. This avoids misunderstandings and also demonstrates that you are listening.

Finally, be sure to separate facts from opinion. The interviewee may say, for example, "We process too many credit card requests." This is an opinion, and it is useful to follow this up with a probing question requesting support for the statement (e.g., "Oh? How many do you process in a day?"). It is helpful to check the facts because any differences between the facts and the interviewee's opinions can point

5-1 INTERVIEW PRACTICE

Interviewing is not as simple as it first appears. This is an exercise for groups of four people that demonstrates how similar interviews can produce different results. Divide the group into two teams of two people each and pick one person in each team to be an interviewer. Each two-person team then conducts a 5-minute interview in which the interviewer collects information about how that person performs some process (e.g., working on a class assignment, making a sandwich, paying bills, getting to class).

After 5 minutes, the interviewers switch teams, and the process is repeated. This time the interviewer conducts the same 5-minute interview but with a different interviewee.

Once the second interview is completed, the teams come together back into a group of four people. The two interviewers then describe each of their two interviews, and the group identifies similarities and differences in the experiences.

out key areas for improvement. Suppose the interviewee complains about a high or increasing number of errors, but the logs show that errors have been decreasing. This suggests that errors are viewed as a very important problem that should be addressed by the new system, even if they are declining.

As the interview draws to a close, be sure to give the interviewee time to ask questions or provide information that he or she thinks is important but was not part of your interview plan. In most cases, the interviewee will have no additional concerns or information, but in some cases this will lead to unanticipated but important information. Likewise, it can be useful to ask the interviewee if there are other people that should be interviewed. Make sure that the interview ends on time. (If necessary, omit some topics or plan to schedule another interview.)

As a last step in the interview, briefly explain what will happen next (see "Follow-Up" below). You don't want to commit to providing certain features in the new system or to deliver a system by a certain date prematurely, but you do want to reassure the interviewee that his or her time was well spent and very helpful to the project.

Follow-Up

After the interview is over, the analyst needs to prepare an *interview report* (Figure 5-4) that describes the information from the interview. The report contains *interview notes,* information that was collected over the course of the interview and is summarized in a useful format. In general, the interview report should be written within 48 hours of the interview because the longer you wait, the more likely you are to forget information.

The interview report is sent to the interviewee with a request to read it and inform the analyst of any clarifications or updates needed. Make sure the interviewee is convinced that you genuinely want his or her corrections to the report. Usually there are few changes, but the need for any significant changes suggests that a second interview will be required. Never distribute someone's information without his or her prior approval.

INTERVIEW REPORT

Interview notes approved by: _____

Person interviewed: _____

Interviewer: _____

Date: _____

Primary purpose:

Summary of interview:

Open items:

Detailed notes:

FIGURE 5-4
Interview Report

JOINT APPLICATION DESIGN

Joint application design (JAD) is an information-gathering technique that allows the project team, users, and management to work together to identify requirements for the system. IBM developed the JAD technique in the late 1970s. It is often the most useful method for collecting information from users.[3] Capers Jones claims that JAD can reduce scope creep by 50%, and it avoids having system requirements that are too specific or too vague, both of which cause trouble during later stages of the SDLC.[4]

JAD is a structured process in which 10 to 20 users meet together under the direction of a *facilitator* skilled in JAD techniques. The facilitator is a person who sets the meeting agenda and guides the discussion but does not join in the discussion as a participant. He or she does not provide ideas or opinions on the topics under discussion, remaining neutral during the session. The facilitator must be an expert in both group process techniques and information systems analysis and design techniques. One or two *scribes* assist the facilitator by recording notes, making copies, and so on. Often the scribes will use computers and computer-aided software engineering (CASE) tools to record information as the JAD session proceeds.

[3] More information on JAD can be found in J. Wood and D. Silver, *Joint Application Design* (New York: John Wiley & Sons, 1989); and Alan Cline, "Joint Application Development for Requirements Collection and Management," http://www.carolla.com/wp-jad.html.

[4] See Kevin Strehlo, "Catching up with the Joneses and 'Requirement' Creep," *InfoWorld,* 18, no. 31 (July 29, 1996), p. 67; and Kevin Strehlo, "The Makings of a Happy Customer: Specifying Project X," *InfoWorld* 18, no. 46 (November 11, 1996), 79 .

The JAD group meets for several hours, several days, or several weeks until all of the issues have been discussed and the needed information is collected. Most JAD sessions take place in a specially prepared meeting room, away from the participants' usual offices, so that they are not interrupted. The meeting room is usually arranged in a U shape so that all participants can easily see each other (Figure 5-5). At the front of the room (the open part of the U), there is a whiteboard, flip chart, and/or overhead projector for use by the facilitator in leading the discussion.

One problem with JAD is that it suffers from the traditional problems associated with groups; sometimes people are reluctant to challenge the opinions of others (particularly their boss), a few people often dominate the discussion, and not everyone participates. In a 15-member group, for example, if everyone participates

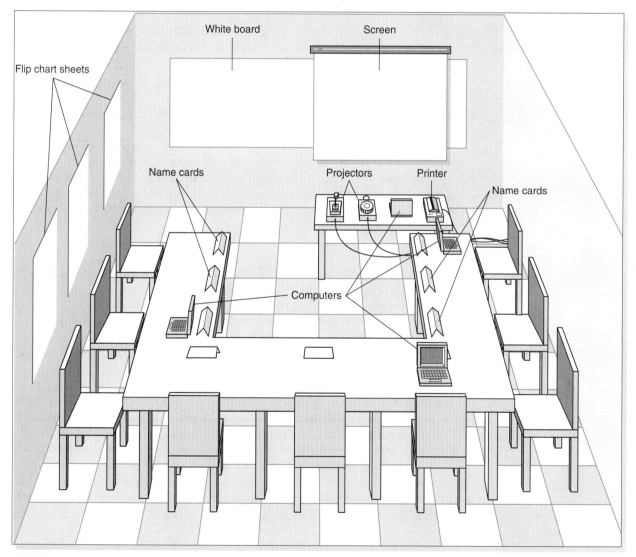

FIGURE 5-5
Joint Application Design Meeting Room

equally, then each person can talk for only 4 minutes each hour and must listen for the remaining 56 minutes—not a very efficient way to collect information.

A new form of JAD called *electronic-JAD* or *e-JAD* attempts to overcome these problems by using groupware. In an e-JAD meeting room, each participant uses special software on a networked computer to send anonymous ideas and opinions to everyone else. In this way, all participants can contribute at the same time, without fear of reprisal from challenging others. Initial research suggests that e-JAD can reduce the time required to run JAD sessions by 50% to 80%.[5]

Selecting Participants

Selecting JAD participants is done in the same basic way as selecting interview participants. Participants are selected on the basis of the information they can contribute, to provide a broad mix of organizational levels, and to build political support for the new system. The need for all JAD participants to be away from their office at the same time can be a major problem. The office may need to be closed or run with a skeleton staff until the JAD sessions are complete. Ideally, the participants who are released from regular duties to attend the JAD sessions should be the very best people in that business unit. However, without strong management support, JAD sessions can fail because those selected to attend the JAD session are those who are less likely to be missed (i.e., the least competent people).

The JAD facilitator should be someone who is an expert in JAD or e-JAD techniques and ideally someone who has experience with the business under discussion. In many cases, the JAD facilitator is a consultant external to the organization because the organization may not have a regular day-to-day need for JAD or e-JAD expertise and developing and maintaining this expertise can be expensive.

Designing the Session

JAD sessions can run from as little as half a day to several weeks, depending on the size and scope of the project. In our experience, most JAD sessions tend to last 5 to 10 days, spread over a 3-week period. Most e-JAD sessions tend to last 1 to 4 days in a 1-week period. JAD and e-JAD sessions usually go beyond the collection of information and move into analysis. In this way, the users themselves participate in analysis tasks and offer different perspectives on the issues than the analysts acting alone.

As with interviewing, success depends on a careful plan. JAD sessions are usually designed and structured using the same principles as interviews. Most JAD sessions are designed to collect specific information from users, which requires the development of a set of questions prior to the meeting. A difference between JAD and interviewing is that *all* JAD sessions are structured—they *must* be carefully planned. The information needed or the analysis technique to be used (e.g., problem analysis, process analysis, breaking assumptions) is identified and put in order. In general, closed-ended questions are seldom used because they do not spark the frank discussion that is typical of JAD. In our experience, it is better to proceed top-down in JAD sessions when gathering information, or from the simplest and least challenging analysis techniques (e.g., such BPA techniques as problem analysis and root cause analysis) to the more innovative techniques (e.g., such BPR techniques as breaking assumptions and process elimination). Typically, 30 minutes is allocated to each separate agenda item, and frequent breaks are scheduled throughout the day because participants tire easily.

Preparing for the Session

As with interviewing, it is important to prepare the analysts and participants for the JAD session. Because the sessions can go beyond the depth of a typical interview and are usually conducted off-site, participants can be more concerned about how to prepare. It is important that the participants understand whether they are going to be providing information about the As-Is system, improvement ideas, or the To-Be system. If the goal of the JAD session, for example, is to develop an As-Is business process model for a noncomputerized system, then participants can bring procedures manuals and documents with them. If the goal is improvement ideas, then they should think about how they would improve the system prior to the JAD session.

Conducting the Session

Most JAD sessions try to follow a formal agenda, and most have formal *ground rules* that define appropriate behavior. Common ground rules include following the schedule, respecting others' opinions, accepting disagreement, and ensuring that only one person talks at once.

The role of the JAD facilitator can be challenging. Many participants come to the JAD session with strong feelings about the business system to be discussed. Channeling these feelings so that the session moves forward in a positive direction and getting participants to recognize and accept—but not necessarily agree on—opinions and situations different from their own requires significant expertise in systems analysis and design, JAD, and interpersonal relations. Few systems analysts attempt to facilitate JAD sessions without being trained in JAD techniques, and most also apprentice with a skilled JAD facilitator before they attempt to lead their first session.

The JAD facilitator performs three key functions. First, he or she ensures that the group sticks to the agenda. The only reason to digress from the agenda is when it becomes clear to the facilitator, project leader, and project sponsor that the JAD session has produced some new information that is unexpected and requires the JAD session (and perhaps the project) to move in a new direction. When participants attempt to divert the discussion away from the agenda, the facilitator must be firm but polite in leading discussion back to the agenda and getting the group back on track.

Second, the facilitator must help the group understand the technical terms and jargon that surround the system development process and help the participants understand the specific analysis techniques used. Participants are experts in their area, their part of the business, but they are not experts in the analysis techniques discussed in the previous chapter. The facilitator must therefore minimize the learning required and teach participants how to effectively provide the information or use the analysis technique.

Third, the facilitator records the group's input on a public display area, which can be a whiteboard, flip chart, or computer display. He or she structures the information that the group provides and helps the group recognize key issues and important solutions. Under no circumstances should the facilitator inject his or her opinions into the discussion. The facilitator *must* remain neutral at all times and simply help the group through the process. The moment the facilitator offers an opinion on an issue, the group will cease seeing him or her as a neutral party but

---

PRACTICAL 5-2 MANAGING PROBLEMS IN JOINT APPLICATION DESIGN SESSIONS

TIP

I have run more than 100 JAD and e-JAD sessions and have learned several standard facilitator tricks. Here are some common problems and some ways to deal with them:

1. **Domination:** The facilitator should ensure that no one person dominates the group discussion. The only way to deal with someone who dominates is head-on. During a break, approach the person, thank him or her for the insightful comments, and ask for help in making sure that others also participate.

2. **Noncontributors:** Drawing out people who have participated very little is challenging because you want to bring them into the conversation so that they will contribute again. The best approach is to ask a direct factual question that you are *certain* they can answer, and it helps to ask the question in a long way to give them time to think. For example, you could say, "Pat, I know you've worked shipping orders a long time. You've probably been in the shipping department longer than anyone else. Could you help us understand exactly what happens when an order is received in Shipping?"

3. **Side discussions:** Sometimes participants engage in side conversations and fail to pay attention to the group. The easiest solution is simply to walk close to the people and continue to facilitate right in front of them. Few people will continue a side conversion when you are 2 feet from them and the entire group's attention is on you and them.

4. **Agenda merry-go-round:** The merry-go-round occurs when a group member keeps returning to the same issue every few minutes and won't let go. One solution is to let the person have 5 minutes to ramble on about the issue while you carefully write down every point on a flip chart or computer file. This flip chart or file is then posted conspicuously on the wall. When the person brings up the issue again, you interrupt, walk to the

paper, and ask him or her what to add. If the person mentions something already on the list, you quickly interrupt, point out it's there, and ask what other information to add. Don't let the person repeat the same point but write any new information.

5. **Violent agreement:** Some of the worst disagreements occur when participants really agree on the issues but don't realize that they agree because they are using different terms. For example, they argue whether a glass is half empty or half full; they agree on the facts but can't agree on the words. In this case, the facilitator has to translate the terms into different words and find common ground so the parties recognize that they really agree.

6. **Unresolved conflict:** In some cases, participants don't agree and can't understand how to determine what alternatives are better. You can help by structuring the issue. Ask for criteria by which the group will identify a good alternative (e.g., "Suppose this idea really did improve customer service. How would we recognize the improved customer service?"). Then once you have a list of criteria, ask the group to assess the alternatives using them.

7. **True conflict:** Sometimes, despite every attempt, participants just can't agree on an issue. The solution is to postpone the discussion and move on. Document the issue as an "open issue" and list it prominently on a flip chart. Have the group return to the issue hours later. Often, the issue will resolve itself by then and you won't have wasted time on it. If the issue cannot be resolved later, move it to the list of issues to be decided by the project sponsor or some other more senior member of management.

8. **Use humor:** Humor is one of the most powerful tools a facilitator has and thus must be used judiciously. The best JAD humor is always in context. Never tell jokes; instead, take the opportunity to find the humor in the situation. *Alan Dennis*

YOUR	**5-2 JOINT APPLICATION DESIGN PRACTICE**
T U R N	

In your class, organize yourselves into groups of 4 to 7 people and pick one person in each group to be the JAD facilitator. Using a blackboard, whiteboard, or flip chart, gather information about how the group performs some process (e.g., working on a class assignment, making a sandwich, paying bills, getting to class).

rather as someone who could be attempting to sway the group into some predetermined solution.

However, this does not mean that the facilitator should not try to help the group resolve issues. For example, if two items appear to be the same to the facilitator, the facilitator should not say, "I think these may be similar." Instead, the facilitator should ask, "Are these similar?" If the group decides they are, the facilitator can combine them and move on. However, if the group decides they are not similar (despite what the facilitator believes), the facilitator should accept the decision and move on. The group is *always* right, and the facilitator has no opinion.

Follow-Up

As with interviews, a JAD *post-session report* is prepared and circulated among session attendees. The JAD post-session report is essentially the same as the interview report shown in Figure 5-4. Since the JAD sessions are longer and provide more information, it usually takes 1 or 2 weeks after the JAD session before the report is complete.

QUESTIONNAIRES

Questionnaires are often used when there is a large number of people from whom information and opinions are needed. In our experience, questionnaires are commonly used for systems intended for use outside the organization (e.g., by customers or vendors) or for systems with business users spread across many geographic locations. Most people automatically think of paper when they think of questionnaires, but today more questionnaires are being distributed in electronic form, either via e-mail or on the Web. Electronic distribution can save a significant amount of money compared with distributing paper questionnaires.

Selecting Participants

As with interviews and JAD sessions, the first step is to select the individuals to whom to send the questionnaire. However, it is not usual to select every person who could provide useful information. The standard approach is to select a *sample* of people who are representative of the entire group. Since sampling guidelines are discussed in most statistics books and most business schools offer courses that cover the topic, we will not discuss it here. The important point, however, in selecting a sample is to realize that not everyone who receives a questionnaire will actually complete it. On average, only 30% to 50% of paper and e-mail questionnaires are returned. Response rates for Web-based questionnaires tend to be significantly lower (often only 5% to 30%).

Designing the Questionnaire

Developing good questions is more important for questionnaires than for interviews and JAD sessions because the information on a questionnaire cannot be immediately clarified for a confused respondent. Questions on questionnaires must be very clearly written and leave little room for misunderstanding, so closed-ended questions tend to be most commonly used. Questions must enable the analyst to clearly separate facts from opinions. Opinion "questions" most often are statements for which the respondents are asked to note the extent to which they agree or disagree (e.g., "Network problems are common"), whereas fact-oriented questions seek more precise values (e.g., "How often does a network problem occur: once an hour, once a day, or once a week?"). Figure 5-6 lists guidelines for questionnaire design.

Perhaps the most obvious issue—but one in our experience that is sometimes overlooked—is to have a clear understanding of how the information collected from the questionnaire will be analyzed and used. You must address this issue before you distribute the questionnaire because it is too late afterward.

Questions should be relatively consistent in style so that the respondent does not have to read instructions for each question before answering it. It is generally good practice to group related questions together to make them simpler to answer. Some experts suggest that questionnaires should start with questions important to respondents, so that the questionnaire immediately grabs their interest and induces them to answer it. Perhaps the most important steps are to have several colleagues review the questionnaire and then to pretest it with a few people drawn from the groups to whom it will be sent. It is surprising how often seemingly simple questions can be misunderstood.

Administering the Questionnaire

The key issue in administering the questionnaire is getting participants to complete the questionnaire and send it back. Dozens of marketing research books have been written about ways to improve response rates. Commonly used techniques include

---

- Begin with nonthreatening and interesting questions.
- Group items into logically coherent sections.
- Do not put important items at the very end of the questionnaire.
- Do not crowd a page with too many items.
- Avoid abbreviations.
- Avoid biased or suggestive items or terms.
- Number questions to avoid confusion.
- Pretest the questionnaire to identify confusing questions.
- Provide anonymity to respondents.

FIGURE 5-6
Good Questionnaire Design

clearly explaining why the questionnaire is being conducted and why the respondent has been selected, stating a date by which the questionnaire is to be returned, offering an inducement to complete the questionnaire (e.g., a free pen), and offering to supply a summary of the questionnaire responses. Systems analysts have additional techniques to improve responses rates inside the organization, such as personally handing out the questionnaire and personally contacting those who have not returned them after 1 or 2 weeks, as well as requesting the respondents' supervisors to administer the questionnaires in a group meeting.

Follow-Up

As with interviews and JAD, it is helpful to process the returned questionnaires and develop a questionnaire report soon after the questionnaire deadline. This ensures that the analysis process proceeds in a timely fashion and that respondents who requested copies of the results receive them promptly.

DOCUMENT ANALYSIS

Document analysis is often used by project teams to understand the As-Is system. Under ideal circumstances, the project team that developed the As-Is system will have produced business process models and data models for the system, and all subsequent projects to improve the system will have updated these models. In this case, the project team can start by reviewing the models and looking at the system itself.

Not all systems, however, are well documented and have high-quality models. Even so, the documents—the forms and reports—used by the As-Is system contain information that may be included in the new system. An obvious starting point, therefore, is to examine documents that are produced by existing systems.

YOUR TURN

5-3 QUESTIONNAIRE PRACTICE

Among your classmates, organize yourselves into groups of four people. Each person develops a short questionnaire to collect information about the frequency with which group members perform some process (e.g., working on a class assignment, making a sandwich, paying bills, getting to class), how long it takes them, how they feel about the process, and opportunities for improving the process. It might help to use the same process you used in the Your Turn 5.1 exercise.

Once all have completed their questionnaire, the first group member reads his or her questionnaire aloud and the other members write their answers on paper, which they then hand to the person. The other members repeat this in turn, until every member has answered all the questionnaires and everyone has a set of answers for his or her questionnaire.

Then each group member prepares a brief summary report of the results from his or her questionnaire (ignoring any information learned from others' questionnaires), which is then shared with the others. You may notice that not all the reports present the same information. Why? What questions were confusing? What problems did you have in analyzing the data?

5-B Publix Credit Card Forms

At my neighborhood Publix grocery store, the cashiers always hand write the total amount of the charge on every credit card charge form, even though it is printed on the form. Why? Because the "back office" staff people who reconcile the cash in the cash drawers with the amount sold at the end of each shift find it hard to read the small print on the credit card forms.

Writing in large print makes it easier for them to add the values up. However, cashiers sometimes make mistakes and write the wrong amount on the forms, which causes problems. *Barbara Wixon*

QUESTION:
How would you improve the system?

The documents (forms, reports, policy manuals) tell only part of the story. They represent the *formal system* that the organization uses. Quite often, the "real" or *informal system* differs from the formal one, and these differences, particularly large ones, give strong indications of what needs to be changed in a new system. For example, forms or reports that are never used likely should be eliminated. Likewise, boxes or questions on forms that are never filled in (or are used for other purposes) should be rethought.

The most powerful indication that the system needs to be changed is when users create their own forms or add more information to existing ones. Such changes clearly demonstrate the need for improvements to existing systems. Thus, it is useful to review both blank and completed forms to identify these deviations. Likewise, when users access multiple reports to satisfy their information needs, it is a clear sign that new information or new information formats would be useful.

Figure 5-7 shows an example of a document used in a veterinary clinic. This document provides a good understanding of the basic information recorded about the clinic's patients (e.g., owner's name, pet's name, address, phone number, and so on). The document also shows some problems with the design of the current form that need to be fixed in the To-Be system. This type of information could also be gathered in an interview, but by analyzing a document, it is less likely that information will be forgotten or overlooked and the users will need to spend less time in interviews. Typically, the analyst will analyze documents first, and then interview users to ensure the documents are still useful and to help clarify any issues that the analyst may observe (e.g., the need to change the document).

OBSERVATION

Observation is a powerful tool for gathering information about the As-Is system because it enables the analyst to see the reality of a situation rather than listen to others describe it in interviews or JAD sessions. Several research studies have shown that many managers really do not remember how they work and how they allocate their time. (Quick—how many hours did you spend last week on each of your courses?) Observation is a good way to check the validity of information gathered from such indirect sources as interviews and questionnaires.

In many ways, the analyst becomes an anthropologist as he or she walks through the organization and observes the business system as it functions. The goal

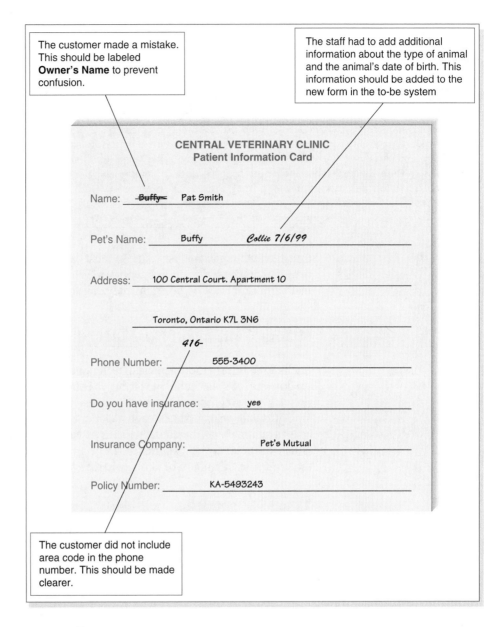

The customer made a mistake. This should be labeled **Owner's Name** to prevent confusion.

The staff had to add additional information about the type of animal and the animal's date of birth. This information should be added to the new form in the to-be system

CENTRAL VETERINARY CLINIC
Patient Information Card

Name: ~~Buffy~~ Pat Smith

Pet's Name: Buffy *Collie 7/6/99*

Address: 100 Central Court. Apartment 10

Toronto, Ontario K7L 3N6

Phone Number: *416-* 555-3400

Do you have insurance: yes

Insurance Company: Pet's Mutual

Policy Number: KA-5493243

The customer did not include area code in the phone number. This should be made clearer.

FIGURE 5-7
Performing a Document Analysis

is to keep a low profile, to not interrupt those working, and to not influence those being observed. Nonetheless, it is important to understand that what analysts observe may not be the normal day-to-day routine, because people tend to be extremely careful in their behavior when they are being watched. Even though normal practice may be to break formal organizational rules, the observer is unlikely to see this. (Remember how you drove the last time a police car followed you?) Thus, what you see may *not* be what you get.

Observation is often used to supplement interview information. The location of a person's office and its furnishings give clues as to the person's power and influence in the organization and can be used to support or refute information given in an interview. For example, an analyst might become skeptical of someone who claims to use the existing computer system extensively if the computer is never

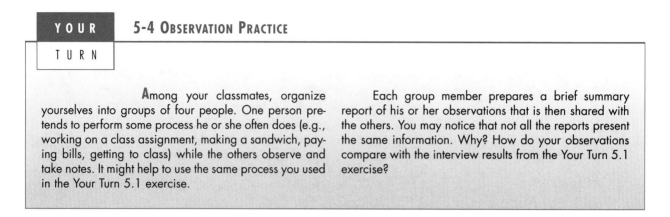

turned on while the analyst visits. In most cases, observation will support the information that users provide in interviews. When it does not, it is an important signal that extra care must be taken in analyzing the business system.

SELECTING THE APPROPRIATE TECHNIQUES

Each of the information-gathering techniques discussed above has its strengths and weaknesses. No one technique is always better than the others, and in practice, most projects use a combination of techniques. Thus, it is important to understand the strengths and weaknesses of each technique and when to use each. Figure 5-8 presents a summary of the characteristics of each technique. One issue not discussed is that of the analyst's experience. In general, document analysis and observation require the least amount of skills, whereas JAD sessions are the most challenging.

Type of Information

The first characteristic is the type of information. Some techniques are more suited for use at different stages of the information-gathering process, whether understanding the As-Is system, identifying improvements, or developing the To-Be system. Interviews and JAD are commonly used in all three stages. By contrast, doc-

	Interviews	Joint Application Design	Questionnaires	Document Analysis	Observation
Type of information	As-Is, improvements, To-Be	As-Is, improvements, To-Be	As-Is, improvements	As-Is	As-Is
Depth of information	High	High	Medium	Low	Low
Breadth of information	Low	Medium	High	High	Low
Integration of information	Low	High	Low	Low	Low
User involvement	Medium	High	Low	Low	Low
Cost	Medium	Low–Medium	Low	Low	Low–Medium

FIGURE 5-8
Table of Information-Gathering Techniques

ument analysis and observation usually are most helpful for understanding the As-Is, although occasionally they provide information about current problems that need to be improved. Questionnaires are often used to gather information about the As-Is system as well as general information about improvements.

Depth of Information

Depth of information refers to how rich and detailed the information is that the technique usually produces and the extent to which the technique is useful at obtaining not only facts and opinions but also an understanding of *why* those facts and opinions exist. Interviews and JAD sessions are very useful for providing a good depth of rich and detailed information and for helping the analyst to understand the reasons behind them. At the other extreme, document analysis and observation are useful for obtaining facts, but not much beyond that. Questionnaires can provide a medium depth of information, soliciting both facts and opinions, but usually little understanding of why.

Breadth of Information

Breadth of information refers to the range of information and information sources that can be easily collected using that technique. Questionnaires and document analysis are both easily capable of soliciting a wide range of information from a large number of information sources. By contrast, both interviews and observation require that the analyst visit each information source individually and take more time. JAD sessions are in the middle, because many information sources are brought together at the same time.

Integration of Information

One of the most challenging aspects of gathering information is the *integration of information* from different sources. Simply put, different people can provide conflicting information. Combining this information and attempting to resolve differences of opinions or facts is usually very time consuming because it means contacting each information source in turn, explaining the discrepancy, and attempting to refine the information. In many cases, the individual wrongly perceives that it is the analyst who is challenging his or her information, when in fact it is another user somewhere else in the organization. This can make the user quite defensive and requires much more time to resolve the differences.

All techniques suffer integration problems to some degree, but JAD sessions are specifically designed to improve integration because all information is integrated when it is collected, not afterward. If two users provide conflicting information, the conflict becomes immediately obvious, as does the source of the conflict. The immediate integration of information is the single most important benefit of JAD that distinguishes it from other techniques. This is why most organizations use JAD for important projects.

User Involvement

User involvement refers to the amount of time and energy the intended users of the new system must devote to the analysis process. It is generally agreed that the more users become more involved in the analysis and design process, the greater the chance that the system will be successful. However, user involvement can have a significant cost, and not all users are willing to contribute valuable time and energy. Questionnaires, document analysis, and observation place the least burden on users, whereas JAD sessions require the greatest effort.

Cost

Cost is an important consideration for every business decision, including the analysis and design process. In general, questionnaires, document analysis, and observation are the lowest-cost techniques (although observation can be quite time consuming). This lower cost does not imply that they are more or less effective than the other techniques. We regard interviews and JAD sessions as having moderate costs. In general, JAD sessions are much more expensive initially because they require many users to be absent from their offices for significant periods of time and often involve highly paid consultants. However, JAD sessions significantly reduce the time spent in information integration and thus result in lower costs in later stages of the information-gathering process.

Combining Techniques

In practice, information gathering and information analysis are often done iteratively and combine a series of different techniques. Most information-gathering plans start with interviews with senior managers to gain an understanding of the project and the big-picture issues. From this, the scope becomes clear as to whether large or small changes are anticipated. These interviews are often followed with analysis of the documents and policies to gain some understanding of the As-Is system. Usually, interviews come next to gather the rest of the information needed regarding the As-Is system.

In our experience, JAD sessions are the most commonly used technique for identifying improvements because the JAD session enables the users and key stakeholders to work together through the analysis techniques (whether they are BPA, BPI, or BPR techniques) and come to a shared understanding of the possibilities for the To-Be system. Occasionally, these JAD sessions are followed by questionnaires sent to a much wider set of users or potential users to see whether the opinions of those who participated in the JAD sessions are widely shared.

The development of the concept for the To-Be system is often done through interviews with senior managers, followed by JAD sessions to make sure the key needs of the new system are well understood.

APPLYING THE CONCEPTS AT CD SELECTIONS

The analysis strategy for CD Selections was presented in the last chapter in Figure 4-8. In this section, we focus on how they translated the strategy into action: gathering information, analyzing it, and developing a vision for the new system. At this point, CD Selections hired an Internet marketing consultant, Dr. Chris Campbell, a professor at the local university. Alec developed an initial information-gathering strategy, which he then discussed with Margaret Mooney, the vice president of marketing, and Chris, the consultant, to help him refine it into a plan.

Since this was a new application, Alec, Margaret, and Chris all agreed that it was important to involve CD Selections' customers in the process to ensure that the system would actually be used. It was decided that the system concept resulting from the analysis would be reviewed in four focus groups run by the marketing department (because of their expertise in customer focus groups) before any behavioral or structural models were developed. Two focus groups would be novice Web users and two focus groups would be expert Web users who had actually purchased at least one CD over the Web. Alec would work with the focus group leader from the marketing department to ensure that information systems got the information it needed. Figure 5-9 summarizes the analysis plan as it was refined to include requirements-gathering activities.

Step	Techniques and Goals	Who and How
Understanding the As-Is system	Document analysis: understand system interface issues	Analysts will review the process and data models for the inventory, accounting, in-store retail systems, and any other system with which the Internet sales system will interact.
	Interviews: understand system interface issues	Alec will interview the senior analysts for the inventory, accounting, in-store retail systems, and senior managers.
	Informal benchmarking: develop a list of "standard" and "advanced" features for Internet sales	Alec will visit Web sites for CDnow, Amazon.com, CD Express, and others.
		Alec will visit Web sites for Land's End and other well-known merchants.
		Alec will review list of features with Chris.
Identifying improvements	Problem analysis: develop a list of features for the Web site and in the system that connects the Web site to the existing information systems	Include problem analysis in JAD session with marketing staff to identify ideas (Alec to facilitate).
		Include this in marketing focus groups.
		Alec will investigate the inventory and accounting systems to identify key processes and data.
		Alec will interview accounting and inventory managers to identify ideas and issues for the system.
	Technology analysis: develop creative ideas for the system that are not commonly offered in the market	Send questionnaire to information systems staff to develop list of technologies.
		Alec will review list of technologies with Chris.
		Include technology analysis in JAD session with marketing staff to identify potential applications for the Internet sales project.
	Outcome analysis: develop a list of alternatives to selling CDs that create business value	Include outcome analysis in JAD session with marketing staff to identify potential ideas.
		Include this in marketing focus groups if time permits.
	Focus groups: validate system concept to ensure customers believe it to be useful	Marketing department will lead customer focus groups with input from Alec.
Developing a To-Be system concept	Basic system concept: produce a bare-bones concept	Analysts will integrate the information from the analyses and produce a very simple description of a To-Be system concept.
	Use case reports: expand and refine the system concept	Conduct a series of JAD sessions to provide the details on how the system will operate.
	Develop behavioral model: Organize use cases into process model	Analysts will develop a behavioral model for To-Be system.
	Develop structural model: identify data	Analysts will develop a structural model for To-Be system.

JAD = joint application design.

FIGURE 5-9
Requirements-Gathering Plan for CD Selections

Understanding the As-Is System

Document Analysis   The first step in the strategy was to understand the As-Is systems with which the Internet sales system had to operate. The project team decided to use document analysis to create the behavioral and structural models for the inventory, accounting, in-store retail systems, and any other system with which the Internet sales system will interact.

Interviews   Alec also decided to interview the senior analysts for the accounting and inventory systems to get a better understanding of how those systems worked. He also thought it would be useful to see if they had any ideas for the new system, as well as to ask about any issues that would need to be addressed.

Identifying Improvements

Informal Benchmarking   The first step in the strategy was to informally benchmark existing CD Web sites and examine Web sites of other leading retailers such as Land's End to identify those features that were commonly offered on the Web as well as those offered only by a few "advanced" merchants. Alec, as the senior analyst on the project, decided to do the benchmarking himself. He planned, once he had the list, to discuss it with Chris, the consultant.

Problem Analysis   Since the marketing staff probably had good ideas for the Web site, Alec decided to simply solicit their general ideas and opinions on features in the JAD sessions. Likewise, the customers in the marketing focus groups should be asked to discuss those features they would like to see in a CD Web site. There were likely some troublesome issues with buying over the Web, so the focus groups should also be asked to discuss any issues that would reduce their chance of buying over the Web.

Technology Analysis   The next step was the technology analysis. Alec decided to e-mail a quick questionnaire (one open-ended question) to all of the staff in the IS department asking them to identify a few "interesting" new technologies on the Internet that might or might not have some interest to CD Selections. He stressed in the questionnaire that he was interested in exciting technologies, even if they did not have any obvious relevance to CD Selections. After he got the questionnaires back, he would organize the answers into a list and discuss them with Chris. He would then develop a short list of 5 to 10 technologies and write a brief description for each.

This technology list would be used in a JAD session with staff from the marketing department to help identify how the technologies could be used to some advantage in the Internet sales project. Alec decided to facilitate the JAD session, because of his JAD experience. He also decided to ask Chris to attend the JAD session because Chris often had ideas that could spark the group to think of novel uses. The results of this JAD session would be a set of ideas about how CD Selections could apply each of the technologies in the Internet sales project. Alec decided to have the group categorize the ideas into three sets: "definite" ideas that would have a good probability of providing business value, "possible" ideas that might add business value, and "unlikely" ideas.

Outcome Analysis   The goal of *outcome analysis* is to promote creative thinking to develop other ways of creating business value on the Internet. Alec decided that this could be an agenda item in the JAD session with the marketing department as well as in the marketing focus groups with customers, provided that there was sufficient time.

The project team would integrate the improvement ideas and develop an outline for the initial system concept. This would be presented to the project sponsor and the marketing staff, who would refine in it into more detail through a series of JAD meetings. The result of these JAD meetings would be a set of use case reports (see Chapter 6) that would describe in some detail how the system would work. From these (and additional interviews as needed), the project team would create behavioral and structural models.

The Results

The requirements-gathering plan was conducted and resulted in a number of important findings. The benchmarking was conducted first, and the key points were shared with the marketing department and discussed with Chris (Figure 5-10). The IS department generated a list of technologies for the technology analysis, and these were shared with the marketing department and Chris as well. Alec reviewed the inventory and accounting system and interviewed the inventory and accounting managers, who provided important ideas.

Alec then conducted the JAD session with the marketing department staff. The one-day meeting was held at a local hotel. Eighteen people participated, including all the managers in the marketing department and a few staff members selected for their insight and Internet knowledge. Margaret also included two managers from the traditional retail stores who were known as stars within the company. The JAD session started with an introduction to the basic concept of the Internet sales project, a demonstration of competitors' Web sites, and a list of the standard and advanced features from the benchmarking analysis. The group then discussed these features and what additional ideas or issues might be important (simple analysis). After lunch, the group did the technology analysis and the outcome analysis to identify a few more ideas (see Figure 5-10). The meeting closed with a discussion of the key points from the inventory and accounting managers, although some issues remained as *open issues,* issues that remained unresolved.

Developing the To-Be System Concept

From this information and analysis, Alec developed the concept for the new system, which was refined through a series of meetings with Margaret and her senior staff in marketing. Given Margaret's desire to have the system operating before the Christmas holidays, Alec recommended a timeboxing approach, where a schedule deadline (the holidays) was used to determine what features could be included (see Chapter 3). They decided to develop the system in three versions rather than attempting to develop a complete system that provided all the features initially (phased development; see Chapter 2). The first version, to be operational well before the holidays, would implement a basic system that would have all the standard features of other Internet retailers, plus a few features unique to CD Selections. The second version, planned for late spring or early summer, would add more advanced features, such as the ability to listen to a sample of music over the Internet, to find similar CDs, and to write reviews. The third version (not yet planned) would investigate and possibly incorporate some of the more innovative ideas, such as the ability to download a CD or individual songs.

Informal benchmarking: develop a list of "standard" and "advanced" features for Internet sales	Standard features
	  Find (or browse) available CDs using such keywords as *category* (e.g., rock, classical, jazz), *artist, CD title, composer* (especially for classical music)
	  Provide information on best-sellers, new, and sale CDs
	  Read reviews of CDs
	Advanced features
	  Listen to a sample of music from the CD
	  Find CDs with "similar" music
	  Enable users to write reviews of CDs

Technology analysis: develop creative ideas for the system that are not commonly offered in the market	Digital audio over the Web (e.g., Real Audio MP3)
	  Listen to a sample of music from the CD
	  Download CD music over the Internet rather than selling the physical CD
	Internet telephony
	  Have a help line to give advice on good CDs
	  CD drives for computers that can write on CDs
	  Download CD music over the Internet rather than selling the physical CD so users can make their own CDs

Outcome analysis: develop a list of alternatives to selling CDs that create business value	Listen to music that I want where I want
	  Enable users to make their own CDs by selecting individual songs, not entire CDs
	  Are digital Walkmans that play computer files rather than CDs widely availablet?

Simple analysis: develop a list of features for the Web site and in the system that connects the Web site to the existing information systems	Marketing
	  Maintain the brand image of our retail stores
	  Pricing same as stores?
	Inventory
	  Do we offer only what we stock in the stores or what we can special-order from suppliers?
	Accounting
	  Do we omit sales tax on CDs shipped out of state?

Focus groups: validate system concept to ensure customers believe it to be useful	Shipping
	  Too expensive; need to find a way to reduce
	  Too long; need to shorten shipping times
	  Cost not immediately displayed; customers select a CD on the basis of price without knowing shipping costs, only to be surprised when they pay; need to develop and post a simple shipping cost table before they select a CD
	Customized CDs
	  Enable users to customize a CD with songs from different CDs
	Recommended "similar" CDs
	  Many of the so-called similar CDs weren't really similar.

FIGURE 5-10
Selected Requirements Gathered by CD Selections

This concept was tested in a series of focus groups run by the marketing department. The basic concept was well received, and some minor improvements were suggested. Three important issues were also identified (see Figure 5-10). First was the issue of high-cost shipping, of the length of time taken in shipping, and of not knowing how much shipping was until the very end of the purchase decision. Many experienced Web users noted that they were frustrated at spending time in the purchase process only to discover that shipping costs were unreasonable, and they canceled the purchase in the end. Second, many experienced Web users had purchased "similar" books or CDs only to discover that they were not really similar. Finally, several customers suggested that they would like to be able to produce customized CDs that included songs from different CDs.

After the focus group sessions, Margaret, Alec, and Chris met and revised the system concept. The ability to present shipping costs early in the ordering process was added to the system features. Margaret decided to retain the shipping options they originally planned to use (the same as those used by the other sites: UPS, FedEx) but to investigate other options for shipping. Likewise, she would investigate the ability to produce customized CDs (which could be incorporated into version 3 of the system) while Alec moved ahead on developing the system concept for the first version.

After some discussion and experimentation on the Internet, they realized that most sites determined similarity of products by looking at purchase patterns; those products bought by the same people were considered similar. There was no link back to whether the products were really similar or just happened to be purchased together. To make matters worse, once a set of products were identified as similar on the Web site, other people bought them on the basis of the recommendation, thus making them appear even more similar, even if they weren't. Margaret suggested that this might be an opportunity for CD Selections to develop a competitive advantage: rather than relying on buying patterns to identify similar products, CD Selections might identify similar products by using recommendations of some of the enthusiastic music lovers that worked in its retail stores. Margaret wasn't sure how this could be fostered but decided to investigate the options.

At this point, the basic system concept was essentially settled (Figure 5-11). The next step was to refine and expand this basic concept into more detailed

Version 1 (October 1)	System will interface with existing systems
	Data on all CDs will be downloaded from the CD Selections distribution system, although the manager will be able to change prices and add new material to help in marketing the CDs over the Internet.
	Orders from the Internet sales system will be sent to the distribution system. This system will process orders and handle accounting in a similar manner as it now processes orders for the individual stores. The Internet sales system will monitor order status to make sure no orders are overlooked.
	On the Web, users should be able to:
	Find (or browse) available CDs using such keywords as *category* (e.g., rock, classical, jazz), *artist, CD title, composer* (especially for classical music)
	Have a shortcut to best-sellers, new, and sale CDs
	Make a purchase with a credit card
	Have CDs shipped via UPS or similar courier
	Read reviews of CDs
	System should:
	Clearly provide shipping costs and total cost
	Link to the inventory database in real time
Version 2 (April 1)	On the Web, users should be able to:
	Listen to a sample of music from the CD
	Find CDs with "similar" music
	Enable users to write reviews of CDs
Possibilities for version 3 (not yet scheduled)	Enable users to make their own CDs by selecting individual songs, not entire CDs
	Download CD music over the Internet rather than selling the physical CD
	Help line for music advice
Open issues	Items still to be resolved:
	Is the pricing same as for our stores?
	Do we offer only what we stock in the stores or what we can special-order from suppliers?
	Are there other shipping alternatives?
	Do we omit sales tax on CDs shipped out of state?
	Can we have our retail staff recommend "similar" CDs?
	Are digital Walkmans that play computer files rather than CDs widely available?

FIGURE 5-11
System Concept for CD Selections' Internet Sales System

descriptions of how the system would work and to develop process and data models. This is described in the next chapter.

SUMMARY

Interviews
The information needed will be the primary determinant of whom to interview, which usually means the key stakeholders from multiple levels of the organization. Unstructured interviews seek a broad and roughly defined set of information, whereas structured interviews seek specific information. Closed-ended questions are those that require a specific answer; open-ended questions leave the type of answer open to the interviewee. Probing questions follow up on what has just been discussed to allow the interviewer to learn more. During the interview, it is critical to carefully record all the information that the interviewee provides. Finally, be sure to separate facts from opinion.

Joint Application Design
Joint application design (JAD) is a structured process in which 10 to 20 users meet together for several hours, days, weeks, or months under the direction of a skilled facilitator. The need for all JAD participants to be away from their office at the same time can be problematic for some organizations. Most JAD sessions try to follow a formal agenda and most have formal ground rules that define appropriate behavior. The facilitator is responsible for conducting the JAD session and making sure the group follows the agenda, for guiding the group through the techniques being used, and for helping the group to organize and structure its information.

Questionnaires
Questionnaires are often used when there is a large number of people from which information and opinions are needed. Most people automatically think of paper when they think of questionnaires, but today more questionnaires are being distributed in electronic form, either via e-mail or on the Web. On average, only 30% to 50% of paper and e-mail questionnaires and 5% to 30% of Web-based questionnaires are returned. Questions on questionnaires must be very clearly written and leave little room for misunderstanding, so closed-ended questions tend to be the most commonly used.

Document Analysis
Document analysis is the examination of the policies, forms, and reports used by the existing information system IS. Quite often, the "real" or informal system differs from the formal one described in the policy manuals. In most cases, the differences are not large, but any differences give indications of what needs to be changed in a new system.

Observation
Observation is a powerful tool for gathering information because it enables the analyst to see the reality of the situation rather than to listen to others describe it in interviews or JAD sessions. In many ways, the analyst becomes a detective or anthropologist as he or she walks through the organization and observes the busi-

ness system as it functions. However, the normal day-to-day routine may change when people are watched, so what is observed may not be "normal."

Selecting the Appropriate Techniques

No one technique is inherently better than another, and in practice, most projects use a combination of techniques. Interviews and JAD sessions are used at all stages of information gathering because they provide a depth of rich and detailed information. Document analysis and observation are commonly used for understanding the As-Is system because they are useful for obtaining facts, but not much beyond that. Questionnaires and document analysis are both capable of soliciting a wide breadth of information, whereas interviews and observation both require the analyst to visit each information source individually. All techniques suffer problems in integrating information from different sources, but JAD sessions are specifically designed to improve integration because all information is combined when it is collected, not afterward. Questionnaires, document analysis, and observation require the least user involvement; JAD sessions require the most. Questionnaires, document analysis, and observation have the lowest cost, JAD sessions have low to moderate costs, and interviews are the most expensive.

KEY TERMS

Bottom-up approach	Integration of information	Post-session report
Breadth of information	Interpersonal skills	Probing question
Closed-ended question	Interview	Questionnaire
Cost	Interview notes	Sample
Depth of information	Interview report	Scribe
Document analysis	Interview schedule	Stakeholder
Electronic JAD (e-JAD)	Joint application design (JAD)	Structured interview
Facilitator	Observation	Top-down approach
Formal system	Open-ended question	Unstructured interview
Ground rules	Open issue	User involvement
Informal system	Outcome analysis	

QUESTIONS

1. Describe the five major steps in conducting interviews.
2. How are participants selected for interviews and joint application design (JAD) sessions?
3. Is it sufficient to ensure that only key managers are interviewed? Explain.
4. Explain the differences between unstructured interviews and structured interviews. When would you use each approach?
5. Explain the difference between top-down and bottom-up interview approaches. When would you use each approach?
6. Explain the difference between a closed-ended question, an open-ended question, and a probing question. When would you use each?
7. How can you differentiate between facts and opinions? How can you use both?

8. Describe the five major steps in conducting JAD sessions.
9. How does a JAD facilitator differ from a scribe?
10. What are the three primary things that a facilitator does in conducting the JAD session?
11. What is e-JAD and why might a company be interested in using it?
12. How does designing questions for questionnaires differ from designing questions for interviews or JAD sessions?
13. What are typical response rates for questionnaires and how can you improve them?
14. Describe document analysis.
15. How does the formal system differ from the informal system?
16. What are the key aspects of using observation in the information-gathering process?
17. Explain the factors that can be used to select information-gathering techniques.
18. What is an open issue?
19. What do you think are three common mistakes that novice analysts make in interviewing?
20. Suppose you are interviewing the project sponsor at the very start of a project. What are the three most important pieces of information you need to gather?
21. Suppose you are interviewing a key mid-level manager at the very start of a project. What are the three most important pieces of information you need to gather?

EXERCISES

A. Who are the key stakeholders involved if your university wanted to change the requirements for your degree (e.g., increasing or decreasing the number and type of courses required)?

B. Find a partner and interview each other about what tasks each of you did in the last job held (full time, part time, past or current). If you haven't held a job, then assume your job is being a student. Before you do this, develop a brief interview plan. After your partner interviews you, identify the type of interview, interview approach, and types of questions used.

C. Design a requirements-gathering plan for exercise D in Chapter 4.

D. Design a requirements-gathering plan for exercise E in Chapter 4.

E. Design a requirements-gathering plan for exercise F in Chapter 4.

F. Find a group of students and run a 60-minute joint application development (JAD) session on improving the alumni relations at your university. First, develop a brief JAD plan and select the techniques to use (three at most) and then develop an agenda. Conduct the session using the agenda and then write your post-session report.

G. Find a questionnaire on the Web that has been created to capture information from Web users.

Describe the purpose of the survey, the way questions are worded, and how the questions have been organized. How can it be improved?

H. Develop an interview plan and conduct a series of three interviews about the important aspects for a system to be developed to help students find permanent and/or part-time jobs. Write brief interview reports.

I. Develop an agenda and conduct a 60-minute JAD session about the important aspects for a system to be developed to help students find permanent and/or part-time jobs. Write a brief post-session report.

J. Develop a questionnaire for requirements-gathering about the important aspects for a system to be developed to help students find permanent and/or part-time jobs. Give the questionnaire to 20 to 50 students and write a brief report.

K. Contact the career services department at your university and find all the pertinent documents designed to help students find permanent and/or part-time jobs. Analyze the documents and write a brief report.

L. Think about the different requirements-gathering strategies used in Exercises H, I, J, and K. What differences are there likely to be in the information gathered by each approach?

MINICASES

1. Barry has recently been assigned to a project team that will be developing a new retail store management system for a chain of submarine sandwich shops. Barry has several years of experience in programming but has not done much analysis in his career. He was a little nervous about the new work he would be doing but was confident he could handle any assignment he was given.

   One of Barry's first assignments was to visit one of the submarine sandwich shops and prepare an observation report on how the store operates. Barry planned to arrive at the store around noon, but he chose a store in an area of town he was unfamiliar with, and due to traffic delays and difficulty in finding the store, he did not arrive until 1:30. The store manager was not expecting him, and refused to let a stranger behind the counter until Barry had him contact the project sponsor (the Director of Store Management) back at company headquarters to verify who he was and what his purpose was.

   After finally securing permission to observe, Barry stationed himself prominently in the work area behind the counter so that he could see everything. The staff had to maneuver around him as they went about their tasks, but there were only minor occasional collisions. Barry noticed that the store staff seemed to be going about their work very slowly and deliberately, but he supposed that was because the store wasn't very busy. At first, Barry questioned each worker about what he or she was doing, but the store manager eventually asked him not to interrupt their work so much—he was interfering with their service to the customers.

   By 3:30, Barry was a little bored. He decided to leave, figuring he could get back to the office and prepare his report before 5:00 that day. He was sure his team leader would be pleased with his quick completion of his assignment. As he drove, he reflected, "There really won't be much to say in this report. All they do is take the order, make the sandwich, collect the payment, and hand over the order. It's really simple!" Barry's confidence in his analytical skills soared as he anticipated his team leader's praise.

   Back at the store, the store manager shook his head, commenting to his staff, "He comes here at the slowest time of day on the slowest day of the week. He never even looked at all the work I was doing in the back room while he was here—summarizing yesterday's sales, checking inventory on hand, making up resupply orders for the weekend . . . plus he never even considered our store opening and closing procedures. I hate to think that the new store management system is going to be built by someone like that. I'd better contact Chuck (the Director of Store Management) and let him know what went on here today." Evaluate Barry's conduct of the observation assignment.

2. Anne has been given the task of conducting a survey of sales clerks who will be using a new order entry system being developed for a household products catalog company. The goal of the survey is to identify the clerks' opinions on the strengths and weaknesses of the current system. There are about 50 clerks who work in three different cities, so a survey seemed an ideal way of gathering the needed information from the clerks.

   Anne developed the questionnaire carefully and pretested it on several sales supervisors who were available at corporate headquarters. After revising it based on their suggestions, she sent a paper version of the questionnaire to each clerk, asking that it be returned within one week. After one week, she had only three completed questionnaires returned. After another week, Anne received just two more completed questionnaires. Feeling somewhat desperate, Anne then sent out an e-mail version of the questionnaire, again to all the clerks, asking them to respond to the questionnaire by e-mail as soon as possible. She received two e-mail questionnaires and three messages from clerks who had completed the paper version expressing annoyance at being bothered with the same questionnaire a second time. At this point, Anne has just a 14% response rate, which she is sure will not please her team leader. What suggestions do you have that could have improved Anne's response rate to the questionnaire?

PLANNING

ANALYSIS

- ☑ **Develop Analysis Plan**
- ☑ **Understand As-Is System**
- ☑ **Identify Improvement Opportunities**
- ☑ **Develop the To-Be System Concept**
- ☐ Develop Use Cases
- ☐ Develop Structural Model
- ☐ Develop Behavioral Model

TASK CHECKLIST

PLANNING → ANALYSIS → DESIGN →

CHAPTER 6
USE-CASE MODELING

A use-case model describes the interaction of an information system and the information system's environment. Use-case models can be used to describe both the current As-Is system and the future To-Be system being developed. This chapter describes use-case modeling as a means to document and understand requirements, and to learn the function or external behavior of the system.

OBJECTIVES

- Understand the rules and style guidelines for use cases and use-case diagrams.
- Understand the process used to create use cases and use-case diagrams.
- Be able to create use cases and use-case diagrams.

CHAPTER OUTLINE

IMPLEMENTATION

INTRODUCTION

The previous chapter discussed the more popular requirements-gathering techniques such as interviewing, JAD, and observation. In this chapter, we discuss how the information that is gathered using these techniques is organized and presented in the form of a use case. Since UML has been accepted as the standard notation by the Object Management Group (OMG), almost all object-oriented development projects today utilize use cases to document and organize the requirements that are obtained during the analysis phase.[1]

A *use case* is a formal way of representing how a business system interacts with its environment. A use case illustrates the activities that are performed by the users of the system. As such, use-case modeling is often thought of as an external view of a business process in that it shows how the users view the process, rather than the internal mechanisms by which the process and supporting systems operate. Use cases can be used to document the current system (i.e., As-Is system) or the new system being developed (i.e., To-Be system).

Use cases are *logical models*—models that describe the business domain's activites without suggesting how they are conducted. Logical models are sometimes referred to as problem domain models. When reading a use case, in principle, you should not be able to tell if an activity is computerized or manual; if a piece of information is collected by paper form or via the Web; or if information is placed in a filing cabinet or a large database. These physical details are defined during the design phase when the logical models are refined into *physical models,* which provide information that is needed to ultimately build the system. By focusing on logical activities first, analysts can concentrate on how the business should run without being distracted by implementation details.

As a first step, the project team gathers requirements from the users about use cases that occur in the business. They then prepare *use-case descriptions* and *use-case diagrams* for each use case. Use cases are the discrete activities that the users perform, such as selling CDs, ordering CDs, and accepting returned CDs from customers. Users work closely with the project team to create the use-case descriptions that contain all the information needed to start modeling the system. Once the use-case descriptions are prepared, the second step is for the systems analysts to transform them into a set of use-case diagrams that formally define the external behavioral or functional view of the business process. Next, the analyst typically creates class diagrams (see the next chapter) to create a structural model of the business problem domain.

In this chapter, we first describe use cases, their elements, and a set of guidelines for creating use cases. We then describe use-case diagrams and how to create them.

USE-CASE DESCRIPTIONS

There are many ways in which the project team can document the requirements for an information system. Use cases are simple descriptions of a system's functions

---

[1] Other similar techniques that are commonly used in non-UML projects are task modeling (see Ian Graham, *Migrating to Object Technology,* Reading, MA: Addison-Wesley, 1995 and Ian Graham, Brian Henderson-Sellers, and Houman Younessi, *The OPEN Process Specification,* Reading, MA: Addison-Wesley, 1997) and scenario-based design (see John M. Carroll, *Scenario-Based Design: Envisioning Work and Technology in System Development,* New York: John Wiley & Sons, 1995).

from the "bird's eye view" of the users.[2] They are functional diagrams in that they describe the basic functions of the system; that is, what the users can do and how the system should respond to the user's actions.[3] Creating use-case diagrams is a two-step process. First, the users work with the project team to write text-based use-case descriptions, and second, the project team translates the use-case descriptions into formal use-case diagrams. *Use-case descriptions* are simply less formal descriptions of the use cases that contain all the information needed to produce the formal use-case diagrams. Although it is possible to skip the use-case description step and move directly into creating use-case diagrams and the other diagrams that follow, users often have difficulty describing their business processes using use-case diagrams without creating use case descriptions first.

Use cases are the primary drivers for all of the UML diagramming techniques. The use case communicates at a high level what the system needs to do. All of the UML diagramming techniques build on this by presenting the use-case functionality in a different way for a different purpose. Use cases are the building blocks by which the system is designed and built.

Use cases capture the typical interaction of the system with the system's users (i.e., end users and other systems). These interactions represent the external, or functional, view of the system from the user's perspective. Each use case describes one and only one function in which users interact with the system,[4] although a use case may contain several "paths" that a user can take while interacting with the system (e.g., when searching for a book in a Web bookstore, the user might search by subject, by author, or by title). Each path through the use case is referred to as a *scenario*.

When creating use cases, the project team must work closely with the users to gather the requirements needed for the use cases. This is often done through interviews, JAD sessions, and observation. Gathering the requirements needed for a use case is a relatively simple process but takes considerable practice. The key thing to remember is that each use case is associated with one and only one *role* that users have in the system. For example, a receptionist in a doctor's office may "play" multiple roles, making appointments, answering the telephone, filing medical

---

[2] For a more detailed description of use-case modeling see Alistair Cockburn, *Writing Effective Use Cases* (Reading, MA: Addison-Wesley, 2001).

[3] Nonfucntional requirements, such as reliability requirements and performance requirements are often documented outside of the use case though more traditional requirements documents. See Gerald Kotonya and Ian Sommerville, *Requirements Engineering,* (Chichester, England: John Wiley & Sons, 1998); Benjamin L. Kovitz, *Practical Software Requirements: A Manual of Content & Style* (Greenwich, CT: Manning, 1999); Dean Leffingwell and Den Widrig, *Managing Software Requirements: A Unified Approach,* (Reading, MA: Addison-Wesley, 2000); and Richard H. Thayer, M. Dorfman, and Sidney C. Bailin, eds., *Software Requirements Engineering,* 2nd Ed. (IEEE Computer Society, 1997).

[4] This is one key difference between traditional structured analysis and design approaches, and object-oriented approaches. If you have experience using traditional structured approaches (or have taken a course on them), then this is an important change for you. If you have no experience with structured approaches, then you can skip this footnote. The traditional structured approach is to start with one overall view of the system and to model processes via functional decomposition—the gradual decomposition of the overall view of the system into the smaller and smaller parts that make up the whole system. However, functional decomposition is *not* used with object-oriented approaches. Instead, each use case documents one individual piece of the system; there is no overall use case that documents the entire system in the same way that a level 0 data flow diagram attempts to document the entire system. By removing this overall view, object-oriented approaches make it easier to decouple the system's objects so they can be designed, developed, and reused independently of the other parts of the system. While this lack of an overall view may prove unsettling initially, it is very liberating over the long term.

records, welcoming patients, and so on. Furthermore, it is possible that multiple users will play the same role. As such, use cases should be associated with the roles "played" by the users and not with the users themselves.

Types of Use Cases

There are many different types of use cases. We suggest two separate dimensions on which to classify a use case based on its purpose and the amount of information it contains: (1) overview versus detail; and (2) essential versus real.

An *overview use case* enables the analyst and user to agree on a high-level overview of the requirements. Typically, it is created very early in the process of understanding the system requirements, and it only documents basic information about the use case such as its name, ID number, primary actor, type, and a brief description.

Once the user and the analyst agree on a high-level overview of the requirements, the overview use cases can be converted to detail use cases. A *detail use case* typically documents, as far as possible, all of the information needed for the use case.

An *essential use case* only describes the minimum essential issues necessary to understand the required functionality. A *real use case* will go further and describe a specific set of steps. For example, an essential use case in a dentist office might say that the receptionist should attempt to "Match the patient's desired appointment times with the available times," while a real use case might say that the receptionist should "Look up the available dates on the calendar using MS Exchange to determine if the requested appointment times are available." The primary difference is that essential use cases are implementation independent, whereas real use cases are detailed descriptions of how to use the system once it is implemented.

Elements of a Use-Case Description

A use-case description contains all the information needed to build the diagrams that follow, but it expresses it in a less formal way that is usually simpler for users to understand. Figure 6-1 shows a sample use case description.[5] The three basic parts of a use-case description are overview information, relationships, and the flow of events.

Overview Information   The overview information identifies the use case and provides basic background information about the use case. The *use-case name* of the use case should be a verb–noun phrase, for example, Make Appointment. The *use-case ID number* provides a unique way to find every use case and enables the team to trace design decisions back to a specific requirement. As stated above, the *use-case type* is either overview or detail and essential or real. The *primary actor* is usually the trigger of the use-case—the person or thing that starts the execution of the use case. The primary purpose of the use case is to meet the goal of the primary

---

[5] Currently, there is no standard set of elements of a use case. The elements described in this section are based on recommendations contained in Alistair Cockburn, *Writing Effective Use Cases,* (Reading, MA: Addison-Wesley, 2001); Craig Larman, *Applying UML and Patterns: An Introduction to Object-Oriented Analysis and Design,* (Englewood Cliffs: NJ: Prentice-Hall, 1998); Brian Henderson-Sellers and Bhuvan Unhelkar, *OPEN Modeling with UML,* (Reading, MA: Addison-Wesley, 2000); Graham, *Migrating to Object Technology,* and Alan Dennis and Barbara Haley Wixom, *Systems Analysis and Design: An Applied Approach,* (New York: John Wiley & Sons, 2000).

actor. The *brief description* is typically a single sentence that describes the essence of the use case.

The *importance level,* located at the top right corner of the form, can be used to prioritize the use cases. Object-oriented development tends to follow a RAD phased development approach, in which some parts of the system are developed first and other parts are only developed in later versions. The importance level enables the users to explicitly prioritize which functions are most important and need to be part of the first version of the system, and which are less important and can wait for later versions if necessary.

The importance level can use a fuzzy scale such as high, medium, and low (e.g., in Figure 6-1 we have assigned an importance level of high to the Make Appointment use case). It can also be done more formally using a weighted average of a set of criteria. For example, Larman[6] suggests rating each use case based on the following criteria using a scale from 0 to 5:

- The use case represents an important business process.
- The use case supports revenue generation or cost reduction.
- Technology needed to support the use case is new or risky and therefore will require considerable research.
- Functionality described in the use case is complex, risky, and/or time-critical. Depending on a use case's complexity, it may be useful to consider splitting its implementation over several different versions.
- The use case could increase understanding of the evolving design relative to the effort expended.

A use case may have multiple stakeholders that have an interest in the use case. As such, each use case lists each of the *stakeholders* with their interest in the use case, for example, Patient and Doctor. The stakeholders list always includes the primary actor, for example, Patient.

Each use case typically has a *trigger*—the event that causes the use case to begin. For example, the "Patient calls and asks for a new appointment or asks to cancel or change an existing appointment." A trigger can be an *external trigger,* such as a customer placing an order or the fire alarm ringing, or it can be a *temporal trigger,* such as book being overdue at the library or time to pay the rent.

Relationships   Use-case *relationships* explain how the use case is related to other use cases and users. The four basic types of relationships are association, extend, include, and generalization. An *association relationship* documents the communication that takes place between the use case and the actors that use the use case. An actor is the UML representation for the *role* that a user plays in the use case. For example, in Figure 6-1, the Make Appointment use case is associated with the Actor-Patient. In this case, a patient makes an appointment. All actors involved in the use case are documented with the association relationship.

An *extend relationship* represents the extension of the functionality of the use case to incorporate optional behavior. In Figure 6-1, the Make Appointment use case uses the Create New Patient use case. This use case is executed only if the patient does not exist in the patient database. As such, it is not part of the normal

---

[6] Larman, *Applying UML and Patterns.*

Use case name:	**Make appointment**		ID:	**2**	Importance level:	**High**
Primary actor:	**Patient**		Use case type:	**Detail, essential**		

Stakeholders and interests:
Patient - wants to make, change, or cancel an appointment
Doctor - wants to ensure patient's needs are met in a timely manner

Brief description: This use case describes how we make an appointment as
                              well as changing or canceling an appointment.

Trigger: Patient calls and asks for a new appointment or asks to cancel or change an existing appointment.

Type: External

Relationships:
      Association: Patient
      Include: Make Payment Arrangements
      Extend: Create New Patient
      Generalization:

Normal flow of events:
1. The Patient contacts the office regarding an appointment.
2. The Patient provides the Receptionist with his or her name and address.
3. The Receptionist validates that the Patient exists in the Patient database.
4. The Receptionist executes the Make Payment Arrangements use case.
5. The Receptionist asks Patient if he or she would like to make a new
   appointment, cancel an existing appointment or change an existing appointment.
       If the patient wants to make a new appointment,
         the S-1: new appointment subflow is performed.
       If the patient wants to cancel a existing appointment,
         the S-2: cancel appointment subflow is performed.
       If the patient wants to change a existing appointment,
         the S-3: change appointment subflow is performed.
6. The Receptionist provides the results of the transaction to the Patient.

Subflows:
  S-1: New Appointment
    1. The Receptionist asks the Patient for possible appointment times.
    2. The Receptionist matches the Patient's desired appointment times
      with available dates and times and schedules the new appointment.
  S-2: Cancel Appointment
    1. The Receptionist asks the Patient for the old appointment time.
    2. The Receptionist finds the current appointment in the appointment
      file and cancels it.
  S-3: Change Appointment
    1. The Receptionist performs the S-2: cancel appointment subflow.
    2. The Receptionist performs the S-1: new appointment subflow.

Alternate/exceptional flows:
  3a: The Receptionist executes the Create New Patient use case.
  S-1, 2a1: The Receptionist proposes some alternative appointment
        times based on what is available in the appointment schedule.
  S-1, 2a2: The Patient chooses one of the proposed times or decides
        not to make an appointment.

FIGURE 6-1
Use-Case Description Example

flow of events and should be modeled with an extend relationship and an alternate/exceptional flow (see below).

An *include relationship* represents the mandatory inclusion of another use case. The include relationship enables *functional decomposition*—the breaking up of a complex use case into several simpler ones. For example, in Figure 6-1, the Make Payment Arrangements was considered to be complex and complete enough to factor out as a separate use case that can be executed by the Make Appointment use case. This also enables parts of use cases to be reused by creating them as a separate use case.

The *generalization relationship* allows use cases to support *inheritance.* For example, the use case in Figure 6-1 could be changed such that a new patient could be associated with a specialized use case called Make New Patient Appointment and an old patient could be associated with a Make Old Patient Appointment. The common or generalized behavior that both the Make New Patient Appointment and Make Old Patient Appointment use cases contain would be placed in the generalized Make Appointment use case, for example, the include relationship to the Make Payment Arrangements use case. In other words, the Make New Patient Appointment and Make Old Patient Appointment use cases would inherit the Make Payment Arrangements use case from the Make Appointment use case. The specialized behavior would be placed in the appropriate specialized use case; for example, the extend relationship to the Create New Patient use case would be placed with the specialized Make New Patient Appointment use case.

Flow of Events   Finally, the individual steps within the business process are described. Three different categories of steps or flows can be documented: normal flow of events, subflows, and alternate or exceptional flows. The *normal flow of events* includes only those steps that normally are executed in a use case. The steps are listed in the order in which they are performed. In Figure 6-1, the Patient and the Receptionist have a conversation regarding the patient's name, address, and action to be performed.

In some cases, it is recommended that the normal flow of events be decomposed into a set of *subflows* to keep the normal flow of events as simple as possible. In Figure 6-1, we have identified three subflows: New Appointment, Cancel Appointment, and Change Appointment. Each of the steps of the subflows is listed. Alternatively, we could replace a subflow with a separate use case that could be included via the include relationships (see above). However, this should be done only if the newly created use case makes sense by itself. For example, in Figure 6-1, does it make sense to factor out a Create New Appointment, Cancel Appointment, and/or Change Appointment use case? If it does, then the specific subflow(s) should be replaced with a call to the related use case and the use case should be added to the include relationship list.

Lastly, we document the *alternate or exceptional flows.* Alternate or exceptional flows are those that happen but are not considered to be the norm. For example in Figure 6-1, we have identified three alternate or exceptional flows. The first one simply addresses the need to create a new patient before the new patient can make an appointment. Normally, the patient would already exist in the patient database. Like the subflows, the primary purpose of separating out alternate or exceptional flows is to keep the normal flow of events as simple as possible.

Optional Characteristics   Yet other characteristics of use cases can be documented, including the level of complexity of the use case, the estimated amount of time it takes to execute the use case, the system with which the use case is associated, specific data flows between the primary actor and the use case, any specific attribute, constraint, or operation associated with the use case, any preconditions that must be satisfied for the use case to execute, or any guarantees that can be made based on the execution of the use case. As we noted at the beginning of this section, there is no standard set of characteristics of a use case that must be captured. In this book, we suggest that the information contained in Figure 6-1 is a minimal amount that should be captured.

Guidelines for Creating Use-Case Descriptions

The essence of a use case is the flow of events. Writing the flow of events in a manner that is useful for later stages of development generally comes with experience. Figure 6-2 provides a set of guidelines that have proven useful.[7]

First, write each individual step in the form Subject–Verb–Direct Object and optionally Preposition–Indirect Object. This form has become known as *SVDPI* sentences. This form of sentence has proven useful in identifying classes and operations (see next chapter). For example, in Figure 6-1, the first step in the normal flow of events, "The Patient contacts the office regarding an appointment," suggests the possibility of three classes of objects: patient, office, and appointment. This approach simplifies the process of identifying the classes in the structural model (described in Chapter 7). Not all steps can use SVDPI sentences, but they should be used whenever possible.

Second, make clear who or what is the initiator of the action and who is the receiver of the action in each step. Normally, the initiator should be the subject of the sentence and the receiver should be the direct object of the sentence. For example, in Figure 6-1, the second step, "Patient provides the Receptionist with his or her name and address," clearly portrays the Patient as the initiator and the Receptionist as the receiver.

Third, write the step from the perspective of an independent observer. Accordingly, you may have to write each step from the perspective of both the initiator and the receiver first. Based on the two points of view, you can then write the "bird's eye view" version. For example, in Figure 6-1, the "Patient provides the Receptionist with his or her name and address." Neither the patient's nor the receptionist's perspective is represented.

---

1. Write each set in the form of Subject–Verb–Direct Object (and sometimes Preposition–Indirect Object).
2. Make sure it is clear who the initiator of the step is.
3. Write the steps from the perspective of the independent observer.
4. Write each step at about the same level of abstraction.
5. Ensure the use case has a sensible set of steps.
6. Apply the KISS principle liberally.
7. Write repeating instructions after the set of steps to be repeated.

FIGURE 6-2

Guidelines for Writing Effective Use-Case Descriptions

---

[7] These guidelines are based on Cockburn, *Writing Effective Use Cases,* and Graham, *Migrating to Object Technology.*

Fourth, write each step at the same level of abstraction. Each step should make about the same amount of progress toward completing the use case as each of the other steps in the use case. On high-level use cases, the amount of progress could be very substantial, whereas on low-level use cases, each step could represent only incremental progress. For example, in Figure 6-1, each step represents about the same amount of effort to complete.

Fifth, ensure that the use case contains a sensible set of actions. Each use case should represent a transaction; therefore, each should be comprised of four parts. (1) The primary actor initiates the execution of the use case by sending a request (and possibly data) to the system. (2) The system ensures that the request (and data) is valid. (3) The system processes the request (and data) and possibly changes its own internal state. (4) The system sends the primary actor the result of the processing. For example, in Figure 6-1, (1) the patient requests an appointment (Steps 1 and 2); (2) the receptionist determines if the patient exists in the database (Step 3); (3) the receptionist sets up the appointment transaction (Steps 4 and 5); and (4) the receptionist gives the time and date of the appointment to the patient (Step 6).

The sixth guideline is the KISS[8] principle. If the use case becomes too complex and/or too long, it should be decomposed into a set of use cases. Furthermore, if the normal flow of events of the use case becomes too complex, subflows should be used. For example, in Figure 6-1, the fifth step in the normal flow of events was sufficiently complex to decompose it into three separate subflows. However, as stated previously, care must be taken to avoid the possibility of decomposing too much. Most decomposition should be done with the classes (see next chapter).

The seventh guideline deals with repeating steps. Normally, in a programming language such as Visual Basic or C, we put loop definition and controls at the beginning of the loop. However, since the steps are written in simple English, it is normally better to simply write "Repeat steps A through E until some condition is met" after step E. It makes the use case more readable to people unfamiliar with programming.

USE-CASE DIAGRAMS

In the previous section, we learned what a use case is and were given a set of guidelines on how to write effective use cases. In this section, we will learn how the use case is the building block for the *use-case diagram,* which summarizes all of the use cases for the part of the system being modeled together in one picture. An analyst can use the use-case diagram to better understand the functionality of the system at a very high level. Typically, the use-case diagram is drawn early on in the SDLC when gathering and defining requirements for the system because it provides a simple, straightforward way of communicating to the users exactly what the system will do. In this manner, the use-case diagram can encourage the users to provide additional requirements that the written use case may not uncover. This section describes the elements of a use-case diagram.

A use-case diagram illustrates in a very simple way the main functions of the system and the different kinds of users that will interact with it. Figure 6-3 describes the basic syntax rules for a use-case diagram. Figure 6-4 presents a use-case diagram for a doctor's office appointment system. We can see from the diagram that patients,

---

[8] Keep It Simple Stupid.

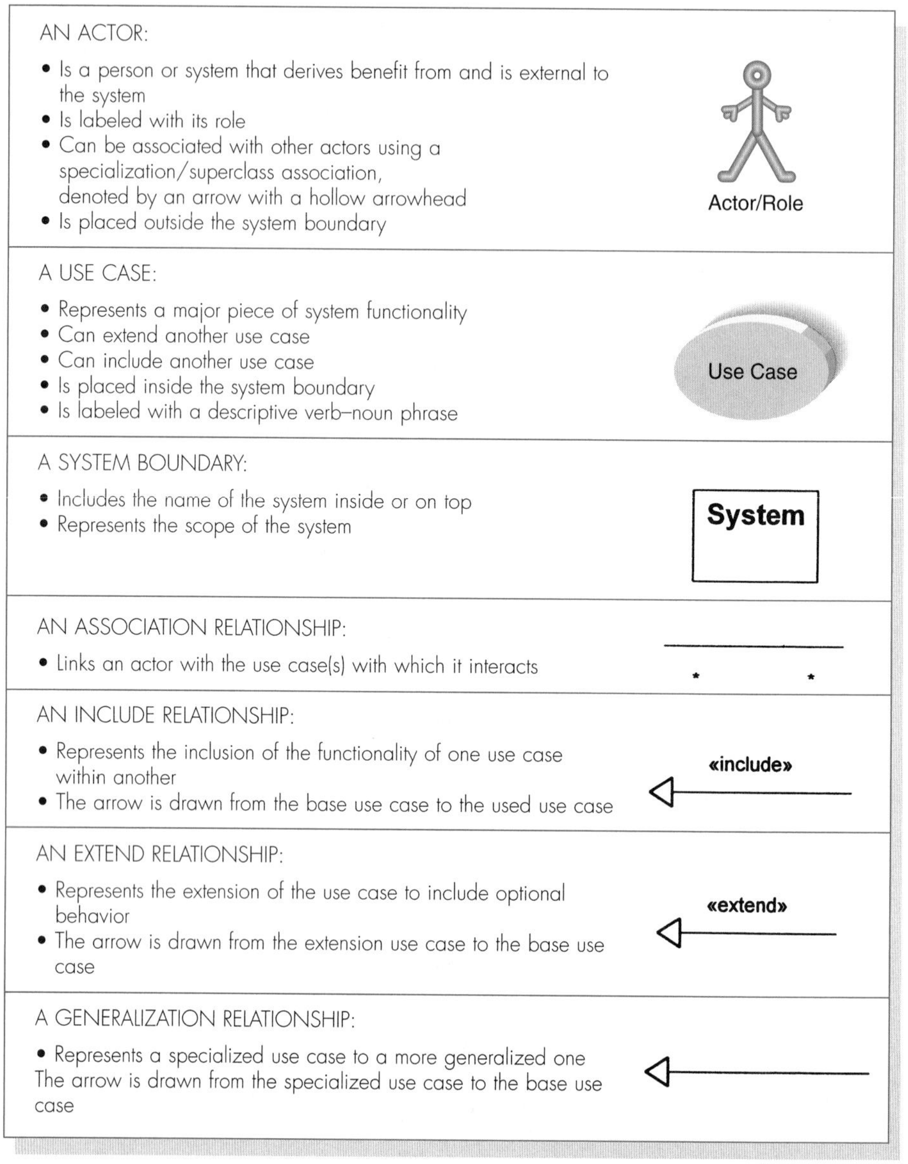

AN ACTOR:

- Is a person or system that derives benefit from and is external to the system
- Is labeled with its role
- Can be associated with other actors using a specialization/superclass association, denoted by an arrow with a hollow arrowhead
- Is placed outside the system boundary

Actor/Role

A USE CASE:

- Represents a major piece of system functionality
- Can extend another use case
- Can include another use case
- Is placed inside the system boundary
- Is labeled with a descriptive verb–noun phrase

Use Case

A SYSTEM BOUNDARY:

- Includes the name of the system inside or on top
- Represents the scope of the system

System

AN ASSOCIATION RELATIONSHIP:

- Links an actor with the use case(s) with which it interacts

*   *

AN INCLUDE RELATIONSHIP:

- Represents the inclusion of the functionality of one use case within another
- The arrow is drawn from the base use case to the used use case

«include»

AN EXTEND RELATIONSHIP:

- Represents the extension of the use case to include optional behavior
- The arrow is drawn from the extension use case to the base use case

«extend»

A GENERALIZATION RELATIONSHIP:

- Represents a specialized use case to a more generalized one
The arrow is drawn from the specialized use case to the base use case

FIGURE 6-3
Syntax for Use-Case Diagram

doctors, and management personnel will use the Appointment System to make appointments, record availability, and produce schedule information, respectively.

Actor

The labeled stick figures on the diagram represent actors (see Figure 6-3). An *actor* is not a specific user but a role that a user can play while interacting with the system. An actor can also represent another system in which the current system interacts. Basically, actors represent the principal elements in the environment in which the system operates. Actors can provide input to the system, receive output from the system, or both. The diagram in Figure 6-4 shows that three actors will interact with the appointment system (a patient, a doctor, and management).

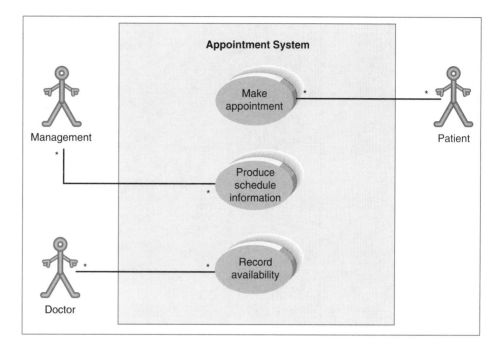

FIGURE 6-4
Use Case Diagram for Appointment System

Sometimes an actor plays a specialized role of a more general type of actor. For example, a new patient may at times interact with the system in a way that is somewhat different than a general patient. In this case, a *specialized actor* (i.e., new patient) can be placed on the model, shown using a line with a hollow triangle at the end of the more general actor (i.e., patient). The specialized actor will inherit the behavior of the more general actor and extend it in some way (see Figure 6-5). Can you think of some ways in which a new patient may behave differently than an existing patient?

Association

Use cases are connected to actors through *association relationships,* which show the use cases with which the actors interact (see Figure 6-3). A line drawn from an actor to a use case depicts an association. The association typically represents two-way communication between the use case and the actor. If the communication is only one-way, then a solid arrowhead can be used to designate the direction of the flow of information. For example, in Figure 6-4 the Patient actor communicates with the Make appointment use case. Since there are no arrowheads on the association, the communication is two-way. Finally, it is possible to represent the multiplicity of the association. Figure 6-4 shows an asterisk (*) at either end of the association between the Patient and the Make appointment use case. This simply designates that an individual patient (instance of the Patient actor) executes the Make appointment use case as many times they wish and that it is possible for the Make appointment use-case appointment to be executed by many different patients. In most cases, this type of many-to-many relationship is appropriate. However, it is possible to restrict the number of patients that can be associated with the Make

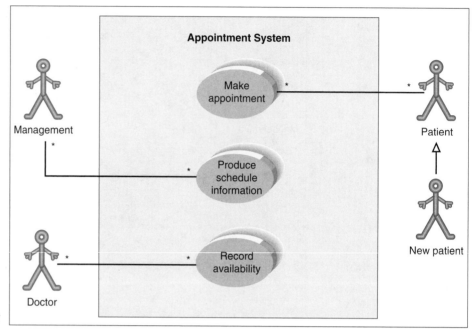

FIGURE 6-5
Use Case Diagram with Specialized Actor

appointment use case. We will discuss the multiplicity issue in detail in the next chapter with regard to class diagrams.

Use Case

A use case, depicted by an oval in the UML, is a major process performed by the system that benefits an actor(s) in some way (see Figure 6-3), and it is labeled using a descriptive verb–noun phrase. Figure 6-4 shows us that the system has three primary use cases: make appointment, produce schedule information, and record availability.

At times a use case includes the functionality, extends the functionality, or generalizes the functionality of another use case on the diagram. This is shown using include, extend, and generalization relationships, respectfully. To help understand a use-case diagram, "higher-level" use cases normally are drawn above the "lower-level" ones. It may be easier to understand these relationships with the help of examples. Let's assume that every time a patient makes an appointment the patient is asked to confirm his or her contact and basic patient information to ensure that the system always contains the most up-to-date information on its patients. However, it is rarely necessary to actually change any of the patient information. Therefore, we may want to have a use case called Update patient information that *extends* the make appointment use case to include the functionality just described. In Figure 6-6, a hollow arrow labeled with "extends" was drawn between Update patient information and Make appointment to denote this special use-case relationship, and the Update patient information use case was drawn lower than the Make appointment use case.

Similarly, a single use case sometimes contains common functions that are used by other use cases. For example, suppose there is a use case called Manage schedule that performs some routine tasks needed to maintain the doctor's office appointment schedule, and the two use cases Record availability and Produce schedule information both perform the routine tasks. Figure 6-6 shows how we can

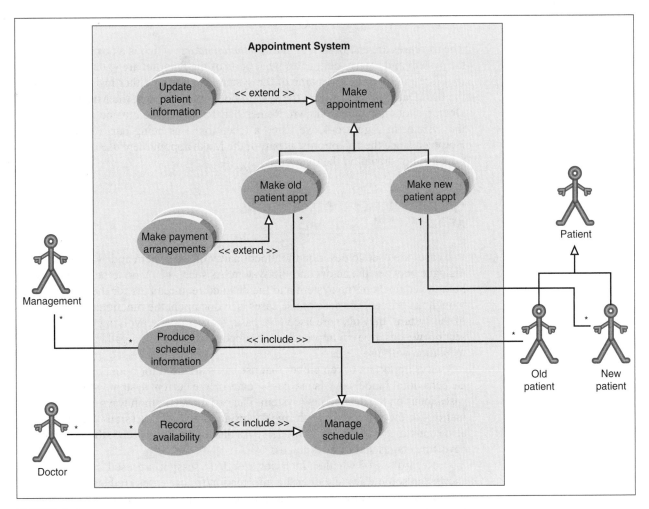

FIGURE 6-6
Extend and Include Relationships

design the system so that Manage schedule is a shared use case that is used by others. A hollow arrow labeled with "include" is used to denote the include relationship, and the *included* use case is drawn below the use cases that contain it.

Finally, at times it makes sense to use a generalization relationship to simplify the individual use cases. For example, in Figure 6-6 the Make appointment use case has been specialized to include a use case for an Old patient and a New patient. The Make old patient appt use case inherits the functionality of the Make appointment use case (including the Update patient information use case extension) and extends its own functionality with the Make payment arrangements use case. The Make new patient appt use case also inherits all of the functionality of the generic Make appointment use case and would include any functionality necessary to capture the information needed to insert the new patient into the patient database. The generalization relationship is represented as an unlabeled hollow arrow, with the more general use case being higher than the lower use cases. Also notice that we have added a second specialized actor, Old patient, and that the Patient actor is now simply a generalization of the old and new patient actors.

System Boundary

The use cases are enclosed within a *system boundary,* which is a box that represents the system and clearly delineates what parts of the diagram are external or internal to it (see Figure 6-3). The name of the system can appear either inside or on top of the box. Depending on where the system boundary is drawn, the Management and Doctor actors may not be drawn. Remember that actors are outside of the scope of the system. In Figure 6-4, we listed a receptionist as being part of the use case. However, since the receptionist is part of the Make appointment use case, the receptionist is not drawn on the diagram.[9]

CREATING USE-CASE DESCRIPTIONS AND USE-CASE DIAGRAMS

Use cases are used to describe the functionality of the system and as a model of the dialogue between the actors and the system. As such, they tend to be used to model both the contexts of the system and the detailed requirements for the system. Even though the primary purpose of use cases is to document the functional requirements of the system, they also are used as a basis for testing the evolving system. In this section we describe an approach that supports requirements gathering and documentation with use cases.

It is important to remember that use cases are used for both the As-Is and To-Be behavioral models. As-Is use cases focus on the current system, while To-Be use cases focus on the desired new system. The two most common ways to gather information for the use cases are through interviews and JAD sessions (observation is also sometimes used for As-Is models). As discussed in Chapter 5, these techniques have advantages and disadvantages.

Regardless of whether interviews or JAD session are used, recent research shows that some ways to gather the information for use cases are better than others. The most effective process has 13 steps to create use-case descriptions[10] and 4 additional steps to create use-case diagrams (see Figure 6-7). These 13 steps are performed in order, but of course the analyst often cycles among them in an iterative fashion as he or she moves from use case to use case.

Identify the Major Use Cases

The first step is to identify the system's boundaries. This helps the analyst to identify the scope of the system. However, as we work through the SDLC, the boundary of the system will change.

---

[9] In other non-UML approaches to object-oriented development, it is possible to represent external actors along with "internal" actors. In this example, the receptionist would be considered an internal actor (see Graham, *Migrating to Object Technology,* and Ian Graham, Brian Henderson-Sellers, and Houman Younessi, *The OPEN Process Specification.*

[10] The approach in this section is based on the work of Cockburn, *Writing Effective Use Cases;* Graham, *Migrating to Object Technology,* George Marakas and Joyce Elam, "Semantic Structuring in Analyst Acquisition and Representation of Facts in Requirements Analysis," *Information Systems Research* 9, no. 1 (1998): 37–63, as well as our own: Alan Dennis, Glenda Hayes, and Robert Daniels, "Business Process Modeling with Group Support Systems," *Journal of Management Information Systems* 15, no. 4, (1999): 115–142.

Identify the Major Use Cases

1. Find the system's boundaries.

2. List the primary actors.

3. List the goals of the primary actors.

4. Identify and write the overviews of the major use cases for the above.

5. Carefully review the current use cases. Revise as needed.

Expand the Major Use Case

6. Choose one of the use cases to expand.

7. Start filling in the details of the chosen use case.

8. Write the normal flow of events of the use case.

9. If the normal flow of events is too complex or long, decompose into subflows.

10. List the possible alternate or exceptional flows.

11. For each alternate or exceptional flow, list how the actor and/or system should react.

Confirm the Major Use Cases

12. Carefully review the current set of use cases. Revise as needed.

13. Start at the top again.

Create the Use-Case Diagram

1. Draw the system boundary.

2. Place the use cases on the diagram.

3. Place the actors on the diagram.

4. Draw the associations.

FIGURE 6-7

Steps for Writing Effective Use-Case Descriptions and Use-Case Diagrams.

The second step is to list the primary actors. The primary actors involved with the system will come from a list of stakeholders and users. However, you should remember that an actor is a role that a stakeholder or user plays, not a specific user, for example, doctor, not Dr. Jones. The first two steps are very intertwined. As actors are identified, the boundary of the system will change.

The third step is to list all of the goals that the actors have in relation to the system. The goals represent the functionality that the system must provide the actor for the system to be a success. Identifying what the tasks are that each actor must perform can facilitate this. For example, does the actor need to create, read, update, or delete (CRUD analysis is described in Chapter 8) any information currently in the system, are there any external changes that an actor must communicate to the system, or is there any information that the system should pass on to the actor?

The fourth step is to identify and write the major use cases, with basic information about each, rather than jump into one use case and describe it completely, that is, only an overview use case. Recall from the previous description of the elements of a use case, that overview use cases only contain the use case name, ID number, type, primary actor, and brief description. Creating overview use cases at this time prevents the users and analysts from forgetting key use cases and helps the users explain the overall set of business processes for which they are responsible. It also helps them understand how to describe the use cases and reduces the chance of overlap between use cases. It is important at this point to understand and define acronyms and jargon so that the project team and others from outside the user group can clearly understand the use cases.

The fifth step is to carefully review the current set of use cases. It may be necessary to split some of them into multiple use cases or merge some of them into a single use case. Also, based on the current set, a new use case may be identified. You should remember that identifying use cases is an iterative process, with users often changing their minds about what a use case is and what it includes. It is very easy to get trapped in the details at this point, so you need to remember that the goal at this step is to identify the *major* use cases and to recognize that you are only creating overview use cases. For example, in the doctor's office example in Figure 6-6, we defined one use case as "Patient makes an appointment." This use case included the cases for both new patients and existing patients, as well as for when the patient changes or cancels the appointment. We could have defined each of these activities (makes an appointment, changes an appointment, or cancels an appointment) as separate use cases, but this would have created a huge set of "small" use cases.

The trick is to select the right "size" so that you end up with three to nine use cases in each system. If the project team discovers many more than eight use cases, this suggests that the use cases are "too small" or that the system boundary is too big. There should be no more than three to nine use cases on the model, so if you identify more than nine, you should group together the use cases into *packages* (i.e., logical groups of use cases) in order to make the diagrams easier to read and keep the models at a reasonable level of complexity. It is simple at that point to sort the use cases and group these "small" use cases together into "larger" use cases that include several small ones or to change the system boundaries.[11]

Expand the Major Use Cases

The sixth is to choose one of the major use cases to expand. Using the importance level of the use case can do this. For example, in Figure 6-1, the Make appointment use case has an importance level of High. As such, it would be one of the earlier use cases to be expanded. You could also use the criteria suggested by Larman[12] to set the prioritization of the use cases as noted earlier.

The seventh step is to start filling in the details on the use-case template. For example, list all of the identified stakeholders and their interests in the use case, indicate the level of importance of the use case, briefly describe the use case, provide the trigger information for the use case, and note the relationships in which the use case participates.

The eighth step is to fill in the steps of the normal flow of events required to describe each use case. The steps focus on what the business process does to complete the use case, as opposed to what actions the users or other external entities take. In general, the steps should be listed in the order in which they are performed, from first to last. Remember to write the steps in a SVDPI form whenever possible.

The ninth step is to ensure that the steps listed in the normal flow of events are not too complex or too long. Each step should be about the same "size" as the others. For example, if we were writing steps for preparing a meal, steps such as "take fork out of drawer" and "put fork on table" are much "smaller" than "prepare cake using mix." If you end up with more than seven steps or steps that vary greatly in size, you should go back and review each step carefully and possibly rewrite the steps.

---

[11] For those familiar with structured analysis and design, packages serve a similar purpose as the leveling and balancing processes used in data flow diagramming. Packages are described in Chapter 9.

[12] Larman, *Applying UML and Patterns.*

One good approach to produce the steps for a use case is to have the users visualize themselves actually performing the use case and to write down the steps as if they were writing a recipe for a cookbook. In most cases, the users will be able to quickly define what they do in the As-Is model. Defining the steps for To-Be use cases may take a bit more coaching. In our experience, the descriptions of the steps change greatly as users work through a use case. Our advice is to use blackboard or white board (or paper with pencil) that can be easily erased to develop the list of steps, and then write it on the use case form. You should write it on the use-case form only after the set of steps is fairly well defined.

The tenth step focuses on identifying the alternate or exceptional flows. Alternate or exceptional flows are those flows of success that represent optional or exceptional behavior. They tend to occur infrequently or as a result of a normal flow failing. They should be labeled in such a way that there is no doubt as to the normal flow of events to which it is related. For example in Figure 6-1, Alternate/Exceptional flow 3a executes when step 3 fails; that is, the Patient does not exist in the Patient database.

The eleventh step is simply to write the description of the alternate or exceptional flow. In other words, if the alternate or exceptional flow is to be executed, describe the response that the actor or system should produce. Like the normal flows and subflows, alternate/exceptional flows should be written in the SVDPI form whenever possible.

Confirm the Major Use Cases

The twelfth step is to carefully review the current set of use cases and to confirm that the use case is correct as written, which means reviewing the use case with the users to make sure each step is correct. The review should look for opportunities to simplify a use case by decomposing it into a set of smaller use cases, merging it with others, looking for common aspects in both the semantics and syntax of the use cases, and identifying new use cases. This is also the time to look into adding the include, extend, or generalization relationships between use cases. The most powerful way to confirm the use case is to ask the user to role-play, or execute the process using the written steps in the use case. The analyst will hand the user pieces of paper labeled as the major inputs to the use case and will have the user follow the written steps like a recipe to make sure that those steps really can produce the outputs defined for the use case using its inputs.

The thirteenth and final step is to *iterate* the entire set of steps. You should continue to iterate until you feel that you have documented a sufficient number of use cases to begin identifying candidate classes for the structural model (see next chapter).

Create the Use-Case Diagram

Basically, drawing the use-case diagram is very straightforward once you have detailed use cases. The actual use-case diagram encourages information hiding. The only information drawn on the use-case diagram includes the system boundary, the use cases themselves, the actors, and the various associations between these components. The major strength of the use-case diagram is that it provides the user with an overview of the detailed use cases. However, you must remember that any time a use case changes, it could impact the use-case diagram.

There are four major steps to drawing the use-case diagram. First, the use-case diagram starts with the system boundary. This forms the border of the system, separating use cases (i.e., the system's functionality) from actors (i.e., the roles of the external users).

Second, the use cases are drawn on the diagram. These are taken directly from the detailed use-case descriptions. Special use-case associations (include, extend, or generalization) are also added to the model at this point. Be careful in laying out the diagram. There is no formal order to the use cases, so they can be placed in whatever fashion is needed to make the diagram easy to read and to minimize the number of lines that cross. It often is necessary to redraw the diagram several times with use cases in different places to make the diagram easy to read. Also, for purposes of understandability, there should be no more than three to nine use cases on the model to keep the diagram as simple as possible. This includes those use cases that have been factored out and now are associated with another use case through the include, extend, or generalization relationships.

Third, the actors are placed on the diagram. They are taken directly from the primary actor element on the detailed use-case description. Like use-case placement, to minimize the number of lines that will cross on the diagram, the actors should be placed near the uses cases with which they are associated.

The fourth and last step is to draw lines connecting the actors to the use cases with which they interact. No order is implied by the diagram, and the items you have added along the way do not have to be placed in a particular order. Therefore, you may want to rearrange the symbols a bit to minimize the number of lines that cross so that the diagram is less confusing.

APPLYING THE CONCEPTS AT CD SELECTIONS

The basic concept for the CD Selections' Internet Sales System was developed in Chapter 5. An expanded version of the concept is presented in Figure 6-8.

Identify the Major Use Cases

The first four steps in writing use cases deal with finding the system boundaries, listing the primary actors and their goals, and identifying and writing overviews of the major use cases based on these results. These use cases will form the basis of the use cases that are contained on the use-case diagram. Take a minute and go back and re-read the description in Figure 6-8 to identify the three to nine major use cases before you continue reading.

To begin with, it seems that the system boundary should be drawn in such a manner that anything that is not part of CD Selections' Internet Sales System, such as the vendors and customers, should be identified as primary actors. Therefore, these are considered to be outside of the scope of the system. The other potential actors identified could be the distribution system and the EM Manager.

Upon closer review of Figure 6-8, it seems that the distribution system should be outside the scope of the Internet Sales System and as such should be identified as a primary actor. In the case of the EM Manager, it seems that the EM Manager should be considered as part of the Internet Sales System and therefore should not be identified as a primary actor. Remember, primary actors are only those that can be viewed as being outside the scope of the system. The decision on whether the EM Manager or the distribution system is inside or outside of the system is somewhat arbitrary. From a customer's perspective, the distribution system could be seen

The Internet Sales System will have a database of basic information about the CDs that it can sell over the Internet, similar to the CD database at each of the retail stores (e.g., title, artist, id number, price, quantity in inventory). Every day, the Internet Sales System will receive an update from the distribution system that will be used to update this CD database. Some new CDs will be added, some will be deleted, and others will be revised (e.g., a new price). The Electronic Marketing (EM) Manager (a position that will need to be created) will also have the ability to update information (e.g., prices for sales).

The Internet Sales System will also maintain a database of marketing materials about each CD that will help Web users learn more about them. Vendors will be encouraged to e-mail marketing materials (e.g., music reviews, links to Web sites, artist information, and sample sound clips) that promote their CDs. The EM Manager will go through the e-mails and determine what information to place on the Web. He or she will add this information to a marketing materials database (or revise it or delete old information) that will be linked to the Web site.

Customers will access the Internet Sales System to look for CDs of interest. Some customers will search for specific CDs or CDs by specific artists, while other customers want to browse for interesting CDs in certain categories (e.g., rock, jazz, classical). When the customer has found all the CDs he or she wants, the customer will "check out" by providing personal information (e.g., name, e-mail, address, credit card), and information regarding the order (e.g., the CDs to purchase, and the quantity for each item). The system will verify the customer's credit card information with an on-line credit card clearance center and either accept the order or reject it.

Every hour or so, the orders will be pulled out of the order database and sent to the distribution system. The distribution system will handle the actual sending of CDs to customers; however, when CDs are sent to customers (via UPS or mail), the distribution system will notify the Internet Sales System, which in turn will e-mail the customer. Weekly reports can be run by the EM manager to check the order status.

FIGURE 6-8
Basic Concept for CD Selections To-Be
Internet Sales System

as being inside of the overall system, and it could be argued that the EM Manager is a primary user of the Internet Sales System.

At this point in the process, it is important to make a decision and to move on. During the process of writing use cases, ample opportunities arise to revisit this decision to determine whether the set of use cases identified is necessary and sufficient to describe the requirements of the Internet Sales System for CD Selections.

As you can see, based on the above considerations, the first two steps of finding the systems boundaries and listing the primary actors are heavily intertwined.

Each of the four paragraphs in the system concept corresponds to one major use case:

1. The distribution system providing information to maintain the basic CD information (e.g., CD name, price, inventory quantity) by adding new CDs, deleting old ones, and updating current ones.
2. The vendors providing information to maintain marketing information about CDs.
3. Customers placing orders over the Web.
4. The distribution system filling orders.

You might have considered adding new CDs to the database, deleting old ones, and changing information about CDs as three separate use cases, which in fact they are. If you think about it, you should see that, in addition to these three, we also need to have use cases for finding CD information and printing reports about CDs. However, our goal at this point is to develop a set of only three to nine *major* use cases. Therefore, we need to put these activities together into one overall "larger" use case.

You should see the same pattern for the marketing materials. We have the same processes for recording new marketing material for a CD, changing it, deleting it, finding it, and printing it. These five activities (creating, changing, deleting, finding, and printing) are the *standard processes* that are usually required anytime you have information to be stored in a database.

The project team at CD Selections identified these same four major use cases. At this point in the process, the project team began writing the overview use cases for these four. Remember that an overview use case only has five pieces of information: use case name, ID number, primary actor, type, and a brief description. At this point in time, we have already identified the primary actors and have associated them with the four use cases. Furthermore, since we are just beginning the development process, the type for all four use cases will be Overview and Essential. Since the ID numbers are simply used for identification purposes (i.e., they act as a key in a database), their values can be assigned in a sequential manner. This only leaves us with two pieces of information for each use case to write. The use-case name should be an action verb–noun phrase combination, for example, Make Appointment. In the CD Selections' Internet Sales situation, Maintain CD Information, Maintain Marketing Information, Place Order, and Fill Order seem to capture the essence of each of the four use cases. Finally, a brief description is written to describe the purpose of the use case or the goal of the primary actor using the use case. This description can range from a sentence to a short essay. The idea is to capture the primary issues in the use case and to make them explicit.

The fifth step is to carefully review the current set of use cases. Take a moment to review the use cases (Figure 6-9) and make sure you understand them. Based on the descriptions of all four use cases, there seems to be an obvious missing use case: Maintain Order. Since the project team did not include language in the brief description of the Place Order use case, it seems that the project team believes that maintaining an order is sufficiently different from placing an order that it should have its own use case to describe it. Furthermore, it does not seem reasonable that a customer would see placing an order and maintaining an order as the same use case. This type of process goes on until the project team feels that they have identified the major use cases for the Internet Sales System.

Expand the Major Use Cases

The sixth step is to choose one of the major use cases to expand. To help the project team to make this choice, they identified the trigger and the stakeholders and their interests for each major use case. The first use case, Maintain CD information, was triggered by the distribution system distributing new information for use in the CD database. Besides the distribution system, another stakeholder was identified: EM Manager. The second use case, Maintain marketing materials, has a similar structure. The receipt of marketing materials from vendors triggers it. And again, the project team believed that the EM Manager was an interested stakeholder. The third use case, Customer places order, is more interesting. The trigger is the customer's actions. Again, the EM Manager is an interested stakeholder. This use case has many more inputs. The fourth use case, Fill order, has a temporal trigger: every hour the Internet Sales System downloads orders into the distribution system. The final use case, Maintain order, is triggered by the customer's action. However, on close review, it seems that it also has much in common with the other maintenance related use cases: Maintain CD information and Maintain marketing information. And like all of the other use cases, the EM Manager has an interest.

Use case name:	Maintain CD information		ID:	1	Importance level:
Primary actor:	Distribution system		Use case type:	Overview, essential	
Stakeholders and interests:					
Brief description:	This adds, deletes, and modifies the basic information about the CDs we have available for sale (e.g., album name, artist(s), price, quantity on hand, etc.).				
Trigger:					

Use case name:	Maintain marketing information		ID:	2	Importance level:
Primary actor:	Vendor		Use case type:	Overview, essential	
Stakeholders and interests:					
Brief description:	This adds, delete, and modifies the additional marketing material available for some CDs (e.g., reviews).				
Trigger:					

Use case name:	Place order		ID:	3	Importance level:
Primary actor:	Customer		Use case type:	Overview, essential	
Stakeholders and interests:					
Brief description:	This use case describes how customers can search the Web site and place orders.				
Trigger:					

Use case name:	Fill order		ID:	4	Importance level:
Primary actor:	Distribution system		Use case type:	Overview, essential	
Stakeholders and interests:					
Brief description:	This describes how orders move from the Internet Sales System into the distribution system and how status information will be updated from the distribution system.				
Trigger:					
	Type:				
Relationships:					
Association:					
Include:					
Extend:					
Generalization:					
Normal flow of events:					
Subflows:					

FIGURE 6-9
Overview Major Use Cases for CD Selections

Based on their review of the major use cases, the project team decided that the Place order use case was the most interesting. The next, seventh, step is to start filling in the details of the chosen use case. At this point in time, the project team had the information to fill in the stakeholders and interests, the trigger, and the Association relationship. Remember that the Association relationship is simply the communication link between the primary actor and the use case. In the Place order use case, the Association relationship is with the Customer.

The project team then had to gather the information needed to define the Place order use case in more detail. Specifically, they needed to begin writing the normal flow of events, that is, they should perform the eighth step for writing effective use cases (see Figure 6-7). This was done based on the results of the earlier analyses described in Chapter 5, as well as through a series of JAD meetings with the project sponsor and the key marketing department managers and staff who would ultimately operate the system.

The goal at this point is to describe how the use case operates. In this example, we will focus on the most complex and interesting use case, Place order. The best way to begin to understand how the customer works through this use case is to visualize yourself placing a CD order over the Web and to think about how other electronic commerce Web sites work—that is, role-play. The techniques of visualizing yourself interacting with the system and of thinking about how other systems work (informal benchmarking) are important, helping analysts and users understand how systems work and how to write the use case. Both techniques (visualization and informal benchmarking) are commonly used in practice. It is important to remember that at this point in the development of the use case, we are only interested in the typical successful execution of the use case. If you try to think of all of the possible combinations of activities, you will never get anything written down.

In this situation, after you connect to the Web site, you probably begin searching, perhaps for a specific CD, perhaps for a category of music, but in any event, you enter some information for your search. The Web site will then present a list of CDs matching your request, along with some basic information about the CDs (e.g., artist, title, and price). If one of the CDs is of interest to you, you might seek more information about it, such as the list of songs, liner notes, and reviews. Once you, the customer, find a CD you like, you will add it to your order and perhaps continue looking for more CDs. Once you are done—perhaps immediately—you will "check out" by presenting your order with information on the CDs you want and giving more information such as mailing address and credit card.

When you write the use case, be sure to remember the seven guidelines described earlier. One might argue that the first step is to present the customer with the home page or a form to fill in to search for an album. This is correct, but this step is usually very "small" compared to other steps that follow.[13] It is analogous to making the first step "hand the user a piece of paper." At this point, we are only looking for the three to seven *major* steps.

The first major step performed by the system is to respond to the customer's search inquiry, which might include a search for a specific album name or albums by a specific artist. Or it might be the customer wanting to see all the classical or alternative CDs in stock. Or it might be a request to see the list of special deals or

---

[13] Since it is so small, it violates the fourth guideline (see Figure 6-2).

CDs on "sale." In any event, the system finds all the CDs matching the request and shows a list of CDs in response. The user will view this response and perhaps will decide to seek more information about one or more CDs. He or she will click on it, and the system will provide additional information. Perhaps the user will also want to see any extra marketing material that is available as well. The user will then select one or more CDs for purchase and will perhaps continue with a new search. These steps correspond to events 1 through 6 in Figure 6-7.

The user may later make changes to the CDs selected, either by dropping some or changing the number ordered. This seems to be similar to an already identified separate use case: Maintain order. As such, we will simple reuse that use case to handle those types of details. At some point/the user will "check-out" by verifying the CDs he or she has selected for purchase and by providing information about him or herself (e.g., name, mailing address, credit card). The system will calculate the total payment and verify the credit card information with the credit card clearance center. At this point in the transaction, the system will send an order confirmation to the customer, and the customer typically leaves the Web site. Figure 6-10 shows the use case at this point. Note that the normal flow of events has been added to the form, but nothing else has changed.

The ninth step for writing uses cases (see Figure 6-7) is basically to try and simplify the current normal flow of events. Currently, we have 11 events, which is a little high. Therefore, we need to see whether there are any steps that we can merge, delete, or rearrange, and whether there are any that we have left out. Based on this type of review, we have decided to create a separate use case that can handle the "checkout" process (events 8, 9, and 10). We also see that event 5 could be viewed as part of the Maintain order use case. As such, we delete event 5 and move event 7 into its place. At this point in time, we have eight events. This seems to be a reasonable number for this particular use case.

The next two steps in writing a use case address the handling of alternate or exceptional flows. If you remember, the normal flow of events only captures the typical set of events that end in a successful transaction. In this case, CD Selections has defined success as a new order being placed. In our current use case, the project team has identified two sets of events that are exceptions to the normal flow. First, event 3 assumes that the list of recommended CDs is acceptable to the customer. However, as one of the team members pointed out, that is an unrealistic assumption. As such, two exceptional flows have been identified and written (3a-1 and 3a-2 in Figure 6-11) to handle this specific situation. Second, a customer may want to abort the entire order instead of going through the checkout process. In this case, exceptional flow 7a was created.[14]

Confirm the Major Use Cases

Once all the use cases had been defined, the final step in the JAD session was to confirm that they were accurate. The project team had the users role-play the use cases. A few minor problems were discovered and easily fixed. However, one major

---

[14] Another approach would be to force the customer through the Maintain order or Checkout use cases. However, the marketing representatives on the project team were concerned with customer frustration. As such, the project team included it in the Place order use case.

Use case name: Place order		ID: ___3___	Importance level: __High__
Primary actor: **Customer**		Use case type: **Detail, essential**	

Stakeholders and interests: **Customer - wants to search Web site to purchase CD**
EM manager - wants to maximize customer

Brief description: **This use case describes how customers can search the Web site and place orders.**

Trigger: **Customer visits Web site and places order**

Type: External

Relationships:
- Association: **Customer**
- Include: **Maintain order**
- Extend:
- Generalization:

Normal flow of events:
1. The Customer submits a search request to the system.
2. The System provides the Customer a list of recommended CDs.
3. The Customer chooses one of the CDs to find out additional information.
4. The System provides the Customer with basic information and reviews on the CD.
5. The Customer adds the CD to their shopping cart.
6. The Customer iterates over 3 through 5 until done shopping.
7. The Customer executes the Maintain Order use case.
8. The Customer logs in to check out.
9. The System validates the Customer's credit card information.
10. The System sends an order confirmation to the Customer.
11. The Customer leaves the Web site.

Subflows:

Alternate/exceptional flows:

FIGURE 6-10
Place Order Use Case after Step 8

problem was discovered: how does a new customer get created? This problem was easily fixed by creating a new use case, Create new customer, and adding it as an extension to the checkout use case. Figure 6-12 shows the revised use cases. At this point in time, the use-case development process can actually start all over again or we can move on to drawing the use-case diagram.

Create the Use-Case Diagram

Creating a use-case diagram from the detailed use-case descriptions is very easy. Since a use-case diagram only shows the system boundary, the use cases themselves, the actors, and the various associations between these components, they support information hiding. Use case diagrams only provide a pictorial overview

| Use case name: | Place order | | | ID: | 3 | Importance level: | High |

Primary actor: Customer   Use case type: Detail, essential

Stakeholders and interests: Customer - wants to search Web site to purchase CD
EM manager - wants to maximize customer satisfaction.

Brief description: This use case describes how customers can search the Web
site and place orders.

Trigger: Customer visits Web site and places order

Type: External

Relationships:
    Association: Customer
    Include: Checkout, Maintain Order
    Extend:
    Generalization:

Normal flow of events:
1. Customer submits a search request to the system.
2. The System provides the Customer a list of recommended CDs.
3. The Customer chooses one of the CDs to find out additional information.
4. The System provides the Customer with basic information and reviews on the CD.
5. The Customer calls the maintain order use case.
6. The Customer iterates over 3 through 5 until done shopping.
7. The Customer executes the Checkout use case.
8. The Customer leaves the Web site.

Subflows:

Alternate/exceptional flows:
    3a-1. The Customer submits a new search request to the system.
    3a-2. The Customer iterates over steps 2 through 3 until satisfied with search results or gives up.
    7a.   The Customer aborts the order.

FIGURE 6-11
Place Order Use Case after Step 11

of the detailed use cases. There are four major steps to drawing a use-case diagram.

Draw the System Boundary First, place a box on the use-case diagram to represent the system and place the system's name either inside or on top of the box. Based on a review of Figure 6-12, be sure to draw the box large enough to contain the use cases. Also, be sure to leave space on the outside of the box for the actors.

Place the Use Cases on the Diagram The next step is to add the use cases. Place the number of use cases inside the system boundary to correspond with the number of functions that the system needs to perform. There should be no more than three to nine use cases on the model. Can you identify the use cases that need to be placed in the system boundary by examining Figure 6-12?

Use case name:	Maintain CD information		ID:	1	Importance level:	High

Primary actor: Distribution system   **Use case type:** Detail, essential

Stakeholders and interests: Distribution System - wants to keep CD information up to date.
EM Manager - wants CD information to be correct to minimize Customer frustration.

Brief description: This adds, deletes, and modifies the basic information
about the CDs we have available for sale (e.g., album
name, artist (s), price).

Trigger: Downloads from the Distribution System

Type: External

Relationships:
Association: Distribution System
Include:

Use case name:	Maintain marketing information		ID:	2	Importance level:	High

Primary actor: Vendor   **Use case type:** Detail, essential

Stakeholders and interests: Vendor - wants to ensure marketing information is
as current as possible
EM Manager - wants marketing information to be
correct to maximize sales.

Brief description: This adds, deletes, and modifies the additional
marketing material available for some CDs (e.g.,
reviews).

Trigger: Materials arrive from vendors, distributors, wholesalers,
record companies, and articles from trade magazines arrive

Type: External

Relationships:
Association: Vendor
Include:

Use case name:	Place order		ID:	3	Importance level:	High

Primary actor: Customer   **Use case type:** Detail, essential

Stakeholders and interests: Customer - wants to search Web site to purchase CD
EM Manager - wants to maximize Customer

Brief description: This use case describes how customers can search the Web
site and place orders.

Trigger: Customer visits Web site and places order.

Type: External

Relationships:
Association: Customer
Include: Checkout, Maintain Order
Extend:
Generalization :

FIGURE 6-12
Revised Major Use Cases for CD Selections

Use case name: Fill order		ID: 4	Importance level: High
Primary actor: Distribution system		Use case type: Detail, essential	
Stakeholders and interests: Distribution System - wants to complete order processing in a timely manner. Customer - wants to receive order in a timely manner. EM Manager - wants to maximize order throughput.			
Brief description: This describes how orders move from the Internet Sales System into the distribution system and how status information will be updated from the distribution system.			
Trigger: Every hour the Distribution System will initiate a trading of information with the Internet Sales System.			
Type: Temporal			
Relationships: Association: Include:			

Use case name: Maintain order		ID: 5	Importance level: High
Primary actor: Customer		Use case type: Detail, essential	
Stakeholders and interests: Customer - wants to modify an order EM manager - wants to ensure high customer satisfaction			
Brief description: This use case describes how a Customer can cancel or modify an open or existing order.			
Trigger: Customer visits Web site and requests to modify a current order.			
Type: External			
Relationships: Association: Customer Include:			

Use case name: Checkout		ID: 6	Importance level: High
Primary actor: Customer		Use case type: Detail, essential	
Stakeholders and interests: Customer - wants to finalize the order. Credit Card Center - wants to provide effective and efficient service to CD selections. EM Manager - wants to maximize order closings.			
Brief description: This use case describes how the customer completes an order including credit card authorization.			
Trigger: Customer signals the system they want to finalize their order.			
Type: External			
Relationships: Association: Customer, Credit Card Center Include: Extend: Create New Customer Generalization:			

FIGURE 6-12 (continued)

Use case name: Create new customer	ID: 7	Importance level: High
Primary actor: Customer	Use case type: Detail, essential	

Stakeholders and interests: Customer - wants to be able to purchase CDs from CD selections.
EM Manager - wants to increase CD selections customer base.

Brief description: This use case describes how a Customer is added to the Customer database.

Trigger: An unknown Customer attempts to check out.

Type: External

Relationships:
    Association: Customer
    Include:
    Extend:
    Generalization :

FIGURE 6-12 *(continued)*

6-1 CD SELECTIONS INTERNET SALES SYSTEM

Complete the detailed use cases for the remaining overview use cases in Figure 6-12.

6-2 CAMPUS HOUSING

Create a set of use cases for the following high-level processes in a housing system run by the campus housing service. The campus housing service helps students find apartments. Owners of apartments fill in information forms about the rental units they have available (e.g., location, number of bedrooms, monthly rent), which are entered into a database. Students can search through this database via the Web to find apartments that meet their needs (e.g., a two-bedroom apartment for $400 or less per month within 1/2 mile of campus). They then contact the apartment owners directly to see the apartment and possibly rent it. Apartment owners call the service to delete their listing when they have rented their apartment(s).

Identify the Actors Once the use cases are placed on the diagram, you will need to place the actors on it. Normally, it is fairly easy to place the actors near their respective use cases. Look at the detailed use cases in Figure 6-12 and see if you can identify the four "actors" that belong on the use-case diagram. Hopefully, you have listed distribution system, customer, vendor, and credit card clearance center. At this point, there are no specialized actors that need to be included.

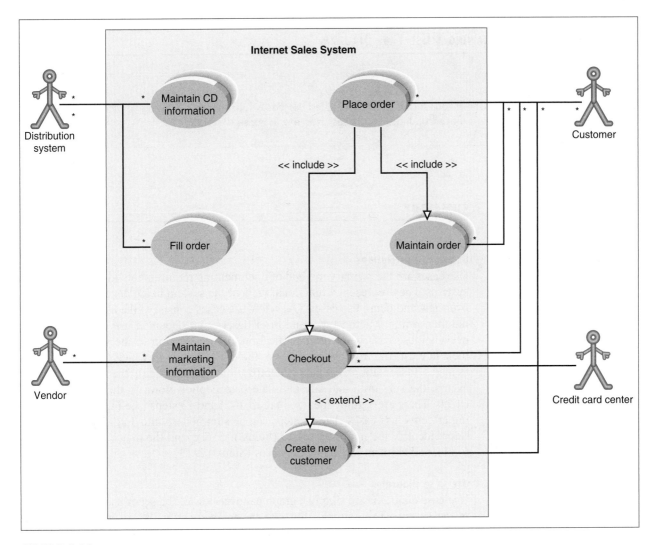

FIGURE 6-13
Use-Case Diagram for the CD Selections
Internet Sales System

Add Associations   The last step is to draw lines connecting the actors with the use cases with which they interact. See Figure 6-13 for the use-case diagram that we have created.

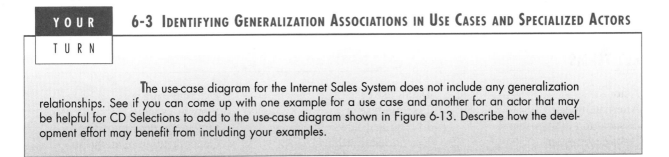

YOUR TURN

6-3 IDENTIFYING GENERALIZATION ASSOCIATIONS IN USE CASES AND SPECIALIZED ACTORS

The use-case diagram for the Internet Sales System does not include any generalization relationships. See if you can come up with one example for a use case and another for an actor that may be helpful for CD Selections to add to the use-case diagram shown in Figure 6-13. Describe how the development effort may benefit from including your examples.

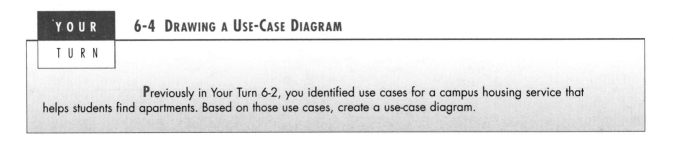

Previously in Your Turn 6-2, you identified use cases for a campus housing service that helps students find apartments. Based on those use cases, create a use-case diagram.

SUMMARY

Use-Case Descriptions

Use cases are the primary method of documenting requirements for object-oriented systems. They represent a functional view of the system being developed. There are overview and detail use cases. Overview use cases consist of the use-case name, ID number, primary actor, type, and brief description. Detailed use cases extend the overview use case with the identification and description of the stakeholders and their interest, the trigger and its type, the relationships in which the use case participates (association, extend, generalization, and include), the normal flow of events, the subflows, and any alternate or exception flows to the normal flow of events. There are 7 guidelines (see Figure 6-2) and 13 steps (see Figure 6-7) to writing effective use cases. The 13 steps can be summed up into 3 aggregate groups of steps: identify the major use cases (Steps 1–5), expand the major use cases (Steps 6–11), and confirm the major use cases (Steps 12–13).

Use-Case Diagrams

Use-case diagrams are simply a graphical overview of the set of use cases contained in the system. They simply illustrate the main functionality of a system and the actors that interact with the system. The diagram includes actors, which are people or things that derive value from the system, and use cases that represent the functionality of the system. The actors and use cases are separated by a system boundary and connected by lines representing associations. At times, actors are specialized versions of more general actors. Similarly, use cases can extend or use other use cases. Creating use-case diagrams is a four-step process (see Figure 6-7) whereby the analyst draws the system boundary, adds the use cases to the diagram, identifies the actors, and finally adds appropriate associations to connect use cases and actors. This process is simple since the written description of the use cases is created first.

KEY TERMS

Actor
Alternate flows
Association relationship
Brief description

Detail use case
Essential use case
Exceptional flows
Extend relationship

External trigger
Flow of events
Functional decomposition
Generalization relationship

Importance level
Include relationship
Inheritance
Iterate
Logical model
Normal flow of events
Overview use case
Packages
Physical model
Primary actor

Real use case
Relationships
Role
Scenario
Specialized actor
Stakeholders
Standard processes
Subflows
SVDPI
System boundary

Temporal trigger
Trigger
Use case
Use-case description
Use-case diagram
Use-case ID number
Use-case name
Use-case type

QUESTIONS

1. How is use-case diagramming related to functional modeling?
2. Explain the following terms. Use layperson's language as though you were describing them to a user: (a) actor; (b) use case; (c) system boundary; (d) relationship.
3. Every association must be connected to at least one _____ and one _____. Why?
4. What is CRUD? Why is this useful?
5. How does a detail use case differ from an overview use case?
6. How does an essential use case differ from a real use case?
7. What are the major elements of an overview use case?
8. What are the major elements of a detail use case?
9. Describe how to create use cases.
10. Why do we strive to have about three to nine major use cases in a business process?
11. Describe how to create use-case diagrams.

12. What are some heuristics for creating a use-case diagram?
13. Why is iteration important in creating use case?
14. What is the viewpoint of a use case, and why is it important?
15. What are some guidelines for designing a set of use cases? Give two examples of the extend associations on a use-case diagram. Give two examples for the uses association.
16. Which of the following could be an actor found on a use-case diagram? Why?

   Ms. Mary Smith
   Supplier
   Customer
   Internet customer
   Mr. John Seals
   Data entry clerk
   Database administrator

EXERCISES

A. Investigate the Web site for Rational Software (www.rational.com) and its repository of information about UML. Write a paragraph news brief on the current state of UML (e.g., the current version and when it will be released; future improvements; etc.).

B. Investigate the Object Management Group. Write a brief memo describing what it is, its purpose, and its influence on UML and the object approach to systems development. (*Hint:* A good resource is: www.omg.org)

C. Create a set of detail use-case descriptions for the process of buying glasses from the viewpoint of the patient, but do not bother to identify the flow of events within each use case. The first step is to see an eye doctor who will give you a prescription. Once you have a prescription, you go to a glasses store, where you select your frames and place the order for your glasses. Once the glasses have been made, you return to the store for a fitting and pay for the glasses.

D. Draw a use-case diagram for the process of buying glasses in Exercise C.

E. Create a set of detail use-case descriptions for the following dentist office system, but do not bother to identify the flow of events within each use case. Whenever new patients are seen for the first time, they complete a patient information form that asks their name, address, phone number, and brief medical history, which are stored in the patient information file. When a patient calls to schedule a new appointment or change an existing appointment, the receptionist checks the appointment file for an available time. Once a good time is found for the patient, the appointment is scheduled. If the patient is a new patient, an incomplete entry is made in the patient file; the full information will be collected when they arrive for their appointment. Because appointments are often made so far in advance, the receptionist usually mails a reminder postcard to each patient two weeks before their appointment.

F. Draw a use-case diagram for the dentist office system in Exercise E.

G. Complete the detail use-case descriptions for the dentist office system in Exercise E by identifying the normal flow of events, subflows, and alternate/exceptional flows within the use cases.

H. Create a set of detail use-case descriptions for an on-line university registration system. The system should enable the staff of each academic department to examine the courses offered by their department, add and remove courses, and change the information about them (e.g., the maximum number of students permitted). It should permit students to examine currently available courses, add and drop courses to and from their schedules, and examine the courses for which they are enrolled. Department staff should be able to print a variety of reports about the courses and the students enrolled in them. The system should ensure that no student takes too many courses and that students who have

any unpaid fees are not permitted to register (assume that a fees file is maintained by the university's financial office which the registration system accesses but does not change).

I. Draw a use-case diagram for the on-line university registration system in Exercise H.

J. Create a set of detail use-case descriptions for the following system. A Real Estate Inc. (AREI) sells houses. People who want to sell their houses sign a contract with AREI and provide information on their house. This information is kept in a database by AREI, and a subset of this information is sent to the citywide multiple listing service used by all real estate agents. AREI works with two types of potential buyers. Some buyers have an interest in one specific house. In this case, AREI prints information from its database, which the real estate agent uses to help show the house to the buyer (a process beyond the scope of the system to be modeled). Other buyers seek AREI's advice in finding a house that meets their needs. In this case, the buyer completes a buyer information form which is entered into a buyer database, and AREI real estate agents use its information to search AREI's database and the multiple listing service for houses that meet their needs. The results of these searches are printed and used to help the real estate agent show houses to the buyer.

K. Draw a use-case diagram for the real estate system in Exercise J.

L. Create a set of detail use-case descriptions for the following system. A Video Store (AVS) runs a series of fairly standard video stores. Before a video can be put on the shelf, it must be catalogued and entered into the video database. Every customer must have a valid AVS customer card in order to rent a video. Customers rent videos for three days at a time. Every time a customer rents a video, the system must ensure that they do not have any overdue videos. If so, the overdue videos must be returned, and an overdue fee must be paid before the customer can rent more videos. Likewise, if the customer has returned overdue videos but has not paid the overdue fee, the fee must be paid before new videos can be rented. Every morning, the store manager prints a report that lists overdue videos. If a video is two or more days overdue, the manager calls the customer as a reminder, to return the video. If a video is returned in damaged condition, the manager removes it from the video database and may sometimes charge the customer.

M. Draw a use-case diagram for the video system in Exercise J.

N. Create a set of detail use-case descriptions for a health club membership system. When members join the health club, they pay a fee for a certain length of time. Most memberships are for one year, but memberships as short as two months are available. Throughout the year, the health club offers a variety of discounts on its regular membership prices (e.g., two memberships for the price of one for Valentine's day). It is common for members to pay different amounts for the same length of membership. The club wants to mail out reminder letters to members asking them to renew their memberships one month before their memberships expire. Some members have become angry when asked to renew at a much higher rate than their original membership contract, so the club wants to track the price paid so that the manager can override the regular prices with special prices when members are asked to renew. The system must track these new prices so that renewals can be processed accurately. One of the problems in the health club industry is the high turnover rate of members. While some members remain active for many years, about half of the members do not renew their memberships. This is a major problem because the health club spends a lot in advertising to attract each new member. The manager wants the system to track each time a member comes into the club. The system will then identify the heavy users and generate a report so that the manager can ask them to renew their memberships early, perhaps offering them a reduced rate for early renewal. Likewise, the system should identify members who have not visited the club in more than a month, so the manager can call them and attempt to re-interest them in the club.

O. Draw a use-case diagram for the system in Exercise N.

P. Create a set of detail use-case descriptions for the following system. Picnics R Us (PRU) is a small catering firm with five employees. During a typical summer weekend, PRU caters 15 picnics with 20 to 50 people each. The business has grown rapidly over the past year, and the owner wants to install a new computer system for managing the ordering and buying process. PRU has a set of 10 standard menus. When potential customers call, the receptionist describes the menus to them. If the customer decides to book a picnic, the receptionist records the customer information (e.g., name, address, phone number, etc.) and the information about the picnic (e.g., place, date, time, which one of the standard menus, total price) on a contract. The customer is then faxed a copy of the contract and must sign and return it along with a deposit (often a credit card or by check) before the picnic is officially booked. The remaining money is collected when the picnic is delivered. Sometimes, the customer wants something special (e.g., birthday cake). In this case, the receptionist takes the information and gives it to the owner who determines the cost; the receptionist then calls the customer back with the price information. Sometimes the customer accepts the price; other times, the customer requests some changes that have to go back to the owner for a new cost estimate. Each week, the owner looks through the picnics scheduled for that weekend and orders the supplies (e.g., plates) and food (e.g., bread, chicken) needed to make them. The owner would like to use the system for marketing as well. It should be able to track how customers learned about PRU and identify repeat customers, so that PRU can mail special offers to them. The owner also wants to track the picnics on which PRU sent a contract, but the customer never signed the contract and actually booked a picnic.

Q. Draw a use-case diagram for the system in Exercise P.

R. Create a set of detail use-case descriptions for the following system. Of-the-Month Club (OTMC) is an innovative young firm that sells memberships to people who have an interest in certain products. People pay membership fees for one year and each month receive a product by mail. For example, OTMC has a coffee-of-the-month club that sends members one pound of special coffee each month. OTMC currently has six memberships (coffee, wine, beer, cigars, flowers, and computer games), each of which costs a different amount. Customers usually belong to just one, but some belong to two or more. When people join OTMC, the telephone operator records the name, mailing address, phone number, e-mail address, credit card information, start date, and membership service(s) (e.g., coffee). Some customers request a double or triple membership (e.g., two pounds of coffee, three cases of beer). The computer game membership operates a bit differently from the others. In this case, the member must also select the type of game (action, arcade, fantasy/science-fiction, educational, etc.) and age level. OTMC is planning to greatly expand the number of memberships it offers (e.g., video games, movies, toys, cheese, fruit, and vegetables), so the system needs to accommodate this future expansion. OTMC is also planning to offer 3-month and 6-month memberships.

S. Draw a use-case diagram for the system in Exercise R.

T. Create a set of detail use-case descriptions for a university library borrowing system (do not worry about catalogue searching, etc.). The system will record the books owned by the library and will record who has borrowed what books. Before people can borrow a book, they must show a valid id card, which is checked to ensure that it is still valid against the student database maintained by the registrar's office (for student borrowers), the faculty/staff database maintained by the personnel office (for faculty/staff borrowers), or against the library's own guest database (for individuals issued a "guest" card by the library). The system must also check to ensure the borrower does not have any overdue books or unpaid fines before he or she can borrow another book. Every Monday, the library prints and mails postcards to those people with overdue books. If a book is overdue by more than two weeks, a fine will be imposed and a librarian will telephone the borrower to remind him or her to return the book(s). Sometimes books are lost or are returned in damaged condition. The manager must then remove them from the database and will sometimes impose a fine on the borrower.

U. Draw a use-case diagram for the system in Exercise T.

MINICASES

1. Williams Specialty Company is a small printing and engraving organization. When Pat Williams, the owner, brought computers into the business office five years ago, the business was very small and very simple. Pat was able to utilize an inexpensive PC-based accounting system to handle the basic information processing needs of the firm. As time has gone on, however, the business has grown and the work being performed has become significantly more complex. The simple accounting software still in use is no longer adequate to keep track of many of the company's sophisticated deals and arrangements with its customers.

   Pat has a staff of four people in the business office who are familiar with the intricacies of the company's record-keeping requirements. Pat recently met with her staff to discuss her plan to hire an IS consulting firm to evaluate their information system needs and recommend a strategy for upgrading their computer system. The staff is excited about the prospect of a new system, since the current system causes them much aggravation. No one on the staff has ever done anything like this before, however, and they are a little wary of the consultants who will be conducting the project.

   Assume that you are a systems analyst on the consulting team assigned to the Williams Specialty Co. engagement. At your first meeting with the Williams staff, you want to be sure that they understand the work that your team will be performing, and how they will participate in that work.

   a. Explain in clear, nontechnical terms, the goals of the Analysis phase of the project.

   b. Explain in clear, nontechnical terms, how use cases and use case diagram will be used by the project team. Explain what these models are, what they represent in the system, and how they will be used by the team.

2. Professional and Scientific Staff Management (PSSM) is a unique type of temporary staffing agency. Many organizations today hire highly skilled, technical employees on a short-term, temporary basis, to assist with special projects or to provide a needed technical skill. PSSM negotiates contracts with its client companies in which it agrees to provide temporary staff in specific job categories for a specified cost. For example, PSSM has a contract with an oil and gas exploration company in which it agrees to supply geologists with at least a master's degree for $5,000 per week. PSSM has contracts with a wide range of companies, and can place almost any type of professional or scientific staff members, from computer programmers to geologists to astrophysicists.

   When a PSSM client company determines that it will need a temporary professional or scientific employee, it issues a staffing request against the contract it had previously negotiated with PSSM. When a staffing request is received by PSSM's contract manager, the contract number referenced on the staffing request is entered into the contract database. Using information from the database, the contract manager reviews the terms and conditions of the contract and determines whether the staffing request is valid. The staffing request is valid if the contract has not expired, the type of professional or scientific employee requested is listed on the original contract, and the requested fee falls within the negotiated fee range. If the staffing request is not valid, the contract manager sends the staffing request back to the client with a letter stating why the staffing request cannot be filled, and a copy of the letter is filed. If the staffing request is valid, the contract manager enters the staffing request into the staffing request database as an outstanding staffing request. The staffing request is then sent to the PSSM placement department.

In the placement department, the type of staff member, experience, and qualifications requested on the staffing request are checked against the database of available professional and scientific staff. If a qualified individual is found, he/she is marked "reserved" in the staff database. If a qualified individual cannot be found in the database, or is not immediately available, the

placement department creates a memo that explains the inability to meet the staffing request, and attaches it to the staffing request. All staffing requests are then sent to the arrangements department.

In the arrangements department the prospective temporary employee is contacted and asked to agree to the placement. After the placement details have been worked out and agreed to, the staff member is marked "placed" in the staff database. A copy of the staffing request and a bill for the placement fee is sent to the client. Finally, the staffing request, the "unable to fill" memo (if any), and a copy of the placement fee bill is sent to the contract manager. If the staffing request was filled, the contract manager closes the open staffing request in the staffing request database. If the staffing request could not be filled the client is notified. The staffing request, placement fee bill, and "unable to fill" memo are then filed in the contract office.

a. Develop a use case for each of the four major scenarios described above.

b. Create a use-case diagram for the system described above.

PLANNING

ANALYSIS

- ☑ **Develop Analysis Plan**
- ☑ **Understand As-Is System**
- ☑ **Identify Improvement Opportunities**
- ☑ **Develop the To-Be System Concept**
- ☑ **Develop Use Cases**
- ☐ Develop Structural Model
- ☐ Develop Behavioral Model

TASK CHECKLIST

PLANNING → **ANALYSIS** → **DESIGN** →

CHAPTER 7
STRUCTURAL MODELING

A structural, or conceptual, model describes the structure of the data that supports the business processes in an organization. During the analysis phase, the structural model presents the logical organization of data without indicating how the data are stored, created, or manipulated so that analysts can focus on the business without being distracted by technical details. Later, during the design phase, the structural model is updated to reflect exactly how the data will be stored in databases and files. This chapter describes Class-Responsibility-Collaboration (CRC) cards, class diagrams, and object diagrams that are used to create the structural model.

OBJECTIVES

- Understand the rules and style guidelines for creating CRC cards, class diagrams, and object diagrams.
- Understand the processes used to create CRC cards, class diagrams, and object diagrams.
- Be able to create CRC cards, class diagrams, and object diagrams.
- Understand the relationship between the structural and use-case models.

CHAPTER OUTLINE

IMPLEMENTATION

INTRODUCTION

During the analysis phase, analysts create use-case models to represent how the business system will behave. At the same time, analysts need to understand the information that is used and created by the business system (e.g., customer information, order information). In this chapter, we discuss how the data underlying the behavior modeled in the use cases are organized and presented.

A *structural model* is a formal way of representing the data that are used and created by a business system. It illustrates people, places, or things about which information is captured, and it shows how they are related to each other. The structural model is drawn using an iterative process in which the model becomes more detailed and less "conceptual" over time. In the analysis phase, analysts draw a *conceptual model,* which shows the logical organization of data without indicating how the data are stored, created, or manipulated. Because this model is free from any implementation or technical details, the analysts can focus more easily on matching the model to the real business requirements of the system.

In the design phase, analysts evolve the conceptual structural model into a design model that reflects how the data will be organized in databases and files. At this point, the model is checked for data redundancy, and the analysts investigate ways to make the data easy to retrieve. The specifics of the design model will be discussed in detail in the design chapters.

In this chapter, we focus on creating a conceptual structural model of the data using CRC cards and class diagrams. Using these techniques, we can show all of the data components of a business system. We first describe structural models and their elements. Then, we focus on CRC cards, class diagrams, and object diagrams. Finally, we discuss how to create structural models using CRC cards and class diagrams and how the structural model relates to the use cases and use-case model that you learned about in Chapter 6.

STRUCTURAL MODELS

Every time a systems analyst encounters a new problem to solve, he or she must learn the underlying application domain. The goal of the analyst is to discover the key data contained in the application domain and to build a structural model of the data. Object-oriented modeling allows the analyst to reduce the "semantic gap" between the underlying application domain and the evolving structural model. However, the real world and the world of software are very different. The real world tends to be messy, whereas the world of software must be neat and logical. As such, an exact mapping between the structural model and the application domain may not be possible. In fact, it may not even be desirable.

One of the primary purposes of the structural model is to create a vocabulary that can be used by both the analyst and users. Structural models represent the things, ideas, or concepts, that is, the objects, contained in the application domain of the problem. They also allow the representation of the relationships between the things, ideas, or concepts. By creating a structural model of the application domain, the analyst creates the vocabulary necessary for the analyst and users to communicate effectively.

One important thing to remember is that, at this stage of development, the structural model does not represent software components or classes in an object-oriented programming language, even though the structural model does contain analysis classes, attributes, operations, and relationships among the analysis classes. The refinement of these initial classes into programming level objects comes later. Nonetheless, the structural model at this point should represent the responsibilities of each class and the collaborations between the classes. Typically, structural models are depicted using CRC cards, class diagrams, and, in some cases, object diagrams. However, before describing CRC cards, class diagrams, and object diagrams, we discuss the basic elements of structural models: classes, attributes, operations, and relationships.

Classes, Attributes, and Operations

As stated in Chapter 1, a *class* is a general template that we use to create specific *instances,* or *objects,* in the application domain. All objects of a given class are identical in structure and behavior but contain different data in their attributes. There are two different general kinds of classes of interest during the analysis phase: concrete and abstract. Normally, when analysts describe the application domain classes, they are referring to concrete classes; that is, *concrete classes* are used to create objects. *Abstract classes,* on the other hand, simply are useful abstractions. For example, from an employee class and a customer class, we may identify a generalization of the two classes and name the abstract class person. We may not actually instantiate the Person class in the system itself, but rather only create and use employees and customers. We will describe generalization in more detail with relationships later in this chapter.

A second classification of classes is the type of real-world thing that it represents. There are application domain classes, user interface classes, data structure classes, file structure classes, operating environment classes, document classes, and various types of multimedia classes. At this point in the development of our evolving system, we are interested only in application domain classes. Later in design and implementation, the other types of classes will become more relevant.

An *attribute* of an analysis class represents a piece of information that is relevant to the description of the class within the application domain of the problem being investigated. An attribute contains information that the analyst or user feels the system should store. For example, a possible relevant attribute of an employee class is employee name, while one that may not be as relevant is hair color. Both describe something about an employee, but hair color probably is not all that useful for most business applications. Only those attributes that are important to task should be included in the class. Finally, one should only add attributes that are primitive or atomic types, that is, integers, strings, doubles, date, time, boolean, and so on. Most complex or compound attributes are really placeholders for relationships between classes. Therefore, they should be modeled as relationships, not attributes.

An *operation,* or service, of an analysis class is where the *behavior* of the class will be defined. In later phases, the operations will be converted to methods (see Chapter 1). However, since methods are more related to implementation, at this point in the development we use the term *operation* to describe the actions to which the instances of the class will be capable of responding. Like attributes, only problem domain-specific operations that are relevant to the problem being investigated should be considered. For example, classes are normally required to provide the means to create instances, delete instances, access individual attribute values, set individual attribute values, access individual relationship values, and remove individual relationship values. However, at this point in the development of the evolving system, the analyst should avoid cluttering up the definition of the class with these basic types of operations and only focus on relevant problem domain-specific operations.

Relationships

Many different types of relationships can be defined, but all can be classified into three basic categories of data abstraction mechanisms: *generalization relationships, aggregation relationships,* and *association relationships.* These data abstraction mechanisms allow the analyst to focus on the important dimensions while ignoring nonessential dimensions. Like attributes, the analyst must be careful to include only relevant relationships.

Generalization Relationships   The generalization abstraction enables the analyst to create classes that inherit attributes and operations of other classes. The analyst creates a *superclass* that contains the basic attributes and operations that will be used in several *subclasses.* The subclasses inherit the attributes and operations of their superclass and can also contain attributes and operations that are unique just to them. For example, a customer class and an employee class can be generalized into a person class by extracting the attributes and operations they have in common and placing them into the new superclass, Person. In this way, the analyst can reduce the redundancy in the class definitions so that the common elements are defined once and then reused in the subclasses. Generalization is represented with the *a-kind-of* relationship, so that we say that an employee is a-kind-of person.

The analyst also can use the flip side of generalization, specialization, to uncover additional classes by allowing new subclasses to be created from an existing class. For example, an employee class can be specialized into a secretary class and an engineer class. Furthermore, generalization relationships between classes can be combined to form generalization hierarchies. Based on the previous examples, a secretary class and an engineer class can be subclasses of an employee class, which in turn could be a subclass of a person class. This would be read as, "a secretary and an engineer are a-kind-of employee and a customer and an employee are a-kind-of person."

The generalization data abstraction is a very powerful mechanism that encourages the analyst to focus on the properties that make each class unique by allowing the similarities to be factored into superclasses. However, to ensure that the semantics of the subclasses are maintained, the analyst should apply the principle of *substitutability.* By this we mean that the subclass should be capable of substituting for the superclass anywhere that uses the superclass (e.g., anywhere we use the employee superclass, we could also logically use its secretary subclass). By focusing on the a-kind-of interpretation of the generalization relationship, the principle of substitutability will be applied.

Aggregation Relationships   Many different types of aggregation or composition relationships have been proposed in data modeling, knowledge representation, and linguistics. For example, a-part-of (logically or physically), a-member-of (as in set membership), contained-in, related-to, and associated-with. However, generally speaking, all aggregation relationships relate *parts* to *wholes* or *parts* to *assemblies.* For our purposes, we use the *a-part-of* or *has-parts* semantic relationship to represent the aggregation abstraction. For example, a door is a-part-of a car, an employee is a-part-of a department, or a department is a-part-of an organization. Like the generalization relationship, aggregation relationships can be combined into aggregation hierarchies. For example, a piston is a-part-of an engine, while an engine is a-part-of a car.

Aggregation relationships are bidirectional in nature. The flip side of aggregation is *decomposition.* The analyst can use decomposition to uncover parts of a class that should be modeled separately. For example, if a door and engine are a-part-of a car, then a car has-parts door and engine. The analyst can bounce around between the various parts to uncover new parts. For example, the analyst can ask, "What other parts are there to a car?" or "To which other assemblies can a door belong?"

Association Relationships   There are other types of relationships that do not fit neatly into a generalization (a-kind-of) or aggregation (a-part-of) framework. Technically speaking, these relationships are usually a weaker form of the aggregation relationship. For example, a patient schedules an appointment. It could be argued that a patient is a-part-of an appointment. However, there is a clear semantic difference between this type of relationship and one that models the relationship between doors and cars or even workers and unions. As such, they are simply considered to be *associations* between instances of classes.

CLASS-RESPONSIBILITY-COLLABORATION CARDS

Class-Responsibility-Collaboration cards, most commonly called *CRC cards,* are used to document the responsibilities and collaborations of a class. We use an extended form of the CRC card to capture all relevant information associated with a class.[1] We describe the elements of our CRC cards below after we explain responsibilities and collaborations.

Responsibilities and Collaborations

The *responsibilities* of a class can be divided into two separate types: knowing and doing. *Knowing responsibilities* are those things that an instance of a class must be capable of knowing. An instance of a class typically knows the values of its attributes and its relationships. *Doing responsibilities* are those things that an instance of a class must be capable of doing. In this case, an instance of a class can execute its operations, or it can request a second instance that it knows about to execute one of its operations on behalf of the first instance.

The structural model describes the data necessary to support the business processes modeled by the use cases. Most use cases involve a set of several classes, not just one class. This set of classes forms a *collaboration.* Collaborations allow the

---

[1] Our CRC cards are based on the work of D. Bellin and S.S. Simone, *The CRC Card Book* (Reading, MA: Addison-Wesley, 1997), I. Graham, *Migrating to Object Technology* (Wokingham, England: Addison-Wesley, 1995), and B. Henderson-Sellers and B. Unhelkar *OPEN Modeling with UML* (Harlow, England: Addison-Wesley, 2000).

analyst to think in terms of clients, servers, and contracts.[2] A *client* object is an instance of a class that sends a request to an instance of another class for an operation to be executed. A *server* object is the instance that receives the request from the client object. A *contract* formalizes the interactions between the client and server objects. For example, a patient makes an appointment with a doctor. This sets up an obligation for both the patient and doctor to appear at the appointed time. Otherwise, consequences, such as billing the patient for the appointment regardless of whether the patient appears, can be dealt out. The contract should also spell out what the benefits of the contract will be, such as a treatment being prescribed for whatever ails the patient and a payment to the doctor for the services provided. Chapter 14 provides a more detailed explanation of contracts and examples of their use.

The analyst can use the idea of class responsibilities and client-server-contract collaborations to help identify the classes, along with the attributes, operations, and relationships, involved with a use case. One of the easiest ways in which to use CRC cards in developing a structural model is through anthropomorphism—pretending that the classes have human characteristics. This is done by having the analyst and/or the user role-play and pretend that they are an instance of the class being considered. They then can either ask questions of themselves or be asked questions by other members of the development team. For example,

> Who or what are you?
> What do you know?
> What can you do?

The answers to the questions are then used to add detail to the evolving CRC cards.

Elements of a CRC Card

The set of CRC cards contains all the information necessary to build a logical structural model of the problem under investigation. Figure 7-1 shows a sample CRC card. Each card captures and describes the essential elements of a class. The front of the card allows the recording of the class's name, ID, type, description, list of associated use cases, responsibilities, and collaborators. The name of a class should be a noun (but not a proper noun such as the name of a specific person or thing). Just like the use cases, in later stages of development, it is important to be able to trace back design decisions to specific requirements. In conjunction with the list of associated use cases, the ID number for each class can be used to accomplish this. The description is simply a brief statement that can be used as a textual definition for the class. The responsibilities of the class tend to be the operations that the class must contain, that is, the doing responsibilities.

The back of a CRC card contains the attributes and relationships of the class. The attributes of the class represent the knowing responsibilities that each instance of the class will have to meet. Typically, the data type of each attribute is listed with the name of the attribute; for example, the amount attribute is double, and the insur-

---

[2] For more information, see K. Beck and W. Cunningham, "A Laboratory for Teaching Object-Oriented Thinking," *Proceedings of OOPSLA, SIGPLAN Notices* 24, no. 10 (1989): 1–6; Henderson-Sellers and Unhelkar, *OPEN Modeling with UML:* C. Larman, *Applying UML and Patterns: An Introduction to Object-Oriented Analysis and Design,* (Englewood Cliffs, NJ; Prentice-Hall, 1998); B. Meyer, *Object-Oriented Software Construction* (Englewood Cliffs, NJ: Prentice Hall, 1994); and R. Wirfs-Brock, B. Wilkerson, and L. Wiener, *Designing Object-Oriented Software* (Englewood Cliffs, NJ: Prentice Hall, 1990).

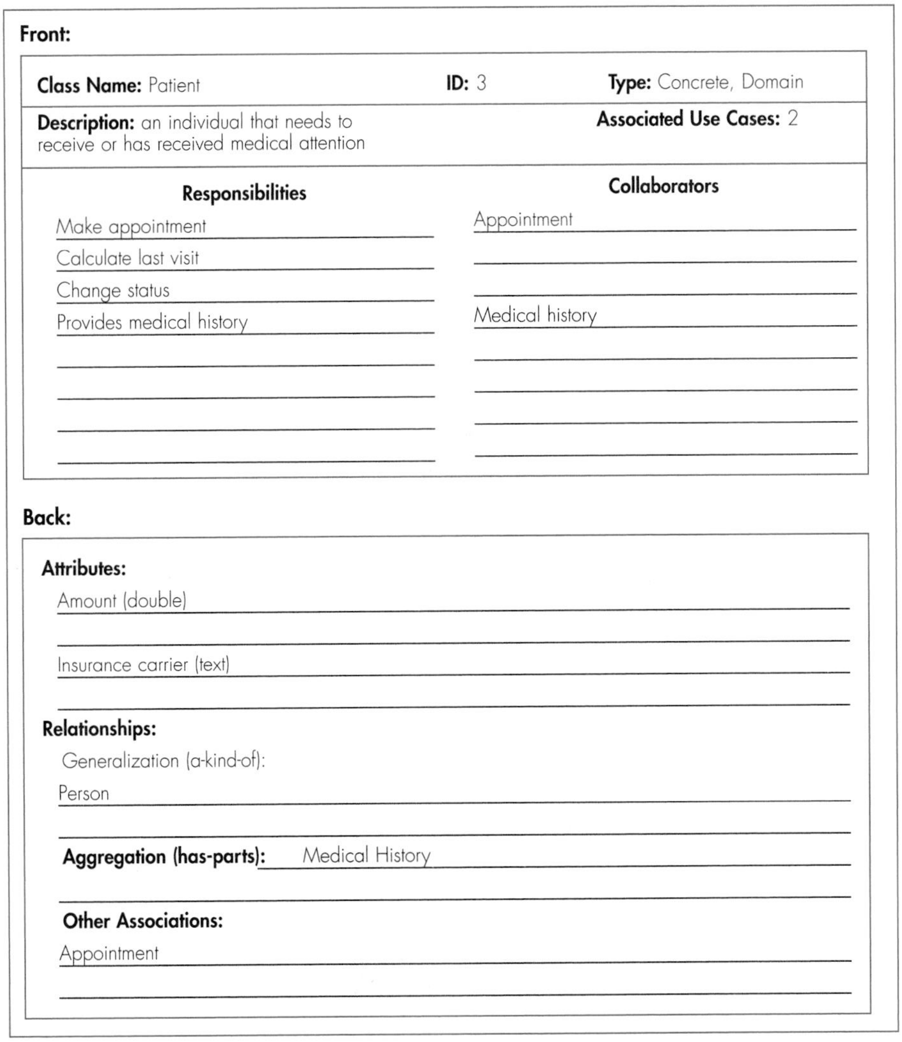

Front:

Class Name: Patient	ID: 3	Type: Concrete, Domain

Description: an individual that needs to receive or has received medical attention

Associated Use Cases: 2

Responsibilities	Collaborators
Make appointment	Appointment
Calculate last visit	
Change status	
Provides medical history	Medical history

Back:

Attributes:

Amount (double)

Insurance carrier (text)

Relationships:

Generalization (a-kind-of):

Person

Aggregation (has-parts):   Medical History

Other Associations:

Appointment

FIGURE 7-1
Example CRC Card

ance carrier is text. Three types of relationships are typically captured at this point in time: generalization, aggregation, and other associations. In Figure 7-1, we see that a Patient is a-kind-of Person and that a Patient is associated with Appointments.

Again, CRC cards are used to document the essential properties of a class. However, once the cards are filled out, the analyst can use the cards and anthropomorphism in role-playing to uncover missing properties.

CLASS DIAGRAMS

The *class diagram* is a *static model* that shows the classes and the relationships among classes that remain constant in the system over time. The class diagram depicts classes, which include both behaviors and states, with the relationships between the classes. The following sections will first present the elements of the class diagram, followed by the way in which a class diagram is drawn.

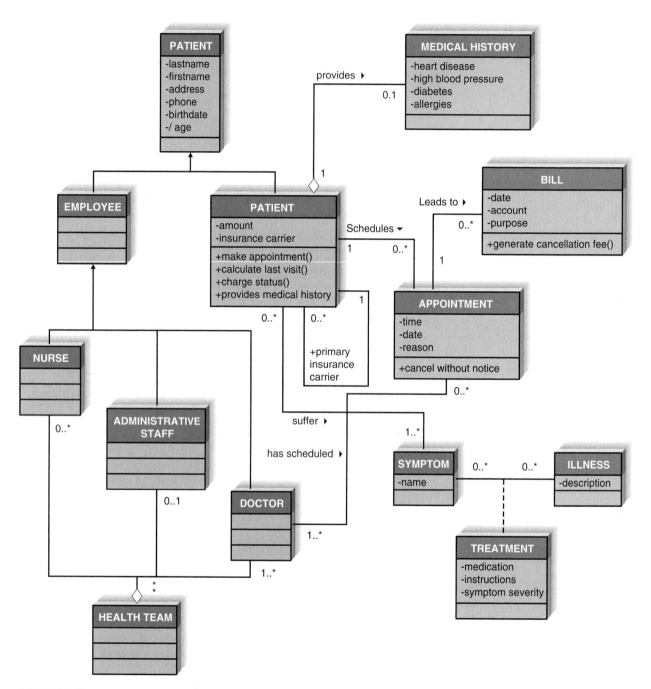

FIGURE 7-2
Example Class Diagram

Elements of a Class Diagram

Figure 7-2 shows a class diagram that was created to reflect the classes and relationships that are needed for the set of use cases that describe the appointment system in Chapter 6.

Class   The main building block of a class diagram is the *class,* which stores and manages information in the system (see Figure 7-3). During analysis, classes refer

A CLASS • Represents a kind of person, place, or thing about which the system will need to capture and store information • Has a name typed in bold and centered in its top compartment • Has a list of attributes in its middle compartment • Has a list of operations in its bottom compartment • Does not explicitly show operations that are available to all classes	<table><tr><td align="center">Class1</td></tr><tr><td>−attribute1</td></tr><tr><td>+operation1()</td></tr></table>
AN ATTRIBUTE • Represents properties that describe the state of an object • Can be derived from other attributes, shown by placing a slash before the attribute's name	attribute name/ derived attribute name
AN OPERATION • Represents the actions or functions that a class can perform • Can be classified as a constructor, query, or update operation • Includes parentheses that may contain parameters or information needed to perform the operation	operation name ()
AN ASSOCIATION • Represents a relationship between multiple classes, or a class and itself • Is labeled using a verb phrase or a role name, whichever better represents the relationship • Can exist between one or more classes • Contains multiplicity symbols, which represent the minimum and maximum times a class instance can be associated with the related class instance	1..* 0..1 verb phrase

FIGURE 7-3
Class Diagram Syntax

to the people, places, events, and things about which the system will capture information. Later, during design and implementation, classes can refer to implementation-specific artifacts like windows, forms, and other objects used to build the system. Each class is drawn using three-part rectangles with the class's name at the top, attributes in the middle, and operations at the bottom. You should be able to identify Person, Medical Staff, Doctor, Patient, Medical History, Appointment, Bill, Diagnosis, Health Problem, and Symptom as classes in Figure 7-2. The attributes of a class and their values define the state of each object that is created from the class, and the behavior is represented by the operations.

Attributes are properties of the class about which we want to capture information (see Figure 7-3). Notice that the Person class in Figure 7-2 contains the attributes lastname, firstname, address, phone, and birthdate. At times, you may want to store *derived attributes,* which are attributes that can be calculated or derived; these special attributes are denoted by placing a slash (/) before the attribute's name. Notice how the Person class contains a derived attribute called "/age," which can be derived by subtracting the patient's birthdate from the current date. It is also possible to show the *visibility* of the attribute on the diagram. *Visibility* relates to the level of information hiding (see Chapter 1) to be enforced for the attribute. The visibility of an attribute can be public (+), protected (#), or private (−). A *public* attribute is one that is not hidden from any other object. As such, other objects can modify its value. A *protected* attribute is one that is hidden from all other classes except its immediate subclasses. A *private* attribute is one that is hidden from all other classes. The default visibility for an attribute is normally private.

Operations are actions or functions that a class can perform (see Figure 7-3). The functions that are available to all classes (e.g., create a new instance, return a value for a particular attribute, set a value for a particular attribute, or delete an instance) are not explicitly shown within the class rectangle. Instead, only those operations that are unique to the class are included, such as the "cancel without notice" operation in the Appointment class and the "generate cancellation fee" operation in the Bill class in Figure 7-2. Notice that both of the operations are followed by parentheses that contain the parameter(s) needed by the operation. If an operation has no parameters, the parentheses are still shown but are empty. As with attributes, the visibility of an operation can be designated as public, protected, or private. The default visibility for an operation is normally public.

A class can contain three kinds of operations: constructor, query, and update. A *constructor operation* creates a new instance of a class. For example, the patient class may have a method called "insert ()" that creates a new patient instance as patients are entered into the system. As we just mentioned, if a method is available to all classes (e.g., create a new instance), it is not explicitly shown on the class diagram, so typically you will not see constructor methods explicitly on the class diagram.

A *query operation* makes information about the state of an object available to other objects, but it will not alter the object in any way. For instance, the "calculate last visit ()" operation that determines when a patient last visited the doctor's office will result in the object being accessed by the system, but it will not make any change to its information. If a query method merely asks for information from attributes in the class (e.g., a patient's name, address, or phone), then it is not shown on the diagram because we assume that all objects have operations that produce the values of their attributes.

An *update operation* will change the value of some or all of the object's attributes, which may result in a change in the object's state. Consider changing the status of a patient from new to current with a method called "change status ()", or associating a patient with a particular appointment with "schedule appointment (appointment)."

Relationships   A primary purpose of the class diagram is to show the relationships, or *associations,* that classes have with one another. These are depicted on the diagram by drawing lines between classes (see Figure 7-3). When multiple classes share a relationship (or a class shares a relationship with itself), a line is drawn and labeled with either the name of the relationship or the roles that the classes play in the relationship. For example, in Figure 7-2 the two classes Patient and Appointment are associated with one another whenever a patient schedules an appointment. Thus, a line labeled "schedules" connects patient and appointment, representing exactly how the two classes are related to each other. Also notice that there is a small solid triangle beside the name of the relationship. The triangle allows a direction to be associated with the name of the relationship. In Figure 7-2, the schedules relationship includes a triangle, indicating that the relationship is to be read as "patient schedules appointment." Inclusion of the triangle simply increases the readability of the diagram.

Sometimes a class is related to itself, as in the case of a patient being the primary insurance carrier for other patients (e.g., their spouse, children). In Figure 7-2, notice that a line was drawn between the Patient class and itself, and is called "primary insurance carrier" to depict the role that the class plays in the relationship. Notice that a plus "+" sign is placed before the label to communicate that it is a role

Exactly one	1	Department — 1 — Boss	A department has one and only one boss.
Zero or more	0..*	Employee — 0..* — Child	An employee has zero to many children.
One or more	1..*	Boss — 1..* — Employee	A boss is responsible for one or more employees.
Zero or one	0..1	Employee — 0..1 — Spouse	An employee can be married to zero or no spouse.
Specified range	2..4	Employee — 2..4 — Vacation	An employee can take between two to four vacations each year.
Multiple, disjoint ranges	1..3, 5	Employee — 1..3, 5 — Committee	An employee is a member of one to three or five committees.

FIGURE 7-4
Multiplicity

as opposed to the name of the relationship. When labeling an association, use either a relationship name or a role name (not both), whichever communicates a more thorough understanding of the model.

Relationships also have *multiplicity*, which documents how an instance of an object can be associated with other instances. Numbers are placed on the association path to denote the minimum and maximum instances that can be related through the association in the format *minimum number . . maximum number* (see Figure 7-4). The numbers specify the relationship from the class at the far end of the relationship line to the end with the number. For example, in Figure 7-2, there is a "0. .*" on the appointment end of the "patient schedules appointment" relationship. This means that a patient can be associated with zero through many different appointments. At the patient end of this same relationship there is a "1," meaning that an appointment must be associated with one and only one (1) patient.

There are times when a relationship itself has associated properties, especially when its classes share a many-to-many relationship. In these cases, a class is formed, called an *association class*, that has its own attributes and methods. It is shown as a rectangle attached by a dashed line to the association path, and the rectangle's name matches the label of the association. Think about the case of capturing information about illnesses and symptoms. An illness (e.g., the flu) can be associated with many symptoms (e.g., sore throat, fever), and a symptom (e.g., sore throat) can be associated with many illnesses (e.g., the flu, strep throat, the common cold). Figure 7-2 shows how an association class can capture information about remedies that change depending on the various combinations. For example, a sore throat caused by the strep bacterium will require antibiotics, whereas treatment for a sore throat from the flu or a cold could be throat lozenges or hot tea. Most often, classes are related through a "normal" association; however, two special cases of an association will appear quite often: generalization and aggregation.

Generalization and Aggregation Generalization shows that one class (subclass) inherits from another class (superclass), meaning that the properties and operations of the superclass are also valid for objects of the subclass. The generalization path is shown with a solid line from the subclass to the superclass and a hollow arrow pointing at the superclass. For example, Figure 7-2 tells us that doctors, nurses, and administrative personnel are all kinds of employees and that those employees and patients are kinds of persons. Remember that the generalization relationship occurs when you need to use words such as "is a-kind-of" to describe the relationship.

Aggregation is used when classes actually comprise other classes. For example, think about a doctor's office that has decided to create health care teams that include doctors, nurses, and administrative personnel. As patients enter the office, they are assigned to a health care team that cares for their needs during their visits. Figure 7-2 shows how this relationship is denoted on the class diagram. A diamond is placed nearest the class representing the aggregation (health care team), and lines are drawn from the arrow to connect the classes that serve as its parts (doctors, nurses, and administrative staff). Typically, you can identify these kinds of associations when you need to use words like "is a part of" or "is made up of" to describe the relationship.

Simplifying Class Diagrams

When a class diagram is fully populated with all of the classes and relationships for a real-world system, the class diagram can become very difficult to interpret, that is, very complex. When this occurs, it is sometimes necessary to simplify the diagram. Using a *view* mechanism can simplify the class diagram. Views were developed originally with relational database management systems and are simply a subset of the information contained in the database. In this case, the view would be a useful subset of the class diagram, such as a use-case view that shows only the classes and relationships that are relevant to a particular use case. A second view could be to show only a particular type of relationship: aggregation, association, or generalization. A third type of view is to restrict the information shown with each class, for example, only show the name of the class, the name and attributes, or the name and operations. These view mechanisms can be combined to further simplify the diagram.

A second approach to simplify a class diagram is through the use of *packages* (i.e., logical groups of classes). To make the diagrams easier to read and to keep the models at a reasonable level of complexity, you can group the classes together into packages. Packages are general constructs that can be applied to any of the elements in UML models. In Chapter 6, we introduced them to simplify use-case diagrams. In the case of class diagrams, it is simple to sort the classes into groups based on the relationships they share.[3]

Object Diagrams

Although class diagrams are necessary to document the structure of the classes, at times a second type of *static structure diagram,* called an *object diagram,* can be useful. An object diagram is essentially an instantiation of all or part of a class diagram. (*Instantiation* means to create an instance of the class with a set of appropriate attribute values.)

---

[3] For those familiar with structured analysis and design, packages serve a similar purpose as the leveling and balancing processes used in data flow diagramming. Packages and package diagrams are described in more detail in Chapter 9.

Object diagrams can be very useful when one is trying to uncover details of a class. Generally speaking, it is easier to think in terms of concrete objects (instances) rather than abstractions of objects (classes). For example, in Figure 7-5, a portion of the class diagram in Figure 7-2 has been copied and instantiated. The top part of the figure simply is a copy of a small view of the overall class diagram. The lower portion is the object diagram that instantiates that subset of classes. By reviewing the actual instances involved, John Doe, Appt1, and Dr. Smith, we may discover additional relevant attributes, relationships, and/or operations or possibly misplaced attributes, relationships, and/or operations. For example, an appointment has a reason attribute. Upon closer examination, the reason attribute may have been better modeled as an association with the Symptom class. Currently, the Symptom class is associated with the Patient class. After reviewing the object diagram, this seems to be in error. As such, we should modify the class diagram to reflect this new understanding of the problem.

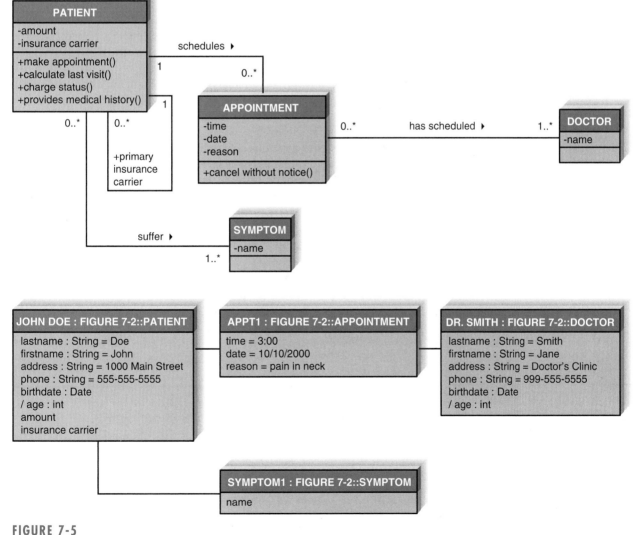

FIGURE 7-5
Example Object Diagram

CREATING CRC CARDS AND CLASS DIAGRAMS

Creating a structural model is an iterative process whereby the analyst makes a roughcut of the model and then refines it over time. Structural models can become quite complex. In fact, some systems have class diagrams that contain hundreds of classes. In this section, we take you through one iteration of creating a structural model, but the model can be expected to change dramatically as you communicate with users and fine-tune your understanding of the system. First, however, we present a set of approaches to identify and refine a set of candidate objects, and provide a set of steps based on these approaches that can be used to create the initial roughcut structural model.

Object Identification

Many different approaches have been suggested to aid the analyst in identifying a set of candidate objects for the structural model. The three most common approaches are textual analysis, common object lists, and patterns. Most analysts use a combination of all three techniques to make sure that no important objects and object attributes, operations, and relationships have been overlooked.

Textual Analysis   *Textual analysis* is an analysis of the text in the use-case descriptions. The analyst starts by reviewing the use-case descriptions and the use-case diagrams. The text in the descriptions is examined to identify potential objects, attributes, operations, and relationships. The nouns in the use case suggest possible classes, whereas the verbs suggest possible operations. Figure 7-6 presents a summary of guidelines we have found useful. The textual analysis of use-case descriptions has been criticized as being too simple, but because its primary purpose is to create an initial roughcut structural model, its simplicity is a major advantage.

Common Object List   As its name implies, a *common object list* is simply a list of objects that are common to the business domain of the system. In addition to look-

---

- A common or improper noun implies a class of objects.
- A proper noun or direct reference implies an instance of a class.
- A collective noun implies a class of objects made up of groups of instances of another class.
- An adjective implies an attribute of an object.
- A doing verb implies an operation.
- A being verb implies a classification relationship between an object and its class.
- A having verb implies an aggregation or association relationship.
- A transitive verb implies an operation.
- An intransitive verb implies an exception.
- A predicate or descriptive verb phrase implies an operation.
- An adverb implies an attribute of a relationship or an operation.

Source: These guidelines are based on Russell J. Abbott, "Program Design by Informal English Descriptions," *Communications of the ACM* 26, no. 11 (1983): 882–894; Peter P-S Chen, "English Sentence Structure and Entity-Relationship Diagrams," *Information Sciences: An International Journal* 29, no. 2–3 (1983): 127–149; and Graham, *Migrating to Object Technology.*

FIGURE 7-6
Textual Analysis Guidelines

ing at the specific use cases, the analyst also thinks about the business separately from the use cases. Several categories of objects have been found to help the analyst in creation of the list, such as physical or tangible things, incidents, roles, and interactions.[4] Analysts should first look for physical or *tangible things* in the business domain. These could include books, desks, chairs, and office equipment. Normally, these types of objects are the easiest to identify. *Incidents* are events that occur in the business domain, such as meetings, flights, performances, or accidents. Reviewing the use cases can readily identify the *roles* that the people "play" in the problem, such as doctor, nurse, patient, or receptionist. Typically, an *interaction* is a transaction that takes place in the business domain, such as a sales transaction. Other types of objects that can be identified include places, containers, organizations, business records, catalogs, and policies. In rare cases, processes themselves may need information stored about them. In these cases, processes may need an object, in addition to a use case, to represent them.

Patterns

The idea of using patterns is relatively new in object-oriented systems development.[5] There have been many definitions of exactly what a pattern is. From our perspective, a *pattern* is simply a useful group of collaborating classes that provide a solution to a commonly occurring problem. Since patterns provide a solution to commonly occurring problems, they are reusable.

An architect, Christopher Alexander, has inspired much of the work associated with using patterns in object-oriented systems development. According to Alexander and his colleagues,[6] very sophisticated buildings can be constructed by stringing together commonly found patterns, rather than creating entirely new concepts and designs. In a very similar manner, it is possible to put together commonly found object-oriented patterns to form elegant object-oriented information systems. For example, many business transactions involve the same type of objects and interactions. Virtually all transactions would require a transaction class, a transaction line item class, an item class, a location class, and a participant class. By simply reusing these existing patterns of classes, we can define the system more quickly and more completely than by starting with a blank piece of paper.

Many different types of patterns have been proposed, ranging from high-level business-oriented patterns to more low-level design patterns. Figure 7-7 lists some common business domains for which patterns have been developed, as well as their source. If you are developing a business information system in one of these business domains, then the patterns developed for that domain may be a very useful starting point in identifying needed classes and their attributes, operations, and relationships.

---

[4] For example, see Larman, *Applying UML and Patterns,* and Sally Shlaer and Stephen J. Mellor, *Object-Oriented Systems Analysis: Modeling the World in Data* (Englewood Cliffs, NJ: Yourdon Press, 1988).

[5] Many books have been devoted to this topic: for example, see Peter Coad, David North, and Mark Mayfield, *Object Models: Strategies, Patterns, & Applications,* 2nd ed., (Englewood Cliffs, NJ: Prentice Hall, 1997); Hans-Erik Eriksson and Magnus Penker, *Business Modeling with UML: Business Patterns at Work,* (New York: John Wiley & Sons, 2000); Martin Fowler, *Analysis Patterns: Reusable Object Models,* (Reading, MA: Addison-Wesley, 1997); Erich Gamma, Richard Helm, Ralph Johnson, and John Vlissides, *Design Patterns: Elements of Reusable Object-Oriented Software,* (Reading, MA: Addison-Wesley, 1995); and David C. Hay, *Data Model Patterns: Conventions of Thought* (New York, NY: Dorset House, 1996).

[6] Christopher Alexander, Sara Ishikawa, Murray Silverstein, Max Jacobson, Ingrid Fiksdahl-King, and Shlomo Angel, *A Pattern Language* (New York: Oxford University Press, 1977).

Business Domains	Sources of Patterns
Accounting	3, 4
Actor-Role	2
Assembly-Part	1
Container-Content	1
Contract	2, 4
Document	2, 4
Employment	2, 4
Financial Derivative Contracts	3
Geographic Location	2, 4
Group-Member	1
Interaction	1
Material Requirements Planning	4
Organization and Party	2, 3
Plan	1, 3
Process Manufacturing	4
Trading	3
Transactions	1, 4

1. Peter Coad, David North, and Mark Mayfield, *Object Models: Strategies, Patterns, & Applications*, 2nd ed., (Englewood Cliffs, NJ: Prentice Hall, 1997).

2. Hans-Erik Eriksson and Magnus Penker, *Business Modeling with UML: Business Patterns at Work*, (New York: John Wiley & Sons, 2000).

3. Martin Fowler, *Analysis Patterns: Reusable Object Models* (Reading, MA: Addison-Wesley, 1997).

4. David C. Hay, *Data Model Patterns: Conventions of Thought* (New York, NY, Dorset House, 1996).

FIGURE 7-7
Useful Patterns

Building CRC Cards and Class Diagrams

It is important to remember that although CRC cards and class diagrams can be used to describe both the As-Is and To-Be structural model of the evolving system, they are most often used for the To-Be model. There are many different ways to identify a set of candidate objects and to create CRC cards and class diagrams. Today most object identification begins with the use cases identified for the problem (see Chapter 6). In this section, we describe a seven-step process used to create the structural model of the problem (see Figure 7-8).

Create CRC Cards The first step is to create CRC cards by performing textual analysis on the use cases. If you recall, the normal flow of events, subflows, and alternate/exceptional flows written as part of the use case were written in a special form called **S**ubject **V**erb **D**irect object **P**reposition **I**ndirect object *(SVDPI)*. By writing use-case events in this form, it is easier to use the guidelines for use-case analysis in Figure 7-6 to identify the objects. Reviewing the primary actors, stakeholders, and interests, and giving a brief description of each use case allow additional candidate objects to be identified. Furthermore, it is useful to go back and

1. Create CRC cards by performing textual analysis on the use cases.
2. Brainstorm additional candidate classes, attributes, operations, and relationships by using the common object list approach.
3. Role-play each use case using the CRC cards.
4. Create the class diagram based on CRC cards.
5. Review the structural model for missing and/or unnecessary classes, attributes, operations, and relationships.
6. Incorporate useful patterns.
7. Review the structural model.

FIGURE 7-8
Steps for Object Identification and Structural Modeling

review the original requirements to look for information that was not included in the text of the use cases. Record all information uncovered for each candidate object on a CRC card.

Examine Common Object Lists   The second step is to brainstorm additional candidate objects, attributes, operations, and relationships using the common object list approach. What are the tangible things associated with the problem? What are the roles played by the people in the problem domain? What incidents and interactions take place in the problem domain? As you can readily see, by beginning with the use cases, many of these questions already have partial answers to them. For example, the primary actors and stakeholders are the roles that are played by the people in the problem domain. However, it is possible to uncover additional roles that were not thought of previously. This obviously would cause the use cases and the use-case model to be modified and possibly expanded. As in the previous step, be sure to record all information uncovered onto the CRC cards. This includes any modifications uncovered for any previously identified candidate objects and any information regarding any new candidate objects identified.

Role-Play the CRC Cards   The third step is to role-play each use case using the CRC cards.[7] Each CRC Card should be assigned to an individual, who will perform the operations for the class on the CRC card. As the "performers" play out their roles, the "system" will tend to break down. When this occurs, additional objects, attributes, operations, or relationships will be identified. Again, like the previous steps, anytime any new information is discovered, new CRC cards are created or modifications to existing CRC cards are made.

Create the Class Diagram   The fourth step is to create the class diagram based on the CRC cards. This is equivalent to creating the use-case diagram from the use cases. Information contained on the CRC cards is simply transferred to the class diagrams. The responsibilities are transferred as operations, and the attributes as attributes, and the relationships are drawn as generalization, aggregation, or association relationships. However, the class diagram also requires that the visibility of the attributes and operations be known. As a general rule of thumb, attributes are private and operations are public. Therefore, unless the analyst has a good reason

---

[7] For more information on role-playing, see Bellin and Simone, *The CRC Card Book* and Dean Leffingwell and Don Widrig, *Managing Software Requirements: A Unified Approach* (Reading, MA: Addison-Wesley, 2000).

to change the default visibility of these properties, then the defaults should be accepted. Finally, the analyst should examine the model for additional opportunities to use aggregation or generalization relationships. These types of relationships can simplify the individual class descriptions. As in the previous steps, any and all changes must be recorded on the CRC cards.

Review the Class Diagram   The fifth step is to review the structural model for missing and/or unnecessary classes, attributes, operations, and relationships. Up until this step, the focus of the process has been on adding information to the evolving model. At this point in time, the focus begins to switch from only adding information to also challenging the reasons for including the information contained in the model.

Incorporate Patterns   The sixth step is to incorporate useful patterns into the evolving structural model. A useful pattern is one that would allow the analyst to more fully describe the underlying domain of the problem being investigated. We can do so by looking at the collection of patterns available (Figure 7-7) and comparing the classes contained in the patterns with those in the evolving class diagram. After identifying the useful patterns, the analyst incorporates the identified patterns into the class diagram and modifies the affected CRC cards. This includes adding and removing classes, attributes, operations, and/or relationships.

Review the Model   The seventh and final step is to review the structural model, including both the CRC cards and the class diagram. This is best accomplished during a formal review meeting using a walk-through approach in which an analyst presents the model to a team of developers and users. The analyst "walks through" the model explaining each part of the model and all of the reasoning that went into the decision to include each class in the structural model. This explanation includes justifications for the attributes, operations, and relationships associated with the classes. Finally, each class should be linked back to at least one use case; otherwise, the purpose of including the class in the structural model will not be understood. It is often best for the review team to also include people outside the development team that produced the model because these individuals can bring a fresh perspective to the model and uncover missing objects.

Applying the Concepts at CD Selections

The previous chapter described the CD Selections Internet Sales System (see Figure 6-8) and also included the use cases derived from the description (see Figures 6-9 through 6-13). In this section, we demonstrate the process of structural modeling using one of the use cases identified: Place Order (see Figure 6-11). Even though we are using just one of the use cases for our structural model, you should remember that to create a complete structural model all use cases should be used.

Create CRC Cards   The first step was to create the set of CRC cards by performing textual analysis on the use cases. To begin with, Alec chose the Place Order use case (see Figure 6-11). He and his team then used the textual analysis rules (see Figure 7-6) to identify the candidate classes, attributes, operations, and relationships. Using these rules on the Normal Flow of Events, they identified Customer, Search Request, CD, List, and Review as candidate classes. They uncovered three different types of search requests: Title Search, Artist Search, and Category Search. By applying the textual analysis rules to the Brief Description, an additional candidate class was discovered: Order. By reviewing the verbs contained in this use case, they saw that a Customer places an Order and that a Customer makes a Search Request.

To be as thorough as possible, Alec and his team also reviewed the original requirements used to create the use case. The original requirements are contained in the third paragraph of Figure 6-8. After reviewing this information, they identified a set of attributes for the Customer (name, address, e-mail, and credit card) and Order (CDs to purchase and quantity) classes and uncovered additional candidate classes: CD Categories and Credit Card Clearance Center. Furthermore, they realized that the Category Search class used the CD Categories class. Finally, they also identified three subclasses of CD Categories: Rock, Jazz, and Classical. Alec's goal, at this point in time, was to be as complete as possible. As such, he realized that they may have identified many candidate classes, attributes, operations, and relationships that may not be included in the final structural model.

Examine Common Object Lists   The second step for Alec and his team was to brainstorm additional candidate classes, attributes, operations, and relationships. He asked the team members to take a minute and think about what information they would like to keep about CDs. The information that they thought of was a set of attributes, for example, title, artist, quantity on hand, price, and category.

He then asked them to take another minute and think about the information that they should store about orders and an order's responsibilities. The responsibilities they identified were a set of operations, including calculate tax, calculate extension price, calculate shipping, and calculate total. Currently, the attributes (CDs to purchase and quantity) of Order implied that a customer should be allowed to order multiple copies of the same CD and allow different CDs to be ordered on the same order. However, the current structural model did not allow this. As such, they created a new class that was associated with both the Order class and the CD class: Order Line Item. This new class only had one attribute, quantity, but it had two relationships: one with Order and the other with CD.

When they reviewed the Customer class, they decided that the name and address attributes needed to be expanded; name should become last name, first name, and middle initial, and address should become street address, city, state, country, and zip code. The updated Customer class and Order class CRC cards are shown in Figures 7-9 and 7-10, respectively.

Role-Play the CRC Cards   The third step was to role-play the classes recorded on the CRC cards. The purpose of this step was to validate the current state of the evolving structural model. Alec handed out the CRC cards to different members of his team. Using the CRC cards, they began executing the different use cases, one at a time, to see if the current structural model could support each use case or whether the use case caused the "system" to crash. Anytime the "system" crashed, there was something

YOUR TURN   **7-1 CD-SELECTIONS INTERNET SALES SYSTEM**

Complete the CRC cards for the remaining identified classes.

Front:

Class Name: Customer	ID: 1	Type: Concrete, Domain
Description: an individual that may or has purchased merchandise from the CD Selections Internet Sales System		Associated Use Cases: 3

Responsibilities	**Collaborators**
_____	_____
_____	_____
_____	_____
_____	_____

Back:

Attributes:

First name _____   Zip code

Last name _____

E-mail _____

Street address _____   Credit card

City _____

State _____

Country _____

Relationships:

Generalization (a-kind-of): _____

_____

Aggregation (has-parts): _____

_____

Other Associations: _____   Order; Search Request

FIGURE 7-9
Customer Class CRC Card

missing: a class, an attribute, a relationship, or an operation. They would then add the missing information to the structural model and try executing the use case again.

First, they decided that the customer had requested the system to perform a search for all of the CDs associated with a specific artist. Based on the current CRC cards, the team felt that the system would produce an accurate list of CDs. They then tried to ask the system for a set of reviews of the CD. At this point in the exercise, the system crashed. The CRC cards did not have the Review class associated with the CD class. Therefore, there was no way to retrieve the requested information. This observation raised another question. Was there other marketing information that should be made available to the customer, for example, artist information and sample clips?

Next, they realized that vendor information should be a separate class that was associated with a CD rather than an additional attribute of a CD. This was because vendors had additional information and operations themselves. If they had modeled the vendor information as an attribute of CD, then the additional information and operations would have been lost. They continued role-playing each of the use cases until they were comfortable with the structural model's ability to support each and every one.

Front:

Class Name: Order   **ID:** 2   **Type:** Concrete, Domain

Description: an order that has been placed by a customer   **Associated Use Cases:** 3
which includes the individual items purchased by the customer

Responsibilities	Collaborators
Calculate tax _____	_____
Calculate shipping _____	_____
Calculate total _____	_____
_____	_____
_____	_____
_____	_____
_____	_____

Back:

Attributes:

   Tax _____   _____

   Shipping _____

   _____

   Total _____

Relationships:

   Generalization (a-kind-of): _____

   _____

   Aggregation (has-parts): _____

   _____

   Other Associations: Order Item; Customer

FIGURE 7-10
Order Class CRC Card

Create the Class Diagram   The fourth step was to create the class diagram from the CRC cards. Figure 7-11 shows the current state of the evolving structural model as depicted in a class diagram based on the Place Order use case.

Review the Class Diagram   The fifth step was to review the structural model for missing and/or unnecessary classes, attributes, operations, and relationships. At this point, the team challenged all components of the model (class, attribute, relationship, or operation) that did not seem to be adding anything useful to the model. If a component could not be justified sufficiently, then they removed it from the structural model. By carefully reviewing the current state of the structural model, they were able to challenge over a third of the classes contained in the class diagram (see Figure 7-11). It seemed that the CD categories, and their subclasses, were not really necessary. There were no attributes or operations for these classes. As such, the idea of CD categories was modeled as an attribute of a CD. The category attribute for the CD class was previously uncovered during the brainstorming step. Also, upon further review of the Search Request class and its subclasses, it was decided that the subclasses were really nothing more than a set of operations of the Search Request class. This was an example of process decomposition creeping into the modeling process. From an object-oriented perspective, we must always be careful to not allow this to occur. However, during the previous steps in the modeling process, Alec wanted to include

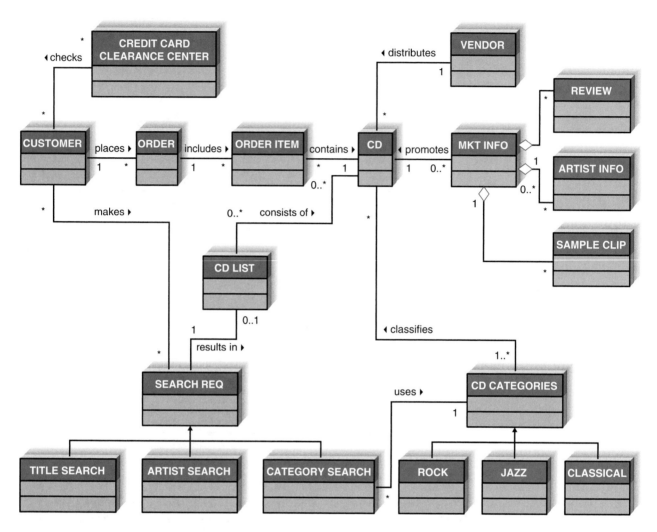

FIGURE 7-11

Preliminary CD Selections Internet Sales System Class Diagram (Place Order Use-Case View)

as much information as possible in the model. It was more beneficial to remove this type of information after it had crept into the model than to take a chance on not capturing the information required to solve the problem.

Incorporate Patterns The sixth step was to incorporate any useful pattern into the structural model. Some of the patterns that could have been useful include the Contract and Transactions patterns listed in Figure 7-7. By reviewing these patterns, Alec and his team uncovered additional classes. For example, there were two different types of Customers: Individual and Organizational. These were uncovered by reviewing the Sales Order contract pattern described in David Hay's *Data Model Patterns* (1996). Furthermore, Peter Coad has identified 12 transaction patterns that could be useful for Alec and his team to investigate further.[8]

---

[8] See Peter Coad, David North, and Mark Mayfield, *Object Models.*

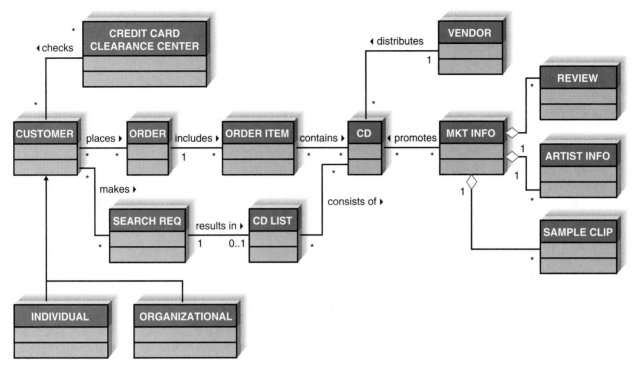

FIGURE 7-12
CD Selections Internet Sales System Class Diagram (Place Order Use-Case View)

Review the Model   The seventh and final step was to carefully review the structural model. Figure 7-12 shows the Place Order use-case view of the structural model as portrayed in a class diagram developed by Alec and his team.

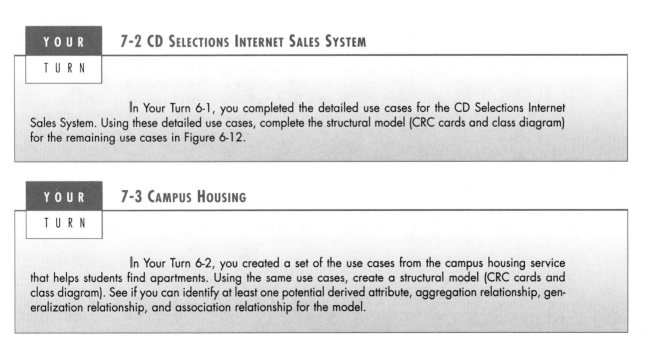

YOUR TURN   7-2 CD Selections Internet Sales System

In Your Turn 6-1, you completed the detailed use cases for the CD Selections Internet Sales System. Using these detailed use cases, complete the structural model (CRC cards and class diagram) for the remaining use cases in Figure 6-12.

YOUR TURN   7-3 Campus Housing

In Your Turn 6-2, you created a set of the use cases from the campus housing service that helps students find apartments. Using the same use cases, create a structural model (CRC cards and class diagram). See if you can identify at least one potential derived attribute, aggregation relationship, generalization relationship, and association relationship for the model.

SUMMARY

Structural Models

Structural models describe the underlying data structure of an object-oriented system. Where use-case models provide an external functional view of the evolving system (i.e., what the system does), structural models provide an internal static view of the evolving system (i.e., how the objects are organized in the system). At this point in the development of the system, the structural model only represents a logical model of the underlying problem domain. One of the primary purposes of the structural model is to create a vocabulary that allows the users and developers to effectively communicate about the problem under investigation. Structural models are comprised of classes, attributes, operations, and relationships. The three basic types of relationships that normally are depicted on structural models are aggregation, generalization, and association. Structural models typically are represented by CRC cards, class diagrams, and, in some cases, object diagrams.

CRC Cards

CRC cards model the classes, their responsibilities, and their collaborations. There are two different types of responsibilities: knowing and doing. Knowing responsibilities are associated mostly with attributes of instances of the class, while doing responsibilities are associated mostly with operations of instances of the class. Collaborations support the concepts of clients, servers, and contracts between objects. CRC cards capture all of the essential elements of the instances of a class. The front of the card allows the recording of the class's name, ID, type, description, list of associated use cases, responsibilities, and collaborators, while the back of the card contains the attributes and relationships.

Class and Object Diagrams

The class diagram is a graphical description of the information contained on the CRC cards. It shows the classes and the relationships between the classes. The visibility of the attributes and operations and the multiplicity of the relationships are additional information that the class diagram portrays, which is not included on the CRC cards. Also, at times a relationship itself contains information. In this case, an associated class is created. There are special areas for each of the relationships (aggregation, generalization, and association) contained in the diagram.

In real-world systems that can have over 100 classes, the class diagram can become overly complicated. To allow for the simplification of the diagram, a view mechanism can be used. A view restricts the amount of information portrayed on the diagram. Some useful views are: hiding all information about the class except for its name and relationships, showing only the classes that are associated with a particular use case, and limiting the relationships included to only one specific type (aggregation, generalization, or association).

When attempting to uncover additional information about the details of a class, it is useful to portray specific instances of a class instead of the classes themselves. An object diagram is used to depict a set of objects that represent an instantiation of all or part of a class diagram.

Creating CRC Cards and Class Diagrams

Creating a structural model is an iterative process. The process described is a seven-step process. The steps include textual analysis of use cases, brainstorming additional objects using common object lists, role-playing each use case using the CRC cards, creating class diagrams, and incorporating useful patterns.

KEY TERMS

A-kind-of
A-part-of
Abstract class
Aggregation relationship
Assemblies
Association relationship
Association class
Attribute
Behavior
Class
Class diagram
Client
Collaboration
Common object list
Conceptual model
Concrete class
Constructor operation
Contract
CRC cards
Decomposition

Derived attribute
Doing responsibility
Generalization relationship
Has-parts
Incidents
Instances
Instantiation
Interactions
Knowing responsibility
Maximum number
Minimum number
Multiplicity
Objects
Object diagram
Operation
Package
Parts
Patterns
Private
Protected

Public
Query operation
Responsibility
Roles
Server
Static model
Static structure diagram
Structural model
Subclass
Substitutability
Superclass
SVDPI
Tangible things
Textual analysis
Update operation
View
Visibility
Wholes

QUESTIONS

1. Describe to a businessperson the multiplicity of a relationship between two classes.
2. Why are assumptions important to a structural model?
3. What is an association class?
4. Contrast the following sets of terms:
   object; class; instance
   property; method; attribute
   superclass; subclass
   concrete class; abstract class
5. Give three examples of derived attributes that may exist on a class diagram. How would they be denoted on the class diagram?
6. What are the different types of visibility? How would they be denoted on a class diagram?
7. Draw the relationships that are described by the following business rules. Include the multiplicities for each relationship.
   A patient must be assigned to only one doctor, and a doctor can have one or many patients.
   An employee has one phone extension, and a unique phone extension is assigned to an employee.
   A movie theater shows at least one movie, and a movie can be shown at up to four other movie theaters around town.

A movie either has one star, two co-stars, or more than 10 people starring together. A star must be in at least one movie.

8. How do you designate the reading direction of a relationship on a class diagram?
9. For what purpose is an association class used in a class diagram? Give an example of an association class that may be found in a class diagram that captures students and the courses they have taken.
10. Give two examples of aggregation, generalization, and association relationships. How is each type of association depicted on a class diagram?
11. Identify the following operations as constructor, query, or update. Which operations would not need to be shown in the class rectangle?
    Calculate employee raise (raise percent)
    Calculate sick days ( )
    Increment number of employee vacation days ( )
    Locate employee name ( )
    Place request for vacation (vacation day)
    Find employee address ( )
    Insert employee ( )
    Change employee address ( )
    Insert spouse ( )

EXERCISES

A. Draw class diagrams for the following classes:
Movie (title, producer, length, director, genre)
Ticket (price, adult or child, showtime, movie)
Patron (name, adult or child, age)

B. Create an object diagram based on the class diagram you drew for Exercise A.

C. Draw a class diagram for the following classes. Consider that the entities represent a system for a patient billing system. Include only the attributes that would be appropriate for this context.
Patient (age, name, hobbies, blood type, occupation, insurance carrier, address, phone)
Insurance carrier (name, number of patients on plan, address, contact name, phone)
Doctor (specialty, provider identification number, golf handicap, age, phone, name)

D. Create an object diagram based on the class diagram you drew for Exercise C.

E. Draw the following relationships:
1. A patient must be assigned to only one patient, and a doctor can have many patients.
2. An employee has one phone extension, and a unique phone extension is assigned to an employee.
3. A movie theater shows many different movies, and the same movie can be shown at different movie theaters around town.

F. Draw a class diagram for the following situations:
1. Whenever new patients are seen for the first time, they complete a patient information form that asks their name, address, phone number, and insurance carrier, which are stored in the patient information file. Patients can be signed up with only one carrier, but they must by signed up to be seen by the doctor. Each time a patient visits the doctor, an insurance claim is sent to the carrier for payment. The claim must contain information about the visit such as the date, purpose, and cost. It would be possible for a patient to submit two claims on the same day.
2. The state of Georgia is interested in designing a system that will track its researchers. Information of interest includes researcher name, title, position; university name, location, enrollment; and research interests. Researchers are associated with one institution, and each researcher has several research interests.
3. A department store has a bridal registry. This registry keeps information about the customer (usually the bride), the products that the store carries, and the

products each customer registers for. Customers typically register for a large number of products, and many customers register for the same products.
4. Jim Smith's dealership sells Fords, Hondas, and Toyotas. The dealership keeps information about each car manufacturer with whom it deals so that it can get in touch with them easily. The dealership also keeps information about the models of cars that it carries from each manufacturer. It keeps information such as list price, the price the dealership paid to obtain the model, and the model name and series (e.g., Honda Civic LX). The dealership also keeps information on all sales that it has made (for instance, it will record the buyer's name, the car bought, and the amount paid for the car). In order to contact the buyers in the future, contact informations is also stored (e.g., address, phone number).

G. Create object diagrams based on the class diagrams that you drew for Exercise F.

H. Examine the class diagrams that you created for Exercise F. How would the models change (if at all) based on these new assumptions?
1. Two patients have the same first and last names.
2. Researchers can be associated with more than one institution.
3. The store would like to keep track of purchase items.
4. Many buyers have purchased multiple cars from Jim over time because he is such a good dealer.

I. Visit a Web site that allows customers to order a product over the Web (e.g., Amazon.com). Create a structural model (CRC cards and class diagram) that the site must need to support its business process. Include classes to show what they need information about. Be sure to include the attributes and operations to represent the type of information they use and create. Finally, draw relationships, making assumptions about how the classes are related.

J. Create a structural model (CRC cards and class diagram) for Exercise C in Chapter 6.

K. Create a structural model for Exercise E in Chapter 6.

L. Create a structural model for Exercise H in Chapter 6.

M. Create a structural model for Exercise J in Chapter 6.

N. Create a structural model for Exercise L in Chapter 6.

O. Create a structural model for Exercise N in Chapter 6.

P. Create a structural model for Exercise P in Chapter 6.

Q. Create a structural model for Exercise R in Chapter 6.

R. Create a structural model for Exercise T in Chapter 6.

MINICASES

1. West Star Marinas is a chain of 12 marinas that offer lakeside service to boaters, service and repair of boats, motors, and marine equipment, and sales of boats, motors, and other marine accessories. The systems development project team at West Star Marinas has been hard at work on a project that eventually will link all the marina's facilities into one unified, networked system.

   The project team has developed a use-case diagram of the current system. This model has been carefully checked. Last week, the team invited a number of system users to role-play the various use cases, and the use cases were refined to the users' satisfaction. Right now, the project manager feels confident that the As-Is system has been adequately represented in the use-case diagram.

   The Director of Operations for West Star is the sponsor of this project. He sat in on the role playing of the use cases, and was very pleased by the thorough job the team had done in developing the model. He made it clear to you, the project manager, that he was anxious to see your team begin work on the use cases for the To-Be system. He was a little skeptical that it was necessary for your team to spend any time modeling the current system in the first place, but grudgingly admitted that the team really seemed to understand the business after going through that work.

   The methodology you are following, however, specifies that the team should now turn its attention to developing the structural models for the As-Is system. When you stated this to the project sponsor, he seemed confused and a little irritated. "You are going to spend even more time looking at the current system? I thought you were done with that! Why is this necessary? I want to see some progress on the way things will work in the future!"

   What is your response to the Director of Operations? Why do we perform structural modeling? Is there any benefit to developing a structural model of the current system at all? How does the use cases and use-case diagram help us develop the structural model?

2. Holiday Travel Vehicles sells new recreational vehicles and travel trailers. When new vehicles arrive at Holiday Travel Vehicles, a new vehicle record is created. Included in the new vehicle record is a vehicle serial number, name, model, year, manufacturer, and base cost.

   When a customer arrives at Holiday Travel Vehicles, he/she works with a salesperson to negotiate a vehicle purchase. When a purchase has been agreed to, a sales invoice is completed by the salesperson. The invoice summarizes the purchase, including full customer information, information on the trade-in vehicle (if any), the trade-in allowance, and information on the purchased vehicle. If the customer requests dealer-installed options, they will be listed on the invoice as well. The invoice also summarizes the final negotiated price, plus any applicable taxes and license fees. The transaction concludes with a customer signature on the sales invoice.

   a. Identify the classes described in the above scenario (you should find six). Create CRC cards for each of the classes.

   Customers are assigned a customer ID when they make their first purchase from Holiday Travel Vehicles. Name, address, and phone number are recorded for the customer. The trade-in vehicle is described by a serial number, make, model, and year. Dealer-installed options are described by an option code, description, and price.

   b. Develop a list of attributes for each of the classes. Place the attributes onto the CRC cards.

   Each invoice will list just one customer. A person does not become a customer until they purchase a vehicle. Over time, a customer may purchase a number of vehicles from Holiday Travel Vehicles.

   Every invoice must be filled out by only one salesperson. A new salesperson may not have sold any vehicles, but experienced salespeople have probably sold many vehicles.

   Each invoice only lists one new vehicle. If a new vehicle in inventory has not been sold, there will be no invoice for it. Once the vehicle sells, there will be just one invoice for it.

   A customer may decide to have no options added to the vehicle, or may choose to add many options. An option may be listed on no invoices, or it may be listed on many invoices.

   A customer may trade in no more than one vehicle on a purchase of a new vehicle. The trade-in vehicle may be sold to another customer, who later trades it in on another Holiday Travel Vehicle.

   c. Based on the above business rules in force at Holiday Travel Vehicles, and CRC cards draw a class diagram and document the relationships with the appropriate multiplicities Remember to update the CRC cards.

PLANNING

ANALYSIS

☑ **Develop Analysis Plan**
☑ **Understand As-Is System**
☑ **Identify Improvement Opportunities**
☑ **Develop the To-Be System Concept**
☑ **Develop Use Cases**
☑ **Develop Structural Model**
☐ Develop Behavioral Models

T A S K C H E C K L I S T

PLANNING   ANALYSIS   DESIGN

CHAPTER 8
BEHAVIORAL MODELING

BEHAVIORAL models describe the internal dynamic aspects of an information system that supports the business processes in an organization. During the analysis phase, behavioral models describe what the internal logic of the processes is without specifying how the processes are to be implemented. Later, in the design and implementation phases, the detailed design of the operations contained in the object are fully specified. In this chapter, we describe three UML models that are used in behavioral modeling: sequence diagrams, collaboration diagrams, and statechart diagrams.

OBJECTIVES

- Understand the rules and style guidelines for sequence, collaboration, and statechart diagrams.
- Understand the processes used to create sequence, collaboration, and statechart diagrams.
- Be able to create sequence diagrams, collaboration diagrams, and statechart diagrams.
- Understand the relationship between the behavioral models and the structural and use-case models.

CHAPTER OUTLINE

Introduction
Behavioral Models
Interaction Diagrams
    Objects, Operations, and
      Messages
    Sequence Diagrams
    Collaboration Diagrams

Statechart Diagrams
    States, Events, Transitions, Actions,
      and Activities
    Elements of a Statechart Diagram
    Building a Statechart Diagram
    Applying Concepts at CD Selections
Summary

IMPLEMENTATION

INTRODUCTION

The previous two chapters discussed use-case models and structural models. Systems analysts utilize use-case models to describe the external behavioral or functional view of an information system, while they use structural models (CRC cards, class diagrams, and object diagrams) to depict the static view of an information system. In this chapter, we discuss how analysts use behavioral models to represent the internal behavior or dynamic view of an information system.

There are two types of *behavioral models.* First is the behavioral model that is used to represent the underlying details of a business process portrayed by a use-case model. In UML, interaction diagrams (sequence and collaboration) are used for this type of behavioral model. Second, there is a behavioral model that is used to represent the changes that occur in the underlying data. UML uses statechart diagrams for this type.

During the analysis phase, analysts use behavioral models to capture a basic understanding of the dynamic aspects of the underlying business process. Traditionally, behavioral models have been used primarily during the design phase where analysts refine the behavioral models to include implementation details (see Chapter 9). For now, our focus is on "what" the dynamic view of the evolving system is and not on "how" the dynamic aspect of the system will be implemented.

In this chapter, we concentrate on creating behavioral models of the underlying business process. Using the interaction diagrams (sequence and collaboration diagrams) and statechart diagrams, we can give a complete view of the dynamic aspects of the evolving business information system. We first describe behavioral models and their components. We then describe each of the diagrams, how they are created, and how they are related to the use-case model and class diagram described in Chapters 6 and 7.

BEHAVIORAL MODELS

When analysts are attempting to understand the underlying application domain of a problem, they must consider both the structural and behavioral aspects of the problem. Unlike other approaches to information systems development, object-oriented approaches attempt to view the underlying application domain in a holistic manner. By viewing the problem domain as a set of *use cases* that are supported by a set of collaborating objects, object-oriented approaches allow the analyst to minimize the "semantic gap" between the real-world set of objects and the evolving object-oriented model of the problem domain. However, as pointed out in the previous chapter, the real world tends to be messy, which makes modeling the application domain perfectly, nearly impossible in software. This is because to work, software must be neat and logical.

One of the primary purposes of behavioral models is to show how the underlying objects in the problem domain will collaborate to support each use case. Whereas structural models represent the objects and the relationships between them, behavioral models depict the internal view of the "business process" that a use case describes. The "process" can be shown by the interaction that takes place between the objects that collaborate to support a use case through the use of interaction (sequence and collaboration) diagrams. It is also possible to show what effect the set of use cases comprising the system has on the objects in the system through the use of statechart diagrams.

Creating behavioral models is an iterative process that not only iterates over the individual behavioral models (e.g., interaction [sequence and collaboration] diagrams and statechart diagrams) but also over the use cases (see Chapter 6) and structural models (see Chapter 7). As the behavioral models are created, it is not unusual to make changes to the use cases and structural models. In the next two sections, we describe interaction and statechart diagrams and when to use each.

INTERACTION DIAGRAMS

One of the primary differences between the class diagram and the interaction diagram, besides the obvious one that the one diagram describes structure and the other behavior, is that the modeling focus on the class diagram is at the class level, while that of the interaction diagram is at the object level. In this section we review objects, operations, and messages, and we cover the two different diagrams (sequence and collaboration) that can be used to model the interactions that take place between the objects in an information system.

Objects, Operations, and Messages

In Chapter 1, we defined an object as an instantiation of a class, that is, an actual person, place, event, or thing about which we want to capture information. If we were building an appointment system for a doctor's office, classes might include doctor, patient, and appointment. The specific patients like Jim Maloney, Mary Wilson, and Theresa Marks are considered objects, that is, *instances* of the Patient class.

Each object has *attributes* that describe information about the object, such as a patient's name, birthdate, address, and phone number. Each object also has *behaviors*. At this point in the development of the evolving system, the behaviors are described by *operations*. An operation is nothing more than an action that an object can perform. For example, an appointment object likely can schedule a new appointment, delete an appointment, and locate the next available appointment.

Each object also can send and receive messages. *Messages* are information sent to objects to tell an object to execute one of its behaviors. Essentially, a message is a function or procedure call from one object to another object. For example, if a patient is new to the doctor's office, the system will send an insert message to the application. The patient object will receive the instruction (i.e., the message) and do what it needs to do to go about inserting the new patient into the system (i.e., the behavior).

Sequence Diagrams

Sequence diagrams are one of two types of interaction diagrams. They illustrate the objects that participate in a use case and the messages that pass between them over time for *one* use case. A sequence diagram is a *dynamic model* that shows the explicit sequence of messages that are passed between objects in a defined interaction. Since sequence diagrams emphasize the time-based ordering of the activity that takes place among a set of objects, they are very helpful for understanding real-time specifications and complex use cases.

The sequence diagram can be a *generic sequence diagram* that shows all possible *scenarios*[1] for a use case, but usually each analyst develops a set of *instance sequence diagrams,* each of which depicts a single scenario within the use case. If you are interested in understanding the flow of control of a scenario by time, you should use a sequence diagram to depict this information. The diagrams are used throughout both the analysis and design phases. However, the design diagrams are very implementation-specific, often including database objects or specific GUI components as the classes. The following section presents the elements of a sequence diagram.

Elements of a Sequence Diagram   Figure 8-1 presents an instance sequence diagram that depicts the objects and messages for the "Make Appointment" use case which describes the process by which a patient creates a new appointment, cancels, or reschedules an appointment for the doctor's office appointment system. In this specific instance, the create appointment process is portrayed.

Actors and *objects* that participate in the sequence are placed across the top of the diagram using actor symbols, from the use-case diagram, and object symbols, from the object diagram (see Figure 8-2). Notice that the actors and objects in Figure 8-1 are aPatient, aReceptionist, Patients, UnpaidBills, Appointments, and anAppt.[2] They are not placed in any particular order, although it is nice to organize

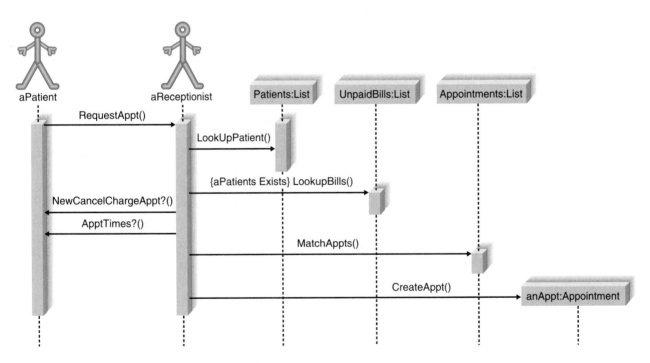

FIGURE 8-1
Instance Sequence Diagram

[1] Remember that a scenario is a single executable path through a use case.

[2] In some versions of the sequence diagram, object symbols are used as surrogates for the actors. However, for purposes of clarity, we recommend using actor symbols for actors instead.

Term and Definition	Symbol

An actor
    Is a person or system that derives benefit
    from and is external to the system
    Participates in a sequence by sending
    and/or receiving messages
    Is placed across the top of the diagram
	 anActor

An object:
    Participates in a sequence by sending
    and/or receiving messages
    Is placed across the top of the diagram
	anObject:aClass

A lifeline:
    Denotes the life of an object during a
    sequence
    Contains an "X" at the point at which the
    class no longer interacts

A focus of control
    Is a long, narrow rectangle placed atop a
    lifeline
    Denotes when an object is sending or
    receiving messages

A message
    Conveys information from one object to
    another one
	aMessage()

Object destruction
    An X is placed at the end of an object's
    lifeline to show that it is going out of
    existence
	X

FIGURE 8-2
Sequence Diagram Syntax

them in some logical way, such as the order in which they participate in the sequence. For each of the objects, the name of the class that they are an instance of is given after the object's name (e.g., Patients: List means that Patients is an instance of the List class that contains individual patient objects).

A dotted line runs vertically below each actor and object to denote the *lifeline* of the actors/objects over time (see Figure 8-1). Sometimes an object creates a *temporary object,* and in this case an X is placed at the end of the lifeline at the point the object is destroyed (not shown). For example, think about a shopping cart object for a Web commerce application. The shopping cart is used for temporarily capturing line items for an order, but once the order is confirmed, the shopping cart is no longer needed. In this case, an X would be located at the point at which the shopping cart object is destroyed. When objects continue to exist in the system after they are used in the sequence diagram, then the lifeline continues to the bottom of the diagram. (This is the case with all of the objects in Figure 8-1.)

A thin rectangular box, called the *focus of control,* is overlaid onto the lifeline to show when the classes are sending and receiving messages (see Figure 8-2). A *message* is a communication between classes that conveys information with the expectation that activity will ensue, and messages passed between classes are shown using solid lines connecting two objects, called *links* (see Figure 8-2). The arrow on the link shows which way the message is being passed, and any argument values for the message are placed in parentheses next to the message's name. The order of messages goes from the top to the bottom of the page, so messages located higher on the diagram represent messages that occur earlier on in the sequence, versus the lower messages that occur later. In Figure 8-1, "LookUpPatient" is a message sent from the actor–aReceptionist to the object Patients, which is a container for the current patients to determine whether the aPatient actor is a current patient.

At times a message is sent only if a *condition* is met. In those cases, the condition is placed between a set of [ ], for example, [aPatient Exists] LookupBills( ). The condition is placed in front of the message name. However, when using a sequence diagram to model a specific scenario, conditions typically are not shown on any single sequence diagram. Instead, conditions are implied only through the existence of different sequence diagram. There are times that a message is repeated. This is designated with an * in front of the message name, for example, *Request CD. An object can send a message to itself; this is known as *self-delegation.* Finally, it is possible to explicitly show the return from a message with a return link, a dashed message. However, adding return links tends to clutter the diagram. As such, unless the return links add a lot of information to the diagram, they should be omitted.

Sometimes an object will create another object. This is shown by the message being sent directly to an object instead of its lifeline. In Figure 8-1, the actor aReceptionist creates an object anAppt.

Building a Sequence Diagram   In this section we describe a six-step process used to create a sequence diagram[3] (see Figure 8-3). The first step in the process is to determine the context of the sequence diagram. The context of the diagram can be a system, a use case, a scenario of a use case, or an operation of a class. Most commonly, it is one use-case scenario.

The second step is to identify the objects that participate in the sequence being modeled, that is, the objects that interact with each other during the use-case scenario. The objects are identified during the development of the structural model (see Chapter 7). These are the classes on which the objects of the sequence diagram for this scenario will be based. However, during this process, it is likely that new classes, and hence new objects, will be uncovered. Don't worry too much about identifying all the objects perfectly; remember that the behavioral modeling process is iterative. Usually the sequence diagrams are revised multiple times during the structural and behavioral modeling processes.

The third step is to set the lifeline for each object. To do this, you need to draw a vertical dotted line below each class to represent the class's existence during the sequence. An X should be placed below the object at the point on the lifeline where the object goes out of existence.

The fourth step is to add the messages to the diagram. This is done by drawing arrows to represent the messages being passed from object to object, with the arrow pointing in the message's transmission direction. The arrows should be placed in order from the first message (at the top) to the last (at the bottom) to show time sequence. Any parameters passed along with the messages should be placed in

---

[3] The approach described in this section is adapted from Grady Booch, James Rumbaugh, and Ivar Jacobson, *The Unified Modeling Language User Guide* (Reading, MA: Addison-Wesley, 1999).

1. Set the context.
2. Identify which objects will participate.
3. Set the lifeline for each object.
4. Lay out the messages from the top to the bottom of the diagram based on the order in which they are sent.
5. Add the focus of control to each object's lifeline.
6. Validate the sequence diagram.

FIGURE 8-3
Steps for Building Sequence Diagrams

parentheses next to the message's name. If a message is expected to be returned as a response to a message, then the return message is not explicitly shown on the diagram.

The fifth step is to place the focus of control on each object's lifeline by drawing a narrow rectangular box over the top of the lifelines to represent when the classes are sending and receiving messages.

The sixth and final step is to validate the sequence diagram. The purpose of this step is to guarantee that the sequence diagram completely represents the underlying process (es). This is done by guaranteeing that the diagram depicts all of the steps in the process.

Apply the Concepts at CD Selections   The best way to learn how to build a sequence diagram is to draw one. In this section we apply the steps described above to the CD Selections case used in the previous chapters.

The first step in the process is to determine the context of the sequence diagram. We will use a scenario[4] from the Place Order use case that was created in Chapter 6 and illustrated in Figure 6-11. (Refer to the original use case for the details.) Figure 8-4 lists the "Normal Flow of Events" that contains the scenario which this sequence diagram will need to describe.

The second step is to identify the objects that participate in the sequence being modeled. The classes associated with the Place Order use case are shown in Figure 7-12. For example, the classes used for the Place Order use case include Customer, CD, Marketing Information, Credit Card Clearance Center, Order, Order Item, Vendor, Search Request, CD List, Review, Artist Information, Sample Clip, Individual, and Organizational.

Normal Flow of Events:
1. Customer submits a search request to the system.
2. The System provides the Customer a list of recommended CDs.
3. The Customer chooses one of the CDs to find out additional information
4. The System provides the Customer with basic information and reviews on the CD
5. The Customer calls the Maintain Order use case.
6. The Customer iterates over 3 through 5 until done shopping.
7. The Customer executes the Checkout use case.
8. The Customer leaves the Web site.

FIGURE 8-4
Normal Flow of Events of the Place Order Use Case

[4] Remember, as stated previously, a scenario is a single executable path through a use case.

During the role-playing of the *CRC cards,* one of the analysts assigned to the CD Selections Internet System Development Team asked whether a "Shopping Cart" class should be included. She stated that every Internet sales site she had visited had a shopping cart that was used to build an order. However, the instance of the Shopping Cart class only existed until either an order was placed or the shopping cart was emptied. Based on this obvious oversight, both the Place Order use case and the class diagram will have to be modified. Another analyst pointed out that the CDs themselves were going to have to be associated with some type of searchable storage. Otherwise, it would be impossible to find the CD in which the Customer was interested. However, it was decided that the CD List class would suffice for both the searchable storage and a temporary instance that is created as a result of a search request. This again is a fairly typical situation in object-oriented development. The more the development team learns about the requirements, the more the models (use case, structural, and behavioral) will evolve. The important thing is to remember that an object-oriented development process is iterative over all of the models.

The objects that will be needed to describe this scenario are instances of the Search Request, CD List, CD, Marketing Information, Customer, Review, Artist Information, Sample Clip, and Shopping Cart classes. We also have an actor (Customer) that interacts with the scenario. To complete this step, you should lay out the objects on the sequence diagram by drawing them, from left to right, across the diagram.

The third step is to set the lifeline for each object. To do this, you should draw a vertical dotted line below each of the objects (aSR, aCDL, CDs, aCD, MI, aR, AI, SC, and anOrder) and the actor (aCustomer). Also, an X should be placed at the bottom of the lifelines for aCDL and aSC since they go away at the end of this process.

The fourth step is to add the messages to the diagram. Examine the steps in Figure 8-4 and see if you can determine the way in which the messages should be added to the sequence diagram. Figure 8-5 shows our results. Notice how we did not include messages back to Customer in response to "create SR" and "add CD." In these cases, it is assumed that the Customer will receive response messages about the requested CD and inserted CD, respectively.

The fifth step is to add the focus of control to each object's and actor's lifeline. This is done by drawing a narrow rectangle box over the top of the lifelines to represent when the objects (actors) are sending and receiving messages. For example, in Figure 8-5 aCustomer is "active" during the entire process, while aSR is only active at the beginning of the process (the top of the diagram).

Lastly, the CD Selections team validated the diagram by ensuring that the diagram accurately and completely represented the underlying process associated with the Place Order use case. Figure 8-5 portrays their completed sequence diagram.

YOUR	8-1 DRAWING A SEQUENCE DIAGRAM
TURN	

In Your Turn 6-2 you were asked to create a set of use cases and a use-case diagram for the campus housing service that helps students find apartments. In Your Turn 7-3 you were asked to create a structural model (CRC cards and class diagram) for those use cases. Select one of the use cases from the use-case diagram and create a sequence diagram that represents the interaction among classes in the use case.

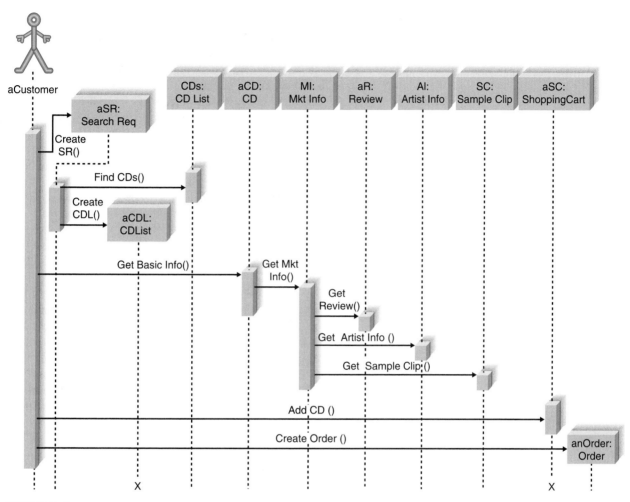

FIGURE 8-5
Sequence Diagram for the Place Order Use Case

Collaboration Diagrams

Collaboration diagrams, like sequence diagrams, essentially provide a view of the dynamic aspects of an object-oriented system. A collaboration diagram is essentially an object diagram that shows message passing relationships instead of aggregation or generalization associations. Collaboration diagrams are very useful to show process patterns, that is, patterns of activity that occur over a set of collaborating classes.

Collaboration diagrams are equivalent to sequence diagrams, but they emphasize the flow of messages through a set of objects, while the sequence diagrams focus on the time ordering of the messages being passed. Therefore, if you are interested in understanding the interactions among a set of collaborating objects, you should use a collaboration diagram. If you are interested in the time ordering of the messages, you should use a sequence diagram. In some cases, you will want to use both to more fully understand the dynamic activity of the system.

Elements of a Collaboration Diagram   A collaboration diagram for the "Make Appointment" use case is portrayed in Figure 8-6. Like the sequence diagram in Figure 8-1, the create appointment process is portrayed.

Actors and objects that collaborate to execute the use case are placed on the collaboration diagram in a manner to emphasize the message passing that takes place between them. Notice that the actors and objects in Figure 8-6 are the same ones in Figure 8-1: aPatient, aReceptionist, Patients, UnpaidBills, Appointments, and anAppt.[5] Again, like the sequence diagram, for each of the objects, the name of the class that they are an instance of is given after the object's name, for example, Patients: List. (See Figure 8-7 for the Collaboration Diagram Syntax.) Unlike the sequence diagram, the collaboration diagram does not have a means to explicitly show an object being deleted or created. It is assumed that when a delete, destroy, or remove message is sent to an object, it will go out of existence and that a create or new message will cause a new object to come into existence. Another difference between the two interaction diagrams is that the collaboration diagram never shows returns from message sends, while the sequence diagram can optionally show them.

An *association* is shown between actors and objects with an undirected line. For example, there is an association shown between the aPatient and aReceptionist actors. Messages are shown as labels on the associations. Included with the labels are lines with arrows showing the direction of the message being sent. For example, in Figure 8-6, the aPatient actor sends the RequestAppt( ) message to the aReceptionist actor, and the aReceptionist actor sends the NewCancelChangeAppt? and the ApptTimes? messages to the aPatient actor. The sequence of sent the message is designated with a sequence number. For example, in Figure 8-6, the RequestAppt( )

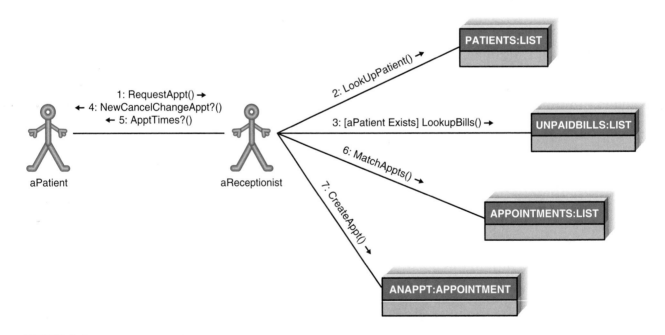

FIGURE 8-6
Example Collaboration Diagram

[5] In some versions of the collaboration diagram, object symbols are used as surrogates for the actors. However, again we recommend using actor symbols for actors instead.

Term and Definition	Symbol
An actor Is a person or system that derives benefit from and is external to the system Participates in a sequences by sending and/or receiving messages	anActor
An object Participates in a sequence by sending and/or receiving messages Is placed across the top of the diagram	anObject:aClass
An association Shows an association between actors and/or objects Messages are sent over associations	_____
A message Conveys information from one object to another one Direction is shown using an arrowhead Sequence is shown by a sequence number	1: aMessage() →

FIGURE 8-7
Collaboration Diagram Syntax

message is the first message sent, whereas the NewCancelChangeAppt? and the ApptTimes? messages are the fourth and fifth messages sent, respectively.

Like the sequence diagram, the collaboration diagram can represent conditional messages. For example in Figure 8-6, the LookupBills() message is sent only if the [aPatient exists] condition is met. If a message is repeatedly sent, an * is placed after the sequence number. Finally, an association that loops onto an object shows self-delegation. The message is shown as the label of the association.

When a collaboration diagram is fully populated with all of the objects, it can become very complex and difficult to understand. When this occurs, it is necessary to simplify the diagram. One approach to simplify a collaboration diagram, such as use-case diagrams (see Chapter 6) and class diagrams (see Chapter 7), is through the use of *packages* (i.e., logical groups of classes). In the case of collaboration diagrams, objects are grouped together based on the messages sent to and received from the other objects.[6]

Building a Collaboration Diagram[7]   Remember that a collaboration diagram is basically an object diagram that shows message passing relationships instead of aggregation or generalization associations. In this section, we describe a five-step process used to build a collaboration diagram (see Figure 8-8).

The first step in the process is to determine the context of the collaboration diagram. Like a sequence diagram, the context of the diagram can be a system, a use case, a scenario of a use case, or an operation of a class.

[6] For those familiar with structured analysis and design, packages serve a similar purpose as the leveling and balancing processes used in data flow diagramming. Packages and Package diagrams are described in Chapter 9.

[7] The approach described in this section is adapted from Booch, Rumbaugh, and Jacobson, *The Unified Modeling Language User Guide.*

1. Set the context.

2. Identify which objects (actors) and the associations between the objects participate in the collaboration.

3. Lay out the collaboration diagram.

4. Add messages.

5. Validate the collaboration diagram.

FIGURE 8-8

Steps for Building Collaboration Diagrams

The second step is to identify the objects (actors) and the associations that link the objects (actors) together that participate in the *collaboration*. Remember: the objects that participate in the collaboration are instances of the classes identified during the development of the structural model (see Chapter 7). Like the sequence diagramming process, it is likely that additional objects, and hence classes, will be discovered. Again, this is normal since the underlying development process is iterative and incremental.

One useful technique to identify potential collaborations is *CRUD analysis*[1]. CRUD analysis uses a CRUD matrix in which each interaction among objects is labeled with a letter for the type of interaction: C for Create, R for Read or Reference, U for update, and D for Delete. In an object-oriented approach, a class by class matrix is used. Each cell in the matrix represents the interaction between instances of the classes. For example in Figure 8-5, an instance of SearchReq created an instance of CDList. As such, assuming a Row:Column ordering, a "C" would be placed in the cell SearchReq:CDList. Also, in Figure 8-5, an instance of CD references an instance of MktInfo. In this case, a "R" would be placed in the CD:MktInfo cell. Figure 8-9 shows the CRUD matrix based on the Place Order use case.

Once a *CRUD matrix* is completed for the entire system, the matrix can be scanned quickly to ensure that each and every class can be instantiated. Furthermore, each type of interaction can be validated for each class. For example, if a class only represents temporary objects, then the column in the matrix should have a D in it somewhere. Otherwise, the instances of the class will never be deleted. Since a data warehouse contains historical data, objects that are to be stored in one should not have any U or D entries in their associated columns. As such, CRUD analysis can be used as away to partially validate the interactions among the objects in an object-oriented system. Finally, the more interactions among a set of classes, the more likely they should be clustered together in a collaboration. However, the number and type of interactions are only an estimate at this point in the development of the system. As such, care should be taken when using this technique to cluster classes together to identify collaborations.

The third step is to lay out the objects (actors) and their associations on the collaboration diagram by placing them together based on the associations that they have with the other objects in the collaboration. By focusing on the associations between the objects (actors) and minimizing the number of associations that cross over one another, we can increase the understandability of the diagram.

The fourth step is to add the messages to the associations between the objects. We do this by adding the name of the message(s) to the association link between the objects and an arrow showing the direction of the message being sent. Further-

---

[8] CRUD analysis is associated with Information Engineering (see James Martin, *Information Engineering, Book II Planning and Analysis,* Englewood Cliffs, NJ: Prentice Hall, 1990).

	Customer	SearchReq	CDList	CD	Mkt Info	Review	Artist Info	Sample Clip	Shopping Cart	Order
Customer		R							U	C
SearchReq			CR							
CDList										
CD					U					
Mkt Info						U	U	U		
Review										
Artist Info										
Sample Clip										
Shopping Cart										
Order										

FIGURE 8-9
CRUD Matrix for the Place Order Use Case

more, each message has a sequence number associated with it to portray the time-based ordering of the message.[9]

The fifth and final step is to validate the collaboration diagram. The purpose of this step is to guarantee that the collaboration diagram faithfully portrays the underlying process(es). This is done by ensuring that all steps in the process are depicted on the diagram.

Apply the Concepts at CD Selections   Like creating sequence diagrams, the best way to learn how to create a collaboration diagram is to draw one. The first step in the process is to determine the context of the collaboration diagram. We will use the same scenario of the Place Order use case that we used above with the description of creating sequence diagrams.

By executing the second step, the CD Selections' team identifies the objects and the associations that link the objects together. Since we are using the same scenario as we did in the sequence diagram above, they need to include the same objects: instances of the Search Request, CD List, CD, Marketing Information, Customer, Review, Artist Information, Sample Clip, and Shopping Cart classes. And again, we have an actor (Customer) that interacts with the scenario. Furthermore, the team identified the associations between the objects, for example, the instances of CD are associated with instances of Mkt Info, which in turn are associated with instances of Review, Artist Info, and Sample Clip.

During the third step, the team placed the objects on the diagram based on the associations that they have with the other objects in the collaboration. This was done to increase the readability, and hence the understandability, of the diagram (see Figure 8-10).

During the fourth step, the CD Selections' team added the messages to the associations between the objects. For example, in Figure 8-10, the Create SR( )

---

[9] However, remember that the sequence diagram portrays the time-based ordering of the messages in a top-down manner. As such, if your focus is on the time-based ordering of the messages, we recommend that you also use the sequence diagram.

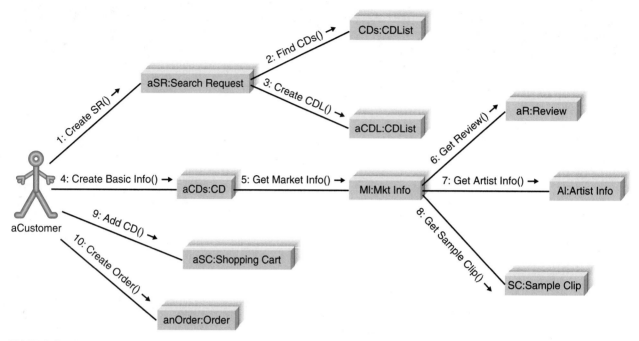

FIGURE 8-10
Collaboration Diagram for the Place Order Use Case

message is the first message sent, and the FindCDs( ) message is the second message sent. The aCustomer actor sends the Create SR( ) message to the aSR object. and the aSR object sends the FindCDs( ) messages to the CDs object.

Finally, the CD Selections' team executed the fifth and final step: validating the diagram. They accomplished this by ensuring that the underlying process associated with the Place Order use case was accurately and completely represented by the diagram. See Figure 8-10 for the completed collaboration diagram for this particular scenario of the Place Order use case.

At this point in time, you should review both Figures 8-5 and 8-10. These two different diagrams are used simply to provide two different views of the same scenario. However, as you can see, it is easier to portray the time ordering of the messages in the sequence diagram than it is in the collaboration diagram, and it is easier to represent how the objects interact together to implement the use case using the collaboration diagram.

YOUR　**8-2 DRAWING A COLLABORATION DIAGRAM**

TURN

In Your Turn 8-1 you were asked to create a sequence diagram for the campus housing service. Draw a collaboration diagram for the same situation.

STATECHART DIAGRAMS

Some of the classes in the *class diagrams* represent a set of objects that are quite dynamic in that they pass through a variety of states over the course of their existence. For example, a patient can change over time from being "new" to "current," "former," and so on, based on his or her status with the doctor's office. A *statechart diagram* is a *dynamic model* that shows the different states that a single object passes through during its life in response to events, along with its responses and actions. Typically, statechart diagrams are not used for all objects, but just to further define complex objects to help simplify the design of algorithms for their methods. CRUD analysis is one technique (see Collaboration Diagram section) that can be used to identify complex objects. The more (C)reate, (U)pdate or (D)elete entries in the column associated with a class, the more likely the instances of the class will have a complex life cycle. As such, these objects are candidates for state modeling with statechart diagrams. The statechart diagram shows the different states of the object and what events cause the object to change from one state to another. In comparison to the interaction diagrams, statechart diagrams should be used if you are interested in understanding the dynamic aspects of a single class and how its instances evolve over time, and not in how a particular use-case scenario is executed over a set of classes. In this section, we describe states, events, transitions, actions, and activities and the use of the statechart diagram to model the state changes through which complex objects pass.

States, Events, Transitions, Actions, and Activities

The *state* of an object is defined by the value of its attributes and its relationships with other objects at a particular point in time. For example, a patient might have a state of "new" or "current" or "former." The attributes or properties of an object affect the state that it is in; however, not all attributes or attribute changes will make a difference. For example, think about a patient's address. Those attributes make very little difference as to changes in a patient's state. However, if states were based on a patient's geographic location (e.g., in-town patients were treated differently than out-of-town patients), changes to the patient's address might influence state changes.

An *event* is something that takes place at a certain point in time and changes a value(s) that describes an object, which in turn changes the object's state. It can be a designated condition becoming true, the receipt of the call for a method by an object, and the passage of a designated period of time. The state of the object determines exactly what the response will be.

A *transition* is a relationship that represents the movement of an object from one state to another state. Some transitions will have a guard condition. A *guard condition* is a boolean expression that includes attribute values, which allows a transition to occur only if the condition is true. An object typically moves from one state to another based on the outcome of an action that is triggered by an event. An *action* is an atomic, nondecomposable process that cannot be interrupted. From a practical perspective, actions take zero time, and they are associated with a transition. In contrast, an *activity* is a nonatomic, decomposable process that can be interrupted. Activities take a long period of time to complete, they can be started and stopped by an action, and they are associated with states.

Elements of a Statechart Diagram

Figure 8-11 presents an example of a statechart diagram representing the patient class in the context of a hospital environment. From this diagram we can tell that a patient enters a hospital and is admitted after checking in. If a doctor finds the patient to be healthy, he or she is released, and is no longer considered a patient after two weeks elapses. If a patient is found to be unhealthy, he or she remains under observation until the diagnosis changes.

A *state* is a set of values that describe an object at a specific point in time, and it represents a point in an object's life in which it satisfies some condition, performs some action, or waits for something to happen (see Figure 8-12). In Figure 8-11 states include entering, admitted, released, and under observation. A state is depicted by a *state symbol,* which is a rectangle with rounded corners with a descriptive label that communicates a particular state. There are two exceptions. An *initial state* is shown using a small solid filled circle, and an object's *final state* is shown as a circle surrounding a small solid filled in circle. These exceptions depict when an object begins and ceases to exist, respectively.

Arrows are used to connect the state symbols, representing the *transitions* between states. Each arrow is labeled with the appropriate event name, and any parameters or conditions that may apply. For example, the two transitions from admitted to released and under observation contain guard conditions.

Building a Statechart Diagram

Statechart diagrams are drawn to depict a single class from a class diagram. Typically, the classes are very dynamic and complex, requiring a good understanding of their states over time and events triggering changes. You should examine your class diagram to identify which classes will need to undergo a complex series of state changes and draw a diagram for each of them. In this section, we describe a five-step process used to build a statechart diagram[10] (see Figure 8-13). Like the other behavioral models, the first step in the process is to determine the context of the statechart diagram. The context of a statechart diagram is usually a class. However, it also could be a set of classes, a subsystem, or an entire system.

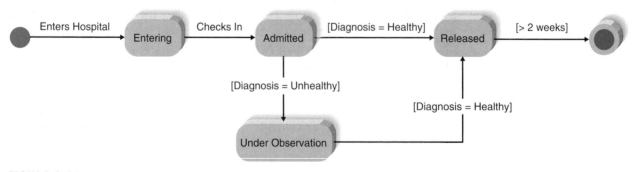

FIGURE 8-11
Example Statechart Diagram

[10] The approach described in this section is adapted from Booch, Rumbaugh, and Jacobson, *The Unified Modeling Language User Guide.*

Term and Definition	Symbol
A state Is shown as a rectangle with rounded corners Has a name that represents the state of an object	 aState
An initial state Is shown as a small filled in circle Represents the point at which an object begins to exist	●
A final state Is shown as a circle surrounding a small solid filled circle (bull's-eye) Represents the completion of activity	 ◉
An event Is a noteworthy occurrence that triggers a change in state Can be a designated condition becoming true, the receipt of an explicit signal from one object to another, or the passage of a designated period of time Is used to label a transition	anEvent
A transition Indicates that an object in the first state will enter the second state Is triggered by the occurrence of the event labeling the transition Is shown as a solid arrow from one state to another, labeled by the event name	⟶

FIGURE 8-12
Statechart Diagram Syntax

The second step is to identify the various states that an order will have over its lifetime. This includes establishing the boundaries of the existence of an object by identifying the initial and final states of an order object. We also must identify the stable states of an object. The information necessary to perform this step is gleaned from reading the use-case descriptions, talking with users, reviewing the CRUD matrix, and relying on the requirements-gathering techniques that you learned about in Chapter 5. An easy way to identify the states of an object is to write the steps of what happens to an object over time, from start to finish, similar to how you would create the "Normal Flow of Events" section of a use-case description.

1. Set the context.
2. Identify the initial, final, and stable states of the object.
3. Determine the order in which the object will pass through the stable states.
4. Identify the events, actions, and guard conditions associated with the transitions.
5. Validate the statechart diagram.

FIGURE 8-13
Steps for Building Statechart Diagrams

The third step is to determine the sequence of the states that an object will pass through during its lifetime. Using this sequence, we can place the states onto the statechart diagram, in a left to right order.

The fourth step is to identify the events, actions, and guard conditions associated with the transitions between the states of the object. The events are the triggers that cause an object to move from one state to the next state. In other words, an event causes an action to execute that changes the value(s) of an object's attribute(s) in a significant manner. The actions are typically operations contained within the object. Also, guard conditions can model a set of test conditions that must be met for the transition to occur. At this point in the process, the transitions are drawn between the relevant states and labeled with the event, action, or guard condition.

The fifth step is to validate the statechart diagram by making sure that each state is reachable and that it is possible to leave all states except for final states. Obviously, if an identified state is not reachable, either a transition is missing or the state was identified in error. Furthermore, only final states can be a dead end from the perspective of an object life cycle.

Applying the Concepts at CD Selections

Basically, the only way to understand the statechart diagramming process is to attempt to draw one. As in the previous example diagrams, we focus our attention on the Place Order use case. In this case, we have picked the order class to investigate.

The second step is to identify the various states that an order will have over its lifetime. To enable the discovery of the initial, final, and stable states of an order, the CD Selections' team interviewed a customer representative that dealt with processing customer orders on a regular basis. Based on this interview, the team uncovered the life of an order (see Figure 8-14) from start to finish, from an order's perspective.

The third step is to determine the sequence of the states that an order object will pass through during its lifetime. Based on the order's life-cycle portrayed in Figure 8-14, the team identified and laid out the states of the order on the statechart diagram.

Next, the team identified the events, actions, and guard conditions associated with the transitions between the states of the order. For example, the event "Order is created" moves the order from the "initial" state to the "In process" state (see Figure 8-15). During the "Processing" state, a credit authorization is requested. The guard condition "Authorization = Approved" prevents the order to move from the "Processing" state to the "Placed" state unless the credit authorization has been approved.

---

1. The customer creates an order on the Web
2. The customer submits the order once he or she is finished
3. The credit authorization needs to be approved for the order to be accepted
4. If denied, the order is returned to the customer for changes or deletion
5. If accepted, the order is placed
6. The order is shipped to the customer
7. The customer receives the order
8. The order is closed

FIGURE 8-14
The Life of an Order

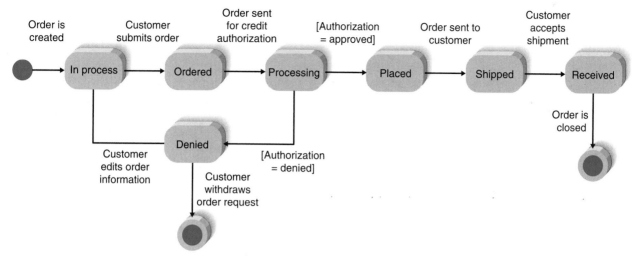

FIGURE 8-15
Statechart Diagram for the Order class

Also, the guard condition "Authorization = Denied" prevents the order to move from the "Processing" state to the "Denied" state unless the credit authorization has been denied. As such, between the two guard conditions, the order is stuck in the processing state until the credit authorization has been either approved or denied.

The team finally validates the statechart by ensuring that each state is reachable and that it is possible to leave all states except for the final states. Furthermore, the team makes sure that all states and transitions for the order have been modeled. At this point in time, one of the analysts on the team suggested that possibly multiple types of orders were being described in the statechart diagram. Specifically, he thought that there were denied and accepted orders. Based on this discovery, he suggested that two new classes, for each subtype of order, be created. However, upon further review by the entire team, it was decided that adding these classes to the class diagram and modifying all of the other diagrams to reflect this change would not add anything to understanding the problem. Therefore, it was decided not to add the classes. However, in many cases, modeling the states that an object will go through during its lifetime may in fact uncover additional useful subclasses. Figure 8-15 illustrates the statechart that the CD Selections' team created for an order object.

8-3 DRAWING A STATECHART DIAGRAM

You have been working with the system for the campus housing service that helps students find apartments. One of the dynamic classes in this system likely is the apartment class. Draw a statechart diagram to show the various states that an apartment class transitions to throughout its lifetime. Can you think of other classes that would make good candidates for a statechart diagram?

SUMMARY

Sequence Diagram

The sequence diagram is a dynamic model that illustrates the classes that participate in a use case and messages that pass between them over time. Classes are placed horizontally across the top of a sequence diagram, each having a dotted line called a lifeline vertically below it. The focus of control, represented by a thin rectangle, is placed over the lifeline to show when the classes are sending or receiving messages. A message is a communication between classes that conveys information with the expectation that activity will ensue, and the messages are shown using an arrow connecting two objects that points in the direction that the message is being passed.

To create a sequence diagram, first identify the context of the diagram: system, use case, scenario, or an operation. Next, identify the classes that participate in the context and then add the messages that pass among them. Finally, you will need to add the lifeline and focus of control. Sequence diagrams are helpful for understanding the time order of the messages.

Collaboration Diagram

Like the sequence diagram, the collaboration diagram provides a dynamic view of an object-oriented system. A collaboration diagram is very similar to an object diagram except that it shows message passing relationships instead of aggregation or generalization associations. The layout of the diagram is done in a manner to accentuate the message passing between the collaborating actors and objects. Undirected lines portray associations that are used to represent the collaborations between the actors and object. Messages are shown as labels on the associations.

When building a collaboration diagram, like a sequence diagram, you should first identify the context of the diagram: system, use case, scenario, or an operation. Next, identify the classes that participate in the context and then identify the associations that represent the collaborations between them. Finally, add the messages to the associations. Collaboration diagrams are helpful for understanding how objects interact together.

Statechart Diagram

The statechart diagram shows the different states that a single class passes through during its life in response to events, along with responses and actions. A state is a set of values that describe an object at a specific point in time; it represents a point in an object's life in which it satisfies some condition, performs some action, or waits for something to happen. An event is something that takes place at a certain point in time and changes a value(s) that describes an object, which in turn changes the object's state. As objects move from state to state, they undergo transitions.

When drawing a statechart diagram, like the other behavioral diagrams, the first thing is to set the context of the diagram: system, subsystem, set of classes, or an individual class. Then, rectangles with rounded corners are placed on the model to represent the various states on which the context will take. Next, arrows are drawn between the rectangles to denote the transitions, and event labels are written above the arrows to describe the event that causes the transition to occur. Typically, statechart diagrams are used to portray the dynamic aspect of a complex class.

KEY TERMS

Action	CRUD matrix	Object
Activity	Dynamic model	Operation
Actor	Event	Packages
Association	Final state	Scenario
Attributes	Focus of control	Self-delegation
Behavior	Generic sequence diagram	Sequence diagram
Behavioral models	Guard condition	State
Class diagram	Initial state	State symbol
Collaboration	Instance	Statechart diagram
Collaboration diagram	Instance sequence diagram	Temporary object
Condition	Lifeline	Transition
CRC cards	Links	Use case
CRUD analysis	Message	

QUESTIONS

1. How is behavioral modeling related to structural modeling?
2. How does a use case relate to a sequence diagram? a collaboration diagram?
3. Contrast the following sets of terms:
   state; behavior
   class; object
   action; activity
   use case; scenario
   method; message
4. Why is iteration important when creating a behavioral model?
5. Describe the main building blocks for the sequence diagram and how they are represented on the model.
6. How do you show that a temporary object is to go out of existence on a sequence diagram?
7. Do lifelines always continue down the entire page of a sequence diagram? Explain.
8. Describe the steps used to create a sequence diagram.
9. Describe the main building blocks for the collaboration diagram and how they are represented on the model.
10. How do you show the sequence of messages on a collaboration diagram?

11. How do you show the direction of a message on a collaboration diagram?
12. Describe the steps used to create a collaboration diagram.
13. Are states always depicted using rounded rectangles on a statechart diagram? Explain.
14. What kinds of events can lead to state transitions on a statechart diagram?
15. What are the steps in building a statechart diagram?
16. How are guard conditions shown on a statechart diagram?
17. Describe the type of class that is best represented by a statechart diagram. Give two examples of classes that would be good candidates for statechart diagrams.
18. Identify the model(s) that contains each of the following components:

actor	focus of control	multiplicity
association	guard condition	object
class	initial state	state
extends association	links	transition
final state	message	update operation

EXERCISES

A. Create a sequence diagram for each of the following scenario descriptions for a video store system. A Video Store (AVS) runs a series of fairly standard video stores.

1. Every customer must have a valid AVS customer card in order to rent a video. Customers rent videos for three days at a time. Every time a customer rents a video, the system must ensure that

the customer does not have any overdue videos. If so, the overdue videos must be returned and an overdue fee paid before the customer can rent more videos.

2. If the customer has returned overdue videos but has not paid the overdue fee, the fee must be paid before new videos can be rented. If the customer is a premier customer, the first two overdue fees can be waived, and the customer can rent the video.

3. Every morning, the store manager prints a report that lists overdue videos; if a video is two or more days overdue, the manager calls the customer to remind them to return the video.

B. Create a collaboration diagram for the video system in Exercise A.

C. Create a sequence diagram for each of the following scenario descriptions for a health club membership system.

1. When members join the health club, they pay a fee for a certain length of time. The club wants to mail out reminder letters to members asking them to renew their memberships one month before their memberships expire. About half of the members do not renew their memberships. These members are sent follow-up surveys to complete about why they decided not to renew so that the club can learn how to increase retention. If the member did not renew because of cost, a special discount is offered to that customer. Typically, 25% of accounts are reactivated because of this offer.

2. Every time a member enters the club, an attendant takes his or her card and scans it to make sure the person is an active member. If the member is not active, the system presents the amount of money it costs to renew the membership. The customer is given the chance to pay the fee and use the club, and the system makes note of the reactivation of the account so that special attention can be given to this customer when the next round of renewal notices are dispensed.

D. Create a collaboration diagram for each of the health club membership scenarios described in Exercise C.

E. Think about sending a first-class letter to an international pen pal. Describe the process that the letter goes through to get from your initial creation of the letter to being read by your friend, from the letter's perspective. Draw a statechart diagram that depicts the states that the letter moves through.

F. Consider the video store that is described in Exercise A. Draw a statechart diagram that describes the various states that a video goes through from the time it is placed on the shelf through the rental and return process.

G. Draw a statechart diagram that describes the various states that a travel authorization can have through its approval process. A travel authorization form is used in most companies to approve travel expenses for employees. Typically, an employee fills out a blank form and sends it to his or her boss for a signature. If the amount is fairly small (<$300), then the boss signs the form and routes it to accounts payable to be input into the accounting system. The system cuts a check that is sent to the employee for the right amount, and after the check is cashed, the form is filed away with the canceled check. If the check is not cashed within 90 days, the travel form expires. When the amount of the travel voucher is a large amount (> $300), then the boss signs the form and sends it to the CFO along with a paragraph explaining the purpose of the travel, and the CFO will sign the form and pass it along to accounts payable. Of course, both the boss and the CFO can reject the travel authorization form if they do not feel that the expenses are reasonable. In this case, the employee can change the form to include more explanation or decide to pay the expenses.

H. Identify the use cases for the following system. Picnics R Us (PRU) is a small catering firm with five employees. During a typical summer weekend, PRU caters 15 picnics with 20 to 50 people each. The business has grown rapidly over the past year, and the owner wants to install a new computer system for managing the ordering and buying process. PRU has a set of 10 standard menus. When potential customers call, the receptionist describes the menus to them. If the customer decides to book a picnic, the receptionist records the customer information (e.g., name, address, phone number, etc.) and the information about the picnic (e.g., place, date, time, which one of the standard menus, total price) on a contract. The customer is then faxed a copy of the contract and must sign and return it along with a deposit (often a credit card or by check) before the picnic is officially booked. The remaining money is collected when the picnic is delivered. Sometimes, the customer wants something special (e.g., birthday cake). In this case, the receptionist takes the information and gives it to the owner who determines the cost; the receptionist then calls the customer back with the price information. Sometimes the customer accepts the price; other times, the customer requests some changes which have to go back to the owner for a new cost estimate. Each week, the owner looks through the picnics scheduled for that weekend and orders the supplies (e.g., plates) and food (e.g., bread, chicken) needed to make them. The owner would like to use the system for marketing as well. It should be able to track how customers learned about PRU, and it should identify repeat customers, so that PRU can mail special offers to them. The owner also wants to track the picnics on which PRU sent a contract, but the customer never signed the contract and actually booked a picnic.

1. Create the use-case diagram for the PRU system.
2. Create a class diagram for the PRU system.
3. Choose one use case and create a sequence diagram.
4. Create a collaboration diagram for the use case chosen in Question 3.
5. Create a statechart diagram to depict one of the classes on the class diagram in Question 2.

I. Identify the use cases for the following system. Of-the-Month Club (OTMC) is an innovative young firm that sells memberships to people who have an interest in certain products. People pay membership fees for one year and each month receive a product by mail. For example, OTMC has a coffee-of-the-month club that sends members one pound of special coffee each month. OTMC currently has six memberships (coffee, wine, beer, cigars, flowers, and computer games), each of which costs a different amount. Customers usually belong to just one, but some belong to two or more. When people join OTMC, the telephone operator records the name, mailing address, phone number, e-mail address, credit card information, start date, and membership service(s) (e.g., coffee). Some customers request a double or triple membership (e.g., two pounds of coffee, three cases of beer). The computer game membership operates a bit differently from the others. In this case, the member must also select the type of game (action, arcade, fantasy/science-fiction, educational, etc.) and age level. OTMC is planning to greatly expand the number of memberships it offers (e.g., video games, movies, toys, cheese, fruit, vegetables), so the system needs to accommodate this future expansion. OTMC is also planning to offer three-month and six-month memberships.

1. Create the use-case diagram for the OTMC system.
2. Create a class diagram for the OTMC system.
3. Choose one use case and create a sequence diagram.
4. Create a collaboration diagram for the use case chosen in Question 3.
5. Create a statechart diagram to depict one of the classes on the class diagram in Question 2.

J. Think about your school or local library and the processes involved in checking out books, signing up new borrowers, and sending out overdue notices from the library's perspective. Describe three use cases that represent these three functions.

1. Create the use-case diagram for the library system.
2. Create a class diagram for the library system.
3. Choose one use case and create a sequence diagram.
4. Create a collaboration diagram for the use case chosen in Question 3.
5. Create a statechart diagram to depict one of the classes on the class diagram in Question 2.

K. Think about the system that handles student admissions at your university. The primary function of the system should be able to track a student from the request for information through the admissions process until the student is either admitted or rejected from attending the school. Write the use-case description that can describe an "admit student" use case.

1. Create the use-case diagram for the one use case. Pretend that students of alumni are handled differently than other students. Also, a generic "update student information" use case is available for your system to use.
2. Create a class diagram for the use case that you selected for Question 1. An admissions form includes the contents of the form, SAT information, and references. Additional information is captured about students of alumni, such as their parents' graduation year, contact information, and college major.
3. Choose one use case and create a sequence diagram. Pretend that the system uses a temporary student object to hold information about people before they send in an admission form. After the form is sent in, these people are considered "students."
4. Create a collaboration diagram for the use case chosen in Question 3.
5. Create a statechart diagram to depict a person as he or she moves through the admissions process.

MINICASES

1. Refer to the use case and use-case diagram you prepared for Professional and Scientific Staff Management (PSSM) for Minicase 2 in Chapter 6. PSSM was so satisfied with your work that they would like you to continue. In particular, they wanted the behavioral models created so that they could understand the interaction that would take place between the users and the system in greater detail.

   a. Create both a sequence and collaboration diagram for each use case.

   b. Based on the sequence and collaboration diagrams, create CRC cards for each class identified and a class diagram.

   c. Choose one of the more complex classes and create a statechart diagram for it.

2. Refer to the structural model that you created for Holiday Travel Vehicles for Minicase 2 in Chapter 7.

   a. Choose one of the more complex classes and create a statechart diagram for it.

   b. Based on the structural model you created, role-play the CRC cards to develop a set of use cases that describe a typical sales process.

   c. Create both a sequence and a collaboration diagram for the use case.

PART THREE
DESIGN PHASE

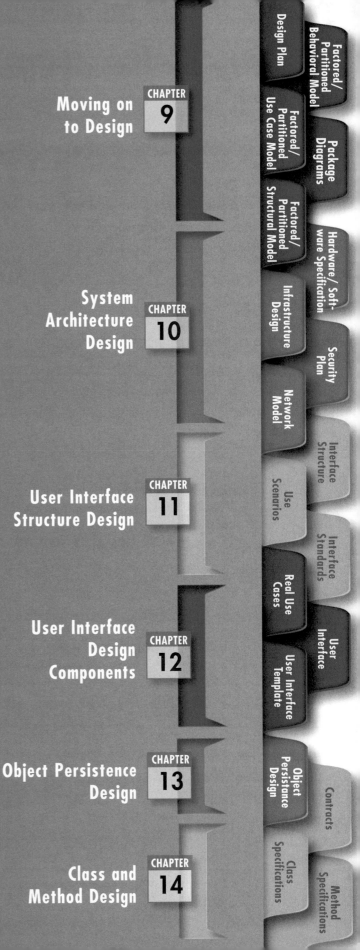

The Design Phase decides *how* the system will operate. First, the project team creates a Design Plan and a set of factored and partitioned Analysis models (Use Case, Structural, and Behavioral). The system architecture design is communicated using an Infrastructure Design, Network Model, Hardware/Software Specification, and Security Plan.

Next, the team designs the interface using Use Scenarios, the Interface Structure (using Windows Navigation Diagrams and Real Use Cases), Interface Standards, User Interface Template, and User Interface Design. Finally, the object persistence and class and method designs are illustrated using the Object Persistence Design, Class Specifications (using CRC Cards and Class Diagrams), Contracts, and Method Specifications. This collection of deliverables is the system specification that is handed to the programming team for implementation. At the end of the Design Phase, the feasibility analysis and project plan are reexamined and revised, and another decision is made by the project sponsor and approval committee about whether to terminate the project or continue.

PROJECT BINDER

Moving on to Design — **CHAPTER 9**

System Architecture Design — **CHAPTER 10**

User Interface Structure Design — **CHAPTER 11**

User Interface Design Components — **CHAPTER 12**

Object Persistence Design — **CHAPTER 13**

Class and Method Design — **CHAPTER 14**

Design Plan
Factored/Partitioned Behavioral Model
Factored/Partitioned Use Case Model
Factored/Partitioned Structural Model
Package Diagrams
Hardware/Software Specification
Infrastructure Design
Security Plan
Network Model
Interface Structure
Use Scenarios
Interface Standards
Real Use Cases
User Interface
User Interface Template
Object Persistence Design
Contracts
Class Specifications
Method Specifications

PLANNING

ANALYSIS

DESIGN

- [ ] **Evolve Analysis Models**
- [ ] **Develop the Design Strategy**
- [ ] **Develop the Design**
- [ ] Develop Infrastructure Design
- [ ] Develop Network Model
- [ ] Develop Hardware/Software Specification
- [ ] Develop Security Plan
- [ ] Develop Use Scenarios
- [ ] Design Interface Structure
- [ ] Design Interface Standards
- [ ] Design User Interface Template
- [ ] Design User Interface
- [ ] Evaluate User Interface
- [ ] Select Object Persistence Format
- [ ] Map Objects to Object Persistence
- [ ] Optimize Object Persistence
- [ ] Size Object Persistence
- [ ] Specify Constraints
- [ ] Develop Contracts
- [ ] Develop Method Specifications

TASK CHECKLIST

PLANNING   ANALYSIS   DESIGN

CHAPTER 9

MOVING ON TO DESIGN

THE design phase in object-oriented system development uses the requirements that were gathered during analysis to create a blueprint for the future system. A successful design builds on what was learned in earlier phases and leads to a smooth implementation by creating a clear, accurate plan of what needs to be done. This chapter describes the initial transition from analysis to design and presents three ways to approach the design for the new system.

OBJECTIVES

- Understand the transition from analysis to design.
- Understand the use of factoring, partitions, and layers.
- Be able to create package diagrams.
- Be familiar with the custom, packaged, and outsource design alternatives.
- Be able to create an alternative matrix.

CHAPTER OUTLINE

IMPLEMENTATION

INTRODUCTION

The purpose of the analysis phase is to figure out *what* the business needs. The purpose of the *design phase* is to decide *how* to build it. The major activity that takes place during this phase is evolving the set of analysis representations into design representations.

Throughout the design phase, the project team carefully considers the new system in respect to the current environment and system that exist within the organization as a whole. Major considerations of the "how" of a system are environmental factors such as integrating with existing systems, converting data from legacy systems, and leveraging skills that exist in-house. Although the planning and analysis phases are undertaken to develop a "possible" system, the goal of the design phase is to create a blueprint for a system that makes sense to implement.

An important initial part of the design phase is to examine several design strategies and decide which will be used to build the system. Systems can be built from scratch, purchased and customized, or outsourced to others, and the project team needs to investigate the viability of each alternative. The decision to make, buy, or outsource influences the design tasks that are accomplished throughout the rest of the phase.

At the same time, technical system architecture decisions are made regarding the hardware and software that will be purchased to support the new system and the way that the processing of the system will be organized. For example, the system can be organized so that its processing is centralized at one location, distributed, or both centralized and distributed, and each solution offers unique benefits and challenges to the project team. Global issues and security need to be considered along with the system's technical architecture because they will influence the implementation plans that are made. Technical system architecture, security, and global issues will be described later in Chapter 10.

The next steps in design include activities such as designing the user interface, system inputs, and system outputs, which involve the ways the user interacts with the system. Chapters 11 and 12 describe these three activities in detail, along with techniques such as storyboarding and prototyping that will help the project team design a system that meets the needs of its users and is satisfying to use.

Finally, Chapters 13 and 14 present object persistence and detailed design activities that are used to map out the nuts and bolts of the system. Techniques like CRC cards, class diagrams, contract specification, method specification, and database design provide the final design details in preparation for the implementation phase, and ensure that programmers have sufficient information to build the right system efficiently.

The many steps of the design phase are highly interrelated and, as with the steps in the analysis phase, the analysts often go back and forth among them. For example, prototyping in the interface design step often uncovers additional information that is needed in the system. Alternatively, a system that is being designed for an organization that has centralized systems may require substantial hardware and software investments if the project team decides to change to a system in which all of the processing is distributed.

In this chapter, we first revisit the object-oriented approach to analysis and design. We then overview the processes that are used to evolve the analysis models into design models. Next, we introduce the use of packages and package diagrams. Finally, we examine the three fundamental approaches to developing new systems: make, buy, or outsource.

9-1 AVOIDING CLASSIC DESIGN MISTAKES

In Chapters 3 and 4, we discussed several classic mistakes and how to avoid them. Here we summarize four classic mistakes in the design phase and discuss how to avoid them.

1. **Reducing Design Time:** If time is short, the temptation is to reduce the time spent in "unproductive" activities such as design so that the team can jump into "productive" programming. This results in missing important details that have to be investigated later at a much higher time cost (usually at least 10 times longer).

   Solution: If time pressure is intense, use timeboxing to eliminate functionality or move it into future versions.

2. **Feature Creep:** Even if you are successful at avoiding scope creep, about 25% of system requirements will still change. And changes—big and small—can significantly increase time and cost.

   Solution: Ensure that all changes are vital and that the users are aware of the impact on cost and time. Try to move proposed changes into future versions.

3. **Silver Bullet Syndrome:** Analysts sometimes believe the marketing claims for some design tools

that purport to solve all problems and magically reduce time and costs. No *one* tool or technique can eliminate overall time or costs by more than 25% (although some can reduce individual steps by this much).

   Solution: If a design tool has claims that appear too good to be true, just say no.

4. **Switching Tools in Mid-Project:** Sometimes analysts switch to what appears to be a better tool during design in the hopes of saving time or costs. Usually, any benefits are outweighed by the need to learn the new tool. This also applies to even "minor" upgrades to current tools.

   Solution: Don't switch or upgrade unless there is a compelling need for *specific* features in the new tool, and then explicitly *increase* the schedule to include learning time.

Source: Adapted from Steve McConnell, *Rapid Development* (Redmond, WA: Microsoft Press, 1996).

REVISITING THE OBJECT-ORIENTED APPROACH TO ANALYSIS AND DESIGN

As we stated in Chapter 1, object-oriented analysis and design (OOAD) approaches use a *phased development RAD* approach (see Figure 1-3). However, based on all of the iterating that goes on over all three sets of models (use case, structural, and behavioral), an actual object-oriented development process tends to be more complex than typical phased development RAD approaches. Also in Chapter 1, we presented three required characteristics of any OOAD approach: *use-case driven, architecture centric,* and *iterative* and *incremental.* In this section, we describe, in more detail, a generic OOAD approach that supports these characteristics.[1]

As you will recall, *architecture centric* means that the underlying software architecture of the evolving system specification drives the specification,

---

[1] The approach described is based on the work of D. Bellin and S.S. Simone, *The CRC Card Book* (Reading, MA: Addison-Wesley, 1997); G. Booch, J. Rumbaugh, and I. Jacobson, *The Unified Modeling Language User Guide* (Reading, MA: Addison-Wesley, 1999); A. Cockburn, *Writing Effective Use Cases,* (Reading, MA: Addison-Wesley, in preparation); I. Graham, *Migrating to Object Technology* (Wokingham, England: Addison-Wesley, 1995); I. Graham, B. Henderson-Sellers, and H. Younessi, *The OPEN Process Specification* (Harlow, England: Addison-Wesley, 1997); B. Henderson-Sellers and B. Unhelkar, *OPEN modeling with UML* (Harlow, England: Addison-Wesley, 2000); I. Jacobson, G. Booch, and J. Rumbaugh, *The Unified Software Development Process* (Reading, MA: Addison-Wesley, 1999); P. Kruchten, *The Rational Unified Process: An Introduction,* 2nd ed. (Reading, MA: Addison-Wesley, 2000), C. Larman, *Applying UML and Patterns: An Introduction to Object-Oriented Analysis and Design,* (Englewood Cliffs, NJ: Prentice-Hall, 1998); and T. Quatrani, *Visual Modeling with Rational Rose 2000 and UML* (Reading, MA: Addison-Wesley, 2000).

construction, and documentation of the system. There are three separate, but inter-related, views of a system: functional, static, and dynamic. The *functional view* describes the external behavior of the system from the perspective of the user. This is supported by use-case modeling utilizing the *use-case model*. The *static view* describes the structure of the systems in terms of attributes, methods, classes, relationships, and messages. This is supported by structural models depicted by CRC cards and the class and object diagrams. Finally, the *dynamic view* describes the internal behavior of the system in terms of messages passed between objects and state changes within an object. The dynamic view is supported by the behavioral models portrayed by sequence, collaboration, and statechart diagrams.

The approach to OOAD that we present in this section is based on the processes expressed in the Unified and Open approaches to object-oriented systems development.[2] However, because of the complexity of the Unified and Open approaches, we have followed a minimalist style[3] of presenting a generic approach to OOAD. As such, we have simplified the OOAD approaches suggested by Unified and Open. Figure 9-1 portrays our modified *phased delivery RAD-based approach.* The solid lines in Figure 9-1 represent information flows from one step in our approach to another step. The dashed lines represent feedback from a later step to an earlier one. For example, plans flow from the Planning step into the Requirements Gathering & Use-Case Development step and feedback from the Requirements Gathering & Use-Case Development step flows back to the Planning step.

The first step of the approach is planning. If you will recall, planning deals with understanding *why* an information system should be built and determining how the project team will go about building it. The second step is gathering requirements and building use cases for the system to be developed. Use cases identify and communicate the high-level business requirements for the system; that is, they provide a functional or external behavioral view of the system. The steps involved in developing use cases were given in Figure 6-7. Use cases and the use-case model drive the remaining steps in our approach; that is, all of the information required from the remaining steps in our approach is derived from the use cases and the use-case model.

Next, the developers of the system perform a *build.* Each build makes incremental progress toward delivering the entire system. The first build is based on the use cases with the highest priority (see Chapter 6). Later builds incorporate additional lower priority use cases. Builds are performed until the system is complete. Each build is comprised of an analysis, design, and implementation step. Each build provides feedback to the Requirements Gathering & Use-Case Development step and delivers a functional system to the next build and a set of patterns that can be included in the library. A build utilizes the idea of *timeboxing.*[4] Typically, in object-oriented development, the time frame for each timebox varies from one to two weeks to one to two months, depending on the size and complexity of the project.

[2] Graham, Henderson-Sellers, and Younessi, *The OPEN Process Specification;* Harlow, England Henderson-Sellers and Unhelkar, *OPEN modeling with UML;* Harlow, England Jacobson, Booch, and Rumbaugh, *The Unified Software Development Process;* Reading, MA and Kruchten, *The Rational Unified Process.* Bosten, MA.

[3] John M. Carrol, *The Nurnberg Funnel: Designing Minimalist Instruction for Practical Computer Skill,* (Cambridge, MA: MIT Press, 1990).

[4] Timeboxing sets a fixed deadline for a project and delivers the system by that deadline no matter what, even if the functionality needs to be reduced. Timeboxing ensures that project teams don't get hung up on the final "finishing touches" that can drag out indefinitely.

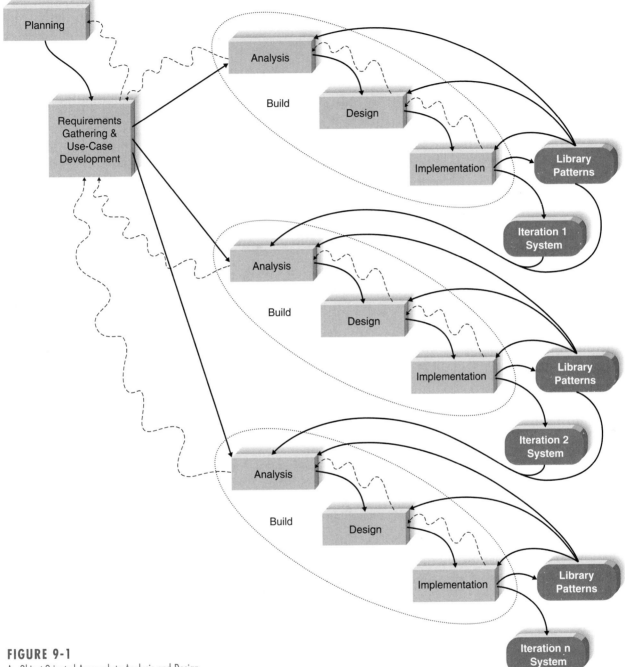

FIGURE 9-1
An Object-Oriented Approach to Analysis and Design

In this section of the book, we only address the second step within a build: design. The analysis steps were described in Chapters 4 through 8. The steps for developing use cases and use-case diagrams were given in Figure 6-7, the steps for object identification and structural modeling were specified in Figure 7-8, and the steps for behavioral modeling were listed in Figures 8-3, 8-8, and 8-13. Figure 9-2 combines and summarizes these steps into a set of steps that can be used to develop

Develop Use Cases (Figure 6-7)

Identify the Major Use Cases
Find the system's boundaries.
List the primary actors.
List the goals of the primary actors.
Identify and write the overviews of the major use cases for the above.
Carefully review the current use cases. Revise as needed.

Expand the Major Use Cases
Choose one of the use cases to expand.
Start filling in the details of the chosen use case.
Write the Normal Flow of Events of the use case.
If the Normal Flow of Events is too complex or too long, decompose into subflows.
List the possible alternate or exceptional flows.
For each alternate or exceptional flow, list how the actor and/or system should react.

Confirm the Major Use Cases
Carefully review the current set of use cases. Revise as needed.
Start at the top again.

Create the Use-Case Diagram
Draw the system boundary.
Place the use cases on the diagram.
Place the actors on the diagram.
Draw the associations.

Develop Structural Model (Figure 7-8)

Create CRC cards by performing textual analysis on use cases.
Brainstorm additional candidate classes, attributes, operations, and relationships.
Role-play each use case using the CRC cards.
Develop behavioral models
If interested in time-based ordering of activities, create sequence diagrams (Figure 8-3).
Set the context.
Identify which objects will participate.
Set the lifeline for each object.
Lay out the messages from the top to the bottom of the diagram based on the order in which they are sent.
Add the focus of control to each object's lifeline.
Validate the sequence diagram.
If interested in patterns of activity over a set of objects, create collaboration diagrams (Figure 8-8).
Set the context.
Identify which objects (actors) and the associations between the objects participate in the collaboration.
Lay out the collaboration diagram.
Add messages.
Validate the collaboration diagram.
For complex classes, create statechart diagrams (Figure 8-13).
Set the context.
Identify the initial, final, and stable states of the object.
Determine the order in which the object will pass through the stable states.
Identify the events, actions, and guard conditions associated with the transitions.
Validate the statechart diagram.
Create class diagrams based on CRC cards.
Review the structural model for missing and/or unnecessary classes, attributes, operations, and relationships.
Incorporate useful patterns.
Review the structural model.

FIGURE 9-2
Object-Oriented Analysis Steps

the use-case, structural, and behavioral models. The details of the implementation steps will be described in the Chapters 15 and 16.

EVOLVING THE ANALYSIS MODELS INTO DESIGN MODELS

The purpose of the analysis models was to represent the underlying business problem domain as a set of collaborating objects. In other words, the analysis activities defined the functional requirements. To successfully model the problem domain, the analysis activities ignored nonfunctional requirements such as performance and the system environment issues (e.g., distributed or centralized processing, user interface issues, and database issues). In contrast, the primary purpose of the design models is to increase the likelihood of successfully delivering a system that implements the functional requirements in a manner that is affordable and easily maintainable.

Therefore, in system design, we address both the functional and nonfunctional requirements. From an object-oriented perspective, system design models simply refine the system analysis models by adding system environment (or solution domain) details to them and refining the problem domain information already contained in the analysis models.

When evolving the analysis model into the design model, you should first carefully review the use cases and the current set of classes (their methods and attributes, and the relationships between them). Are all of the classes necessary, or are there any missing classes? Are the classes fully defined? Are there any missing attributes or methods? Do the classes have any unnecessary attributes and methods? Is the current representation of the evolving system optimal?

In the following sections, we introduce factoring, partitions and collaborations, and layers as a way to evolve problem domain-oriented analysis models into optimal solution domain-oriented design models.

Factoring

Factoring is the process of separating out a *module* into a standalone module in of itself. The new module can be a new *class* or a new *method.* For example, when reviewing a set of classes, it may be discovered that they have a similar set of attributes and methods. As such, it may make sense to factor out the similarities into a separate class. Depending on whether or not the new class should be in a superclass relationship to the existing classes, the new class can be related to the existing classes through *generalization (a-kind-of)* or possibly through *aggregation (has-parts)* relationship. For example, using the appointment system example in the previous chapters (see Figure 7-2), if the Employee class had not been identified, we could possibly identify it at this stage by factoring out the similar methods and attributes from the Nurse, Administrative Staff, and Doctor classes. In this case, we would relate the new class (Employee) to the existing classes using the generalization (a-kind-of) relationship.

Abstraction and *refinement* are two processes closely related to factoring. Abstraction deals with the creation of a "higher" level idea from a set of ideas. The above example of identifying the Employee class is an example of abstracting from a set of lower classes to a higher one. In some cases, the abstraction process will identify *abstract classes,* while in other situations, it will identify additional *concrete classes.*[5] The refinement process is the opposite of the abstraction process. In the appointment system example in the previous chapters (see Figure 7-2), we could identify additional subclasses of the Administrative Staff class, such as receptionist, secretary, and bookkeeper. Of course, we would only add the new classes if there were sufficient differences between them. Otherwise, the more general class, Administrative Staff, would suffice.

Partitions and Collaborations

Based on all of the factoring, refining, and abstracting that can take place to the evolving system, the sheer size of the system representation can overload both the user and the developer. At this point in the evolution of the system, it may make sense to split the representation into a set of *partitions*. A partition is the object-oriented equivalent of a subsystem,[6] in which a subsystem is a decomposition of a larger system into its component systems. For example, an accounting information system could be functionally decomposed into an accounts payable system, an accounts receivable system, a payroll system, and so on. From an object-oriented perspective, partitions are based on the pattern of activity (messages sent) among the objects in an object-oriented system.

A good place to look for potential partitions is the *collaborations* modeled in UML's collaboration diagrams (see Chapter 8). As noted earlier, one useful way to identify collaborations is to create a collaboration diagram for each use case. However, since an individual class may support multiple use cases, an individual class can participate in multiple use-case-based collaborations. In those cases where classes are supporting multiple use cases, the collaborations should be merged together.[7]

Depending on the complexity of the merged collaboration, it may be useful to decompose the collaboration into multiple partitions. In this case, in addition to having collaborations between objects, it is possible to have collaborations among partitions. The general rule of thumb is the more messages sent between objects, the more likely the objects belong in the same partition. The fewer messages sent, the less likely the two objects belong together.

Another useful approach to identify potential partitions is to model each collaboration between objects in terms of clients, servers, and contracts, where a *client* is an instance of a class that sends a *message* to an instance of another class for a *method* to be executed, a *server* is the instance of a class that receives the message, and a *contract* is the specification that formalizes the interactions between the client and server objects (see Chapters 7 and 14). This allows the developer to build up potential partitions by looking at the contracts that have been specified between objects. In this case, the more contracts there are between objects, the more often than not the objects belong in the same partition. The fewer contracts, the chances are the two classes do not belong in the same partition.

---

[5] See Chapter 7 for the differences between abstract and concrete classes.

[6] Some authors refer to partitions as subsystems (e.g., see Rebecca Wirfs-Brock, Brian Wilkerson, and Lauren Weiner, *Designing Object-Oriented Software,* (Englewood Cliffs, NJ: Prentice-Hall, 1990), while others refer to them as layers (e.g., see Graham, *Migrating to Object Technology*). However, we have chosen to use the term *partition* (from Larman, *Applying UML and Patterns*) to minimize confusion between subsystems in a traditional system development approach and layers associated with *Rational's* Unified Process (Kruchten, *The Rational Unified Process).*

[7] CRUD analysis (see Chapter 8) can be used to identify potential classes on which to merge collaborations.

Layers

Up to this point in the development of our system, we have focused only on the problem domain and have totally ignored the system environment (system architecture, user interface, and data management). To successfully evolve the analysis model of the system into a design model of the system, we must add the system environment information. One useful way to do this, without overloading the developer, is to use *layers*. A layer represents an element of the software architecture of the evolving system. We have concentrated only on one layer in the evolving software architecture: the problem domain layer. There should be a layer for each different element of the system environment (e.g., system architecture, user interface, and data management).

The idea of separating the different elements of the system architecture into separate layers can be traced back to the MVC architecture of *Smalltalk*.[8] When Smalltalk was first created,[9] the authors decided to separate the application logic from the logic of the user interface. In this manner, it was possible to easily develop different user interfaces that worked with the same application. To accomplish this, they created the *Model-View-Controller* (MVC) architecture in which *Models* implemented the application logic (problem domain), and *Views* and *Controllers* implemented the logic for the user interface: Views handled the output and Controllers handled the input. Since graphical user interfaces were first developed in the Smalltalk language, the MVC architecture served as the foundation for virtually all graphical user interfaces that have been developed today (including the Mac interfaces, the Windows family, and the various Unix-based GUI environments).

Based on Smalltalk's innovative MVC architecture, many different software layers have been proposed.[10] Based on these proposals, we suggest the following layers on which to base a software architecture: foundation, system architecture, human–computer interaction, data management, and problem domain (see Figure 9.3). Each layer limits the types of classes that can exist on it; for example, only user interface classes may exist on the human computer interaction layer. Next we give a short description of each layer.

Foundation   The *foundation layer* is, in many ways, a very uninteresting layer. It contains classes that are necessary for any object-oriented application to exist. It includes classes that represent fundamental data types, such as integers, real numbers, characters, and strings, classes that represent fundamental data structures

---

[8] See Simon Lewis, *The Art and Science of Smalltalk: An Introduction to Object-Oriented Programming Using VisualWorks.* (Englewood Cliffs, NJ: Prentice-Hall, 1995).

[9] Smalltalk was first invented in the early 1970s by a software development research team at Xerox PARC. It introduced many new ideas into the area of programming languages, for example, object-orientation, windows-based user interfaces, reusable class library, and the idea of a development environment. In many ways, Smalltalk is the father (or mother) of all object-based and object-oriented languages, such as Visual Basic, C++, and Java.

[10] For example, Problem Domain, Human Interaction, Task Management, and Data Management (Peter Coad and Edward Yourdon, *Object-Oriented Design,* Englewood Cliffs, NJ: Yourdon Press, 1991); Domain, Application, and Interface (Graham, *Migrating to Object Technology*); Domain, Service, and Presentation (Larman, *Applying UML and Patterns*); Business, View, and Access (Ali Bahrami, *Object-Oriented Systems Development Using the Unified Modeling Language,* New York: McGraw-Hill, 1999); Application-Specific, Application-General, Middleware, System-Software (Jacobson, Booch, and Rumbaugh, *The Unified Software Development Process*); and Foundation, Architecture, Business, and Application (Meilir Page-Jones, *Fundamentals of Object-Oriented Design in UML,* Reading, MA: Addison-Wesley, 2000).

Layers	Examples	Relevant Chapters
Foundation	Date Enumeration	9, 14
System Architecture	ServerSocket URLConnection	10, 14
Human–Computer Interaction	Button Panel	11, 12, 14
Data Management	DataInputStream FileInputStream	13, 14
Problem Domain	Employee Customer	6, 7, 8, 9, 14

FIGURE 9-3
Layers and Sample Classes

(sometimes referred to as container classes), such as lists, trees, graphs, sets, stacks, and queues, and classes that represent useful abstractions (sometimes referred to as utility classes) such as data, time, and money. Today, the classes found on this layer typically are included with the object-oriented development environments.

System Architecture The *system architecture layer* addresses how the software will execute on specific computers and networks. As such, this layer includes classes that deal with communication between the software and the computer's operating system and the network. For examp!e, classes that address how to interact with the various ports on a specific computer would be part of this layer. Also included are classes that would interact with so-called *middleware* applications such as the OMG's *CORBA* and Microsoft's *DCOM* architectures that handle distributed objects.

Unlike the foundation layer, many design issues must be addressed before choosing the appropriate set of classes for this layer. Among these design issues are the choice of a computing or network architecture (such as the various client-server architectures), the actual design of a network, hardware and server software specification, global/international issues (such as multilingual requirements), and security issues. A complete description of all the issues related to system architecture is beyond the scope of this book; entire courses are dedicated to this subject. However, we do present the basic issues in Chapter 10.

Human–Computer Interaction The *human–computer interaction layer* contains classes associated with the View and Controller idea from Smalltalk. The primary purpose of this layer is to keep the specific user interface implementation separate from the application or problem domain classes. This increases the portability of the evolving system. Typical classes found on this layer include classes that can be used to represent buttons, windows, text fields, scroll bars, check boxes, drop-down lists, and many other classes that represent user interface elements.

Many issues must be addressed when one is designing the user interface for an application. For example, how important is consistency across different user interfaces? What about differing levels of user experience? How is the user expected to be able to navigate through the system? What about help systems and online manuals? What types of input elements should be included (e.g., text box, radio buttons, check boxes, sliders, drop-down list boxes, etc.)? What types of output elements should be included (e.g., text, tables, graphs, etc.)? Like the System architecture layer, a complete description of all the issues related to human–computer interaction is beyond the scope of this book.[11] However, from the user's per-

[11] One of the best books on user interface design is Ben Schneiderman, *Designing the User Interface: Strategies for Effective Human Computer Interaction,* 3rd ed. (Reading, MA: Addison-Wesley, 1998).

spective, the user interface is the system. As such, we present the basic issues in user interface design in Chapters 11 and 12.

Data Management   The *data management layer* addresses the issues involving the persistence of the objects contained in the system. The types of classes that appear in this layer deal with how objects can be stored and retrieved. Like the human–computer interface layer, the classes contained in this layer allow the problem domain classes to be independent of the storage utilized, and hence increase the portability of the evolving system. Some of the issues related to this layer include choice of the storage format (such as relational, object/relational, and object databases) and storage optimization (such as clustering and indexing). A complete description of all the issues related to the data management layer also is beyond the scope of this book.[12] However, we do present the fundamentals in Chapter 13.

Problem Domain   The previous layers dealt with classes that represent elements from the system environment and that will be added to the evolving system specification. The *problem domain layer* is what we have focused our attention on up until now. At this stage of the development of our system, we will need to further detail the classes so that it will be possible to implement them in an effective and efficient manner.

Many issues need to be addressed when designing classes, no matter on which layer they appear. For example, some issues relate to factoring, cohesion and coupling, connascence, encapsulation, proper use of inheritance and polymorphism, constraints, contract specification, and detailed method design. These issues are discussed in Chapter 14.

PACKAGES AND PACKAGE DIAGRAMS

In UML, collaborations, partitions, and layers can be represented by a higher level construct: a package.[13] A *package* is a general construct that can be applied to any element in UML models. In Chapter 6, we introduced the idea of packages as a way to group use cases together to make the use-case diagrams easier to read and to keep the models at a reasonable level of complexity. In Chapters 7 and 8, we did the same thing for class and collaboration diagrams, respectively. In this sections, we describe a package diagram: a diagram that is composed only of packages. A *package diagram* is effectively a class diagram that shows packages only.

The symbol for a package is similar to a tabbed folder (see Figure 9-4). Depending on where a package is used, packages can participate in different types of relationships. For example, in a class diagram, packages represent groupings of classes; therefore, aggregation and association relationships are possible.

In a package diagram, it is useful to depict a new relationship, a *dependency relationship*. A dependency relationship is portrayed by a dashed arrow (see Figure 9-4).

---

[12] Many good database design books are relevant to this layer; see, for example, Fred R. McFadden, Jeffrey A. Hoffer, and Mary B. Prescott, *Modern Database Management,* 4th ed. (Reading, MA: Addison-Wesley, 1998); Michael Blaha and William Premerlani, *Object-Oriented Modeling and Design for Database Applications,* (Englewood Cliffs, NJ: Prentice Hall, 1998), and Robert J. Muller, *Database Design for Smarties: Using UML for Data Modeling,* (San Mateo, CA: Morgan Kaufmann, 1999).

[13] This discussion is based on material in Chapter 7 of Martin Fowler with Kendal Scott, *UML Distilled: A Brief Guide to the Standard Object Modeling Language,* 2nd ed., (Reading, MA: Addison Wesley, 2000).

A PACKAGE A logical grouping of UML elements Used to simplify UML diagrams by grouping related elements into a single higher-level element.	Package
A DEPENDENCY RELATIONSHIP Represents a dependency between packages; that is, if a package is changed, the dependent package may also have to be modified. The arrow is drawn from the dependent package toward the package on which it is dependent.	┈┈┈┈┈▶

FIGURE 9-4
Syntax for Package Diagram

This relationship shows that a modification dependency exists between two packages. That is, a change in one package potentially could cause a change to be required in another package. Figure 9-5 portrays the dependencies among the different layers (foundation, system architecture, human–computer interaction, data management, and problem domain). For example, a change in the problem domain layer will most likely cause changes to occur in the human–computer interaction layer. However, the reverse is not true.

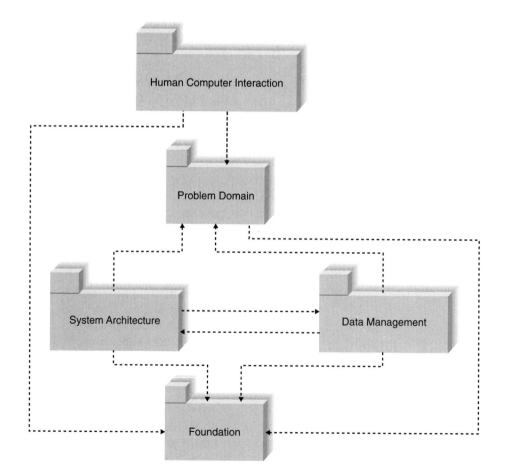

FIGURE 9-5
Package Diagram of Dependency
Relationships among Layers

At the class level, many causes for dependencies among classes are possible. For example, a change in the protocol for a method will cause the interface for all objects of this class to change. Therefore, all classes that have objects that send messages to the instances of the modified class could have to be modified. Capturing dependency relationships among the classes and packages helps maintain systems.

As stated above, collaborations, partitions, and layers are modeled as packages in UML. Furthermore, collaborations are normally factored into a set of partitions, which are typically placed on a layer. In addition, partitions can be made up of other partitions. It is also possible to have classes in partitions, which are contained in another partition and placed on a layer. All of these groupings are represented using packages in UML. Remember that a package is simply a generic grouping construct used to simplify UML models through the use of composition.[14]

A simple package diagram, based on the appointment system example from the previous chapters, is shown in Figure 9-6. This diagram portrays only a very small portion of the entire system. In this case, we see that the Patient UI, DM-Patient, and Patient Table classes are dependent on the Patient class. Furthermore, the DM-Patient class is dependent on the Patient Table class. The same can be seen with the classes dealing with the actual appointments. By isolating the Problem Domain classes (such as the Patient and Appt classes) from the actual object persistence classes (such as the Patient Table and Appt Table classes) through the use of the intermediate Data Management classes (DM-Patient and DM-Appt classes), we isolate the Problem Domain classes from the actual storage medium.[15] This greatly simplifies the maintenance and increases the reusability of the Problem Domain classes. Of course, in a complete description of a real system, there would be many more dependencies.

Identifying Packages and Creating Package Diagrams

In this section, we describe a simple five-step process to create package diagrams (see Figure 9-7). The first step is to set the context for the package diagram. Remember: packages can be used to model partitions and/or layers. Revisiting the appointment system again, let's set the context as the problem domain layer.

The second step is to cluster the classes together into partitions based on the relationships that the classes share. The relationships include generalization, aggregation, the various associations, and the message sending that takes place between the objects in the system. To identify the packages in the appointment system, we should look at the different analysis models, for example, class diagrams (see Figure 7-2), sequence diagrams (see Figure 8-1), and collaboration diagrams (see Figure 8-6). Any classes in a generalization hierarchy should be kept together in a single partition.

The third step is to place the clustered classes together in a partition and model the partitions as packages. Figure 9-8 portrays five packages: PD Layer, Person Pkg, Patient Pkg, Appt Pkg, and Treatment Pkg.

The fourth step is to identify the dependency relationships among the packages. In this case, we review the relationships that cross the boundaries of the packages to uncover potential dependencies. In the appointment system, we see association

---

[14] For those familiar with traditional approaches, such as structured analysis and design, packages serve a similar purpose as the leveling and balancing processes used in data flow diagramming.

[15] These issues are described in more detail in Chapter 13.

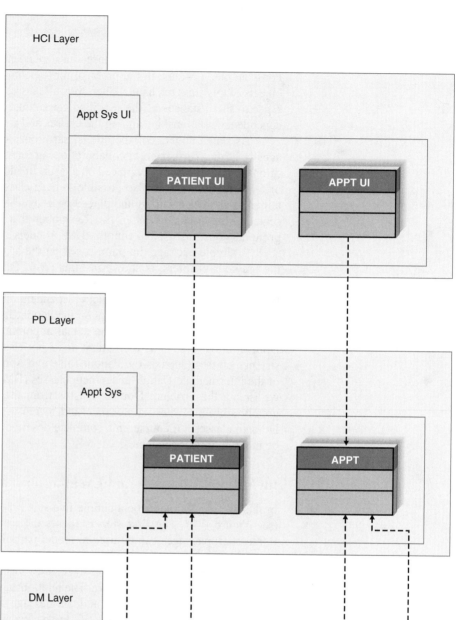

FIGURE 9-6
Package Diagram of Appointment
System

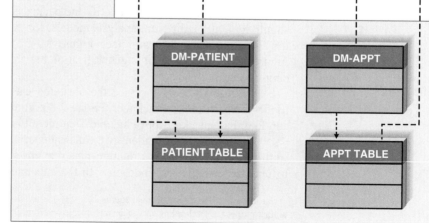

1. Set the context.
2. Cluster classes together based on shared relationships.
3. Model clustered classes as a package.
4. Identify dependency relationships among packages.
5. Place dependency relationships between packages.

FIGURE 9-7
Steps for Identifying Packages and
Building Package Diagrams

relationships that connect the Person Pkg with the Appt Pkg (via the association between the Doctor class and the Appointment class), and the Patient Pkg, which is contained within the Person Pkg, with the Appt Pkg (via the association between the Patient and Appointment classes) and the Treatment Pkg (via the association between the Patient and Symptom classes).

The fifth step is to place the dependency relationships on the evolved package diagram. In the case of the Appointment system, dependency relationships exist between the Person Pkg and the Appt Pkg, and the Person Pkg and the Treatment Pkg. To increase the understandability of the dependency relationships among the different packages, a pure package diagram that only shows the dependency relationships among the packages is created (see Figure 9-9).

Applying the Concepts at CD Selections

The CD Selections Internet Sales System was described in the previous chapters. In Chapter 6, the basic concept for the system was given in Figure 6-8, and the use cases and use-case diagram were given in Figures 6-9 through 6-13. The structural model comprised of the CRC cards and class diagram was given in Chapter 7 (see Figures 7-9 through 7-12). Finally, behavioral models developed for the Place Order use case and the Order class were shown in Chapter 8 (see the sequence diagram in Figure 8-5, the collaboration diagram in Figure 8-10, and the statechart diagram in Figure 8-15). In this section, we demonstrate the creation of a package diagram for the CD Selections Internet Sales System.

As detailed above, the first thing to do when identifying packages and creating package diagrams is to set the context. At this point in the development of the Internet Sales System for CD Selections, Alec wants the development team to concentrate solely on the problem domain layer.

The second step, cluster classes together, is accomplished by reviewing the relationships among the different classes (see Figures 7-12, 8-5, and 8-10). Through this review process, we see that there are generalization, aggregation, various associations, and message sending relationships. Since we always want to keep classes in a generalization hierarchy together, we automatically cluster the Customer, Individual, and Organization classes together to form a partition. It is also preferred to keep together the classes that participate in aggregation relationships. Based on aggregation relationships, we cluster the Mkt Info, Review, Artist Info, and Sample Clip classes together in a partition. Based on the association relationship and the message sending pattern of activity (contained in the collaboration diagram in Figure 8-10) between the CD and Mkt Info classes, the development team concluded that these classes should be in the same partition. Furthermore, since the Vendor class is related only to the CD class (see the class diagram in Figure 7-12), the development team decided to place it in the same

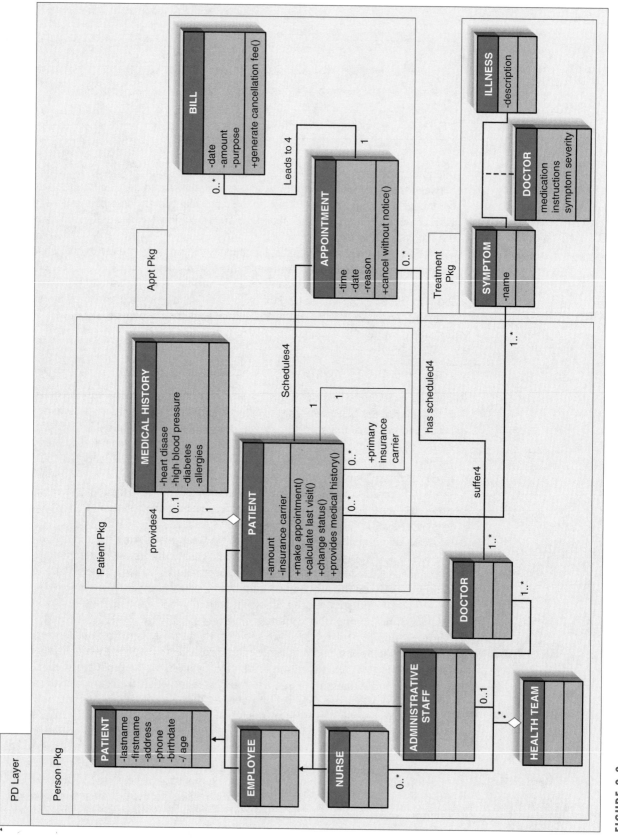

FIGURE 9-8
Package Diagram of the PD Layer for the Appointment System

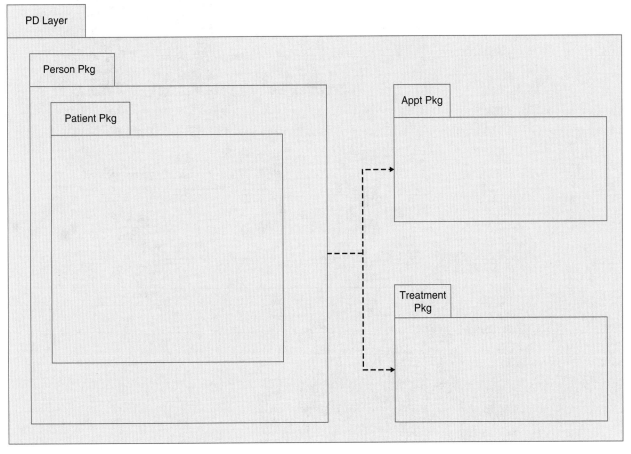

FIGURE 9-9
Overview Package Diagram of the PD Layer for the Appointment System

partition. Finally, the development team placed the Order and Order Item classes together and the Search Req and CD List classes together in their own partitions.

The third step was to model each of these partitions as packages. Figure 9-10 shows the classes being contained in their respective packages. Observe that the Credit-Card Clearance Center currently is not contained in any package.

Identifying dependency relationships among the packages is the fourth step. In this case, Alec was able to quickly identify four associations among the different packages: the Customer Package and the Order Package, the Customer Package and the Search Package, the Order Package and the CD Package, and the Search Package and the CD Package. He also identified an association between the Credit Card Clearance Center class and the Customer Package. Based on these associations, five dependency relationships were identified.

The fifth and final step is to place the dependency relationships on the package diagram. Again, to increase the understandability of the dependency relationships among the different packages, Alec decided to create a pure package diagram that only depicted the highest level packages (and in this case the Credit Card Clearance Center class) and the dependency relationships (see Figure 9-11).

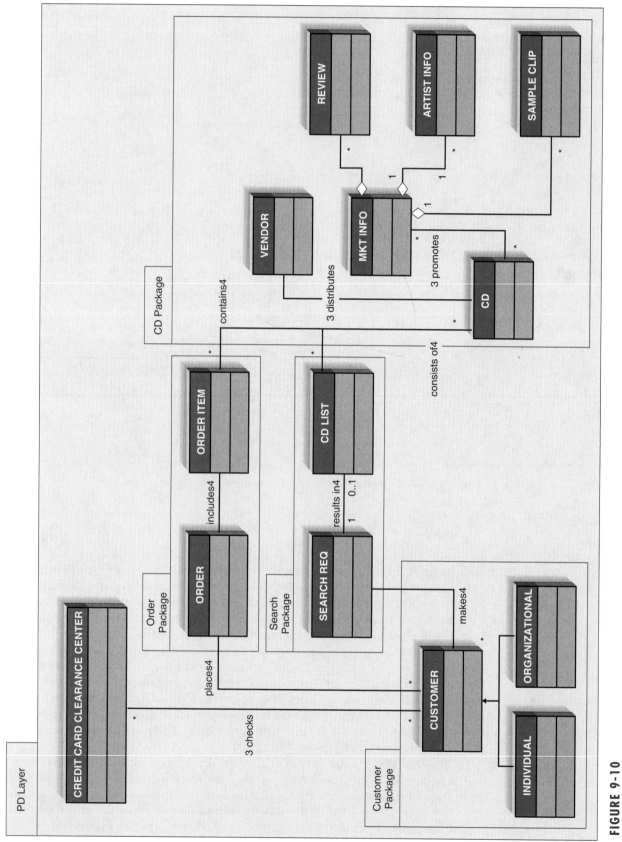

FIGURE 9-10
Package Diagram of the PD Layer of CD Selections Internet Sales System

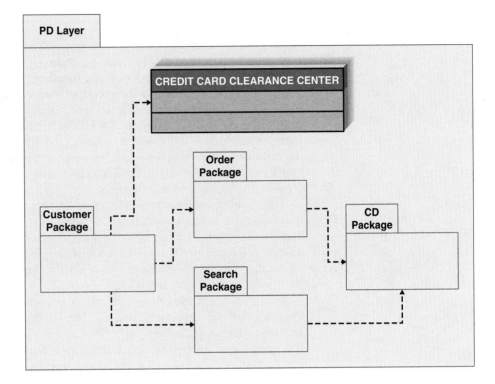

FIGURE 9-11
Overview Package Diagram of the PD
Layer of CD Selections Internet Sales
System

DESIGN STRATEGIES

Until now, we have assumed that the system will be built and implemented by the project team. However, there are actually three ways to approach the creation of a new system: develop a custom application in-house; buy a packaged system and customize it; and rely on an external vendor, developer, or service provider to build the system. Each of these choices has its strengths and weaknesses, and each is more appropriate in different scenarios. The following sections describe each design choice in turn, and then we present criteria that you can use to select one of the three approaches for your project.

YOUR	9-1 CAMPUS HOUSING
TURN	

You have been working with the system for the campus housing service over the previous three chapters. In Chapter 6, you were asked to create a set of use cases (Your Turn 6-2) and a use-case diagram (Your Turn 6-4). In Chapter 7, you created a structural model (CRC cards and class diagram) for the same situation. In Chapter 8, you created a sequence diagram (Your Turn 8-1) and a collaboration diagram (Your Turn 8-2) for one of the use cases you had identified in Chapter 6. Finally, you also created a statechart diagram for the apartment class in Chapter 8 (Your Turn 8-3).

Now, you should identify a set of partitions for the problem domain layer and model them in a package diagram.

Custom Development

Many project teams assume that *custom development,* or building a new system from scratch, is the best way to create a system. For one thing, teams have complete control over how the system looks and functions. Let's consider the order-taking process for CD Selections. If the company wants a Web order-taking feature that links tightly with its existing distribution system, the project may involve a complex, highly specialized program. Or CD Selections might have a technical environment in which all information systems are built using standard technology and interface designs so that they are consistent and easier to update and support. In both cases, it could be very effective to create a new system from scratch that meets these highly specialized requirements.

Custom development also allows developers to be flexible and creative in the way they solve business problems. CD Selections may envision the Web interface that takes customer orders as an important strategic enabler. The company may want to use the information from the system to better understand their customers who order over the Web, and it may want the flexibility to evolve the system to incorporate technology, such as data mining software and geographic information systems to perform marketing research. A custom application would be easier to change to include components that take advantage of current technologies that can support such strategic efforts.

Building a system in-house also builds technical skills and functional knowledge within the company. As developers work with business users, their understanding of the business grows and they become better able to align IS with strategies and needs. These same developers climb the technology learning curve so that future projects applying similar technology become much easier.

CONCEPTS **9-A BUILDING A CUSTOM SYSTEM—WITH SOME HELP**

IN ACTION

I worked with a large financial institution in the Southeast that suffered serious financial losses several years ago. A new CEO was brought in to change the strategy of the organization to being more "customer-focused." The new direction was quite innovative, and it was determined that custom systems, including a data warehouse, would have to be built to support the new strategic efforts. The problem was that the company did not have the in-house skills for these kinds of custom projects.

The company now has one of the most successful data warehouse implementations because of its willingness to use outside skills and because of its focus on project management. To supplement skills within the company, eight sets of external consultants, including hardware vendors, system integrators, and business strategists, were hired to take part in the data warehouse project and transfer critical skills to internal employees.

An in-house project manager coordinated the data warehouse implementation full-time, and her primary goals were to clearly set expectations, define responsibilities, and communicate the interdependencies that existed among the team members.

This company shows that successful custom development can be achieved, even when the company may not start off with the right skills in-house. But this kind of project is not easy to pull off. It takes a talented project manager to keep the project moving along and to transition the skills to the right people over time. *Barbara Wixom*

QUESTIONS

What are the risks in building a custom system without the right technical expertise? Why did the company select a project manager from within the organization? Would it have been better to hire an external professional project manager to coordinate the project?

Custom application development, however, also includes a dedicated effort that includes long hours and hard work. Many companies have a development staff that already is overcommitted to filling huge backlogs of systems requests, and so they simply do not have time for another project. Also, a variety of skills—technical, interpersonal, functional, project management, and modeling—all have to be in place for the project to move ahead smoothly. IS professionals, especially highly skilled individuals, are quite difficult to hire and retain.

The risks associated with building a system from the ground up can be quite high, and there is no guarantee that the project will succeed. Developers could be pulled away to work on other projects, technical obstacles could cause unexpected delays, and the business users could become impatient with a growing timeline.

Packaged Software

Many business needs are not unique, and because it makes little sense to "reinvent the wheel," many organizations buy *packaged software*—a software program that has already been written—rather than developing their own custom solution. Thousands of commercially available software programs have already been written to serve a multitude of purposes. Think about your own need for a word processor. Did you ever consider writing your own word processing software? That would be very silly considering the number of good software packages that are available for a relatively inexpensive cost.

Similarly, most companies have needs that can be met quite well by packaged software such as payroll or accounts receivable. It can be much more efficient to buy programs that have already been created, tested, and proven, and a packaged system can be bought and installed in a relatively short period of time, when compared with a custom system. In addition, packaged systems incorporate the expertise and experience of the vendor who created the software.

Let's examine the Order Taking process once again. Certain programs are available, called shopping cart programs, that allow a company to sell products on the Internet by keeping track of customers' selections, totaling them, and then e-mailing the order to a mailbox. You can easily install the program on an existing Web page, and it allows you to take orders. Some shopping cart programs are even available over the Internet for free. Therefore, CD Selections could decide that acquiring a ready-made order-taking application might be much more efficient than creating one from scratch.

Packaged software can range from reusable components (such as ActiveX and Javabeans) to small single-function tools (like the shopping cart program) to huge, all-encompassing systems such as *enterprise resource planning (ERP)* applications that are installed to automate an entire business. Implementing ERP systems is a growing trend in which large organizations spend millions of dollars installing packages by companies like SAP, Peoplesoft, Oracle, and Baan and then change their businesses accordingly. Installing ERP software is much more difficult than installing small application packages because the benefits it brings can be harder to realize and the problems it involves are much more serious.

One problem is that companies buying packaged systems must accept the functionality that the system provides, and rarely is there a perfect fit. If the packaged system is large in scope, its implementation could mean a substantial change in the way the company does business. Letting technology drive the business can be a dangerous way to go.

Most packaged applications allow for *customization,* or the manipulation of system parameters to change the way certain features work. For example, the package might have a way to accept information about your company or the company logo that would then appear on input screens. Or an accounting software package could offer a choice of ways to handle cash flow or inventory control so that it could support the accounting practices in different organizations. If the amount of customization is not enough, and the software package has a few features that don't quite work the way the company needs it to work, the project team can create workarounds.

A *workaround* is a custom-built add-on program that interfaces with the packaged application to handle special needs. It can be a nice way to create needed functionality that does not exist in the software package. But workarounds should be a last resort for several reasons. First, they are not supported by the vendor who supplied the packaged software, so when upgrades are made to the main system, they may make the workaround ineffective. Also, if problems arise, vendors have a tendency to make the workaround the culprit and so they refuse to provide support.

Although choosing a packaged software system is simpler than custom development, it too can benefit from following a formal methodology just as if you were building a custom application.

Systems integration refers to the process of building new systems by combining packaged software, existing legacy systems, and new software written to integrate these together. Many consulting firms specialize in systems integration, so it is not uncommon for companies to select the packaged software option and then outsource the integration of a variety of packages to a consulting firm. (Outsourcing is discussed in the next section.)

The key challenge in systems integration is finding ways to integrate the data produced by the different packages and legacy systems. Integration often hinges on taking data produced by one package or system and reformatting it for use in another package/system. The project team starts by examining the data produced by and needed by the different packages/systems and identifying the transformations that must occur to move the data from one to the other. In many cases, this involves "fooling" the different packages/systems into thinking that the data was produced by an existing program module that the package/system expects to produce the data rather than the new package/system that is being integrated.

For example, CD Selections needs to integrate the new Internet Sales System with existing legacy systems such as the distribution system. The distribution system was written to support a different application—the distribution of CDs to retail stores—and it currently exchanges data with the system that supports the CD Selections retail stores. The Internet Sales System will need to produce data in the same format as this existing system so that the distribution system will think the data is from the same system as always. Conversely, the project team may need to revise the existing distribution system, so that it can accept data from the Internet Sales System in a new format. A third approach is through the use of an *object wrapper.*[16] An object wrapper essentially is an object that "wraps around" a legacy system, enabling an object-oriented system to send messages to the legacy system. Effectively, object wrappers create an application program interface (API) to the legacy system. The creation of an object wrapper allows the protection of the corporation's investment in the legacy system.

[16] Ian Graham, *Object-Oriented Methods: Principles & Practice,* 3rd ed., (Harlow, England: Addison-Wesley, 2001).

Outsourcing

The design choice that requires the least amount of in-house resources is *outsourcing*—hiring an external vendor, developer, or service provider to create the system. Outsourcing has become quite popular in recent years. Some estimate that as many as 50% of companies with IT budgets over $5 million are currently outsourcing or evaluating the approach. Although these figures include outsourcing for all kinds of systems functions, this section focuses on outsourcing a single development project.

There can be great benefit to having someone else develop your system. They may be more experienced in the technology or have more resources, like experienced programmers. Many companies embark on outsourcing deals to reduce costs, while others see it as an opportunity to add value to the business. For example, instead of creating a program that handles the order-taking process or buying a pre-existing package, CD Selections may decide to let a Web service provider provide commercial services for them.

For whatever reason, outsourcing can be a good alternative for a new system; however, it does not come without costs. If you decide to leave the creation of a new system in the hands of someone else, you could compromise confidential information or lose control over future development. In-house professionals are not benefiting from the skills that could be learned from the project, and instead the expertise is transferred to the outside organization. Ultimately, important skills can walk right out the door at the end of the contract.

Most risks can be addressed if you decide to outsource, but two are particularly important. First, assess the requirements for the project thoroughly—you should never outsource what you don't understand. If you have conducted rigorous planning and analysis, then you should be well aware of your needs. Second, carefully choose a vendor, developer, or service with a proven track record with the type of system and technology that your system needs.

Three primary types of contracts can be drawn to control the outsourcing deal. A *time and arrangements* deal is very flexible because you agree to pay for whatever time and expenses are needed to get the job done. Of course, this agreement could result in a large bill that exceeds initial estimates. This works best when you and the outsourcer are unclear about what it is going to take to finish the job.

You will pay no more than expected with a *fixed-price* contract because if the outsourcer exceeds the agreed-upon price, it will have to absorb the costs. Outsourcers are much more careful about defining requirements clearly up front, and there is little flexibility for change.

The type of contract gaining in popularity is the *value-added* contract whereby the outsourcer reaps some percentage of the completed system's benefits. You have very little risk in this case but expect to share the wealth once the system is in place.

Creating fair contracts is an art because you need to carefully balance flexibility with clearly defined terms. Needs often change over time, so you don't want the contract to be so specific and rigid that alterations can't be made. Think about how quickly technology like the World Wide Web changes. It is difficult to foresee how a project may evolve over a long period of time. Short-term contracts help leave room for reassessment if needs change or if relationships are not working out the way both parties expected. In all cases, the relationship with the outsourcer should be viewed as a partnership in which both parties benefit and communicate openly.

Managing the outsourcing relationship is a full-time job. Thus, someone needs to be assigned full-time to manage the outsourcer, and the level of that person should be appropriate for the size of the job (a multimillion dollar outsourcing engagement should be handled by a high-level executive). Throughout the relationship, progress should be tracked and measured against predetermined goals. If you do embark on an outsourcing design strategy, be sure to get more information. Many books have been written that provide much more detailed information on the topic.[17] Figure 9-12 summarizes some guidelines for outsourcing.

Selecting a Design Strategy

Each of the design strategies we have discussed has its strengths and weaknesses, and no one strategy is inherently better than the others. Thus, it is important to understand the strengths and weaknesses of each strategy and when to use each. Figure 9-13 presents a summary of the characteristics of each strategy.

Business Need   If the business need for the system is common, and technical solutions already exist in the marketplace that can meet the business need of the system, it makes little sense to build a custom application. Packaged systems are good

---

CONCEPTS   **9-B EDS VALUE-ADDED CONTRACT**

IN ACTION

Value-added contracts can be quite rare—and very dramatic. They exist when a vendor is paid a percentage of revenue generated by the new system, which reduces the up-front fee, sometimes to zero. The City of Chicago and EDS (a large consulting and systems integration firm) agreed to reengineer the process by which the city collects the fines on 3.6 million parking tickets per year and signed the landmark deal of this type three years ago. At the time, because of clogged courts and administrative problems, the city collected on only about 25% of all tickets issued. It had a $60 million backlog of uncollected tickets.

Dallas-based EDS invested an estimated $25 million in consulting and new systems in exchange for the right to up to 26% of the uncollected fines, a base pro-cessing fee for new tickets, and software rights. To date, EDS has taken in well over $50 million on the deal, analysts say. The deal has come under some fire from various quarters as an example of an organization giving away too much in a risk/reward-sharing deal. City officials, however, counter that the city has pulled in about $45 million in previously uncollected fines and has improved its collection rate to 65% with little up-front investment.

Source: Datamation, February 1, 1995, pp. 58–61.

QUESTION
Do you think the City of Chicago got a good deal from this arrangement? Why or why not?

---

[17] For more information on outsourcing, we recommend M. Lacity, and R. Hirschheim, *Information Systems Outsourcing: Myths, Metaphors, and Realities* (New York: John Wiley, 1993); and L. Willcocks and G. Fitzgerald, *A Business Guide to Outsourcing Information Technology* (London: Business Intelligence, 1994).

Outsourcing Guidelines

Keep the lines of communication open between you and your outsourcer.

Define and stabilize requirements before signing a contact.

View the outsourcing relationship as a partnership.

Select the vendor, developer, or service provider carefully.

Assign a person to manage the relationship.

Don't outsource what you don't understand.

Emphasize flexible requirements, long-term relationships, and short-term contracts.

FIGURE 9-12
Outsourcing Guidelines

alternatives for common business needs. A custom alternative will need to be explored when the business need is unique or has special requirements. Usually, if the business need is not critical to the company, then outsourcing is the best choice—someone outside of the organization can be responsible for the application development.

In-house Experience   If in-house experience exists for all of the functional and technical needs of the system, it will be easier to build a custom application than if these skills do not exist. A packaged system may be a better alternative for companies that do not have the technical skills to build the desired system. For example, a project team that does not have Web commerce technology skills may want to acquire a Web commerce package that can be installed without many changes. Outsourcing is a good way to bring in outside experience that is missing in-house so that skilled people are in charge of building the system.

	Use Custom Development When...	Use a Packaged System When...	Use Outsourcing When...
Business need	The business need is unique	The business need is common	The business need is not core to the business
In-house experience	In-house functional and technical experience exists	In-house functional experience exists	In-house functional or technical experience does not exist
Project skills	There is a desire to build in-house skills	The skills are not strategic	The decision to outsource is a strategic decision
Project management	The project has a highly skilled project manager and a proven methodology	The project has a project manager who can coordinate vendor's efforts	The project has a highly skilled project manager at the level of the organization that matches the scope of the outsourcing deal
Time frame	The time frame is flexible	The time frame is short	The time frame is short or flexible

FIGURE 9-13
Selecting a Design Strategy

Project Skills   The skills that are applied during projects are either technical (e.g., Java, SQL) or functional (e.g., electronic commerce), and different design alternatives are more viable depending on how important the skills are to the company's strategy. For example, if certain functional and technical expertise that relates to Internet sales applications and to Web commerce application development are important to the organization because it expects the Internet to play an important role in its sales over time, then it makes sense for the company to develop Web commerce applications in-house, using company employees so that the skills can be developed and improved. On the other hand, some skills like network security may be either beyond the technical expertise of employees or not of interest to the company's strategists—it is just an operational issue that needs to be addressed. In this case, packaged systems or outsourcing should be considered so that internal employees can focus on other business critical applications and skills.

Project Management   Custom applications require excellent project management and a proven methodology. There are so many things that can push a project off-track, including funding obstacles, staffing holdups, and overly demanding business users. Therefore, the project team should choose to develop a custom application only if they are certain that the underlying coordination and control mechanisms will be in place. Packaged and outsourcing alternatives also need to be managed. However, they are more shielded from internal obstacles because the external parties have their own objectives and priorities (e.g., it may be easier for an outside contractor to say "no" to a user than a person within the company). The latter alternatives typically have their own methodologies, which can benefit companies that do not have an appropriate methodology to use.

Time Frame   When time is a factor, the project team should probably start looking for a system that is already built and tested. In this way, the company will have a good idea of how long the package will take to put in place and what the final result will contain. The time frame for custom applications is hard to pin down, especially when you consider how many projects end up missing important deadlines. If you must choose the custom development alternative and the time frame is very short, consider using techniques like timeboxing to manage this problem. The time to produce a system using outsourcing really depends on the system and the outsourcer's resources. If a service provider has services in place that can be used to support the company's needs, then a business need could be implemented quickly. Otherwise, an outsourcing solution could take as long as a custom development initiative.

YOUR	**9-2 CHOOSE A DESIGN STRATEGY**
TURN	

   Suppose that your university is interested in creating a new course registration system that can support Web-based registration. What should the university consider when determining whether to invest in a custom, packaged, or outsourced system solution?

DEVELOPING THE ACTUAL DESIGN

Once the project team has a good understanding of how well each design strategy fits with the project's needs, they must begin to understand exactly *how* to implement these strategies. For example, what tools and technology would be used if a custom alternative were selected? What vendors make packaged systems that address the project needs? What service providers would be able to build this system if the application were outsourced? This information can be obtained from people working in the IS department and from recommendations by business users. Or the project team can contact other companies with similar needs and investigate the types of systems that they have put in place. Vendors and consultants usually are willing to provide information about various tools and solutions in the form of brochures, product demonstrations, and information seminars.

The project team likely will identify several ways that the system could be constructed after weighing the specific design options. For example, the project team may have found three vendors that make packaged systems that potentially could meet the project's needs. Or the team may be debating over whether to develop a system using Java as a development tool and the database management system from Oracle; or to outsource the development effort to a consulting firm like Accenture or American Management Systems. Each alternative will have pros and cons associated with it that need to be considered, and only one solution can be selected in the end.

Alternative Matrix

An *alternative matrix* can be used to organize the pros and cons of the design alternatives so that the best solution will be chosen in the end. This matrix is created using the same steps as the feasibility analysis, which was presented in Chapter 2. The only difference is that the alternative matrix combines several feasibility analyses into one matrix so that the alternatives can be easily compared. The alternative matrix is a grid that contains the technical, budget, and organizational feasibilities for each system candidate, pros and cons associated with adopting each solution, and other information that is helpful when making comparisons. Sometimes weights are provided for different parts of the matrix to show when some criteria are more important to the final decision.

To create the alternative matrix, draw a grid with the alternatives across the top and different criteria (e.g., feasibilities, pros, cons, and other miscellaneous criteria) along the side. Next, fill in the grid with detailed descriptions about each alternative. This becomes a useful document for discussion because it clearly presents the alternatives being reviewed and comparable characteristics for each one.

Suppose that your company is thinking about implementing a packaged financial system like Oracle Financials or Microsoft's Great Plains, but there is not enough expertise in-house to create a thorough alternative matrix. This situation is quite common—often the alternatives for a project are unfamiliar to the project team, so outside expertise is needed to provide information about the alternatives' criteria.

One helpful tool is the *Request for Proposals* (RFP). A RFP is a document that solicits proposals to provide alternative solutions from a vendor, developer, or service provider. Basically, the RFP explains the system that you are trying to build and the criteria that you will use to select a system. Vendors, then, respond by

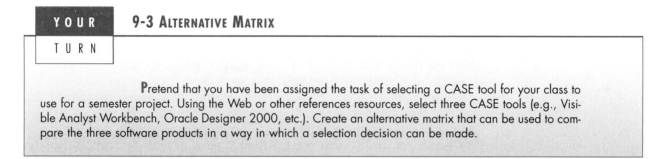

describing what it would mean for them to be part of the solutions. They communicate the time, cost, and exactly how their product or services will address the needs of the project.

There is no formal way to write an RFP, but it should include basic information, such as the description of the desired system, any special technical needs or circumstances, evaluation criteria, instructions for how to respond, and the desired schedule. An RFP can be a very large document (i.e., hundreds of pages) because companies try to include as much detail as possible about their needs so that the respondent can be just as detailed in the solution that would be provided. Thus, RFPs typically are used for large projects rather than small ones because they take a lot of time to create, and even more time and effort for vendors, developers, and service providers to develop high-quality responses. Only a project with a fairly large price tag would be worth the time and cost to develop a response for the RFP.

A less effort-intensive tool is a *Request for Information* (RFI) that includes the same format as the RFP. The RFI is shorter and contains less detailed information about a company's needs. It requires general information from respondents that communicates the basic services that they can provide.

The final step, of course, is to decide which solution to design and implement. The decision should be made by a combination of business users and technical professionals after the issues involved with the different alternatives are well understood. Once the decision is finalized, the design phase can continue as needed, based on the selected alternative.

Applying the Concepts at CD Selections

Alec could use one of three different approaches with the new system: he could develop the entire system using development resources from CD Selections; he could buy a commercial Internet sales packaged software program (or a set of different packages and integrate them); or he could hire a consulting firm or service provider to create the system. Immediately, Alec ruled out the third option. Building Internet applications, especially sales systems, was becoming increasingly important to the CD Selections' business strategy. By outsourcing the Internet Sales System, CD Selections would not develop Internet application development skills and business skills within the organization.

Instead, Alec decided that a custom development project using the company's standard Web development tools would be the best choice for CD Selections. In this way, the company would be developing critical technical and business skills in-house, and the project team would be able to have a high level of flexi-

	Alternative 1: Shop With Me	Alternative 2: WebShop	Alternative 3: Shop-N-Go
Technical feasibility	Developed using C; very little C experience in-house Orders sent to company using e-mail files	Developed using C and Java; would like to develop in-house Java skills Flexible export features for passing order information to other systems	Developed using Java; would like to develop in-house Java skills Orders saved to a number of file formats
Economic feasibility	$150 initial charge	$700 up-front charge, no yearly fees	$200/year
Organizational feasibility	Program used by other retail music companies	Program used by other retail music companies	Brand-new application; few companies have experience with Shop-N-Go to date
Other benefits	Very simple to use	Tom in Information Systems support has had limited but positive experience with this program Easy to customize	
Other limitations			The interface is not easily customized

FIGURE 9-14
Alternative Matrix for Shopping
Cart Program

bility and control over the final product. Also, Alec wanted the new Internet Sales System to directly interface with the existing distribution system, and there was a chance that a packaged solution would not be able to integrate as well into the CD Selections environment.

There was one part of the project that could be handled using packaged software: the shopping cart portion of the application. Alec realized that a multitude of programs have been written and are available (at low prices) to handle a customer's order transaction over the Web. These programs allow customers to select items for an order form, input credit card and billing information, and finalize the order transaction. Alec believed that the project team should at least consider some of these packaged alternatives so that less time would be spent writing a program that handled basic Web tasks, and more time could be devoted to innovative marketing ideas and custom interfaces with the distribution system.

To help better understand some of the shopping cart programs that were available in the market and how their adoption could benefit the project, Alec created an alternative matrix that compared three different shopping cart programs to one another (see Figure 9-14). Although all three alternatives had positive points, Alec saw Alternative B (WebShop) as the best alternative for handling the shopping cart functionality for the new Internet Sales System. WebShop was written in Java, the tool that CD Selections selected as its standard Web development language; the expense was reasonable, with no hidden or recurring costs; and there was a person in-house who had some positive experience with the program. Alec made a note to look into acquiring WebShop as the shopping cart program for the Internet Sales System.

SUMMARY

The design phase is the phase of the SDLC in which the blueprint for the new system is developed. It contains many steps that guide the project team through planning out exactly how the system needs to be constructed. The requirements that were identified and the models that were created in the analysis phase serve as the primary inputs for the design activities. In Object-Oriented design, the primary activity is to evolve the analysis models into design models by optimizing the problem domain information that is already contained in the analysis models and adding system environment details to them.

Object-Oriented Approach to Analysis and Design

Our approach to OOAD is based on the processes described in the Unified and Open approaches to object-oriented systems development. It uses a modified phased delivery RAD approach to its life cycle. It is also use-case driven, architecture centric, and iterative and incremental. It supports three different generic views of the evolving system: functional, static, and dynamic. It utilizes time-boxes to manage the creation of builds of the system. Finally, it employs use-case modeling (use cases and use-case models) to support the functional view of the system, structural modeling (CRC cards, class diagrams, and object diagrams) to support the static view of the system, and behavioral modeling (sequence diagrams, collaboration diagrams, and statechart diagrams) to support the dynamic view of the system.

Evolving the Analysis Models into Design Models

When evolving the analysis models into design models, you should first carefully review the analysis models: use cases, use-case diagrams, CRC cards, class and object diagrams, sequence diagrams, collaboration diagrams, and statechart diagrams. During this review, factoring, refinement, and abstraction processes can be used to polish the current models. The analysis models may become overly complex during this polishing. If this occurs, then the models should be partitioned based on the interactivity (message sending) and relationships (generalization, aggregation, and association) shared among the classes. The more a class has in common with another class, that is, the more relationships shared, the more likely they belong in the same partition.

The second thing to do to evolve the analysis mode is to add the system environment (system architecture, user interface, and data management) information to the problem domain information already contained in the model. To accomplish this and to control the complexity of the models, layers are used. A layer represents an element of the software architecture of the system. Layers evolved from the MVC architecture of Smalltalk. We recommend the use of five different layers: foundation, system architecture, human–computer interaction, data management, and problem domain. Each layer only supports certain types of classes; for example, database manipulation classes would only be allowed on the data management layer.

Packages and Package Diagrams

A package is a general UML construct used to represent collaborations, partitions, and layers. The primary purpose is to support the logical grouping of other

UML constructs together, for example, use cases and classes, by the developer and user to simplify and increase the understandability of a UML diagram. In some instances a diagram that only contains packages is useful. A package diagram contains packages and dependency relationships. A dependency relationship represents the possibility of a modification dependency existing between two packages; that is, changes in one package could cause changes in the dependent package.

Identifying packages and creating a package diagram are accomplished using a five-step process (see Figure 9-7): setting the context, clustering similar classes, placing the clustered classes into a package, identifying dependency relationships among the packages, and placing the dependency relationship on the package diagram.

Design Strategies

During the design phase, the project team also needs to consider three approaches to creating the new system, including developing a custom application in-house, buying a packaged system and customizing it, and relying on an external vendor, developer, or system provider to build and/or support the system.

Custom development allows developers to be flexible and creative in the way they solve business problems, and it builds technical and functional knowledge within the organization. But many companies have a development staff that is already overcommitted to filling huge backlogs of systems requests, and they just don't have time to devote to a project where a system is built from scratch. It can be much more efficient to buy programs that have been created, tested, and proven, and a packaged system can be bought and installed in a relatively short period of time, when compared with a custom solution. Workarounds can be used to meet the needs that are not addressed by the packaged application.

The third design strategy is to outsource the project and pay an external vendor, developer, or service provider to create the system. It can be a good alternative for how to approach the new system. However, it does not come without costs. If a company decides to leave the creation of a new system in the hands of someone else, the organization could compromise confidential information or lose control over future development.

Each of the design strategies has its strengths and weaknesses, and no one strategy is inherently better than the others. Thus, it is important to consider such issues as the uniqueness of business need for the system, the amount of in-house experience that is available to build the system, and the importance of the project skills to the company. The existence of good project management and the amount of time available to develop the application also play a role in the selection process.

Developing the Actual Design

Ultimately, the decision must be made regarding the specific type of system that needs to be designed. An alternative matrix can help make this decision by presenting feasibility information for several candidate solutions in a way in which they can be compared easily. Both the Request for Proposal and the Request for Information are two ways to gather accurate information regarding the alternatives.

KEY TERMS

A-kind-of
Abstract classes
Abstraction
Aggregation
Alternative matrix
Architecture centric
Build
Class
Client
Collaboration
Concrete classes
Contract
Controller
CORBA
Custom development
Customization
DCOM
Data Management layer
Dependency relationship
Design phase
Dynamic view

Enterprise resource planning (ERP)
Factoring
Fixed-price contract
Foundation layer
Functional view
Generalization
Has-parts
Human–computer interaction layer
Incremental
Iterative
Layer
Message
Method
Middleware application
Model
Model-View-Controller (MVC)
Module
Network model
Object wrapper
Outsourcing
Package

Package diagram
Packaged software
Partition
Phased delivery RAD
Phased development RAD
Problem domain layer
Refinement
Request for Information
Request for Proposals
Server
Smalltalk
Static view
System architecture layer
Systems integration
Time and arrangements contract
Timeboxing
Use-case driven
Use-case model
Value-added contract
View
Workaround

QUESTIONS

1. What is meant by use-case driven?
2. Compare and contrast the typical phased delivery RAD SDLC with the modified phased delivery RAD SDLC associated with the OOAD approach described in this chapter.
3. What are the different views supported by the OOAD approach described in this chapter?
4. What diagrams and models support the different views identified in the previous question?
5. What is a build?
6. What are the different levels of abstraction in an object-oriented system?
7. What are the main activities that happen during the design phase of the SDLC?
8. How do analysis activities lead into design tasks?

9. What is the primary difference between an analysis model and a design model?
10. What does factoring mean? How is it related to abstraction and refinement?
11. What is a partition? How does a partition relate to a collaboration?
12. What is a layer? Name the different layers.
13. What is a package? How are packages related to partitions and layers?
14. What is a dependency relationship? How do you identify them?
15. What are the five steps for identifying packages and creating package diagrams?
16. What situations are most appropriate for a custom development design strategy?

17. What are some problems with using a packaged software approach to building a new system? How can these problems be addressed?
18. Why do companies invest in ERP systems?
19. What are the pros and cons of using a workaround?
20. When is outsourcing considered a good design strategy? When is it not appropriate?
21. What are the differences between the time and arrangements, fixed-price, and value-added contracts for outsourcing?
22. How are the alternative matrix and feasibility analysis related?
23. What is an RFP? How is this different from an RFI?

EXERCISES

A. Based on the use cases and use-case diagrams you created for Exercises E, F, and G in Chapter 6 and the structural model you created in Exercise K in Chapter 7, draw a package diagram for the problem domain layer for the dentist office system.

B. Based on the use cases and use-case diagrams you created for Exercises J and K in Chapter 6 and the structural model you created in Exercise M in Chapter 7, draw a package diagram for the problem domain layer for the real estate system.

C. Based on the use cases and use-case diagrams you created for exercises L and M in Chapter 6, the structural model you created in Exercise N in Chapter 7, and the behavioral models you created in exercises A and B in Chapter 8, draw a package diagram for the problem domain layer for the video store system.

D. Based on the use cases and use-case diagrams you created for exercises N and O in Chapter 6, the structural model you created in Exercise O in Chapter 7, and the behavioral models you created in exercises C and D in Chapter 8, draw a package diagram for the problem domain layer for the health club membership system.

E. Based on the use cases and use-case diagrams you created for exercises T and U in Chapter 6, the structural model you created in Exercise R in Chapter 7, and the behavioral models you created in exercise J in Chapter 8, draw a package diagram for the problem domain layer for the university library system.

F Based on the use cases and use-case diagrams you created for exercises P and Q in Chapter 6, the structural model you created in Exercise P in Chapter 7, and the behavioral models you created in exercise H in Chapter 8, draw a package diagram for the problem domain layer for the catering system.

G. Based on the use cases and use-case diagrams you created for exercises R and S in Chapter 6, the structural model you created in Exercise Q in Chapter 7, and the behavioral models you created in exercise I in Chapter 8, draw a package diagram for the problem domain layer for the system for the Of-the-Month Club.

H. Which design strategy would you recommend for the construction of this system in:
   1. Exercise A. Why?
   2. Exercise B. Why?
   3. Exercise C. Why?
   4. Exercise D. Why?
   5. Exercise E. Why?
   6. Exercise F. Why?
   7. Exercise G. Why?

I. Pretend that you are leading a project that will implement a new course enrollment system for your university. You are thinking about either using a packaged course enrollment application or outsourcing the job to an external consultant. Create a Request for Proposal to which interested vendors and consultants could respond.

J. Pretend that you and your friends are starting a small business painting houses in the summertime. You need to buy a software package that handles the financial transactions of the business. Create an alternative matrix that compares three packaged systems (e.g., Quicken, MS Money, Quickbooks). Which alternative appears to be the best choice?

MINICASES

1. Susan, president of MOTO, Inc., a human resources management firm, is reflecting on the client management software system her organization purchased four years ago. At that time, the firm had just gone through a major growth spurt, and the mixture of automated and manual procedures that had been used to manage client accounts became unwieldy. Susan and Nancy, her IS department head, researched and selected the package that is currently used. Susan had heard about the software at a professional conference she attended, and at least initially, it worked fairly well for the firm. Some of their procedures had to change to fit the package, but they expected that and were prepared for it.

Since that time, MOTO, Inc. has continued to grow, not only through an expansion of the client base, but also through the acquisition of several smaller employment-related businesses. MOTO, Inc. is a much different business than it was four years ago. Along with expanding to offer more diversified human resource management services, the firm's support staff has also expanded. Susan and Nancy are particularly proud of the IS department they have built up over the years. Using strong ties with a local university, an attractive compensation package, and a good working environment, the IS department is well-staffed with competent, innovative people, plus a steady stream of college interns that keeps the department fresh and lively. One of the IS teams pioneered the use of the Internet to offer MOTO's services to a whole new market segment, an experiment that has proven very successful.

It seems clear that a major change is needed in the client management software, and Susan has already begun to plan financially to undertake such a project. This software is a central part of MOTO's operations, and Susan wants to be sure that a quality system is obtained this time. She knows that the vendor of their current system has made some revisions and additions to its product line. There are also a number of other software vendors who offer products that may be suitable. Some of these vendors did not exist when the purchase was made four years ago. Susan is also considering Nancy's suggestion that the IS department develop a custom software application.

a. Outline the issues that Susan should consider that would support the development of a custom software application in-house.

b. Outline the issues that Susan should consider which would support the purchase of a software package.

c. Within the context of a systems development project, when should the decision of "make-versus-buy" be made? How should Susan proceed? Explain your answer.

2. Refer to Minicase 1 in Chapter 7. After completing all of the analysis models (both the As-Is and To-Be models) for West Star Marinas, the Director of Operations finally understood why it was important to understand the As-Is system before delving into the development of the To-Be system. However, you now tell him that the To-Be models are only the problem domain portion of the design. To say the least, he is now very confused. After explaining to him the advantages of using a layered approach to developing the system, he informs you that "I don't care about reusability or maintenance. I only want the system to be implemented as soon as possible. You IS types are always trying to pull a fast one on the users. Just get the system completed."

What is your response to the Director of Operations? Do you jump into implementation as he seems to want? What do you do next?

3. Refer to the analysis models that you created for Professional and Scientific Staff Management (PSSM) for Minicase 2 in Chapter 6 and for Minicase 1 in Chapter 8.
   a. Add packages to your use-case diagram to simplify it.
   b. Use the collaboration diagram to identify logical partitions in your class diagram. Add packages to the diagram to represent the partitions.

4. Refer to the analysis models that you created for Holiday Travel Vehicles for Minicase 2 in Chapter 7 and for Minicase 2 in Chapter 8.
   a. Add packages to your use-case diagram to simplify it.
   b. Use the collaboration diagram to identify logical partitions in your class diagram. Add packages to the diagram to represent the partitions.

PLANNING

ANALYSIS

DESIGN

- ✔ **Evolve Analysis Models**
- ✔ **Develop the Design Strategy**
- ✔ **Develop the Design**
- ☐ **Develop Infrastructure Design**
- ☐ **Develop Network Model**
- ☐ **Develop Hardware/Software Specification**
- ☐ **Develop Security Plan**
- ☐ Develop Use Scenarios
- ☐ Design Interface Structure
- ☐ Design Interface Standards
- ☐ Design User Interface Template
- ☐ Design User Interface
- ☐ Evaluate User Interface
- ☐ Select Object Persistence Format
- ☐ Map Objects to Object Persistence
- ☐ Optimize Object Persistence
- ☐ Size Object Persistence
- ☐ Specify Constraints
- ☐ Develop Contracts
- ☐ Develop Method Specifications

TASK CHECKLIST

PLANNING

ANALYSIS

DESIGN

CHAPTER 10

SYSTEM ARCHITECTURE DESIGN

AN **IMPORTANT** component of the design phase is the system architecture design, which describes the proposed technical environment for the new system. This technical environment contains the hardware, software, and communications infrastructure on which the new system will be created, and the methods for supporting the system's security needs and global requirements. The deliverable from the system architecture design contains the network model, the hardware and software specification, and the plan for security and global support.

OBJECTIVES

- Understand the differences among server-based, client-based, and client server computing.
- Be familiar with distributed objects computing.
- Be able to create a network model.
- Be able to create a hardware/software specification.
- Become familiar with global and security issues.

CHAPTER OUTLINE

Introduction
Computing Architectures
  Server-Based Computing
  Client-Based Architectures
  Client-Server Architectures
  Client-Server Tiers
  Distributed Objects Computing
  Selecting a Computing Architecture
Infrastructure Design
  The Network Model
  Hardware and Software
    Specification

Global Issues
  Multilingual Requirements
  Local versus Centralized Control
  Unstated Norms
  24-7 Support
  Communications Infrastructure
Security
  Identifying Threats to the System
  Assessing the Risk of Each Threat
  Creating Controls
Applying the Concepts at CD Selections
Summary

IMPLEMENTATION

INTRODUCTION

An important step of the design phase is creation of the design of the system architecture layer: the *system architecture design.* The system architecture design is the plan for the hardware, software, and communications infrastructure for the new system, security, and global support. The first step in system architecture design is to determine whether the system will have a *server-based, client-based,* or *client-server architecture.* A server-based architecture builds the processing for the system into the server, whereas client-based architecture relies on personal computers to support the processing of the new application. This decision influences other components of the system architecture design, so it typically is determined early on in the design phase.

Next, a *network model* is created to show where the major components of the system (e.g., servers, personal computers) will be located and how the components will be connected to one another. A *hardware and software specification* is then developed to describe in detail the components that are pictured on the network model diagram. Details include the descriptions of the actual hardware and software products. The specification provides guidance to the people in the organization who are responsible for purchasing and acquiring the hardware and software for the new system. Typically, it takes quite some time to receive hardware and software from the vendors, so it is best to place important (or large) orders sooner rather than later.

Two additional factors affect system architecture design—the global needs of the application and system security—and the project team creates a global support and security plan that explains how the project will address both. Global issues result when companies develop systems for users in areas around the world, and they can affect things like standards, application support, culture, and the decision to have centralized or distributed coordination. Security is threatened by problems arising from system disruption, data destruction, and disaster as well as unauthorized access.

Every part of the system architecture design—the technical environment, security, and global support—is designed based on the business needs that were specified earlier in the analysis phase. The project team does not create the system architecture design deliverable in a vacuum. Rather, they develop it to closely match with the business requirements that were uncovered before the design phase began.

This chapter first describes server-based, client-based, and client-server computing, followed by steps for creating the network model. The hardware and software specification is presented next, along with security and global issues that need to be considered for a new project. Finally, a system architecture design example is presented for CD Selections.

COMPUTING ARCHITECTURES

There are three fundamental computing architectures. In *server-based* computing, the *server* performs virtually all of the work. In *client-based* computing, the *client computers* are responsible for most of the application functions. In *client-server* computing, the work is shared between the servers and clients.

The work done by any application system can be divided into four general functions. The first is *data storage* (located on the data management layer). Most application programs require data to be stored and retrieved, whether it is a small file such as a memo produced by a word processor, or a large database that stores an organization's accounting records. These are the data documented in the structural model (CRC cards and class diagrams). The second function is *data access logic* (located on the data management layer)—the processing required to access data, which often means database queries in SQL. The third function is the *application logic* (located on the problem domain layer) which can be simple or complex depending on the application. This is the logic that is documented in the use cases and behavioral models (sequence, collaboration, and statechart diagrams). The fourth function is the *presentation logic* (located on the human–computer interaction layer)—the presentation of information to the user and the acceptance of the user's commands (the user interface). These four functions (data storage, data access logic, application logic, and presentation logic) are the basic building blocks of any application.

The components of the three computing architectures include servers, clients, and networks. *Servers* can come in several flavors: mainframes, minicomputers, and microcomputers. A *mainframe* is a very large general-purpose computer usually costing millions of dollars that is capable of performing many simultaneous functions. A second type of server is a *minicomputer,* which is a large general-purpose computer costing tens of thousands to hundreds of thousands of dollars and is often used in support of database management systems and in manufacturing for process control. Third, the *microcomputer* (or *personal computer*) can range from a small desktop machine that you might use to a machine costing $50,000 or more. These categories (i.e., mainframe, minicomputer, and microcomputer) were developed years ago to describe the development of hardware technology. The actual delineation among these three categories of computing architectures has become blurred however, and today the terms are used quite loosely.

The *client,* which is the input/output hardware device that is used by the user, probably is the one piece of equipment with which you are most familiar. The three major categories of clients are terminals, personal computers, and special-purpose terminals. *Terminals* are used for display and data-entry purposes only whereas *personal computers* are "intelligent" and can handle a great number of processing tasks. *Special-purpose terminals* include machines like ATMs, kiosks, Palm Pilots, and many other interfaces with which users can communicate with an application.

A complete description of networks is beyond the scope of this book.[1] Indeed, entire courses are dedicated to the subject. For the purposes of this chapter, it will be sufficient to understand a *network* as the hardware and software that are used to connect clients and servers to one another. A network supports the communication between the client and server components that make up the computing architecture.

Server-Based Computing

The very first computing architectures were server-based, with the server (usually a central mainframe computer) performing all four application functions. The clients (usually terminals) enabled users to send and receive messages to and from the server computer. The clients merely captured keystrokes and sent them to the server for processing, accepting instructions from the server on what to display (see Figure 10-1).

---

[1] For information on networking and data communications, see Jerry FitzGerald and Alan Dennis, *Business Data Communications and Networking,* 6th ed. (New York: John Wiley & Sons, 1999).

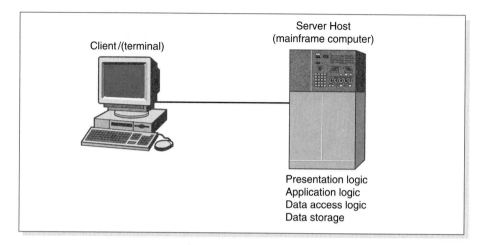

FIGURE 10-1
Server-Based Architecture

This very simple architecture often works very well. Application software is developed and stored on one computer, and all data are on the same computer. There is one point of control, because all messages flow through the one central server.

The fundamental problem associated with server-based networks is that the server must process all messages. As the demands for more and more applications grow, many server computers become overloaded and unable to quickly process all the users' demands. Response time becomes slower, and network managers are required to spend increasingly more money to upgrade the server computer. Unfortunately, upgrades come in large increments and are expensive (e.g., $500,000); it is difficult to upgrade "a little."

Client-Based Architectures

In the mid-1980s, there was an explosion in the use of personal computers. Part of this trend was fueled by a number of low-cost, highly popular applications such as word processors, spreadsheets, and presentation graphics programs. It was also fueled in part by managers' frustrations with application software on server mainframe computers. Most mainframe software is not as easy to use as personal computer software, is far more expensive, and can take years to develop. In the mid-1980s many large organizations had application development backlogs of two to three years; that is, getting any new mainframe application program written would take a very long time. In contrast, managers could buy personal computer packages or develop personal computer-based applications in a few months.

With client-based architectures, the clients are personal computers on a local area network, and the server computer is a server on the same network. The application software on the client computers is responsible for the presentation logic, the application logic, and the data access logic; the server simply stores the data (see Figure 10-2).

This simple architecture often works very well. However, as the demands for more and more network applications grow, the network circuits can become overloaded. The fundamental problem in client-based networks is that all data on the server must travel to the client for processing. For example, suppose the user wishes to display a list of all employees with company life insurance. All the data in the

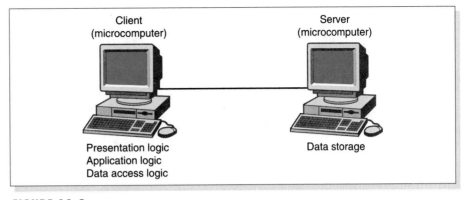

FIGURE 10-2
Client-based Architectures

database must travel from the server where the database is stored over the network to the client, which then examines each record to see whether it matches the data requested by the user. This can overload both the network and the power of the client computers.

Client-Server Architectures

Most organizations today are moving to client-server architectures, which attempt to balance the processing between the client and the server by having both do some of the application functions. In these architectures, the client is responsible for the presentation logic, while the server is responsible for the data access logic and data storage. The application logic may reside on either the client or the server, or be split between both (see Figure 10-3). The client shown in Figure 10-3 can be referred to as a *fat client* if it contains the bulk of application logic. A recent trend is to create client-server architectures using *thin clients* because there is less overhead and maintenance in supporting thin-client applications. For example, many Web-based systems are designed with the Web browser performing presentation,

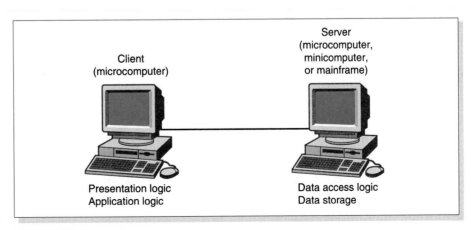

FIGURE 10-3
Client-Server Architecture

with only minimal application logic using programming languages like Java, and the Web server having the application logic, data access logic, and data storage.

Client-server architectures have some important benefits. First and foremost, they are *scalable.* That means it is easy to increase or decrease the storage and processing capabilities of the servers. If one server becomes overloaded, you simply add another server and move some of the application logic or data storage to it. The cost to upgrade is much more gradual, and you can upgrade in smaller steps (e.g., $5,000) rather than spending hundreds of thousands to upgrade a mainframe server.

Client-server architectures can support many different types of clients and servers. You are not locked into one vendor, as is often the case in server-based networks. Similarly, it is possible to connect computers that use different operating systems so that users can choose which type of computer they prefer (e.g., combining both IBM personal computer and Apple Macintoshes on the same network).

Finally, because no single server computer supports all the applications, the network is generally more reliable. There is no central point of failure that will halt the entire network if it fails, as there is in a server-based computing. If any one server fails in a client-server environment, the network can continue to function using all the other servers (but, of course, any applications that require the failed server will not work).

Client-server computing also has some critical limitations, the most important of which is its complexity. All applications in client-server computing have two parts, the software on the client, and the software on the server. Writing this software is more complicated than writing the traditional all-in-one software used in server-based computing. Programmers often must learn new programming languages and new programming techniques, which requires retraining. Updating the network with a new version of the software is more complicated, too. In server-based computing, there is one place in which application software is stored; to update the software, you simply replace it there. With client-server computing, you must update all clients and all servers.

Much of the debate between server-based and client-server computing has centered on cost. One of the great claims of server-based networks in the 1980s was that they provided economies of scale. Manufacturers of big mainframes claimed it was cheaper to provide computer services on one big mainframe than on a set of smaller computers. The personal computer revolution changed this thinking. Over the past 20 years, the costs of personal computers have continued to drop, while their performance has increased significantly. Today, personal computer hardware is more than 1,000 times cheaper than mainframe hardware for the same amount of computing power.

With cost differences like these, it is easy to see why there has been a sudden rush to microcomputer-based client-server computing. The problem with these cost comparisons is that they ignore the total cost of ownership, which includes factors other than obvious hardware and software costs. For example, many cost comparisons overlook the increased complexity associated with developing application software for client-server networks. Most experts believe that it costs four to five times more to develop and maintain application software for client-server computing than it does for server-based computing.

Client-Server Tiers

The application logic can be partitioned between the client and the server in many ways. The example in Figure 10-3 is one of the most common. In this case, the server is responsible for the data and the client, the application and presentation. This is called a *two-tiered architecture* because it uses only two sets of computers, clients and servers.

A *three-tiered architecture* uses three sets of computers as shown in Figure 10-4. In this case, the software on the client computer is responsible for presentation logic, an application server(s) is responsible for the application logic, and a separate database server(s) is responsible for the data access logic and data storage.

An *n-tiered architecture* uses more than three sets of computers. In this case, the client is responsible for presentation, a database server(s) is responsible for the data access logic and data storage, and the application logic is spread across two or more different sets of servers. Figure 10-5 shows an example of an n-tiered architecture of a groupware product called TCBWorks developed at the University of Georgia. TCBWorks has four major components. The first is the Web browser on the client computer that a user employs to access the system and enter commands (presentation logic). The second component is a Web server that responds to the user's requests, either by providing HTML pages and graphics (application logic), or by sending the request to the third component (a set of 28 programs written in the C programming language) on another application server that performs various functions (application logic). The fourth component is a database server that stores all the data (data access logic and data storage). Each of these four components is separate, making it easy to spread the different components on different servers and to partition the application logic on two different servers.

The primary advantage of an n-tiered client-server architecture compared to a two-tiered architecture (or a three-tiered to a two-tiered) is that it separates out the processing that occurs to better balance the load on the different servers; it is more "scalable." In Figure 10-5, we have three separate servers, which provide more power than if we had used a two-tiered architecture with only one server. If we discover that the application server is too heavily loaded, we can simply replace it with a more powerful server, or even put in two application servers. Conversely, if we discover the database server is underused, we could store data from another application on it.

There are two primary disadvantages to an n-tiered architecture compared to a two-tiered architecture (or a three-tiered to a two-tiered). First, it puts a greater

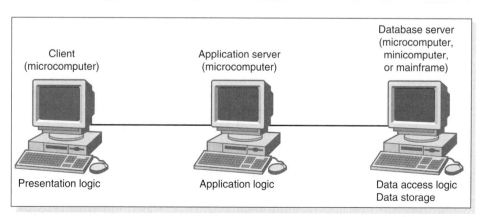

FIGURE 10-4
A Three-tier Client-Server Architecture

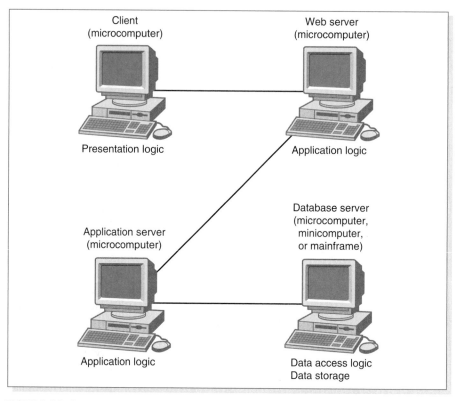

FIGURE 10-5
A Four-tier Client-Server Architecture

load on the network. If you compare Figures 10-3, 10-4, and 10-5, you will see that the n-tiered model requires more communication among the servers; it generates more network traffic, so you need a higher capacity network. Second, it is much more difficult to program and test software in n-tier architectures than two-tiered architectures because more devices have to communicate to complete a user's transaction.

Distributed Objects Computing

From an object-oriented perspective, *distributed objects computing (DOC)* is the next version of client-server computing. DOC represents a software layer that goes between the clients and servers; hence, it is known as *middleware*. Middleware supports the interaction between objects in a distributed computing environment. Furthermore, middleware treats the actual physical network architecture in a transparent manner. This allows the developer to focus on the development of the application and ignore the peculiarities of a specific distributed environment.

From a practical perspective, DOC allows the developer to simply concentrate on the users, objects, and methods of an application instead of worrying about which server contains which set of objects. The client object simply requests the "network" to locate and execute the server object's method. This basically allows the physical location of the server object to become irrelevant from the client object's perspective. Therefore, since servers are no longer being addressed directly on the client, servers can be added and subtracted from the network without having to update the client code. Only the middleware must be made aware of the new addresses of the server objects. This can greatly reduce the maintenance in a client-server environment. However, since middleware adds an additional layer between the clients and servers, it can reduce the efficiency of the application. Currently, there are two primary competing DOC standards: the Object Management Group's (OMG) CORBA and Microsoft's DCOM.

OMG's *Common Object Request Broker Architecture (CORBA)*[2] is a platform-independent standard for *Object Request Brokers (ORBs)*. ORBs allow objects to be distributed over a set of clients and servers, regardless of the underlying platform (e.g., Windows or Unix) or language (e.g., Java or C++). In other words, it increases the interoperability among applications in a heterogeneous, distributed environment. This includes allowing objects on the client to communicate with nonobject applications on the servers. This is accomplished in a manner similar to using *object wrappers* (see Chapter 9); client objects only interact with client broker objects, client broker objects only interact with server broker objects, and server broker objects only interact with server objects. Practically, the system architecture layer becomes a CORBA layer. CORBA-based ORBs accomplish this transparency through the use of CORBA's Interface Definition Language (IDL) and an interface definition repository. IDL looks a lot like C++, but it only allows the definition of interfaces. As such, it is not a full-blown programming language. Through the use of IDL, CORBA allows client and server objects to be maintained in a platform-independent manner.

Even though Microsoft is involved in OMG, Microsoft has its own competing approach to support DOC: the *distributed component object model (DCOM)*.[3] Basically, DCOM is nothing more than Microsoft's *component object model* (COM) and *component object model+* (COM+) for a distributed environment. These models support a binary standard for interfaces between objects. This should, in principle, support interoperability. However, unlike CORBA, DCOM interfaces must follow the same memory layout as C++ virtual function tables.[4] This can seriously compromise the goal of supporting distributed object computing in a platform-independent manner. At this point in time, CORBA and DCOM are competitors to support DOC; it is not clear which approach will win in the marketplace.[5]

---

[2] Many books have been written on CORBA (e.g., Robert Orfali and Dan Harkey, *Client/Server Programming with Java and CORBA,* 2nd ed., New York: John Wiley & Sons, 1998; Alan Pope, *The Corba Reference Guide: Understanding the Common Object Request Broker Architecture,* Reading, MA: Addison-Wesley, 1998; Zahir Tari and Omran Bukhres, *Distributed Object Computing: A CORBA Perspective,* New York: John Wiley & Sons, 2001). However, since it is an evolving standard, you should go to OMG's CORBA Web site (www.corba.org/) for up-to-date information on CORBA.

[3] Many books have been written on COM/DCOM (e.g., Guy Eddon and Henry Eddon, *Inside Distributed Com.* Redmond, WA: Microsoft Press, 1998; Jim Maloney, *Distributed COM Application Development Using Visual Basic 6.0 and MTS,* Englewood Cliffs, NJ: Prentice-Hall, 1999; and Nathan Wallace, *COM/DCOM Blue Book,* Scottsdole, AZ: The Coriolis Group, 1999). For up-to-date information on DCOM, you should go to Microsoft's COM Web site (www.microsoft.com/com).

[4] CORBA supports multiple-language bindings and does not specify how any specific ORB must layout memory. However, CORBA is simply a standard, while DCOM is an actual product.

[5] For head-to-head comparisons of CORBA and DCOM, see Jason Pritchard, *COM and CORBA Side by Side: Architectures, Strategies, and Implementations* (Reading, MA: Addison-Wesley, 1999) and to see how to integrate the two approaches, see Michael Rosen, David Curtis, and Dan Foody, *Integrating Corba and Com Applications* (New York: John Wiley & Sons, 1998).

Selecting a Computing Architecture

Most systems are built to use the existing infrastructure in the organization, so often the current infrastructure restricts the choice of architecture. For example, if the new system is to be built for a mainframe-centric organization, a server-based architecture may be the best option. Other factors like corporate standards, existing licensing agreements, and product/vendor relationships also can mandate what architecture the project team needs to design. However, many organizations now have a variety of infrastructures available or are openly looking for pilot projects to test new architectures and infrastructures, enabling a project team to select an architecture based on other important factors.

Each of the computing architectures has its strengths and weaknesses, and no one architecture is inherently better than the others. Thus, it is important to understand the strengths and weaknesses of each computing architecture and when to use each. Figure 10-6 presents a summary of the important characteristics of each architecture.

Cost of Infrastructure   One of the strongest driving forces to client-server architectures is the cost of infrastructure (the hardware, software, and networks that will support the application system). Simply put, personal computers are more than 1,000 times cheaper than mainframes for the same amount of computing power. The personal computers on our desks today have more processing power, memory, and hard disk space than the typical mainframe of the past, and the cost of the personal computers is a fraction of the cost of the mainframe.

Therefore, the cost of client-server architectures is low compared to server-based architectures that rely on mainframes. Client-server architectures also tend to be cheaper than client-based architectures because they place less of a load on networks and thus require less network capacity.

Cost of Development   The cost of developing systems is an important factor when considering the financial benefits of client-server architectures. Developing application software for client-server computing is extremely complex, and most experts believe that it costs four to five times more to develop and maintain application software for client-server computing than it does for server-based computing. Developing application software for client-based architectures is usually cheaper still because there are many GUI development tools for simple stand-alone computers that communicate with database servers (e.g., Visual Basic, Access).

The cost differential may change as more companies gain experience with client-server applications; new client-server products are developed and refined;

	Server-Based	Client-Based	Client–Server
Cost of infrastructure	Very high	Medium	Low
Cost of development	Medium	Low	High
Ease of development	Low	High	Low–medium
Interface capabilities	Low	High	High
Control and security	High	Low	Medium
Scalability	Low	Medium	High

FIGURE 10-6
Characteristics of Computing Architectures

and client-server standards mature. However, given the inherent complexity of client-server software and the need to coordinate the interactions of software on different computers, a cost difference is likely to remain.

Ease of Development   In most organizations today, there is a huge backlog of mainframe applications, systems that have been approved but that lack the staff to implement them. This backlog signals the difficulty in developing server-based systems. The tools for mainframe-based systems often are not user-friendly and require highly specialized skills (e.g., COBOL)—skills that new graduates often don't have and aren't interested in acquiring. In contrast, client-based and client-server architectures can rely on GUI development tools that can be intuitive and easy to use. The development of applications for these architectures can be fast and painless. Unfortunately, the applications for client-server can be very complex because they must be built for several layers of hardware (e.g., database servers, Web servers, client workstations) that need to communicate effectively with each other. Project teams often underestimate the effort involved in creating secure, efficient client-server applications.

Interface Capabilities   Typically, server-based applications contain plain, character-based interfaces. For example, think about airline reservation systems like SABRE, which can be quite difficult to use unless the operator is well trained in the commands and hundreds of codes that are used to navigate through the system. Today, most users of systems expect a *graphical user interface* (GUI) or a Web-based interface, which they can operate using a mouse and graphical objects (e.g., push-buttons, drop-down lists, icons, etc.). GUI and Web development tools typically are created to support client-based or client-server applications; rarely can server-based environments support these types of applications.

Control and Security   The server-based architecture was originally developed to control and secure data, and it is much easier to administer because all of the data is stored in a single location. In contrast, client-server computing requires a high degree of coordination among many components, and the chance for security holes or control problems is much more likely. Also, the hardware and software that are used in client-server computing are still maturing in terms of security. When an organization has a system that absolutely must be secure (e.g., an application used by the Department of Defense), then the project team may be more comfortable with the server-based alternative on highly secure and control-oriented mainframe computers.

Scalability   Scalability refers to the ability to increase or decrease the capacity of the computing infrastructure in response to changing capacity needs. The most scal-

YOUR TURN   **10-1 COURSE REGISTRATION SYSTEM**

Think about the course registration system in your university. What computing architecture does it use? If you had to create a new system today, would you use the current computing architecture or change to a different one? Describe the criteria that you would consider when making your decision.

able architecture is client-server computing because servers can be added to (or removed from) the architecture when processing needs change. Also, the types of hardware that are used in client-server (e.g., minicomputers) typically can be upgraded at a pace that most closely matches the growth of the application. In contrast, server-based architectures rely primarily on mainframe hardware that needs to be scaled up in large, expensive increments, and client-based architectures have ceilings above which the application cannot grow because increases in use and data can result in increased network traffic to the extent that performance is unacceptable.

INFRASTRUCTURE DESIGN

In most cases, a system is built for an organization that has a hardware, software, and communications infrastructure already in place. Thus, project teams are usually more concerned with how an existing infrastructure needs to be changed or improved to support the requirements that were identified during the analysis phase, as opposed to how to design and build an infrastructure from scratch. Further, the coordination of infrastructure components is very complex, and it requires highly skilled technical professionals. As a project team, it is best to leave changes to the computing infrastructure to the infrastructure analysts. In this section we will summarize key elements of infrastructure design to give you a basic understanding of what it includes. We will describe the network model, the hardware and software specification, and close with a discussion of global and security issues.

The Network Model

The *network model* is a diagram that shows the major components of the information system (e.g., servers, communication lines, networks) and their geographic locations throughout the organization. There is no one way to depict a network model, and in our experience analysts create their own standards and symbols, using presentation applications (e.g., Microsoft PowerPoint) or diagramming tools (e.g., Visio).

The purpose of the network model is twofold: to convey the complexity of the system and to show how the system's components will fit together. The diagram also helps the project team develop the hardware and software specification that is described in the next section.

The components of the network model are the various clients (e.g., personal computers, kiosks), servers (e.g., database, network, communications, printer), network equipment (e.g., wires, dial-up connections, satellite links), and external systems or networks (e.g., Internet service providers) that support the application. *Locations* are the geographic sites related to these components. For example, if a company created an application for users at four of its plants in Canada and eight plants in the United States, and it used one external system to provide Internet service, the network model to depict this would contain 12 locations (4 + 8 = 12).

Creating the network model is a top-down exercise whereby you first graphically depict all of the locations where the application will reside. This is accomplished by placing symbols that represent the locations for the components on a diagram, and then connecting them with lines that are labeled with the approximate amount of data or types of network circuits between the separated components.

Companies seldom build networks to connect distant locations by buying land and laying cable (or sending up their own satellites). Instead, they usually

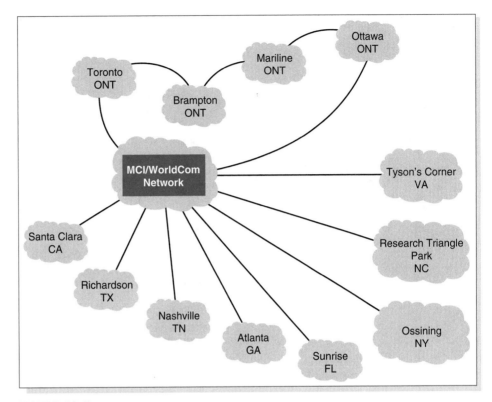

FIGURE 10-7
Top-Level Network Model with 13 Locations

lease services provided by large telecommunications firms such as AT&T, Sprint, and WorldCom/MCI. Figure 10-7 shows a typical network. The "clouds" in the diagram represent the networks at different locations (e.g., Toronto, Atlanta). The lines represent network connections between specific points (e.g., Toronto to Brampton). In other cases, a company may lease connections from many points to many others, and rather than trying to show all the connections, a separate "cloud" may be drawn to represent this many-to-many type of connection (e.g., the cloud in the center of this Figure 10-7 represents a network of many-to-many connections provided by a telecom firm such as MCI/WorldCom).

This high-level diagram has several purposes. First, it shows the locations of the components that are needed to support the application. Therefore, the project team can get a good understanding of the geographic scope of the new system and how complex and costly the communications infrastructure will be to support. (For example, an application that supports one site likely will have less communications costs as compared to a more complex application that will be shared all over the world.) The diagram also indicates the external components of the system (e.g., customer systems, supplier systems), which may impact security or global needs (discussed later in this chapter).

The second step in the network model is to create low-level network diagrams for each of the locations shown on the top-level diagram. First, hardware is drawn on the model in a way that depicts how the hardware for the new system will be placed throughout the location. It usually helps to use symbols that resemble the

10-A Nortel's Network

Nortel, the giant Canadian manufacturer of network switches, recently implemented a network using its own equipment. Prior to implementing it, Nortel had used a set of more than 100 circuits that was proving to be more and more expensive and unwieldy to manage. Nortel started by identifying the 20% of its sites that accounted for 80% of the network traffic. These were the first to move to the new network. By the time the switchover is complete, more than 100 sites will be connected.

There are two primary parts to the new network (see Figure 10-7). The largest part uses a public network provided by WorldCom/MCI. Nortel also operates a private network in Ontario. Both parts of the network use 44 Mbps circuits. Nortel is planning to move to 155 Mbps circuits when costs drop.

Question:

What are the advantages and disadvantages of Nortel's phased implementation of the new network?

Source: "A Case for ATM," Network World, June 2, 1997. Vol. 14, No. 22 pp. 1, 90.

hardware that will be used. For example, if you use Microsoft PowerPoint, you can place clip art that represents workstations, mainframes, printers, and the like, on the diagram to clearly communicate the various hardware components at that location. The amount of detail to include on the network model depends on the needs of the project. Some low-level network models contain text descriptions below each of the hardware components that describe in detail the proposed hardware configurations and processing needs. Others only include the number of users that are associated with the different parts of the diagram.

Next, lines are drawn connecting the components that will be physically attached to each other, and other symbols like "lightning bolts" and satellite dishes are used to communicate dial-up and satellite connections. In terms of software, some network, models list the required software for each network model component right on the diagram, whereas other models describe the software in a memo

10-B The Department of Education's Financial Aid Network

Each year, the U.S. Department of Education (DOE) processes more than 10 million student loans, such as Pell Grants, Guaranteed Student Loans, and college work-study programs. The DOE has more than 5,000 employees that need access to more than 400 gigabytes of student data.

The financial aid network, located in one Washington, DC building, has two major parts (i.e., backbones) each of which uses high-speed 155 Mbps circuits to connect four local area networks and three high-performance database servers (see Figure 10-8). These two backbones are connected to another network backbone that connects the financial aid building using 100 Mbps circuits to five other DOE buildings in Washington that have similar networks. The two backbones are also connected to the ten regional DOE offices across the United States and to other U.S. government agencies via a set of lower speed 1.5 Mbps connections.

Source: Network Computing, May 15, 1996.

Question

Why does each part of the network have different data rates?

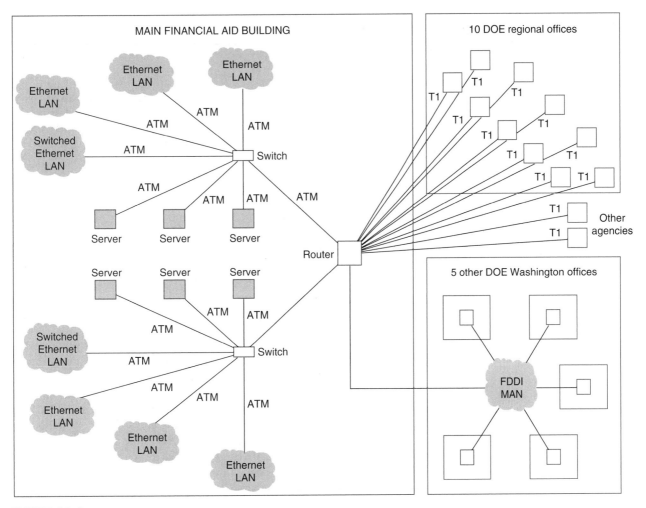

FIGURE 10-8
Low-Level Network Model

that is attached to the network model and stored in the project binder. Also, with UML, the system architecture layer can be described with class and package diagrams that show the network interface and how the other layers interact with it.[6] Figure 10-8 shows an example of a network model for a single location.

Our experiences have shown that most project teams create a memo for the project files that provides additional detail about the network model. This information is helpful to the people who are responsible for creating the hardware and software specification (described next) and who will work more extensively with the infrastructure development. This memo can include special issues that impact communications, requirements for hardware and software that may not be obvious from the network model, or specific hardware or software vendors or products that should be acquired.

[6] The other layers include foundation, human–computer interaction, data management, and problem domain. Layers, packages, and package diagrams were described in Chapter 9.

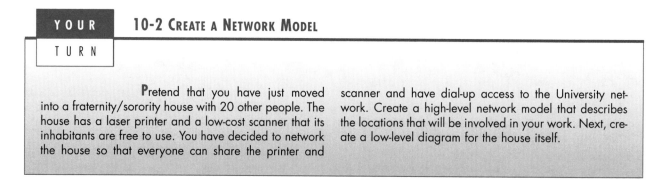

10-2 CREATE A NETWORK MODEL

Pretend that you have just moved into a fraternity/sorority house with 20 other people. The house has a laser printer and a low-cost scanner that its inhabitants are free to use. You have decided to network the house so that everyone can share the printer and scanner and have dial-up access to the University network. Create a high-level network model that describes the locations that will be involved in your work. Next, create a low-level diagram for the house itself.

Again, the primary purpose of the network model diagram is to present the proposed infrastructure for the new system. The project team can use the diagrams to understand the scope of the system, the complexity of its structure, any important communication issues that may affect development and implementation, and the actual components that will need to be acquired or integrated into the environment.

Hardware and Software Specification

The design phase is the time to begin acquiring the hardware and software that will be needed for the future system. In many cases, the new system will simply run on the existing equipment in the organization. Other times, however, new hardware and software (usually for servers) must be purchased. The *hardware and software specification* is a document that describes the hardware and software needed to support the application. The actual acquisition of hardware and software should be left to the Purchasing Department or the area in the organization that handles capital procurement; however, the project team can use the hardware and software specification to communicate the project needs to the appropriate people. There are several steps involved in creating the document.

First, you must create a list of the hardware that is needed to support the future system. The low-level network model provides a good starting point for recording the project's hardware needs because each of the components on the diagram corresponds to an item on this list. In general, the list can include things like database servers, network servers, peripheral devices (e.g., printers, scanners), backup devices, storage components, and any other hardware component that is needed to support an application. At this time, you also should note the quantity of each item that will be needed.

The second step is to describe in as much detail as possible the minimum requirements for each piece of hardware. Typically, the project team must convey requirements like the amount of processing capacity, the amount of storage space, and any special features that should be included. This step becomes easier with experience; however, there are some hints that can help you describe hardware needs accurately. For example, consider the hardware standards within the organization or those recommended by vendors. Talk with experienced system developers or other companies with similar systems. Finally, think about the factors that impact hardware performance, such as the response time expectations of the users, data volumes, software memory requirements, the number of users accessing the system, the number of external connections, and growth projections. Figure 10-9 presents a sample hardware specification for a client workstation.

Client Workstation

Minimum requirements:
- Pentium III Processor
- 128 megabytes of RAM
- 40-gigabyte hard disk
- CD-ROM drive
- Large-screen monitor

Number needed:
200

Software requirements:
- Windows 2000 operating system—current version
- Netscape Navigator—current version
- Microsoft Office—current version

Other acquisition information:
- Extended warrantees should be purchased for all components of the client workstations.
- Site licenses (as opposed to individual copies) are available for Microsoft Office software.
- We will need to send the 1 or 2 employees who will be supporting these workstations to Windows 2000 training.

FIGURE 10-9
Hardware and Software Specification for Client Workstation

Next, you will need to write down the software that will run on each hardware component as well as any additional costs. For example, technical training, maintenance, extended warranties, and licensing agreements (e.g., a site license for a software package) are hardware- and software-related costs that should be considered during the acquisition process. Again, the needs that you list are influenced by decisions that are made in the other design phase activities. Figure 10.9 shows the specification with software and other cost information.

CONCEPTS
IN ACTION

10-C THE IMPORTANCE OF INFRASTRUCTURE PLANNING

At the end of 1997, Oxford Health Plans, Inc. posted a $120 million loss to its books. The company's unexpected growth was its undoing because the system, which was originally planned to support the company's 217,000 members, had to meet the needs of a membership that now exceeded 1.5 million. System users found that processing a new member sign-up took 15 minutes instead of the proposed 6 seconds. Also, the computer problems left Oxford unable to send out bills to many of its customer accounts and rendered it unable to track payments to hundreds of doctors and hospitals. In less than a year, uncollected payments from customers tripled to more than $400 million, while the payments owed to caregivers amounted to more than $650 million. Mistakes in infrastructure planning can cost far more than the cost of hardware, software, and network equipment alone.

Source: The Wall Street Journal, December 11, 1997.

QUESTION
If you had been in charge of the Oxford project, how would you have avoided these problems?

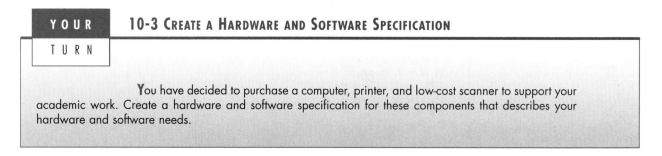

YOUR TURN

10-3 CREATE A HARDWARE AND SOFTWARE SPECIFICATION

You have decided to purchase a computer, printer, and low-cost scanner to support your academic work. Create a hardware and software specification for these components that describes your hardware and software needs.

GLOBAL ISSUES

Yet another important part of the design of the system's architecture is the way in which the project team will design global support. In today's global business environment, organizations are expanding their systems to reach users around the world. Although this can make great business sense, its impact on application development should not be underestimated. Typically, a memo is included in the Project Binder that describes the relevant global support issues for the project and exactly how each of them will be addressed by the design of the system. The following sections describe several concepts that you can incorporate in a global support memo when creating a system that will be used in multiple countries.

Multilingual Requirements

The first and most obvious difference between applications used in one region and those designed for global use is language. Global applications often have to support users who speak different languages and write using non-English letters (e.g., accents, Cyrillic, Japanese). One of the most challenging aspects in designing global systems is getting a good translation of the original language messages into a new language. Words often have similar meanings but can convey subtly different meanings when they are translated, so it is important to use translators skilled in translating technical words.

The other challenge is often screen space. In general, English-language messages usually take 20 to 30% fewer letters than their French or Spanish counterparts. Designing global systems requires allocating more screen space to message than might be used in the English-language version.

Some systems are designed to handle multiple languages on the fly so that users in difference countries can use different languages concurrently; that is, the same system supports several different languages simultaneously (a *concurrent multilingual system*). Other systems contain separate parts that are written in each language and must be reinstalled before a specific language can be used; that is, each language is provided by a different version of the system so that any one installation will only use one language (i.e., a *discrete multilingual system*). Either approach can be effective, but this functionality must be designed into the system well in advance of implementation.

Local versus Centralized Control

The project team will need to give some thought to how much of the application will be controlled by a central group and how much of it will be managed locally.

For example, some companies allow groups of users to customize the application by permitting the omission or addition of certain features. This decision has trade-offs between flexibility and control because customization often makes it more difficult for the project team to create and maintain the application. It also means that training can differ between different parts of the organization, and customization can create problems when staff move from one location to another.

Unstated Norms

Many countries have unstated norms that are not shared internationally. It is important for the application designer to make these assumptions explicit because they can lead to confusion otherwise. In the United States, the unstated norm for entering a date is the date format MM/DD/YYYY; however, in Canada and most European countries, the implicit format is DD/MM/YYYY. When designing global systems, it is critical to recognize these unstated norms and make them explicit so that users in different countries do not become confused.

Currency is the other item often overlooked in system design. Global application systems must specify the currency in which information is being entered and reported. Being explicit can also address inconsistencies in the way business is conducted across the world.

24-7 Support

A 40-hour workweek mentality has little place in a project whose goal is to build an application to be used by people around the world. Instead, project team members need to consider how the application can be developed, maintained, and used *24-7* (i.e., 24 hours a day, 7 days a week). This 24-7 requirement means that users may need help or have questions at any time, and a support desk that is available eight hours a day will not be sufficient support. Also, it is more difficult to predict

---

CONCEPTS **10-D DEVELOPING MULTILINGUAL SYSTEMS**

IN ACTION

I've had the opportunity to develop two multilingual systems. The first was a special-purpose decision support system to help schedule orders in paper mills called *BCW-Trim.* The system was installed by several dozen paper mills in Canada and the United States, and it was designed to work in either English or French. All messages were stored in separate files (one set English, one set French), with the program written to use variables initialized either to the English or French text. The appropriate language files were included when the system was compiled to produce either the French or English version.

The second program was a groupware system called *GroupSystems,* for which I designed several modules. The system has been translated into dozens of different languages including French, Spanish, Portuguese, German, Finnish, and Croatian. This system enables the user to switch between languages at will by storing messages in simple text files. This design is far more flexible because each individual installation can revise the messages at will. Without this approach, it is unlikely that there would have been sufficient demand to warrant the development of versions to support less commonly used languages (e.g., Croatian). *Alan Dennis*

QUESTION
How would you decide how much to support users who speak languages other than English? Would you create multilingual capabilities for any application that would be available to non-English speaking people? Think about Web sites that exist today.

the peaks and valleys in usage of the system. Typically, applications are backed up on weekends or late evenings when users are no longer accessing the system. Such maintenance activities will have to be rethought if the users are continuously using the application. The development of Web interfaces in particular has escalated the need for 24-7 support because by default the Web can be accessed by anyone at anytime. For example, the developers of a Web application for outdoor gear and clothing retailer Orvis were surprised when the first order after going live came from Japan!

Communications Infrastructure

The communications infrastructure typically is more complex for applications that are accessed worldwide because the infrastructure includes more components and more connections that may not be commonly available worldwide. The project team must pay careful attention to how the components can effectively communicate information quickly and accurately. This may mean distributing the data in the application across servers in different locations to improve performance or investing in permanent, high-speed communications connections between geographic sites. The network model is a helpful tool when planning the communications infrastructure for a global application. Communications technologies are particularly susceptible to problems in going global because different continents often favor different technologies.

SECURITY

The last piece to include in the system architecture design is a security plan that addresses how to keep the application and its data secure. Like all design activities, the security plan should be driven by requirements that were identified during analysis, not by the newest or least expensive security features available. Three main steps are needed to create the plan: identifying threats to the system, assessing the risk of each threat, and creating controls that maintain security.

Identifying Threats to the System

A *threat* to the system is any potentially adverse occurrence that can do harm to the application or its data, such as a computer virus, a hacker, or an unexpected natural disaster. Figure 10-10 lists the most commonly occurring threats. As you can see, about 87% of organizations suffer losses from viruses each year, while only 30% experience losses due to an external hacker; organizations are also much more likely to be attacked by an internal hacker—one of their own employees—than by someone from outside.

In general, application security threats can be classified into one of two categories: disruption, destruction, and disaster; and unauthorized access. *Disruptions* occur when users cannot use the application for a short period of time. For example, a network wire may be cut, causing part of the network to cease functioning until the failed component can be replaced. Some users may be affected, but others are not. Some disruptions may also be caused by or result in the *destruction* of data. For example a virus may destroy files, or the "crash" of a hard disk may cause files to be destroyed. Other disruptions may be catastrophic. Natural (or

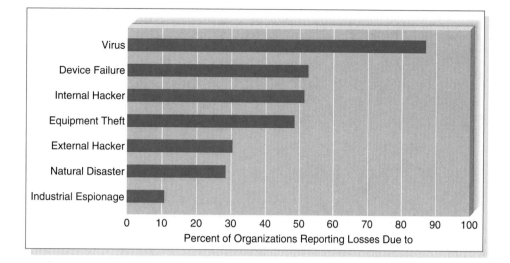

FIGURE 10-10
Most Common Threats

man-made) *disasters* may occur that destroy servers or large sections of the network. For example, fires, floods, earthquakes, mudslides, tornadoes, or terrorist attacks can destroy large parts of the buildings and networks in their path.

    Unauthorized access is often viewed as hackers gaining access to an application; however, most unauthorized access incidents involve employees. Unauthorized access may have only minor effects. A curious intruder may simply explore the system, gaining knowledge that has little value. A more serious intruder may be a competitor bent on industrial espionage who could attempt to gain access to information on products under development, or the details and price of a bid on a large contract. Worse still, the intruder could change files to commit fraud or theft, or destroy information to injure the organization.

Assessing the Risk of Each Threat

Once the threats are identified, they are listed across the top of a *control spreadsheet,* with all of the *system components* listed down the side[7] (see Figure 10-11). A system component is one of the individual pieces that compose the system, including servers (e.g., database servers and Web servers), clients (e.g., personal computers and kiosks), networks, and application software.

    Each threat should be ranked according to its importance and listed in order from most to least important. Importance can be based on number of criteria, such as dollar loss, embarrassment, liability, and probability of occurrence.

Creating Controls

Finally, all of the controls that will be incorporated into the new system are listed in the center of the spreadsheet. A *control* is something that mitigates or stops a threat. It also restricts, safeguards, or protects a component. During this step, you identify the controls and put them into each cell for each threat and component.

    Begin by considering the system component and the specific threat, and then describe each control that prevents, detects, or corrects that threat. The description

---

[7] Risk assessment was described in Chapter 3.

Threats	Disruption, Destruction, Disaster					Unauthorized Access		
Components	Fire	Flood	Power loss	Circuit Failure	Virus	External Intruder	Internal Intruder	Eavesdrop
Servers	1,2	1,3	4	1,5,6	7,8	9,10,11,12	9,10	
Client Computers								
Communications Circuits								
Network Devices								
Network Software								
People								

Controls
1. Disaster recovery plan
2. Halon fire system in host computer room; sprinklers in rest of building
3. Host computer room on 5th floor
4. Uninterruptable Power Supply (UPS) on all major network servers
5. Contract guarantees from interexchange carriers
6. Extra backbone fiber cable laid in different conduits between major servers
7. Virus checking software present on the network
8. Extensive user training on virurses and reminders in monthly newsletter
9. Strong password software
10. Extensive user training on password security and reminders in monthly newsletter
11. Call-back modem system
12. Application layer firewall

FIGURE 10-11
Security Plan

of the control (and its role) is placed in a numerical list, and the control's number is placed in the cell. For example, assume 12 controls have been identified. Each one is described, named, and numbered consecutively. The numbered list of controls has no ranking attached to it: the first control is number 1 just because it is the first control identified (see Figure 10-11).

Controls for Disruption, Destruction, and Disaster   The key principle in preventing disruption, destruction, and disaster—or at least reducing their impact—is *redundancy*. Redundant hardware that automatically recognizes failure and intervenes to replace the failed component can mask a failure that would otherwise result in a service disruption. For example, a special-purpose *fault-tolerant server* contains many redundant components to prevent failure. With redundancy, if the primary device fails, the redundant device automatically takes over, with no observable effects on the application.

Disasters are different. In this case, an entire site can be destroyed. Even if redundant components are present, often the scope of the loss is such that returning the network to operation is extremely difficult. The best solution is to have the application and its data stored in a separate location so that it can be retrieved in the case of an emergency. All organizations need a *disaster recovery plan* that describes how operations will be restored in the event of a disaster. Most organizations have a separate computer facility (either one that they operate or one operated by a disaster recovery firm) to which they can move computer operations should their main computer facility be destroyed.

Special attention also must be paid to preventing computer *viruses.* Viruses cause unwanted events, some are harmless (such as nuisance messages), some serious (such as the destruction of data). Many antivirus software packages are available to check disks and files to make sure they are virus-free. Always check all diskettes and files for viruses before using them (even those from friends!). Researchers estimate that six new viruses are developed every day, so it is important to frequently update the virus information files that are provided by the antivirus software.

Controls for Unauthorized Access   Some people will try to gain unauthorized access to the system. Some will do no harm, while others attempt to access a system for personnel gain or to sabotage the system in some way. The key principle in preventing unauthorized access is to be proactive. There are three general approaches to preventing unauthorized access: developing a security policy, securing application access, and training. A combination of all techniques is best to ensure strong security.

A *security policy* is critical and should clearly define the important application components to the safeguarded components and the important controls needed to do that. It should contain a section devoted to what users of the system should and should not do. It should also contain a clear plan for routinely training users—particularly those with little computer expertise—on key security rules and a clear plan for routinely testing and improving the security controls in place.

It is important to screen and classify both users and data. Some organizations, especially those in government, assign different security clearance levels to users as well as to data, thus permitting users to see only what they "need to know." Adequate user training on the application should be provided through self-teaching manuals, newsletters, policy statements, and short courses. A well-publicized security campaign may deter potential intruders.

Several technologies are also available to increase security. The use of *passwords* and *encryption* are central to any type of secure system. Passwords restrict

---

CONCEPTS   **10-E POWER OUTAGE COSTS A MILLION DOLLARS**

IN ACTION

Lithonia Lighting, located just outside of Atlanta, is the world's largest manufacturer of light fixtures with more than $1 billion in annual sales. One afternoon, the power transformer at its corporate headquarters exploded, leaving the entire office complex, including the corporate data center, without power. The data center's backup power system immediately took over and kept critical parts of the data center operational. However, it was insufficient to power all systems, so the system supporting sales for all of Lithonia Lighting's North American agents, dealers, and distributors had to be turned off.

The transformer was quickly replaced and power restored. However, the three-hour shutdown of the sales system cost $1 million in potential sales lost. Unfortunately, it is not uncommon for the cost of a disruption to be hundreds or thousands of times the cost of the failed components.

QUESTION
How could these losses have been avoided?

access to those people who know the password, while encryption prevents someone who does not have the encryption key from decoding and understanding data, whether the data is in transit on the Internet or stored on a database server.

Firewalls are software programs designed to prevent unauthorized access to organizational networks and are often used to protect a network from hackers on the Internet. There are several types of firewalls that provide different levels of security. Some firewalls, for example, permit Web traffic but prevent people from logging in or transferring files. Other types of firewalls only allow traffic if a user applies a password or uses a special format. For this reason, organizations often use two firewalls: one between their Web server and the Internet that lets Web traffic in, and a second firewall between the Web server and the main networks that have a higher level of security.

APPLYING THE CONCEPTS AT CD SELECTIONS

Alec realized that the hardware, software, and communications that would support the new application would need to be integrated into the current infrastructure at CD Selections. Therefore, he set up a meeting with the project team and the information systems manager who was responsible for designing and maintaining the infrastructure at the company.

During the meeting he learned that CD Selections had been moving towards a client-server environment over the past few years, although a central mainframe still existed as the primary server for many server-based applications.

The group discussed the Internet sales application and unanimously agreed that it should be built using a multitier (probably three-tier) client-server architecture. They felt that most Web-based applications were best suited for the client-server platform because of the high availability of development tools and the high scalability. It was hard to know at this point exactly how much traffic this Web site would get and how much power the system would require, but a client-server architecture would allow CD Selections to easily scale up the system as needed.

By the end of the meeting, it was agreed that a three-tiered client-server architecture was the best configuration for the Internet portion of the Internet Sales System (i.e., the Place Order process in Figures 6-11 and 6-13). Customers will use their personal computers running a Web browser as the client. A database server will store the Internet System's databases, while an application server will have Web server software and the application software to run the system.

A separate two-tier client server system will maintain the CD and Marketing Material information (i.e., the Maintain CD Information and Maintain Marketing Information processes in Figure 6-13). This system will have an application for the personal computers of the staff working in the Internet sales group that communicates directly with the database server and enables staff to update the information. The database server will have a separate program to enable it to exchange data with CD Selections' distribution system on the company mainframe.

Next, Alec created a network model to show the major components of the Internet Sales System (see Figure 10-12). The Internet Sales System is on a separate network segment divided from the CD Selections' main network by a firewall that separates the network from the Internet while granting access to the Web and database servers. The Internet Sales System has two parts. A firewall is used to connect the Web/Application server to the Internet, while another firewall further

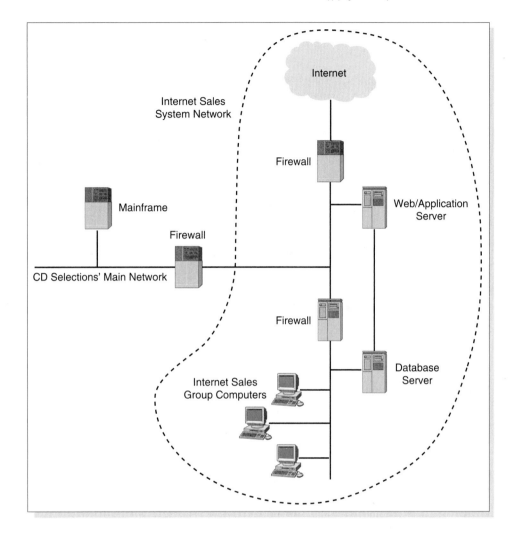

FIGURE 10-12
Network Model for CD Selections
Internet Sales System

protects the Internet Sales group's client computers and database server from the Internet. In order to improve response time, a direct connection is made from the Web/Application server to the database server because these will exchange a lot of data.

After examining the network model, Alec decided that the only components that need to be acquired for the project are a database server and a Web server, and three client computers for the Internet Sales group. He developed a hardware and software specification for these components and handed them off to the Purchasing Department to start the acquisition process.

Finally, Alec put together plans for global support and security. Given the project's short time frame, he did not want to complicate the system with global support, so he planned to target the system for English-speaking users and operators. But he made a note to investigate global issues in later versions of the system. Given that the Web interface could reach a geographically dispersed group, Alec did want to plan for 24-7 system support. He scheduled a meeting to talk with the CD Selections' systems support group about how they might be able to support the Internet Sales System outside of standard working hours.

Alec developed a control spreadsheet listing security threats and controls to file as the system's security plan. He included items like the firewalls that would be set up to prevent unauthorized access to the system; standard backup and recovery procedures as laid out by CD Selections' company policy to avoid destruction of data; and a password strategy for the internal (e.g., marketing users) and external (e.g., customers) users of the system. He distributed the spreadsheet to the project team to review before the next staff meeting. At that time he planned to include additional threats that other team members identify, along with the ways in which the system can be designed to handle them.

SUMMARY

An important component of the design phase is the creation of the system architecture design, which includes the hardware, software, and communications infrastructure for the new system and the way in which security and global support will be provided. The system architecture design is described in a deliverable, which contains the network model, the hardware and software specification, and plans for how the system will address global requirements and security.

Computing Architectures

There are three fundamental computing architectures. In server-based computing, the server performs virtually all of the work. In client-based computing, the client computers are responsible for most of the application functions. In client-server computing, the work is shared between the servers and clients, and it is distributed across two components (two-tiered architecture), three components (three-tiered architecture), or more than three components (n-tiered architecture). From an object-oriented point of view, the future trend in client-server computing is distributed object computing (DOC). Currently, there are two competing approaches to support DOC: OMG's CORBA and Microsoft's DCOM.

Each of the computing architectures has its strengths and weaknesses, and no one architecture is inherently better than the others. The choice that the project team makes should be based on several criteria, including cost of development, ease of development, need for GUI applications, network capacity, central control and security, and scalability. The project team may also want to consider the existing architecture in place in the organization and special software requirements of the project.

Network Model

The network model is a diagram that shows the technical components of the information system (e.g., servers, personal computers, networks) and their geographic locations throughout the organization. The components of the network model are the various clients (e.g., personal computers, kiosks), servers (e.g., database, network, communications, printer), network equipment (e.g., wire, dial-in connections, satellite links), and external systems or networks (e.g., Internet service provider) that support the application. Creating a network model is a top-down process whereby a high-level diagram is first created to show the geographic sites, or locations, that house the various components of the future system. Next, a low-level diagram is created to describe each location in detail, and it shows the system's hardware components and how they are attached to one another.

Hardware and Software Specification

The hardware and software specification is a document that describes what hardware and software are needed to support the application. To create a specification document, the hardware that is needed to support the future system is listed and then described in as much detail as possible. Next, the software to run on each hardware component is written down, along with any additional associated costs, such as technical training, maintenance, extended warrantees, and licensing agreements. Although the project team may suggest specific products or vendors, ultimately the hardware and software specification is turned over to the people who are in charge of procurement.

Global Issues

More and more applications are being built to support users around the world. Project teams need to design the system to handle the effects of this trend on norms that are associated with the system, the multilanguage requirements, and whether the control for the application should be centralized or controlled by local groups. Also, this global requirement forces project teams to plan how the application will be developed, maintained, and used in a 24-7 environment and to pay close attention to the communications infrastructure that is in place to support the unique global requirements.

Security

In general, an application is threatened in two main ways: disruption, destruction, and disaster; and unauthorized access. To be sure that the application has the necessary protection against both types of threats, the project team should follow a three-step process: identifying threats to the system, assessing their risks, and developing controls to address each threat. Controls include things like security plans, password protection, and firewalls.

KEY TERMS

24-7
Application logic
Client computer
Client-based architecture
Client-server architecture
Common Object Request Broker
  Architecture (CORBA)
Component object model
  (COM)
Concurrent multilingual system
Control
Control spreadsheet
Data access logic
Data storage
Destruction
Disaster
Disaster recovery plan
Discrete multilingual system
Disruptions

Distributed component object
  model (DCOM)
Distributed objects computing
  (DOC)
Encryption
Fat client
Fault-tolerant server
Firewalls
Graphical user interface (GUI)
Hardware and software specification
Locations
Mainframe
Microcomputer
Middleware
Minicomputer
Network
Network model
N-tiered architecture
Object Request Broker (ORB)

Object wrappers
Passwords
Personal computer
Presentation logic
Redundancy
Risk analysis
Scalable
Security policy
Server
Server-based architecture
Special-purpose terminal
System architecture design
System component
Terminal
Thin client
Threat
Three-tiered architecture
Two-tiered architecture
Unauthorized access
Viruses

QUESTIONS

1. What are the four general functions of any application?
2. What are the three main components of any computing architecture?
3. Name two examples of a server.
4. Name two examples of a client.
5. What is the biggest problem with server-based computing?
6. What is the biggest problem with client-based computing?
7. Is client-server a less expensive alternative than server-based? Why or why not?
8. Describe three benefits and three limitations of client-server computing.
9. Describe the differences among two-tiered, three-tiered, and n-tiered computing.
10. What are distributed object computing, CORBA, and DCOM?
11. Define scalable. Why is this term important to system developers?
12. What six criteria are helpful to use when comparing the appropriateness of computing alternatives?
13. Why should the project team consider the existing computing architecture in the organization and special software requirements of the project?
14. What does the network model communicate to the project team?
15. What are the differences between the top-level network model and the low-level network model?
16. Describe the steps for creating a network model for a system.
17. How is information from the network model used to create the hardware and software specification?
18. Describe the steps to creating a hardware and software specification.
19. What additional hardware- and software-associated costs may need to be included on the hardware and software specification?
20. Who ultimately is in charge of acquiring hardware and software for the project?
21. Name four issues that project teams must think about when building global applications.
22. What are two ways to address multilanguage requirements for a system?
23. Describe the three-step process for creating a security plan.
24. What are the two kinds of threats to a system? Provide examples for each kind.
25. Describe three controls that can be used to address the threats from Question 24.

EXERCISES

A. Using the Web (or past issues of computer industry magazines such as *Computerworld*), locate a system that runs in a server-based environment. Based on your reading, why do you think the company chose that computing environment?
B. Using the Web (or past issues of computer industry magazines such as *Computerworld*), locate a system that runs in a client-server environment. Based on your reading, why do you think the company chose that computing environment?
C. Using the Web, locate examples of a mainframe component, a minicomputer component, and a microcomputer component. Compare the components in terms of price, speed, available memory, and disk storage. Did you find large differences in prices when the performances of the components are considered?
D. You have been selected to find the best client-server computing architecture for a Web-based order entry system that is being developed for LL Bean. Write a short memo that describes to the project manager your reason for selecting an n-tiered architecture over a two-tiered architecture. In the memo, give some idea as to what the different components of the architecture you would include.
E. You have been selected to determine whether your firm should invest in CORBA- or DCOM-based technologies to support distributed objects computing. Write a short memo that describes to the steering committee the advantages and disadvantages of both approaches. (*Note:* You probably should check out the Web sites associated with both.)
F. Think about the system that your university currently uses for career services and pretend that you are in charge of replacing the system with a new one. Describe how you would decide on the computing architecture for the new system using the criteria presented in this chapter. What information

will you need to find out before you can make an educated comparison of the alternatives?

G. Locate a consumer products company on the Web and read its company description (so that you get a good understanding of the geographic locations of the company). Pretend that the company is about to create a new application to support retail sales over the Web. Create a high-level network model that depicts the locations that would include components that support this application.

H. An energy company with headquarters in Dallas, Texas, is thinking about developing a system to track the efficiency of its oil refineries in North America. Each week, the 10 refineries as far as Valdez, Alaska, and as close as San Antonio, Texas, will upload performance data via satellite to the corporate mainframe in Dallas. Production managers at each site will use a personal computer to dial into an Internet service provider and access reports via the Web. Create a high-level network model that depicts the locations that have components supporting this system.

I. Create a low-level network diagram for the building that houses the computer labs at your university. Choose an application (e.g., course registration, student admissions) and only include the components that are relevant to that application.

J. Pretend that your mother is a real estate agent, and she has decided to automate her daily tasks using a laptop computer. Consider her potential hardware and software needs, and create a hardware and software specification that describes them. The specification should be developed to help your mother buy her hardware and software on her own.

K. Pretend that the admissions office in your university has a Web-based application so that students can apply for admission on-line. Recently, there has been a push to admit more international students into the university. What do you recommend that the application include to ensure that it will support this global requirement?

MINICASES

1. The system development project team at Birdie Masters golf schools has been working on defining the system architecture design for the system. The major focus of the project is a networked school location operations system, allowing each school location to easily record and retrieve all school location transaction data. Another system element is the use of the Internet to enable current and prospective students to view class offerings at any of the Birdie Masters' locations, schedule lessons and enroll in classes at any Birdie Masters location, and maintain a student progress profile—a confidential analysis of the student's golf skill development.

   The project team has been considering the globalization issues that should be factored into the architecture design. The school's plan for expansion into the golf-crazed Japanese market is moving ahead. The first Japanese school location is tentatively planning to open about six months after the target completion date for the system project. Therefore, it is important that issues related to the international location be addressed now during design.

   Assume that you have been given the responsibility of preparing a summary memo for the Project Binder on the globalization issues that should be factored into the design. Prepare this memo discussing the global-

ization issues that are relevant to Birdie Masters' new system.

2. Jerry is a relatively new member of a project team that is developing a retail store management system for a chain of sporting goods stores. Company headquarters is in Las Vegas, and the chain has 27 locations throughout Nevada, Utah, and Arizona. Several cities have multiple stores.

   The new system will be a networked, client-server architecture. Stores will be linked to one of three regional servers, and the regional servers will be linked to corporate headquarters in Las Vegas. The regional servers also link to each other. Each retail store will be outfitted with similar configurations of two PC-based point-of-sale terminals networked to a local file server. Jerry has been given the task of developing a network model that will document the geographic structure of this system. He has not faced a system of this scope before and is a little unsure how to begin.

   a. Prepare a set of instructions for Jerry to follow in developing this network model.

   b. Draw a network model for this organization.

   c. Prepare a set of instructions for Jerry to follow in developing a hardware and software specification.

PLANNING

ANALYSIS

DESIGN

- ✔ **Evolve Analysis Models**
- ✔ **Develop the Design Strategy**
- ✔ **Develop the Design**
- ✔ **Develop Infrastructure Design**
- ✔ **Develop Network Model**
- ✔ **Develop Hardware/Software Specification**
- ✔ **Develop Security Plan**
- ☐ Develop Use Scenarios
- ☐ Design Interface Structure
- ☐ Design Interface Standards
- ☐ Design User Interface Template
- ☐ Design User Interface
- ☐ Evaluate User Interface
- ☐ Select Object Persistence Format
- ☐ Map Objects to Object Persistence
- ☐ Optimize Object Persistence
- ☐ Size Object Persistence
- ☐ Specify Constraints
- ☐ Develop Contracts
- ☐ Develop Method Specifications

T A S K C H E C K L I S T

PLANNING　　　ANALYSIS　　　DESIGN

CHAPTER 11
USER INTERFACE STRUCTURE DESIGN

A user interface is the part of the system with which the users interact. It includes the screen displays that provide navigation through the system, the screens and forms that capture data, and the reports that the system produces (whether on paper, on the screen, or via some other media). This chapter introduces the basic principles and processes of interface design and discusses how to design the interface structure and standards.

OBJECTIVES

- Understand several fundamental user interface design principles.
- Understand the process of user interface design.
- Understand how to design the user interface structure.
- Understand how to design the user interface standards.

CHAPTER OUTLINE

IMPLEMENTATION

INTRODUCTION

Interface design is the process of defining how the system will interact with external entities (e.g., customers, suppliers, other systems). In this chapter we focus on the design of *user interfaces,* but it is also important to remember that there are sometimes *system interfaces* that exchange information with other systems (e.g., in the case of CD Selections, the Internet Sales System must exchange data with the distribution system). System interfaces are typically designed as part of a systems integration effort. They are defined in general terms as part of the system architecture and data management layers, and they are designed specifically during the data storage design (see Chapter 13) and class and method design (see Chapter 14).

The user interface design defines the way in which the users will interact with the system and the nature of the inputs and outputs that the system accepts and produces; that is, it deals with the human computer interface layer. The user interface includes three fundamental parts. The first is the *navigation mechanism,* the way in which the user gives instructions to the system and tells it what to do (e.g., buttons, menus). The second is the *input mechanism,* the way in which the system captures information (e.g., forms for adding new customers). The third is the *output mechanism,* the way in which the system provides information to the user or to other systems (e.g., reports, Web pages). Each of these is conceptually different, but all are closely intertwined. All computer displays contain navigation mechanisms, and most contain input and output mechanisms. Therefore, navigation design, input design, and output design are tightly coupled.

This chapter introduces several fundamental design principles and provides an overview of the design process for the human–computer interaction layer. The next chapter continues this process by presenting a series of specific techniques for designing navigation, inputs, and outputs.

PRINCIPLES FOR USER INTERFACE DESIGN

In many ways, user interface design is an art. The goal is to make the interface pleasing to the eye and simple to use, while minimizing the effort the users need to accomplish their work. The system is never an end in itself; it is merely a means to accomplish the "business" of the organization.

In our experience, the greatest problem facing experienced designers is using space effectively. Simply put, often too much information needs to be presented on a screen or report or form than will fit comfortably. Analysts must balance the need for simplicity and pleasant appearance against the need to present the information across multiple pages or screens, which decreases simplicity. In this section, we discuss some fundamental interface design principles, which are common for navigation design, input design, and output design[1] (see Figure 11-1).

Layout

The first element of design is the basic *layout* of the screen, form, or report. Most software designed for personal computers follows the standard Windows or Macintosh

[1] A good book on the design of interfaces is Susan Weinschenk, Pamela Jamar, and Sarah Yeo, *GUI Design Essentials* (New York: John Wiley & Sons, 1997).

Principle	Description
Layout	The interface should be a series of areas on the screen that are used consistently for different purposes—for example, a top area for commands and navigation, a middle area for information to be input or output, and a bottom area for status information.
Content awareness	Users should always be aware of where they are in the system and what information is being displayed.
Aesthetics	Interfaces should be functional and inviting to users through careful use of white space, colors, and fonts. There is often a tradeoff between including enough white space to make the interface look pleasing without losing so much space that important information does not fit on the screen.
User experience	Although ease of use and ease of learning often lead to similar design decisions, there is sometimes a tradeoff between the two. Novice users or infrequent users of software will prefer ease of learning, whereas frequent users will prefer ease of use.
Consistency	Consistency in interface design enables users to predict what will happen before they perform a function. It is one of the most important elements in ease of learning, ease of use, and aesthetics.
Minimal user effort	The interface should be simple to use. Most designers plan on having no more than three mouse clicks from the starting menu until users perform work.

FIGURE 11-1
Principles of User Interface Design

approach for screen design. The screen is divided into three boxes (see Figure 11-2). The top box is the navigation area through which the user issues commands to navigate through the system. The bottom box is the status area, which displays information about what the user is doing. The middle—and largest—box is used to display reports and present forms for data entry.

In many cases (particularly on the Web), multiple layout areas are used. Figure 11-3 shows a screen with four navigation areas, each of which is organized to provide different functions and navigation within different parts of the system. The top area provides the standard Netscape navigation and command controls that change the contents of the entire system. The navigation area on the left edge navigates between sections and changes all content to its right. The navigation areas at the top and bottom of the page provide identical controls in graphical (top) and text (bottom) format, and are used to navigate within a specific section.

This use of multiple layout areas for navigation also applies to inputs and outputs. Data areas on reports and forms are often subdivided into subareas, each of which is used for different types of information. These areas are almost always rectangular in shape, although sometimes space constraints will require odd shapes. Nonetheless, the margins on the edges of the screen should be consistent. Each of the areas within the report or form is designed to hold different information. For example, on an order form (or order report), one part may be used for customer information (e.g., name, address), one part for information about the order in general (e.g., date, payment information), and one part for the order details (e.g., how many units of which items at what price each). Each area is self-contained, so that information in one area does not run into another.

The areas and information within areas should have a natural intuitive flow to minimize the users' movement from one area to the next. People in westernized nations (e.g., the United States, Canada, Mexico) tend to read top-to-bottom left-to-

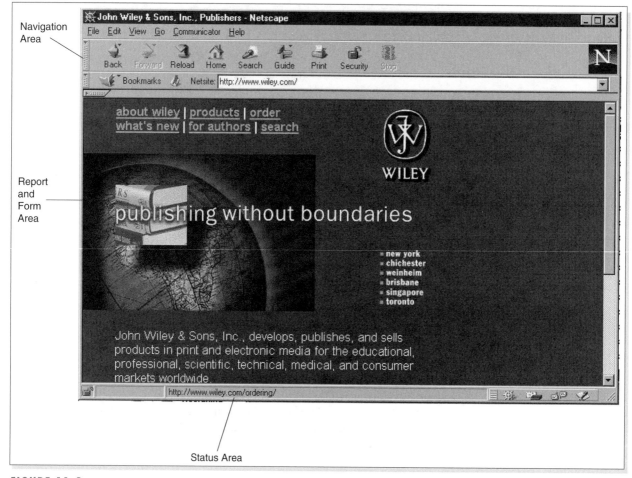

Navigation Area

Report and Form Area

Status Area

FIGURE 11-2
The Three Principal Screen Layout Areas

right; related information should therefore be placed so that it is used in this order (e.g., address lines, followed by city, state/province, and then zip code/postal code.) Sometimes the sequence is in chronological order, or from the general to the specific, or from most frequently to least frequently used. In any event, before the areas are placed on a form or report, the analyst should have a clear understanding of what arrangement makes the most sense for how the form or report will be used. The flow between sections should also be consistent, whether horizontal or vertical (see Figure 11-4). Ideally, the areas will remain consistent in size, shape, and placement for the forms used to enter information (whether paper or on-screen) and the reports used to present it.

Content Awareness

Content awareness refers to the ability of an interface to make the user aware of the information it contains with the least amount of effort on the user's part. All parts of the interface, whether navigation, input, or output, should provide as

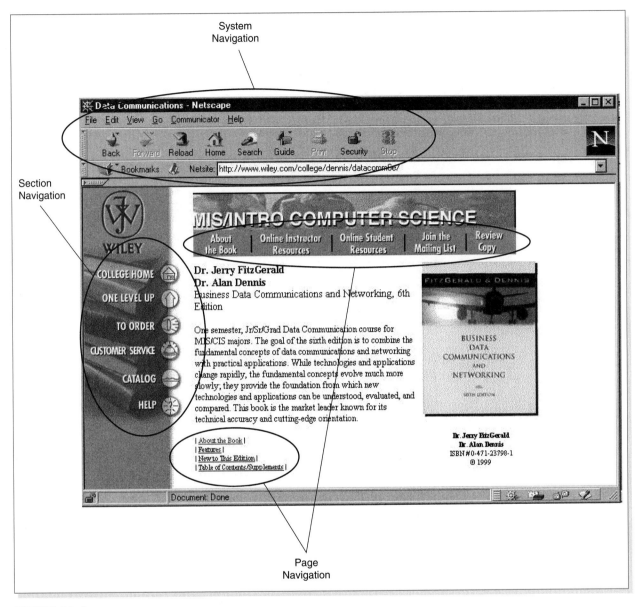

FIGURE 11-3
Layout with Multiple Navigation Areas

much content awareness as possible, but it is particularly important for forms or reports that are used quickly or irregularly (e.g., a Web site).

Content awareness applies to the interface in general. All interfaces should have titles (on the screen frame, for example). Menus should show where you are and, if possible, where you came from to get there.

Content awareness also applies to the areas within forms and reports. All areas should be clear and well defined (with titles if space permits) so that it is difficult for the user to become confused about the information in any area. Then users can quickly locate the part of the form or report that is likely to contain the information they need. Sometimes the areas are marked using lines, colors, or headings

FIGURE 11-4
Flow between Interface Sections
(a) Vertical (b) Horizontal

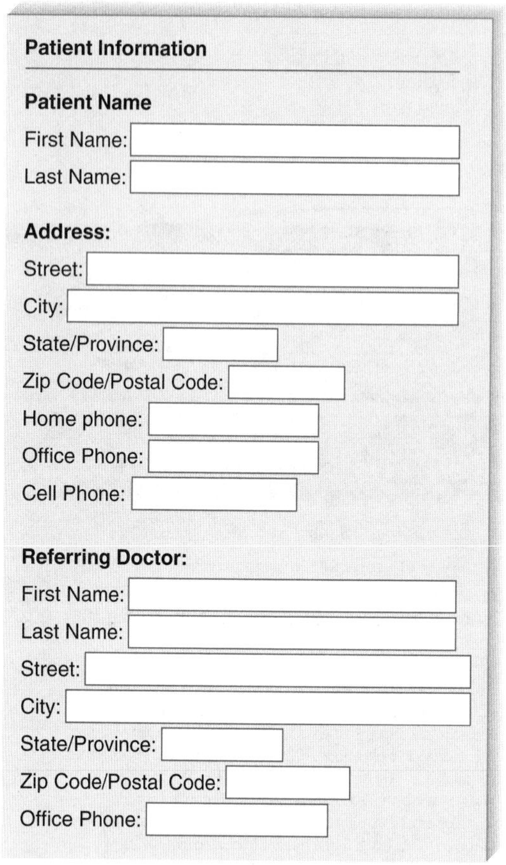

Patient Information

Patient Name

First Name:

Last Name:

Address:

Street:

City:

State/Province:

Zip Code/Postal Code:

Home phone:

Office Phone:

Cell Phone:

Referring Doctor:

First Name:

Last Name:

Street:

City:

State/Province:

Zip Code/Postal Code:

Office Phone:

(*a*) Vertical flow

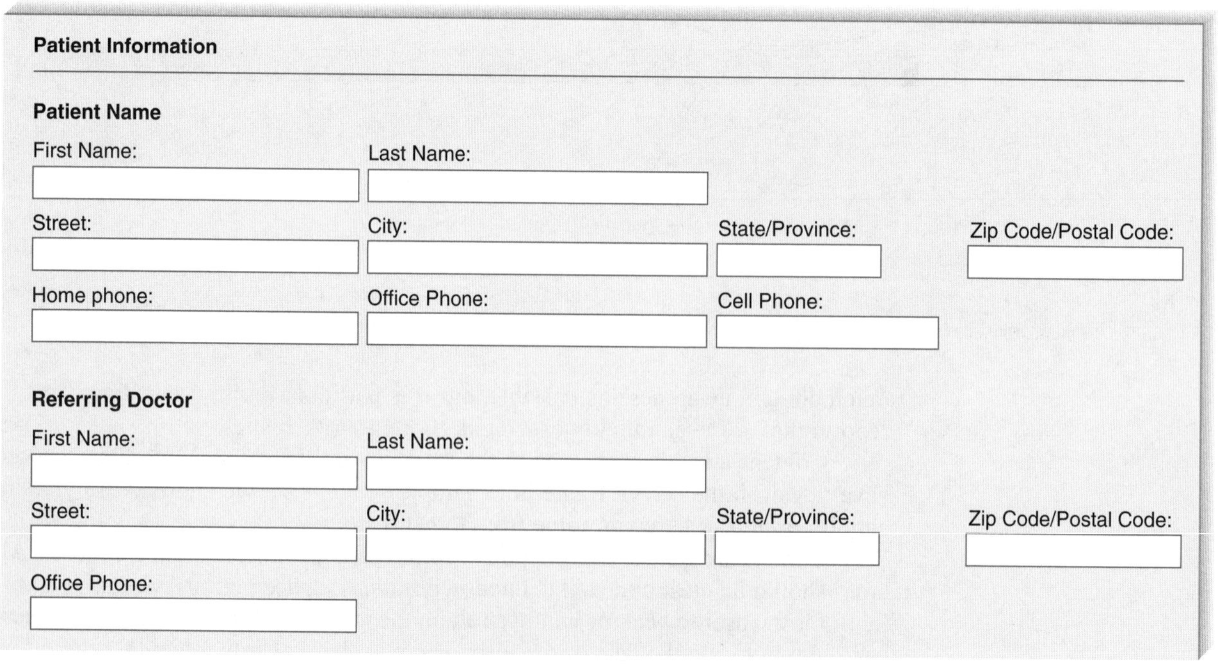

Patient Information

Patient Name

First Name: Last Name:

Street: City: State/Province: Zip Code/Postal Code:

Home phone: Office Phone: Cell Phone:

Referring Doctor

First Name: Last Name:

Street: City: State/Province: Zip Code/Postal Code:

Office Phone:

(*b*) Horizontal Flow

(e.g., the section area in Figure 11-3), whereas in other cases the areas are only implied (e.g., the page controls at the bottom of Figure 11-3).

Content awareness also applies to the *fields* within each area. Fields are the individual elements of data that are input or output. The *field labels* that identify the fields on the interface should be short and specific—objectives that often conflict. There should be no uncertainty about the format of information within fields, whether for entry or display. For example, a date of 10/5/99 is different depending on whether you are in the United States (October 5, 1999) or in Canada (May 10, 1999). Any fields for which there is the possibility of uncertainty or multiple interpretations should provide explicit explanations.

Content awareness also applies to the information that a form or report contains. In general, all forms and reports should contain a preparation date (i.e., the date printed or the data completed) so that the age of the information is obvious. Similarly, all printed forms and software should provide version numbers so that users, analysts, and programmers can identify outdated materials.

Figure 11-5 shows a form from the University of Georgia. This form illustrates the logical grouping of fields into areas with an explicit box (top left), as well as an implied area with no box (lower left). The address fields within the address area follow a clear, natural order. Field labels are short where possible (see the top left) but long where more information is needed to prevent misinterpretation (see the bottom left).

Aesthetics

Aesthetics refers to designing interfaces that are pleasing to the eye. They do not have to be "works of art," but they do need to be functional and inviting to use.

Space is usually at a premium on forms and reports, and often there is the temptation to squeeze as much information as possible onto a page or a screen. Unfortunately, this can make a form or report so unpleasant that users do not want to use it. In general, all forms and reports need a minimum amount of *white space* that is intentionally left blank.

What was your first reaction when you looked at Figure 11-5? This is the most unpleasant form at the University of Georgia, according to staff members. Its *density* is too high; it has too much information packed into too small of a space, with too little white space. Although it may be efficient in saving paper by using one page, not two, it is not effective for many users.

In general, novice or infrequent users of an interface, whether on a screen or on paper, prefer interfaces with low density, often one with a density of less than 50% (i.e., less than 50% of the interface occupied by information). More experienced users prefer higher densities, sometimes approaching 90% occupied because they know where information is located and high densities reduce the amount of physical movement through the interface. We suspect that the form in Figure 11-5 was designed for the experienced staff in the personnel office who use it daily rather than for the clerical staff in academic departments with less personnel experience who use the form only a few times a year.

The design of text is equally important. As a general rule, all text should be in the same font and about the same size. Fonts should be no less than 8 point in size, but 10 point is often preferred, particularly if the interface will be used by older people. Changes in font and size are used to indicate changes in the type of

EMPLOYEE PERSONNEL REPORT

UNIVERSITY OF GEORGIA

PAY TYPE

DOCUMENT NO. PAGE DATE FY DEPARTMENT PHONE COLLEGE OR DIVISION

UGA EMPLOYMENT HISTORY
☐ (C) CURRENT ☐ (P) PREVIOUS

ACTION
MO DA YR

DEPARTMENT/PROJECT

PRI-DEPT HIGH DEGREE INSTITUTION YEAR DATE

UGA % TIME

SOC.SEC.NUM. LAST NAME FIRST NAME/INITIAL MIDDLE INITIAL/NAME SUF

☐ (1) REGULAR ☐ (3) TEMPORARY ☐ (T) TIPPED
☐ (2) UGA STUDENT ☐ (4) NR-ALIEN

STREET OR ROUTE NO. (LINE 1) BIRTH DATE SPOUSE'S NAME CHAIR

☐ (E) EXEMPT ☐ (N) NON-EXEMPT ☐ (Y) FACULTY-RANK
☐ (M) MALE ☐ (S) SINGLE ☐ (N) NON-FACULTY

STREET OR ROUTE NO. (LINE 2) UNIVERSITY PHONE CITIZEN OF 1-9 VISA COUNTY

☐ (F) FEMALE ☐ (M) MARRIED ☐ (5) HISPANIC
☐ (1) WHITE ☐ (3) ORIENTAL/ASIAN ☐ (6) MULTIRACIAL

CITY STATE ZIP + 4 UNIVERSITY BUILDING NAME BLDG-NO/FLOOR/ROOM

☐ (2) BLACK ☐ (4) AMERICAN INDIAN ☐ (9)

| FOR PAYROLL DEPT USE ONLY | COUNTY MONEY (PER PAY PERIOD) | COOP. EXT. EMPLOYEES ONLY | PAYROLL PAYMENT DISTRIBUTION |
| FED EXM STATE EXM | | UGA SALARY | ☐ (1) SEND TO DEPT (DIST CODE) |

OASDI RETIRE GDCP
HI EIC

COUNTY MONEY ☐ (2) DIRECT DEPOSIT(SEND PR105) TO PAYROLL
TOTAL ☐ (3) PICK UP AT PAYROLL WINDOW

TRX	HOME DEPT	SHORT TITLE	POSN NO.	APPT. BEGIN MO DA YR HR	APPT. END MO DA YR HR	JOBCLASS CODE	POSITION TITLE	POS % TIME	C N	FULL TIME ANNUAL SALARY	C N	S C	SUPPLEMENT AMOUNT

MO DA YR HR MO DA YR HR MO DA YR HR MO DA YR HR

PAYROLL AUTHORIZATION

TRX	HOME DEPT	SHORT TITLE	POSN NO	ACCOUNT	FISCAL YEAR BUDGET	EFT

TOTALS

FROM
THRU
AMOUNT PER PAY PERIOD OR HOURLY RATE

☐ (A) NEW UGA EMPLOYEE ☐ (B) LATERAL TRANSFER
☐ (D) REPLACEMENT POSN-NAME OF LAST INCUMBENT
☐ (E) APPOINTMENT TO NEW POSITION
☐ (F) CHANGE % TIME EMPLOYED FROM ____ TO ____
☐ (G) CONTINUATION WITHIN EXISTING BUDGET POSITION
☐ (H) REVISE DISTRIBUTION OF SALARY
☐ (I) TRANSFER FROM DEPT ____ TO ____
☐ (J) CHANGE PAY TYPE FROM ____ TO ____

☐ (C) PROMOTION

☐ (K) CHANGE TITLE FROM ____ TO ____
☐ (L) CHANGE NAME FROM ____ TO ____
☐ (M) CHANGE SSN FROM ____ TO ____
☐ (N) LEAVE W/O PAY FROM ____ TO ____
☐ (O) CHG COUNTY $ FROM ____ TO ____
☐ (P) TERMINATION-REASON
☐ (Q) OTHER (SPECIFY)

REMARKS

DEPARTMENT HEAD DATE VICE PRESIDENT DATE BUDGET REVIEW DATE BUDGET OFFICE DATE

DEAN/DIRECTOR DATE FACULTY RECORDS DATE CONTRACTS & GRANTS DATE PERSONNEL DATE

FIGURE 11-5
Form Example

information that is presented (e.g., headings, status indicators). In general, italics and underlining should be avoided because they make text harder to read.

Serif fonts (i.e., those having letters with serifs or tails (e.g., Times Roman) such as the font you are reading right now) are the most readable for printed reports, particularly for small letters. San-serif fonts (i.e., those without serifs (e.g., Helvetica or Arial) such as the ones used for the chapter titles in this book) are the most readable for computer screens and are often used for headings in printed reports. Never use all capital letters, except possibly for titles.

Color and patterns should be used carefully and sparingly and only when it serves a purpose. (About 10% of men are color blind, so the improper use of color can impair their ability to read information.) A quick trip around the Web will demonstrate the problems caused by indiscriminate use of colors and patterns. Remember, the goal is pleasant readability, not art; color and patterns should be used to strengthen the message, not overwhelm it. Color is best used to separate and categorize items, such as showing the difference between headings and regular text, or to highlight important information. Therefore, colors with high contrast should be used (e.g., black and white). In general, black text on a white background is the most readable, with blue on red the least readable. (Most experts agree that background patterns on Web pages should be avoided.) Color has been shown to affect emotion, with red provoking intense emotion (e.g., anger) and blue provoking lowered emotions (e.g., drowsiness).

User Experience

Two types of users will use most computer systems: those with experience and those without. Interfaces should be designed for both types of users. Novice users are usually most concerned with ease of learning—how quickly they can learn new systems. Expert users are usually most concerned with ease of use—how quickly they can use the system, once they have learned how to use it. Often these two are complementary and lead to similar design decisions, but sometimes there are tradeoffs. Novices, for example, often prefer menus that show all available system functions because these promote ease of learning. Experts, on the other hand, sometimes prefer fewer menus that are organized around the most commonly used functions.

Systems that will end up being used by many people on a daily basis are more likely to have a majority of expert users (e.g., order entry systems). Although interfaces should try to balance ease of use and ease of learning, these types of systems should put more emphasis on ease of use rather than ease of learning. Users should be able to access the commonly used functions quickly, with few keystrokes or a small number of menu selections.

In many other systems (e.g., decision support systems), most people will remain occasional users for the lifetime of the system. In this case, greater emphasis may be placed on ease of learning than on ease of use.

Ease of use and ease of learning often go hand-in-hand, but sometimes they do not. Research shows that expert and novice users may have different requirements and behavior patterns. For example, novices virtually never look at the bottom area of a screen that presents status information, whereas experts refer to the status bar when they need information. Most systems should be designed to support frequent users, except for those systems that are meant to be used infrequently or when many new users or occasional users are expected (e.g., the Web). Similarly, systems that contain functionality that is used only occasionally must possess a highly intuitive interface, or an interface that contains explicit guidance regarding its use.

The balance of quick access to commonly used and well-known functions, as well as guidance through new and less well-known functions, is challenging to the interface designer, and this balance often requires elegant solutions. Microsoft Office, for example, addresses this issue through the use of the Office Assistant (also known as "the paper clip guy"[2]) and the "show me" functions that demonstrate the menus and buttons for specific functions. These features remain in the background until they are needed by novice users (or even by experienced users when they are using an unfamiliar part of the system).

Consistency

Consistency in design is probably the single most important factor in making a system simple to use because it enables users to predict what will happen. When interfaces are consistent, users can interact with one part of the system and then know how to interact with the rest, aside of course, from elements unique to those parts. Consistency usually refers to the interface within one computer system, so that all parts of the same system work in the same way. Ideally, however, the system should also be consistent with other computer systems in the organization and with commercial software that is used (e.g., Windows). For example, many users are familiar with the Web, so the use of Web-like interfaces can reduce the amount of learning required by the user. In this way, the user can reuse Web knowledge, thus significantly reducing the learning curve for a new system. Many software development tools support consistent system interfaces by providing standard interface objects (e.g., list boxes, pull-down menus, and radio buttons).

Consistency occurs at many different levels. Consistency in the *navigation controls* conveys how actions in the system should be performed. For example, using the same icon or command to change an item clearly communicates how changes are made throughout the system. Consistency in *terminology* is also important. This refers to using the same words for elements on forms and reports (e.g., not "customer" in one place and "client" in another). We also believe that consistency in *report and form design* is important, although a recent study suggests that being *too* consistent can cause problems.[3] When reports and forms are very similar except for very minor changes in titles, users sometimes mistakenly use the wrong form, and either enter incorrect data or misinterpret its information. The implication for design is to make the reports and forms similar, but to give them some distinctive elements (e.g., color, size of titles) that enable users to immediately detect differences.

Minimize Effort

Finally, interfaces should be designed to minimize the amount of effort needed to accomplish tasks. This means using the fewest possible mouse clicks or keystrokes

---

[2] The paper clip guy usually evokes a strong reaction; people either like him or hate him. We think the concept is useful, but his appearance can be irritating after a while, an example of humor that quickly wears off. He was designed by an undergraduate computer science student working at Microsoft as a summer intern.

[3] John Satzinger and Lorne Olfman, "User Interface Consistency Across End-User Application: The Effects of Mental Models," *Journal of Management Information Systems* (Spring 1998): 167–193.

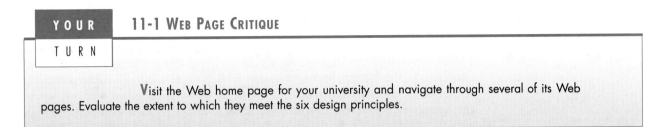

YOUR

TURN

11-1 WEB PAGE CRITIQUE

Visit the Web home page for your university and navigate through several of its Web pages. Evaluate the extent to which they meet the six design principles.

to move from one part of the system to another. Most interface designers follow the *three clicks rule:* users should be able to go from the start or main menu of a system to the information or action they want in no more than three mouse clicks or three keystrokes.

USER INTERFACE DESIGN PROCESS

User interface design is a five-step process that is iterative—analysts often move back and forth between steps rather than proceeding sequentially from step one to step five (see Figure 11-6). First, the analysts examine the use cases (see Chapter 6) and sequence diagrams (see Chapter 8) developed in the analysis phase. Then they interview users in order to develop *use scenarios* that describe commonly employed patterns of actions that the users will perform so the interface enables users to quickly and smoothly perform these scenarios. Second, the analysts develop the *window navigation diagram* (WND) that defines the basic structure of the interface.

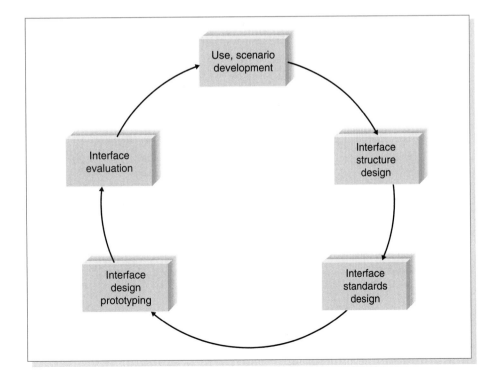

FIGURE 11-6
User Interface Design Process

These diagrams show all the interfaces (e.g., screens, forms, and reports) in the system and how they are connected. Third, the analysts design *interface standards,* which are the basic design elements on which interfaces in the system are based. Fourth, the analysts create an *interface design prototype* for each of the individual interfaces in the system, such as navigation controls (including the conversion of the *essential use cases* to *real use cases*), input screens, output screens, forms (including preprinted paper forms), and reports. Chapter 12 contains sections with specific guidelines for the design of these navigation controls, inputs, and outputs and the development of real use cases. Finally, the individual interfaces are subjected to *interface evaluation* to determine whether they are satisfactory and how they can be improved.

Interface evaluations almost always identify improvements, so the interface design process is repeated in a cyclical process until no new improvements are identified. In practice, most analysts interact closely with the users during the interface design process, so that users have many chances to see the interface as it evolves, rather than waiting for one overall interface evaluation at the end of the interface design process. It is better for all concerned (both analysts and users) if changes are identified sooner rather than later. For example, if the interface structure or standards need improvements, it is better to identify changes before most of the screens that use the standards have been designed.

Use Scenario Development

A *use scenario* is an outline of the steps that the users perform to accomplish some part of their work. A use scenario is one path through an essential use case. For example, Figure 6-13 shows the use-case diagram for the Web section of the Internet Sales System. This figure shows that the Maintain Order use case is distinct from the Checkout use case. We model the two use cases separately because they represent two separate processes that are included in the Place Order use case.

The use-case diagram was designed to model all possible uses of the system—that is, its complete functionality or all possible paths through the use case at a fairly high level of abstraction. In one use scenario, a user will browse for many CDs, much like someone browsing through a real music store looking for interesting CDs. He or she will search for a CD, read the marketing materials for it, perhaps add it to the shopping cart, browse for more, and so on. Eventually, the user will want to place the order, perhaps removing some things from the shopping cart beforehand.

In another use scenario, a user will want to buy one specific CD. He or she will go directly to the CD, price it, and buy it immediately, much like someone running into a store, making a beeline for the one CD he or she wants, and immediately paying

11-3 Use Scenario Development for an ATM

Pretend you have been charged with the task of redesigning the interface for the ATM at your local bank. Develop two use scenarios for it.

and leaving the store. This user will enter the CD information in the search portion of the system, look at the resulting cost information, and immediately place an order. Anything that slows this reader down will risk loss of the sale. For this use scenario, we need to ensure that the path through the use case as presented by the interface is short and simple, with very few menus and mouse clicks.

Use scenarios are presented in a simple narrative description that is tied to the essential use cases developed during the analysis phase (see Chapter 6). Figure 11-7 shows the two use scenarios we have described. The key point to remember when using use cases for interface design is *not* to document all possible use scenarios within a use case. The goal is to document two or three of the *most common* use scenarios so that the interface can be designed to enable the most common uses to be performed simply and easily.

Use scenario: The Browsing Shopper User is not sure what he or she wants to buy and will browse for several CDs	Use scenario: The Hurry-up Shopper User knows exactly what he or she wants and wants it quickly
1. User may search for a specific artist or browse through a music category (1). 2. User will likely read the basic information for several CDs, as well as the marketing material for some. He or she will likely listen to music samples and browse related CDs (after we implement these) (3). 3. User will put several CDs in the shopping cart and will continue browsing (5). 4. Eventually, the user will want to place the order but will probably want to look through the shopping cart, possibly discarding some CDs first (7).	1. User will search for a specific artist or CD (1). 2. User will look at the price and perhaps other information (3). 3. User will want to place the order (7) or do another search (1) or surf on the other Web sites (3a-2).

The numbers in parentheses refer to specific events in the essential use case.

FIGURE 11-7

Use Scenarios

Interface Structure Design

The interface structure defines the basic components of the interface and how they work together to provide functionality to users. A *window navigation diagram (WND)*[4] is used to show how all the screens, forms, and reports used by the system are related and how the user moves from one to another. Most systems have several WNDs, one for each major part of the system.

A WND is very similar to a statechart diagram (see Chapter 8) in that they both model state changes. A statechart typically models the *state* changes of an object, whereas a WND models the state changes of the user interface. In a WND, each state that the user interface may be in is represented as a box. Furthermore, a box typically corresponds to a user interface component such as a window, form, button, or report. For example, in Figure 11-8, there are 13 separate states: Main Menu, Menu A, Menu B, Menu C, and so on.

Transitions are modeled as either a single-headed or double-headed arrow. A single-headed arrow indicates that a return to the calling state is not required, whereas double-headed arrow means a required return. For example, in Figure 11-8, the transition from the Main Menu state to the Menu A state does not require a return; however, the transition from the Menu A state to the Form A state does. The arrows are labeled with the action that causes the user interface to move from one state to another. For example, in Figure 11-8, to move from the Main Menu state to the Menu A state, the user must "Select Menu A" from the Main Menu.

The last item to be described in a window navigation diagram is the *stereotype*. The stereotype is modeled as a text item surrounded by "<< >>" symbols. It represents the type of user interface component of a box on the diagram. For example, the Main Menu is a window, whereas Form A is a form.

The basic structure of the interface follows the basic structure of the business process itself as defined in the use cases and behavioral model. The analyst starts with the essential use cases and develops the fundamental flow of control of the system as it moves from object to object. The analyst then examines the use scenarios to see how well the WND supports them. Quite often, the use scenarios identify paths through the WND that are more complicated than they should be. The analyst then reworks the WND to better support the use scenarios, sometimes by making major changes to the menu structure, sometimes by adding shortcuts.

YOUR	**11-4 INTERFACE STRUCTURE DESIGN**
TURN	

Pretend you have been charged with the task of redesigning the interface for the ATM at your local bank. Design an interface structure design using a WND that shows how a user would navigate among the screens.

---

[4] The window navigation diagram is actually an adaptation of the statechart and object diagrams (see Meilir Page-Jones, *Fundamentals of Object-Oriented Design in UML* (Boston: Addison-Wesley, 2000).

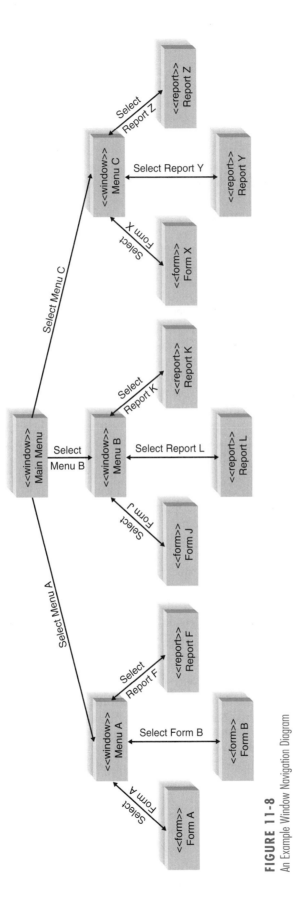

FIGURE 11-8
An Example Window Navigation Diagram

Interface Standards Design

The *interface standards* are the basic design elements that are common across the individual screens, forms, and reports within the system. Depending on the application, there may be several sets of interface standards for different parts of the system (e.g., one for Web screens, one for paper reports, one for input forms). For example, the part of the system used by data-entry operators may mirror other data-entry applications in the company, while a Web interface for displaying information from the same system may adhere to some standardized Web format. Likewise, each individual interface may not contain all of the elements in the standards (e.g., a report screen may not have an "edit" capability), and they may contain additional characteristics beyond the standard ones, but the standards serve as the touchstone that ensures the interfaces are consistent across the system.

Interface Metaphor   First, the analysts must develop the fundamental *interface metaphor(s)* that defines how the interface will work. An interface metaphor is a concept from the real world that is used as a model for the computer system. The metaphor helps the user understand the system and enables the user to predict what features the interface might provide, even without actually using the system. Sometimes systems have one metaphor, while in other cases, there are several metaphors in different parts of the system.

In many cases, the metaphor is explicit. Quicken, for example, uses a checkbook metaphor for its interface, even to the point of having the users type information into an on-screen form that looks like a real check. In other cases, the metaphor is implicit or unstated, but it is there nonetheless. Many Windows systems use the paper form or table as a metaphor.

In some cases, the metaphor is so obvious that it requires no thought. The CD Selections Internet Sales System, for example, will use the music store as the metaphor (e.g., shopping cart). In other cases, a metaphor is hard to identify. In general, it is better not to force a metaphor that really doesn't fit a system because an ill-fitting metaphor will confuse users by promoting incorrect assumptions.

Interface Objects   The template specifies the names that the interface will use for the major *interface objects,* the fundamental building blocks of the system such as the classes. In many cases, the object names are straightforward, such as calling the shopping cart the "shopping cart." In other cases, it is not so simple. For example, CD Selections sells both CDs and tapes. When users search for items to buy, should they search for "CDs" or "CDs & Tapes" or "CDs/Tapes" or "Music" or "Albums" or something else? Obviously, the object names should be easily understood and help promote the interface metaphor.

In general, in cases of disagreements between the users and the analysts over names, whether for objects or actions (see below), the users should win. A more "understandable" name always beats a more "precise" or more "accurate" one.

Interface Actions   The template also specifies the navigation and command language style (e.g., menus), and grammer (e.g., object-action order; see the navigation design section in Chapter 12). It gives names to the most commonly used *interface actions* in the navigation design (e.g., "buy" versus "purchase" or "modify" versus "change").

Interface Icons

The interface objects and actions as well as their status (e.g., "deleted" or "over-drawn") may be represented by *interface icons*. Icons are pictures that will appear on command buttons as well as in reports and forms to highlight important information. Icon design is very challenging because it means developing a simple picture less than half the size of a postage stamp that needs to convey an often complex meaning. The simplest and best approach is simply to adopt icons developed by others (e.g., a blank page to indicate "create a new file," a diskette to indicate "save"). This has the advantage of quick icon development, and the icons may already be well understood by users because they have seen them in other software.

Commands are actions that are especially difficult to represent with icons because they are in motion, not static. Many icons have become well known from widespread use, but icons are not as well understood as was first believed. Use of icons can sometimes cause more confusion than insight. (For example, did you know that a picture of a sweeping broom (paintbrush) in Microsoft Word means "format painter"?) Icon meanings become clearer with use, but sometimes a picture is not worth even one word. When in doubt, use a word, not a picture.

Interface Templates

The *interface template* defines the general appearance of all screens in the information system, and the paper-based forms and reports that are used. The template design, for example, specifies the basic layout of the screens (e.g., where the navigation area(s), status area, and form/report area(s) will be placed) and the color scheme(s) that will be applied. It defines whether windows will replace one another on the screen or will cascade over the top of each other. The template defines a standard placement and order for common interface actions (e.g., "File Edit View" rather than "File View Edit"). In short, the template draws together the other major interface design elements: metaphors, objects, actions, and icons.

Interface Design Prototyping

An *interface design prototype* is a "mock-up" or a simulation of a computer screen, form, or report. A prototype is prepared for each interface in the system to show the users and the programmers how the system will perform. In the "old days," an interface design prototype was usually specified on a paper form that showed what would be displayed on each part of the screen. Paper forms are still used today, but more and more interface design prototypes are being built using computer tools instead of paper. The three most common approaches to interface design prototyping are storyboarding, HTML prototyping, and language prototyping.

YOUR **11-5 INTERFACE STANDARDS DEVELOPMENT**

TURN

Pretend you have been charged with the task of redesigning the interface for the ATM at your local bank. Develop an interface standard that includes metaphors, objects, actions, icons, and a template.

Storyboard At its simplest, an interface design prototype is a paper-based *storyboard*. The storyboard shows hand-drawn pictures of what the screens will look like and how they flow from one screen to another, in the same way a storyboard for a cartoon shows how the action will flow from one scene to the next (see Figure 11-9). Storyboards are the simplest technique because all they require is paper (often a flip chart) and a pen—and someone with some "artistic" ability.

HTML Prototype One of the most common types of interface design prototypes used today is the *HTML prototype*. As the name suggests, an HTML prototype is built using Web pages created in HTML (hypertext mark-up language). The designer uses HTML to create a series of Web pages that shows the fundamental parts of the system. The users can interact with the pages by clicking on buttons and entering pretend data into forms (but because there is no "system" behind the pages the data is never processed). The pages are linked together so that as the user clicks on buttons, the requested part of the system appears. HTML prototypes are superior to storyboards in that they enable users to interact with the system and gain a better sense

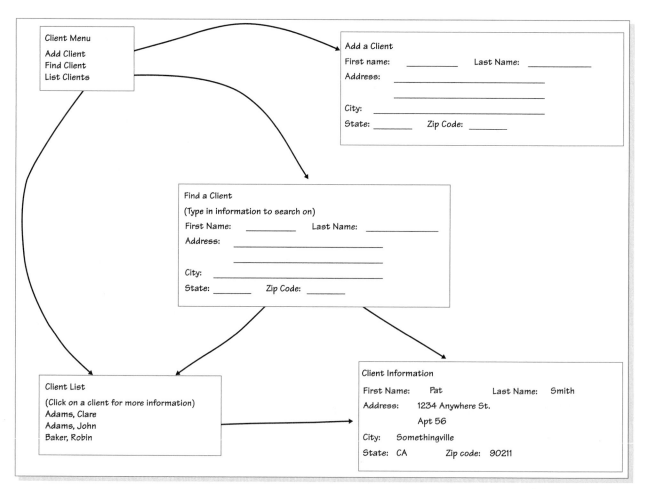

FIGURE 11-9
An Example Storyboard

of how to navigate among the different screens. However, HTML has limitations—the screens shown in HTML will never appear exactly like the real screens in the system (unless, of course, the real system will be a Web system in HTML).

Language Prototype A *language prototype* is an interface design prototype built using the actual language or tool that will be used to build the system. Language prototypes are designed in the same ways as HTML prototypes (they enable the user to move from screen to screen but perform no real processing). For example, in Visual Basic, it is possible to create and view screens without actually attaching program code to the screens. Language prototypes take longer to develop than storyboards or HTML prototypes but have the distinct advantage of showing *exactly* what the screens will look like. The user does not have to guess about the shape or position of the elements on the screen.

Selecting the Appropriate Techniques Projects often use a combination of different interface design prototyping techniques for different parts of the system. Storyboarding is the fastest and least expensive, but provides the least amount of detail. Language prototyping is the slowest, most expensive, and most detailed approach. HTML prototyping falls between the two extremes. Therefore, storyboarding is used for parts of the system in which the interface is well understood and when more expensive prototypes are thought to be unnecessary. HTML prototypes and language prototypes are used for parts of the system that are critical, yet not well understood.

Interface Evaluation

The objective of *interface evaluation* is to understand how to improve the interface design before the system is complete. Many organizations save interface evaluation for the very last step in the systems development life cycle (SDLC) before the system is installed. Ideally, however, interface evaluation should be performed while

---

CONCEPTS **11-A INTERFACE DESIGN PROTOTYPES FOR A DSS APPLICATION**

IN ACTION

I was involved in the development of several decision support systems (DSS) while working as a consultant. On one project, a future user was frustrated because he could not imagine what a DSS looked like and how one would be used. He was a key user, but the project team had a difficult time involving him in the project because of his frustration. The team used SQL Windows (one of the most popular development tools at the time) to create a language prototype that demonstrated the future system's appearance, proposed menu system, and screens (with fields but no processing).

The team was amazed at the user's response to the prototype. He appreciated being given a context with which to visualize the DSS, and he soon began to recommend improvements to the design and flow of the system, and to identify some important information that was overlooked during the analysis phase. Ultimately, the user became one of the strongest supporters of the system, and the project team felt sure that the prototype led to a much better product in the end. *Barbara Wixom*

QUESTION
Why do you think the team chose to use a language prototype rather than a storyboard or HTML prototype? What tradeoffs were involved in the decision?

the system is being designed—before it is built—so that any major design problems can be identified and corrected before the time and cost of programming has been spent on a weak design. It is not uncommon for the system to undergo one or two major changes after the users see the first interface design prototype because they identify problems that are overlooked by the project team.

As with interface design prototyping, interface evaluation can take many different forms, each with different costs and different amounts of detail. Four common approaches are heuristic evaluation, walk-through evaluation, interactive evaluation, and formal usability testing. As with interface design prototyping, the different parts of a system can be evaluated using different techniques.

Heuristic Evaluation   A *heuristic evaluation* examines the interface by comparing it to a set of heuristics or principles for interface design. The project team develops a checklist of interface design principles — from the list at the start of this chapter, for example, as well as the list of principles in the navigation, input, and output design sections in Chapter 12. At least three members of the project team then individually work through the interface design prototype examining each and every interface to ensure it satisfies each design principle on a formal checklist. After each has gone through the prototype separately, they meet as a team to discuss their evaluation and identify specific improvements that are required.

Walk-Through Evaluation   An interface design *walk-through evaluation* is a meeting conducted with the users who will ultimately have to operate the system. The project team presents the prototype to the users and walks them through the various parts of the interface. The project team shows the storyboard or actually demonstrates the HTML or language prototype and explains how the interface will be used. The users identify improvements to each of the interfaces that are presented.

Interactive Evaluation   With an *interactive evaluation,* the users themselves actually work with the HTML or language prototype in a one-person session with member(s) of the project team (an interactive evaluation cannot be used with a storyboard). As the user works with the prototype (often by going through the use scenarios, using the real use cases (see Chapter 12), or just navigating at will through the system), he or she tells the project team member(s) what he or she likes and doesn't like, and what additional information or functionality is needed. As the user interacts with the prototype, team member(s) record the cases when the user appears to be unsure what to do, makes mistakes, or misinterprets the meaning of an interface component. If several of the users participating in the evaluation show a pattern of uncertainty, mistakes, or misinterpretations, it is a clear indication that those parts of the interface need improvement.

Formal Usability Testing   Formal *usability testing* is commonly done with commercial software products and products developed by large organizations that will be widely used through the organization. As the name suggests, it is a very formal—almost scientific—process that can be used only with language prototypes (and systems that have been completely built awaiting installation or shipping).[5] As with

---

[5] A good source for usability testing is Jakob Nielsen and Robert Mack (eds.), *Usability Inspection Methods* (New York: John Wiley & Sons, 1994). See also www.useit.com/papers.

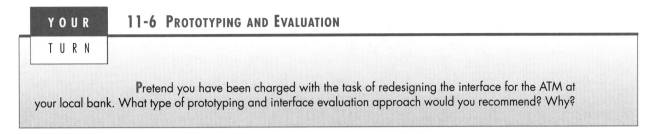

YOUR

TURN

11-6 PROTOTYPING AND EVALUATION

Pretend you have been charged with the task of redesigning the interface for the ATM at your local bank. What type of prototyping and interface evaluation approach would you recommend? Why?

interactive evaluation, usability testing is done in one-person sessions in which a user works directly with the software. However, it is typically done in a special lab equipped with video cameras and special software that records each and every keystroke and mouse operation so they can be replayed to understand exactly what the user did.

The user is given a specific set of tasks to accomplish (usually the use scenarios), and after some initial instructions, the project teams member(s) are not permitted to interact with the user to provide assistance. The user must work with the software without help, which can be hard on the users if they become confused with the system. It is critical that users understand that the goal is to test the interface, not their abilities, and if they are unable to complete the task, the interface—not the user—has failed the test.

Formal usability testing is very expensive because each one-user session (which typically lasts one to two hours) can take one to two days to analyze due to the volume of detail collected in the computer logs and videotapes. Most usability testing involves five to ten users because fewer than five users makes the results depend too much on the specific individual users who participated, and more than ten users is often too expensive to justify (unless you work for a large commercial software developer).

APPLYING THE CONCEPTS AT CD SELECTIONS

In the CD Selections example, there are four different high-level use cases in the Internet Sales System (see Figure 6-13 in Chapter 6): Maintain CD Information, Fill Order, Place Order, and Maintain Marketing Information. There are also three additional use cases, Maintain Order, Checkout, and Create New Customer, associated with the Place Order use case. In this section we focus only on the Place Order and its associated use cases, that is, the Web portion used by customers.

Use Scenario Development

The first step in the interface design process was to develop the key use scenarios for the Internet Sales System. Alec Adams, senior systems analyst at CD Selections and project manager for the Internet Sales System, began by examining the essential use cases and thinking about the types of users and how they would interact with the system. As discussed previously, Alec identified two use scenarios: the browsing shopper and the hurry-up shopper (see Figure 11-7).[6] Alec also thought

---

[6] Of course, it may be necessary to modify the original essential use scenarios in light of these new subtypes of customer. Furthermore, the structural and behavioral models may have to be modified. Remember that object-oriented systems analysis and design follows a RAD-based approach to developing systems; as such, additional requirements can be uncovered at any time.

of several other use scenarios for the Web site in general but omitted them because they were not relevant to the Internet Sales portion. Likewise, he thought of several use scenarios that did not lead to sales (e.g., fans looking for information about their favorite artists and albums), and omitted them as well.

Interface Structure Design

Next, Alec created a window navigation diagram (WND) for the Web system. He began with the essential use cases to ensure that all functionality defined for the system was included in the WND. Figure 11-10 shows the WND for the Web portion of the Internet Sales System. The system will start with a home page that contains the main menu for the sales system. Based on the essential use cases, Alec identified four basic operations that he felt made sense to support on the main menu: search the CD catalog, search by music category, review the contents of the shopping cart, and actually place the order. Each of these was modeled as a hyperlink on the home page.

Alec decided to model the full search option as a pop-up search menu that allowed the customer to choose to search the CD catalog based on artist, title, or composer. He further decided that a textbox would be required to allow the customer to type in the name of the artist, title, or composer, depending on the type of search requested. Finally, he chose to use a button to submit the request to the system. After the "submit" button is pressed, the system produces a report composed of hyperlinks to the individual information on each CD. A CD report containing the basic information is generated by clicking on the hyperlink associated with the CD. On the basic report, Alec added buttons for choosing to find out additional information on the CD and to add the CD to the shopping cart. If the "Detail" button is pressed, a detailed report containing the marketing information on the CD is produced. Finally, Alec decided to include a button on this report to add the CD to the shopping cart.

The second basic operation supported on the home page was to allow the user to search the CD catalog by category of music. Like the previous operation, Alec chose to model the category search with a pop-up search menu. In this case, once the customer chose the category, the system would produce the report with the hyperlinks to the individual information on each CD. From that point on, the navigation would be identical to the previous searches.

The third operation supported was to review the contents of the shopping cart. In this case, Alec decided to model the shopping cart as a report that contained three types of hyperlinks; one for removing an individual CD from the shopping cart, one for removing all CDs from the shopping cart, and one for placing the order. The removal hyperlinks would remove the individual CD (or all of the CDs) from the shopping cart if the user would confirm the operation. The place order link would send the customer to an order form. Once the customer filled out the order form, the customer would press the order button. The system would then respond with a order confirmation message box.

The fourth operation supported on the home page was to directly allow the customer to place an order. Upon review, Alec decided that the place order and review cart operations were identical. As such, he decided to force the user to have the place order and review shopping cart operations go through the same process.

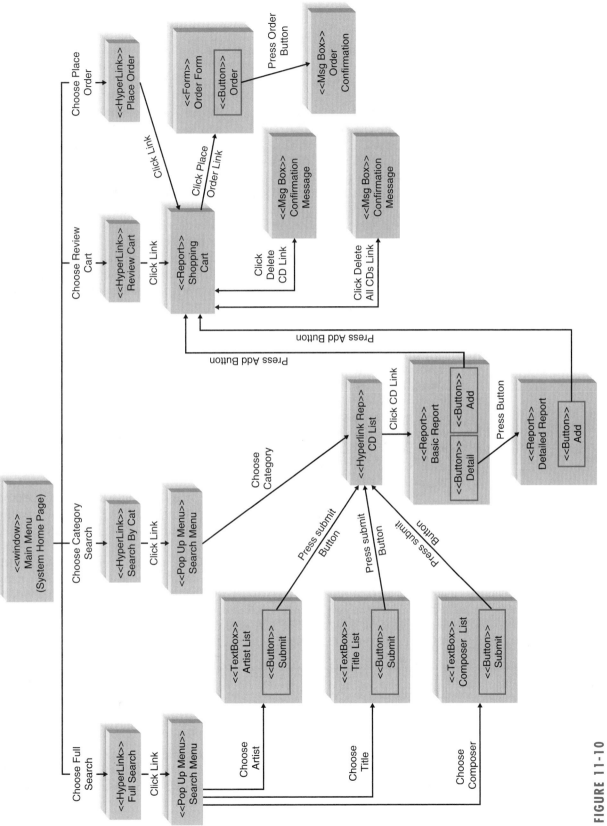

FIGURE 11-10
CD-Selections Initial WND for the Web Portion of the Internet Sales System

Alec also envisioned that by using frames, the user would be able to return to the home page from any screen. Documenting these would give the WND too many lines, so Alec simply put a note describing it with the WND.

The Revised WND Alec then examined the use scenarios to see how well the initial WND enabled different types of users to work through the system. He started with the Browsing Shopping use scenario and followed it through the WND, imagining what would appear on each screen and pretending to navigate through the system. He found the WND to work well but noticed a couple of minor issues related to the shopping cart. First, he decided that it would make sense to allow the customer to retrieve the information related to the CDs contained in the shopping cart. As such, he changed the stereotype of the user interface component from Report to HyperLink Rep and added a hyperlink from the Shopping Cart to the Basic Report created by the different search requests. Second, he noticed that the Shopping Cart was using hyperlinks to link to the removal and place order processes. However, in all the other elements of the WND, he was using buttons to model the equivalent ideas. As such, he decided to change the shopping cart component to model these connections as buttons. Of course, this forced him to modify the transitions as well.

Alec next explored the Hurry-up Shopper use scenario. In this case, the WND did not work as well. Moving from the home page to the search page, to the list of matching CDs, to the CD page with price and other information takes three mouse clicks. This falls within the three clicks rule, but for someone in a hurry, this may be too many. Alec decided to add a "quick-search" option to the home page that would enable the user to enter one search criterion (e.g., just artist name or title rather than a more detailed search as would be possible on the search page) that with one click would take the user to the one CD that matched the criteria or to a list of CDs if there were more than one. This would enable an impatient user to get to the CD of interest in one or two clicks.

Once the CD is displayed on the screen, the Hurry-up Shopper use scenario would suggest that the user would immediately purchase the CD, do a new search, or abandon the Web site and surf elsewhere. This suggested two important changes. First, there had to be an easy way to go to the place order screen. As the WND stands (see Figure 11-10), the user must add the item to the shopping cart and then click on the link on the HTML frame to get to the place order screen. While the ability of users to notice the place order link in the frame would await the interface evaluation stage, Alec suspected, based on past experience, that a significant number of users would not see it. Therefore, he decided to add a button to the Basic Report screen and the Detailed Report screen called "Buy" (see Figure 11-11).

Second, since the Hurry-up Shopper might want to search for another CD instead of buying the CD, Alec decided to include the quick-search item from the home page on the frame. This would make all searches immediately available from anywhere in the system. As a result, all functionality on the home page would now be carried on the frame. Alec updated the note on the bottom attached to the WND to reflect the change.

Finally, upon review of the WND, Alec decided to remodel the Artist List, Title List, and Composer Lists as window stereotypes instead of textbox stereotypes. He then added a Search Item textbox to each of these elements. Figure 11-11 shows the revised WND for the Web portion of the Place Order use case. All changes are highlighted.

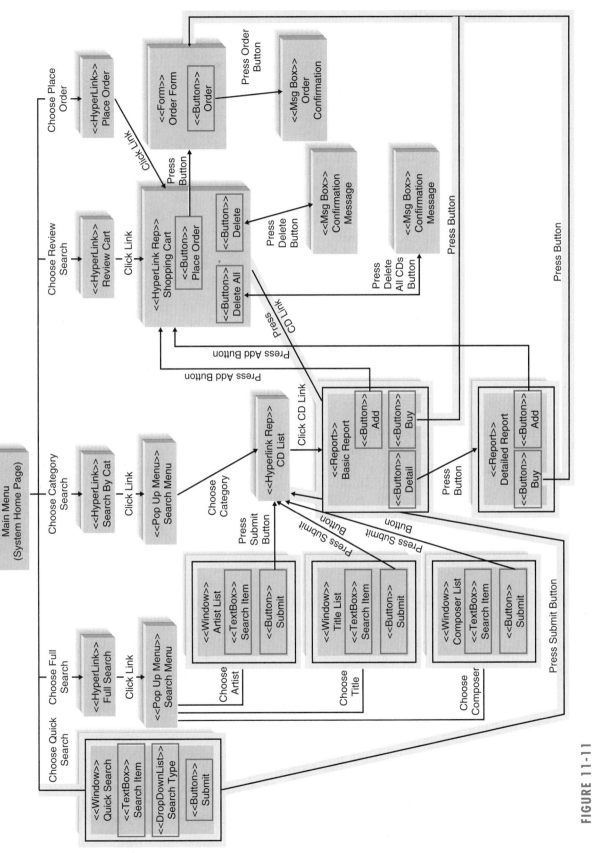

FIGURE 11-11
CD-Selections Revised WND for the Web Portion of the Internet Sales System

Interface metaphor: A CD Selections music store

Interface objects:
- *Album:* All music items, whether CD or tape
- *Artist:* Person or group who records the CD
- *Title:* Title or name of CD
- *Composer:* Person or group who wrote the music for the CD (primarily used for classical music)
- *Music category:* Type of music; current categories include rock, jazz, classical, country, alternative, sound tracks, rap, folk, gospel
- *CD list:* List of CD(s) matching the specified criteria
- *Shopping cart:* Place to store selected CDs until they are ordered

Interface actions:
- *Search for:* Displays a CD list that matches specified criteria
- *Browse:* Displays a CD list sorted in order by some criteria
- *Buy:* Authorizes payment by a credit card for specific CDs

Interface icons:
CD Selections logo will be used on all screens

FIGURE 11-12
CD Selections Interface Standards

Interface Standards Design

Once the WND was complete, Alec moved on to develop the interface standards for the system. The interface metaphor was straightforward: a CD Selections music store. The key interface objects and actions were equally straightforward, as was the use of the CD Selections logo icon (see Figure 11-12).

At this point, the next step is to design the interface template and then move on to design the individual interfaces. However, this requires an understanding of the basic interface navigation, input, and output elements, which are discussed in Chapter 12. So, we'll stop here and continue the example in the next chapter.

SUMMARY

User Interface Design Principles
The first element of the user interface design is the layout of the screen, form, or report, which is usually depicted using rectangular shapes with a top area for navigation, a central area for inputs and outputs, and a status line at the bottom. The design should help the user be aware of content and context, both between different parts of the system as they navigate through it and within any one form or report. All interfaces should be aesthetically pleasing (not works of art) and need to include significant white space, use colors carefully, and be consistent with fonts. Most interfaces should be designed to support both novice/first-time users and experienced users. Consistency in design (both within the system and across other systems employed by the users) is important for the navigation controls, terminology, and layout of forms and reports. Finally, all interfaces should attempt to minimize user effort, for example, by requiring no more than three clicks from the main menu to perform an action.

The User Interface Design Process

First, analysts develop use scenarios that describe commonly used patterns of actions that the users will perform. Second, they design the interface structure via a window navigation diagram (WND) based on the essential use cases. The WND is then tested with the use scenarios to ensure that it enables users to perform these scenarios quickly and smoothly. Third, analysts define the interface standards in terms of interface metaphor(s), objects, actions, and icons. These elements are drawn together by the design of a basic interface template for each major section of the system. Fourth, the designs of the individual interfaces are prototyped, either through a simple storyboard, an HTML prototype, or a prototype using the development language of the system itself (e.g., Visual Basic). Finally, interface evaluation is conducted using heuristic evaluation, walk-through evaluation, interactive evaluation, or formal usability testing. This evaluation almost always identifies improvements, so the interfaces are redesigned and evaluated further.

KEY TERMS

Aesthetics	Interface evaluation	Storyboard
Consistency	Interface icon	System interface
Content awareness	Interface metaphor	Three clicks rule
Density	Interface object	Transition
Essential use case	Interface standards	Usability testing
Field	Interface template	Use scenario
Field label	Language prototype	User experience
Heuristic evaluation	Layout	User interface
HTML prototype	Navigation mechanism	Walk-through evaluation
Input mechanism	Output mechanism	White space
Interactive evaluation	Real use case	Window navigation diagram
Interface action	State	(WND)
Interface design prototype	Stereotype	

QUESTIONS

1. Explain three important user interface design principles.
2. What are three fundamental parts of most user interfaces?
3. Why is content awareness important?
4. What is white space and why is it important?
5. Under what circumstances should densities be low? high?
6. How can a system be designed to be used by experienced and first-time users?
7. Why is consistency in design important? Why can too much consistency cause problems?
8. How can different parts of the interface be consistent?

9. Describe the basic process of user interface design.
10. What are use scenarios and why are they important?
11. What is a window navigation diagram (WND) and why is it used?
12. Why are interface standards important?
13. Explain the purpose and contents of interface metaphors, interface objects, interface actions, interface icons, and interface templates.
14. Why do we prototype the user interface design?
15. Compare and contrast the three types of interface design prototypes.
16. Why is it important to perform an interface evaluation before the system is built?

17. Compare and contrast the four types of interface evaluation.
18. Under what conditions is heuristic evaluation justified?

19. What type of interface evaluation did you perform in the Your Turn Box 11-1?

EXERCISES

A. Develop two use scenarios for a Web site that sells some retail products (e.g., books, music, clothes).
B. Draw a WND for a Web site that sells some retail products (e.g., books, music, clothes).
C. Describe the primary components of the interface standards for a Web site that sells some retail products (metaphors, objects, actions, icons, and template).
D. Develop two use scenarios for the use-case diagram in Exercise O in Chapter 6.
E. Develop the interface standards (omitting the interface template) for the use-case diagram in Exercise O in Chapter 6.
F. Draw a WND for the use-case diagram in Exercise O in Chapter 6.
G. Design a storyboard for Exercise O in Chapter 6.
H. Develop two use scenarios for the use-case diagram in Exercise Q in Chapter 6.
I. Develop the interface standards (omitting the interface template) for the use-case diagram in Exercise Q in Chapter 6.

J. Draw a WND for the use-case diagram in Exercise Q in Chapter 6.
K. Design a storyboard for Exercise Q in Chapter 6.
L. Develop two use scenarios for the use-case diagram in Exercise S in Chapter 6.
M. Develop the interface standards (omitting the interface template) for the use-case diagram in Exercise S in Chapter 6.
N. Draw a WND for the use-case diagram in Exercise S in Chapter 6.
O. Design a storyboard for Exercise S in Chapter 6.
P. Develop two use scenarios for the use-case diagram in Exercise U in Chapter 6.
Q. Develop the interface standards (omitting the interface template) for the use-case diagram in Exercise U in Chapter 6.
R. Draw a WND for the use-case diagram in Exercise U in Chapter 6.
S. Design a storyboard for Exercise U in Chapter 6.

MINICASES

1. Tots to Teens is a catalog retailer specializing in children's clothing. A project has been under way to develop a new order entry system for the company's catalog clerks. The old system had a character-based user interface that corresponded to the system's COBOL underpinnings. The new system will feature a graphical user interface more in keeping with up-to-date PC products in use today. The company hopes that this new user interface will help reduce the turnover they have experienced with their order entry clerks. Many newly hired order entry staff found the old system very difficult to learn and were overwhelmed by the numerous mysterious codes that had to be used to communicate with the system.

A user interface walk-through evaluation was scheduled for today to give the user a first look at the new system's interface. The project team was careful to invite several key users from the order entry department. In

particular, Norma was included because of her years of experience with the order entry system. Norma was known to be an informal leader in the department; her opinion influenced many of her associates. Norma had let it be known that she was less than thrilled with the ideas she had heard for the new system. Due to her experience and good memory, Norma worked very effectively with the character-based system and was able to breeze through even the most convoluted transactions with ease. Norma had trouble suppressing a sneer when she heard talk of such things as "icons" and "buttons" in the new user interface.

Cindy was also invited to the walk-through because of her influence in the order entry department. Cindy has been with the department for just one year, but she quickly became known because of her successful organization of a sick child day-care service for the children of the department workers. Sick children are the number-one cause of absenteeism in the department, and many of the workers could not afford to miss workdays. Never one to keep quiet when a situation needed improvement, Cindy has been a vocal supporter of the new system.

a. Drawing upon the design principles presented in the text, describe the features of the user interface that will be most important to experienced users like Norma.

b. Drawing upon the design principles presented in the text, describe the features of the user interface that will be most important to novice users like Cindy.

2. The members of a systems development project team have gone out for lunch together, and as often happens, the conversation turns to work. The team has been working on the development of the user interface design, and so far, work has been progressing smoothly. The team should be completing work on the interface prototypes early next week. A combination of storyboards and language prototypes has been used in this project. The storyboards depict the overall structure and flow of the system, but the team developed language prototypes of the actual screens because they felt that seeing the actual screens would be valuable for the users.

CHRIS (the youngest member of the project team): I read an article last night about a really cool way to evaluate a user interface design. It's called usability testing, and it's done by all the major software vendors. I think we should use it to evaluate our interface design.

HEATHER (system analyst): I've heard of that, too, but isn't it really expensive?

MARK (project manager): I'm afraid it is expensive, and I'm not sure we can justify the expense for this project.

CHRIS: But we really need to know that the interface works. I thought this usability testing technique would help us prove we have a good design.

AMY (systems analyst): It would, Chris, but there are other ways too. I assumed we'd do a thorough walk-through with our users and present the interface to them at a meeting. We can project each interface screen so that the users can see it and give us their reaction. This is probably the most efficient way to get the users' response to our work.

HEATHER: That's true, but I'd sure like to see the users sit down and work with the system. I've always learned a lot by watching what they do, seeing where they get confused, and hearing their comments and feedback.

RYAN (systems analyst): It seems to me that we've put so much work into this interface design that all we really need to do is review it ourselves. Let's just make a list of the design principles we're most concerned about and check it ourselves to make sure we've followed them consistently. If we have, we should be fine. We want to get moving on the implementation, you know.

MARK: These are all good ideas. It seems like we've all got a different view of how to evaluate the interface design. Let's try and sort out the technique that is best for our project.

Develop a set of guidelines that can help a project team like the one above select the most appropriate interface evaluation technique for their project.

PLANNING

ANALYSIS

DESIGN

☑ **Evolve Analysis Models**
☑ **Develop the Design Strategy**
☑ **Develop the Design**
☑ **Develop Infrastructure Design**
☑ **Develop Network Model**
☑ **Develop Hardware/Software Specification**
☑ **Develop Security Plan**
☑ **Develop Use Scenarios**
☑ **Design Interface Structure**
☑ **Design Interface Standards**
☐ Design User Interface Template
☐ Design User Interface
☐ Evaluate User Interface
☐ Select Object Persistence Format
☐ Map Objects to Object Persistence
☐ Optimize Object Persistence
☐ Size Object Persistence
☐ Specify Constraints
☐ Develop Contracts
☐ **Develop Method Specifications**

T A S K C H E C K L I S T

PLANNING → ANALYSIS → DESIGN →

CHAPTER 12

USER INTERFACE DESIGN COMPONENTS

THE user interface is the part of the system with which the users interact. It includes the screen displays that provide navigation through the system, the screens and forms that capture data, and the reports that the system produces (whether on paper, on the screen, or via some other media). This chapter introduces the basic principles and techniques for navigation design, input design, and output design.

OBJECTIVES

- Understand commonly used principles and techniques for navigation design.
- Understand commonly used principles and techniques for input design.
- Understand commonly used principles and techniques for output design.
- Be able to design a user interface.

CHAPTER OUTLINE

IMPLEMENTATION

INTRODUCTION

Interface design is the process of defining how the users will interact with the system and the nature of the inputs and outputs that the system accepts and produces. The interface includes three fundamental parts. The first is the *navigation mechanism,* the way in which the user gives instructions to the system and tells it what to do (e.g., buttons, menus). The second is the *input mechanism,* the way in which the system captures information (e.g., forms for adding new customers). The third is the *output mechanism,* the way in which the system provides information to the user or to other systems (e.g., reports, Web pages). Each of these is conceptually different, but they all are intertwined closely. All computer displays contain navigation mechanisms, and most contain input and output mechanisms. Therefore, navigation design, input design, and output design are tightly coupled.

This chapter provides an overview of the navigation, input, and output components that are used in interface design. This chapter focuses on the design of Web-based interfaces and *graphical user interfaces* (GUI) that use windows, menus, icons, and a mouse (e.g., Windows, Macintosh).[1] Although text-based interfaces are still commonly used on mainframes and Unix systems, GUI interfaces are probably the most common type of interfaces that you will use, with the possible exception of printed reports.[2]

NAVIGATION DESIGN

The navigation component of the interface enables the user to enter commands to navigate through the system and perform actions to enter and review information it contains. The navigation component also presents messages to the user about the success or failure of his or her actions. The goal of the navigation system is to make the system as simple as possible to use. A good navigation component is one the user never really notices. It simply functions the way the user expects, and thus the user gives it little thought.

Basic Principles

One of the most difficult things about using a computer system is learning how to manipulate the navigation controls to make the system do what you want. Analysts usually need to assume that users have not read the manual, have not attended training, and do not have external help readily at hand. All controls should be clear and understandable, and placed in an intuitive location on the screen. Ideally, the controls should anticipate what the user will do and simplify his or her efforts. For example, many setup programs are designed so that for a "typical" installation the user can simply keep pressing the "next" button.

---

[1] Many people attribute the origin of GUI interfaces to Apple or Microsoft. Some people know that Microsoft copied from Apple, which in turn "borrowed" the whole idea from a system developed at the Xerox Palo Alto Research Center (PARC) in the 1970s. Very few know that the Xerox system was based on a system developed by Doug Englebart of Stanford, which was first demonstrated at the Western Computer Conference in 1968. Around the same time, Englebart also invented the mouse, desktop video conferencing, groupware, and a host of other things we now take for granted. Doug is a legend in the computer science community and has won too many awards to count, but he is relatively unknown by the general public.

[2] A good book on GUI design is Susan Fowler, *GUI Design Handbook* (New York: McGraw-Hill, 1998).

Prevent Mistakes   The first principle of designing navigation controls is to prevent the user from making mistakes. A mistake costs time and creates frustration. Worse still, a series of mistakes can cause the user to discard the system. Mistakes can be reduced by labeling commands and actions appropriately and by limiting choices. Too many choices can confuse the user, particularly when they are similar and hard to describe in the short space available on the screen. When there are many similar choices on a menu, consider creating a second level of menu or a series of options for basic commands.

Never display a command that cannot be used. For example, many Windows applications "gray out" commands that cannot be used; they are displayed on pull-down menus in a very light-colored font, but they cannot be selected. This shows that they are available (and keeps all menu items in the same place), but they cannot be used in the current context.

When the user is about to perform a critical function that is difficult or impossible to undo (e.g., deleting a file), it is important to confirm the action with the user (and make sure the selection was not made by mistake). This is usually done by having the user respond to a confirmation message that explains what the user has requested and asks the user to confirm that this action is correct.

Simplify Recovery from Mistakes   No matter what the system designer does, users will make mistakes. The system should make it as easy as possible to correct these errors. Ideally, the system will have an "undo" button that makes mistakes easy to override; however, writing the software for such buttons can be very complicated.

Use Consistent Grammar Order   One of the most fundamental decisions is the *grammar order*. Most commands require the user to specify an object (e.g., file, record, word) and the action to be performed on that object (e.g., copy, delete). The interface can require the user to choose first the object and then the action (an *object-action order*), or first the action and then the object (an *action-object order*). Most Windows applications use an object-action grammar order (e.g., think about copying a block of text in your word processor).

The grammar order should be consistent throughout the system, both at the data element level and at the overall menu level. Experts debate about the advantages of one approach over the other, but because most users are familiar with the object-action order, most systems today are designed using that approach.

Types of Navigation Controls

Two traditional hardware devices can be used to control the user interface: the keyboard and a pointing device such as a mouse, trackball, or touch screen. In recent years, voice recognition systems have made an appearance, but they are not yet common. The three basic software approaches for defining user commands are languages, menus, and direct manipulation.

Languages   With a *command language,* the user enters commands using a special language developed for the computer system (e.g., DOS and SQL both use command languages). Command languages sometimes provide greater flexibility than other approaches because the user can combine language elements in ways that are not predetermined by developers. However, they put a greater burden on users because users must learn syntax and type commands rather than select from a well-defined, limited number of choices. Systems today use command languages sparingly, except in cases where there are an extremely large number of command combinations that make it impractical to try to build all combinations into a menu (e.g., SQL queries for databases).

Natural language interfaces are designed to understand the user's own language (e.g., English, French, Spanish). These interfaces attempt to interpret what the user means, and often they present back to the user a list of interpretations from which to choose. An example of the use of natural language is Microsoft's Office Assistant, which enables users to ask free-form questions for help.

Menus The most common type of navigation system today is the *menu*. A menu presents the user with a list of choices, each of which can be selected. Menus are easier to learn than languages because a limited number of available commands are presented to the user in an organized fashion. Clicking on an item with a pointing device or pressing a key that matches the menu choice (e.g., a function key) takes very little effort. Therefore, menus are usually preferred to languages.

Figure 12-1 shows a traditional text-based menu on a Unix system. The user can select menu items by typing the letter in front of the item (e.g., C to compose a message) or by using the arrow keys to move the highlighted bar to the item and pressing the enter key.

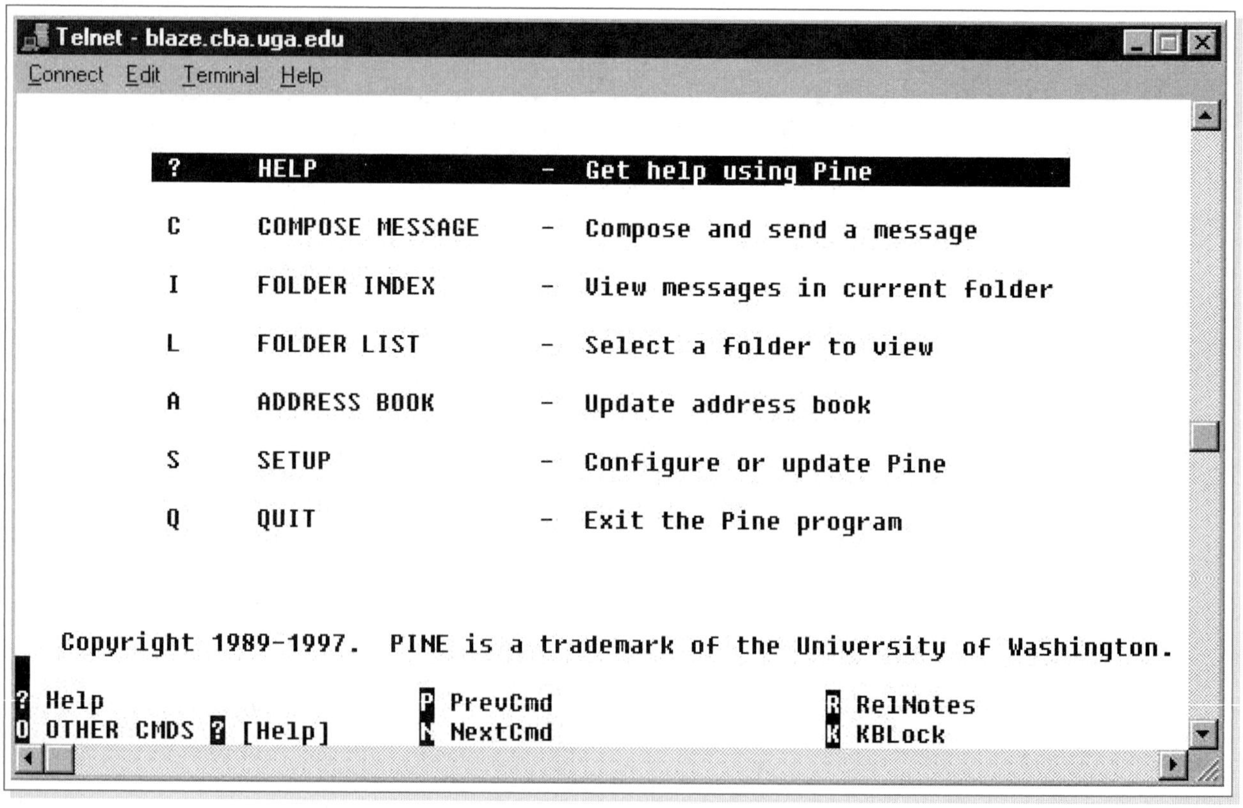

FIGURE 12-1
A Traditional Menu in a Unix System

Menu design needs to be done with care because the submenus behind a main menu are hidden from users until they click on the menu item. It is better to make menus broad and shallow (i.e., each menu containing many items with only one or two layers of menus) rather than narrow and deep (i.e., each menu containing only a few items, but each leading to three or more layers of menus). A broad and shallow menu presents the user with the most information initially so that he or she can see many options and requires only a few mouse clicks or keystrokes to perform an action. A narrow and deep menu makes users hunt and seek for items hidden behind menu items and requires many more clicks or keystrokes to perform an action.

Research suggests that in an ideal world, any one menu should contain no more than eight items, and it should take no more than two mouse clicks or keystrokes from any menu to perform an action (or three from the main menu that starts a system).[3] However, analysts must sometimes break this guideline in the design of complex systems. In this case, menu items are often grouped together and separated by a horizontal line (see Figure 12-2). Often menu items have *hot keys* that enable experienced users to quickly invoke a command with keystrokes in lieu of a menu choice (e.g., Ctrl-F in Word invokes the Find command or Alt-F opens the file menu).

Menus should put together like items so that the user can intuitively guess what each menu contains. Most designers recommend grouping menu items by interface objects (e.g., customers, purchase orders, inventory) rather than by interface actions (e.g., new, update, format), so that all actions pertaining to one object are in one menu, all actions for another object are in a different menu, and so on. However, this is highly dependent on the specific interface. As Figure 12-2 shows, Microsoft Word groups menu items by interface objects (e.g., File, Table, Window) *and* by interface actions (e.g., Edit, Insert, Format) on the same menu.

Figures 12-2 and 12-3 illustrate some other types of menus. Figure 12-4 summarizes several commonly used types of menus and when to use them.

Direct Manipulation   With *direct manipulation,* the user enters commands by working directly with interface objects. For example, users can change the size of objects in Microsoft PowerPoint by clicking on them and moving their sides, or they can move files in Windows Explorer by dragging the filenames from one folder to another. Direct manipulation can be simple, but it suffers from two problems. First, users familiar with language- or menu-based interfaces don't always expect it. Second, not all commands are intuitive (e.g., how do you copy [not move] files in Win-

---

YOUR	12-1 DESIGN A NAVIGATION SYSTEM
TURN	

Design a navigation system for a system into which users must enter information about customers, products, and orders. For all three, users will want to change, delete, find one specific record, and list all records.

---

[3] Kent L. Norman, *The Psychology of Menu Selection* (Norwood, NJ: Ablex Publishing Corp., 1991).

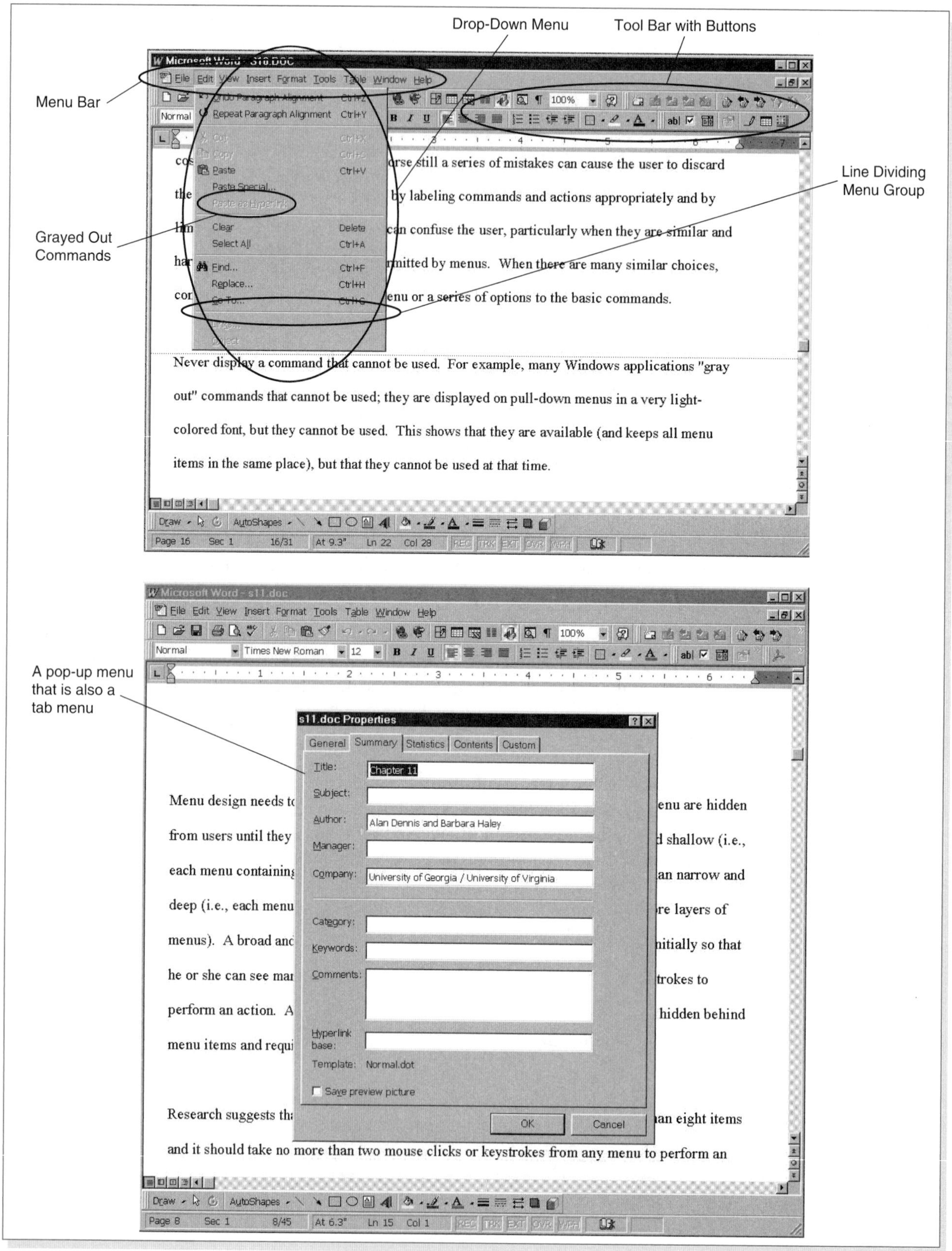

Drop-Down Menu   Tool Bar with Buttons

Menu Bar

W Microsoft Word - s16.DOC

File Edit View Insert Format Tools Table Window Help

Undo Paragraph Alignment   Ctrl+Z
Repeat Paragraph Alignment   Ctrl+Y

Normal

Cut   Ctrl+X
Copy   Ctrl+C
Paste   Ctrl+V
Paste Special...
Paste as Hyperlink
Clear   Delete
Select All   Ctrl+A
Find...   Ctrl+F
Replace...   Ctrl+H
Go To...   Ctrl+G

Grayed Out Commands

Line Dividing Menu Group

cos   orse still a series of mistakes can cause the user to discard

the   by labeling commands and actions appropriately and by

lim   can confuse the user, particularly when they are similar and

har   mitted by menus. When there are many similar choices,

cor   enu or a series of options to the basic commands.

Never display a command that cannot be used. For example, many Windows applications "gray

out" commands that cannot be used; they are displayed on pull-down menus in a very light-

colored font, but they cannot be used. This shows that they are available (and keeps all menu

items in the same place), but that they cannot be used at that time.

Draw ▸ AutoShapes ▸
Page 16   Sec 1   16/31   At 9.3"   Ln 22   Col 28   REC TRK EXT OVR WPH

W Microsoft Word - s11.doc

File Edit View Insert Format Tools Table Window Help

Normal   Times New Roman   12   B I U

A pop-up menu that is also a tab menu

s11.doc Properties

General | Summary | Statistics | Contents | Custom

Title: Chapter 11
Subject:
Author: Alan Dennis and Barbara Haley
Manager:
Company: University of Georgia / University of Virginia

Category:
Keywords:
Comments:

Hyperlink base:

Template: Normal.dot

☐ Save preview picture

OK   Cancel

Menu design needs to   enu are hidden

from users until they   d shallow (i.e.,

each menu containing   an narrow and

deep (i.e., each menu   re layers of

menus). A broad and   initially so that

he or she can see man   rokes to

perform an action. A   hidden behind

menu items and requi

Research suggests that   an eight items

and it should take no more than two mouse clicks or keystrokes from any menu to perform an

Draw ▸ AutoShapes ▸
Page 8   Sec 1   8/45   At 6.3"   Ln 15   Col 1   REC TRK EXT OVR WPH

FIGURE 12-2
Common Types of Menus

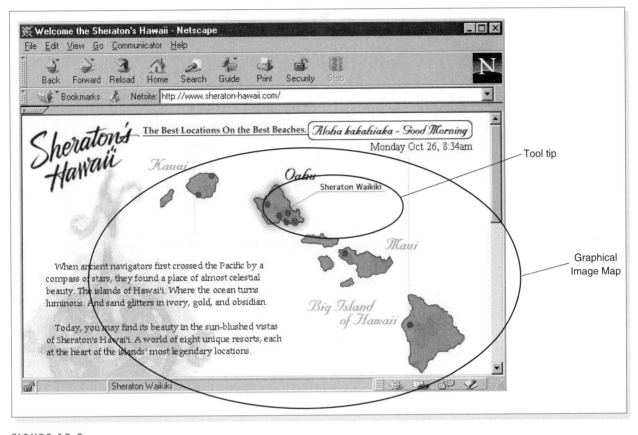

FIGURE 12-3
Image Map

dows Explorer? On the Macintosh, why does moving a folder to the trash delete the file if it is on the hard disk but eject the diskette if the file is on a diskette?).

Messages

Messages are the way in which the system responds to a user and informs him or her of the status of the interaction. In general, messages should be clear, concise, and complete, which are sometimes conflicting objectives. All messages should be grammatically correct and free of jargon and abbreviations (unless they are the users' jargon and abbreviations). Avoid negatives because they can be confusing (e.g., replace "Are you sure you do not want to continue?" with "Do you want to quit?"). You should also avoid humor because it wears off quickly after the same message appears dozens of times.

Messages should require the user to acknowledge them (by clicking, for example) rather than being displayed for a few seconds and then disappearing. The exceptions are messages that inform the user of delays in processing, which should disappear once the delay has passed. In general, messages are text, but sometimes standard icons are used. For example, Windows displays an hourglass when the system is busy.

There are many different types of messages, such as *error messages, confirmation messages, acknowledgment messages, delay messages,* and *help messages* (see Figure 12-5). All messages should be carefully crafted, but error messages and

Type of Menu	When to Use	Notes
Menu bar List of commands at the top of the screen; always on-screen	Main menu for system	Use the same organization as the operating system and other packages (e.g., File, Edit, View). Menu items are always one word, never two. Menu items lead to other menus rather than perform action. Never allow users to select actions they can't perform (instead, use grayed-out items).
Drop-down menu Menu that drops down immediately below another menu; disappears after one use	Second-level menu, often from menu bar	Menu items are often multiple words. Avoid abbreviations. Menu items perform action or lead to another cascading drop-down menu, pop-up menu, or tab menu.
Pop-up menu Menu that pops up and floats over the screen; disappears after one use	As a shortcut to commands for experienced users	Pop-up menus often (not always) invoked by a right click in Windows-based systems. These menus are often overlooked by novice users, so usually they should duplicate functionality provided in other menus.
Tab menu Multipage menu with one tab for each page that pops up and floats over the screen; remains on-screen until closed	When user needs to change several settings or perform several related commands	Menu items should be short to fit on the tab label. Avoid more than one row of tabs, because clicking on a tab to open it can change the order of the tabs and in virtually no other case does selecting from a menu rearrange the menu itself.
Toolbar Menu of buttons (often with icons) that remains on-screen until closed	As a shortcut to commands for experienced users	All buttons on the same toolbar should be the same size. If the labels vary dramatically in size, then use two different sizes (small and large). Buttons with icons should have a *tool tip*, an area that displays a text phrase explaining the button when the user pauses the mouse over it.
Image map Graphic image in which certain areas are linked to actions or other menus	Only when the graphic image adds meaning to the menu	The image should convey meaning to show which parts perform action when clicked. Tool tips can be helpful.

FIGURE 12-4
Types of Menus

help messages require particular care. For example, error messages should *first* explain the problem, *second* describe how to correct it, and *third* provide button(s) for the user response. Compare Figure 12-6a and 12-6b.

Messages (and especially error messages) should always explain the problem (what the user did incorrectly) and corrective action as clearly and as explicitly as possible so that the user knows exactly what needs to be done. In the case of complicated errors, the error message should display what the user entered, suggest probable causes for the error, and propose possible user responses. When in doubt, provide more information than the user needs or give the user the ability to get additional information (see Figure 12-6c).

Messages should be polite. Impolite error messages, like impolite behavior, can provoke an emotional response. Polite error messages won't reduce the error, but they may diffuse anger. Politeness also means that the system should take the blame for any mistakes on the user's part[4] (see Figure 12-6d).

Finally, messages should provide a message number. Message numbers are not intended for users but may make it simpler for help desks and customer support

---

[4] It's not fair to the computer, but computers don't buy software.

Type of Messages	When to Use	Notes
Error message Informs the user that he or she has attempted to do something to which the system cannot respond	When user does something that is not permitted or not possible	Always explain the reason and suggest corrective action. Traditionally, error messages have been accompanied by a beep, but many applications now omit it or permit users to remove it.
Confirmation message Asks users to confirm that they really want to perform the action they have selected	When user selects a potentially dangerous choice, such as deleting a file	Always explain the cause and suggest possible action. Often include several choices other than "OK" and "cancel."
Acknowledgment message Informs the user that the system has accomplished what it was asked to do	Seldom or never. Users quickly become annoyed with all the unnecessary mouse clicks	Acknowledgment messages are typically included because novice users often like to be reassured that an action has taken place. The best approach is to provide acknowledgment information without a separate message on which the user must click. For example, if the user is viewing items in a list and adds one, then the updated list on the screen showing the added item is sufficient acknowledgment.
Delay message Informs the user that the computer system is working properly	When an activity takes more than seven seconds	Should permit the user to cancel the operation in case he or she does not want to wait for its completion. Should provide some indication of how long the delay may last.
Help message Provides additional information about the system and its components	In all systems	Help information is organized by table of contents and/or keyword search. *Context-sensitive help* provides information that is dependent on what the user was doing when help was requested. Help messages and on-line documentation are discussed in Chapter 15.

FIGURE 12-5
Types of Messages

lines to identify problems and help users because many messages use similar wording. They also make it easier for programmers to locate and change messages (see Figure 12-6e).

Navigation Design Documentation

Navigation for a system is designed through the use of *window navigation diagrams* (WND) and real use cases. WNDs were described in Chapter 11. In this section we describe real use cases.

Real use cases are derived from the essential use cases (see Chapter 6), *use scenarios* (see Chapter 11), and WNDs. Recall that an *essential use case* is one that only describes the minimum essential issues necessary to understand the required functionality. A real use case describes a specific set of steps that a user performs to use a specific part of a system. As such, real use cases are implementation dependent; that is; they are detailed descriptions of how to use the system once it is implemented.

To evolve an essential use case into a real use case, two changes must be made. First, the use-case type must be changed from essential to real. Second, all events must be specified in terms of the actual user interface. Therefore, the Normal Flow of Events, Subflows, and Alternate/Exceptional Flows must be modified. The Normal Flow of Events, Subflows, and Alternate/Exceptional Flows for the real use case

Stating corrective action before the problem explanation can be confusing.

Stating the problem before the corrective action is more intuitive.

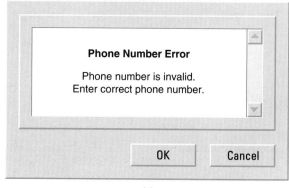

(a)

(b)

The problem should be stated in as much detail as possible.

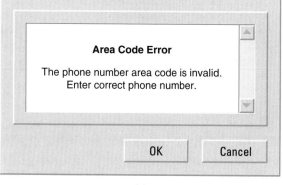

The error message should be polite.

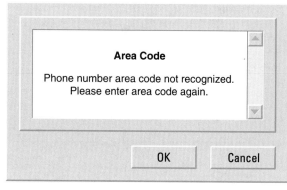

(c)

(d)

Error messages should include a message number.

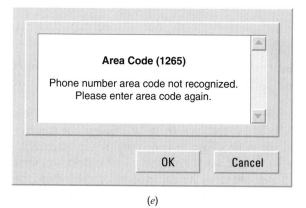

(e)

FIGURE 12-6
Crafting Messages

associated with the storyboard user interface prototype given in Figure 11-9 is shown in Figure 12-7. For example, Step 2 of the Normal Flow of Events states that "The System provides the Sales Rep with the Main Menu for the System" which allows the Sales Rep to interact with the Maintain Client List aspect of the system.

Use-Case Name: *Maintain Client List*	ID: *12*	Importance Level: *High*

Primary Actor: *Sales Rep* **Use-Case Type:** *Detail, Real*

Stakeholders and Interests: *Sales Rep - wants to add, find or list clients*

Brief Description: *This use case describes how sales representatives can search and maintain the client list.*

Trigger: *Patient calls and asks for a new appointment or asks to cancel or change an existing appointment.*

Type: External

Relationships:
    Association: *Sales Rep*
    Include:
    Extend:
    Generalization:

Normal Flow of Events:

1. *The Sales Rep starts up the system.*
2. *The System provides the Sales Rep with the Main Menu for the System.*
3. *The System asks Sales Rep if he or she would like to Add a client, Find an existing client, or to List all existing clients.*
    If the Sales Rep wants to add a client, they click on the Add Client Link and execute S-1: New Client.
    If the Sales Rep wants to find a client, they click on the Find Client Link and execute S-2: Find Client.
    If the Sales Rep wants to list all clients, they click on the List Client Link and execute S-3: List Clients.
4. *The System returns the Sales Rep to the Main Menu of the System.*

Subflows:

S-1: New Client
  1. *The System asks the Sales Rep for relevant information.*
  2. *The Sales Rep types in the relevant information into the Form*
  3. *The Sales Rep submits the information to the System.*
S-2: Find Client
  1. *The System asks the Sales Rep for the search information.*
  2. *The Sales Rep types in the search information into the Form*
  3. *The Sales Rep submits the information to the System.*
  4. *If the System finds a single client that meets the search information, the System produces a Client Information report and returns the Sales Rep to the Main Menu of the System*
    Else If the System finds a list of clients that meet the search information, the System executes S-3: List Clients.
S-3: List Clients
  1. *If this Subflow is executed from Step 3*
    The System creates a List of All clients
    Else
    The System creates a List of clients that matched the S-2: Find Client search criteria.
  2. *The Sales Rep selects a client.*
  3. *The System produces a Client Information report.*

Alternate/Exceptional Flows:
  S-2 4a. The System produces an Error Message.

FIGURE 12-7
Example Real Use Case

INPUT DESIGN

Inputs facilitate the entry of data into the computer system, whether highly structured data, such as order information (e.g., item numbers, quantities, costs) or unstructured information (e.g., comments). Input design means designing the screens used to enter the information, as well as any forms on which users write or type information (e.g., timecards, expense claims).

Basic Principles

The goal of the input mechanism is to capture accurate information for the system both simply and easily. The fundamental principles for input design reflect the nature of the inputs (whether batch or on-line) and ways to simplify their collection.

On-line versus Batch Processing   There are two general formats for entering inputs into a computer system: on-line processing and batch processing. With *on-line processing* (sometimes called *transaction processing*), each input item (e.g., a customer order, a purchase order) is entered into the system individually, usually at the same time as the event or transaction prompting the input. For example, when you take a book out from the library, buy an item at the store, or make an airline reservation, the computer system that supports that process uses on-line processing to immediately record the transaction in the appropriate database(s). On-line processing is most commonly used when it is important to have *real-time information* about the business process. For example, when you reserve an airline seat, the seat is no longer available for someone else to use.

With *batch processing,* all the inputs collected over some time period are gathered together and entered into the system at one time in a batch. Some business processes naturally generate information in batches. For example, most hourly payrolls are done using batch processing because time cards are gathered together in batches and processed at once. Batch processing is also used for transaction processing systems that do not require real-time information. For example, most stores send sales information to district officers so that new replacement inventory can be ordered. This information could be sent in real time as it is captured in the store, so that the district offices are aware within a second or two that a product is sold. If stores do not need this up-to-the-second real time data, they will collect sales data throughout the day and transmit it every evening in a batch to the district office. This batching simplifies the data communications process and often saves communications costs. However, it means that inventories are accurate not in real time, but only at the end of the day after the batch has been processed.

Capture Data at the Source   Perhaps the most important principle of input design is to capture the data in an electronic format at its original source or as close to the original source as possible. In the early days of computing, computer systems replaced traditional manual systems that operated on paper forms. As these business processes were automated, many of the original paper forms remained, either because no one thought to replace them or because it was too expensive to do so. Instead, the business process continued to contain manual forms that were taken to the computer center in batches to be typed into the computer system by a *data-entry operator.*

Many business processes still operate this way today. For example, most organizations have expense claim forms that are completed by hand and submitted to an accounting department, which approves them and enters them into the system in batches. This approach presents: three problems: First, it is expensive because it duplicates work (the form is filled out twice, once by hand, once by keyboard[5]). Second, it increases processing time because the paper forms must be physically moved through the process. Third, it increases the cost and probability of error because it separates the entry from the processing of information; someone may misread the handwriting on the input form, data may be entered incorrectly, or the original input may contain an error that invalidates the information.

Most transaction processing systems today are designed to capture data at its source. *Source data automation* refers to using special hardware devices to automatically capture data without requiring anyone to type it. Stores commonly use *bar code readers* that automatically scan products and that enter data directly into the computer system. No intermediate formats such as paper forms are used. Similar technologies include *optical character recognition,* which can read printed numbers and text (e.g., on checks), *magnetic stripe readers,* which can read information encoded on a stripe of magnetic material similar to a diskette (e.g., credit cards), and *smart cards* that contain microprocessors, memory chips, and batteries (much like credit-card-sized calculators). As well as reducing the time and cost of data entry, these systems reduce errors because they are far less likely to capture data incorrectly. Today, portable computers and scanners allow data to be captured at the source even when in mobile settings (e.g., air courier deliveries, use of rental cars).

These automatic systems are not capable of collecting a lot of information, so the next best option is to capture data immediately from the source using a trained entry operator. Many airline and hotel reservations, loan applications, and catalog orders are recorded directly into a computer system while the customer provides the operator with answers to questions. Some systems eliminate the operator altogether and allow users to enter their own data. For example, several universities (e.g., the Massachusetts Institute of Technology [MIT]) no longer accept paper-based applications for admissions; all applications are typed by students into electronic forms.

The forms for capturing information (on a screen, on paper, etc.) should support the data source. That is, the order of the information on the form should match the natural flow of information from the data source, and data-entry forms should match paper forms used to initially capture the data.

Minimize Keystrokes   Another important principle is to minimize keystrokes. Keystrokes cost time and money, whether they are performed by a customer, user, or trained data-entry operator. The system should never ask for information that can be obtained in another way (e.g., by retrieving it from a database or by performing a calculation). Similarly, a system should not require a user to type information that can be selected from a list; selecting reduces errors and speeds entry.

In many cases, some fields have values that often reoccur. These frequent values should be used as the *default value* for the field so that the user can simply

---

[5] Or in the case of the University of Georgia, three times: first by hand on an expense form; a second time when it is typed onto a new form for the "official" submission because the accounting department refuses handwritten forms; and finally when it is typed into the accounting computer system.

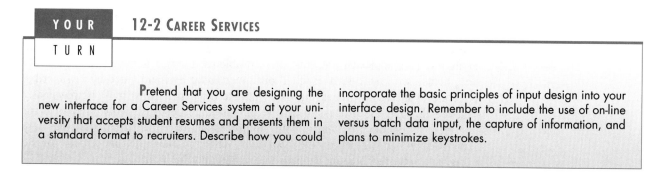

accept the value and not have to retype it time and again. Examples of default values are the current date, the area code held by the majority of a company's customers, and a billing address that is based on the customer's residence. Most systems permit changes to default values to handle data-entry exceptions as they occur.

Types of Inputs

Each data item that has to be input is linked to a *field* on the form into which its value is typed. Each field also has a *field label,* which is the text beside, above, or below the field that tells the user what type of information belongs in the field. Often the field label is similar to the name of the data element, but they do not have to have identical words. In some cases, a field will display a template over the entry box to show the user exactly how data should be typed. There are many different types of inputs, in the same way that there are many different types of fields (see Figure 12-8).

Text   As the name suggests, a *text box* is used to enter text. Text boxes can be defined to have a fixed length or can be scrollable and accept a virtually unlimited amount of text. In either case, boxes can contain single or multiple lines of textual information. Never use a text box if you can use a selection box (see below).

Text boxes should have field labels placed to the *left* of the entry area, their size clearly delimited by a box (or a set of underlines in a non-GUI interface). If there are multiple text boxes, their field labels and the left edges of their entry boxes should be aligned. Text boxes should permit standard GUI functions such as cut, copy, and paste.

Numbers   A *number box* is used to enter numbers. Some software can automatically format numbers as they are entered, so that 3452478 becomes $34,524.78. Dates are a special form of numbers that sometimes have their own type of number box. Never use a number box if you can use a selection box.

Selection Box   A *selection box* enables the user to select a value from a predefined list. The items in the list should be arranged in some meaningful order, such as alphabetical for long lists or in order of most frequently used. The default selection value should be chosen with care. A selection box can be initialized as "unselected," or better still, start with the most commonly used item already selected.

Six commonly used types of selection boxes are *check boxes, radio buttons, on-screen list boxes, drop-down list boxes, combo boxes,* and *sliders* (see Figure 12-9).

FIGURE 12-8
Types of Input Boxes

The choice among the types of text selection boxes generally comes down to one of screen space and the number of choices the user can select. If screen space is limited and only one item can be selected, then a drop-down list box is the best choice because not all list items need to be displayed on the screen. If screen space is limited but the user can select multiple items, an on-screen list box that displays only a few

12-A ERP USER INTERFACES DRIVE USERS NUTS

It used to take workers at Hydro Agri's Canadian Fertilizer stores about 20 seconds to process an order. After installing SAP, it now takes 90 seconds. Entering an order requires users to navigate through six screens to find the data fields that were on one screen in the old system. The problem became so critical during the spring planting rush that the project team installing the SAP system was pressed into service to take telephone orders.

Many other customers have complained about similar problems in SAP and the other leading ERP systems. Ontario-based Algoma Steel uses PeopleSoft and now has to use a dozen screens to enter employee data that were contained in two screens in their old custom-built personnel system. A-dec, a dental equipment maker based in Oregon, discovered the hard way that its Baan inventory system was still counting products that had been shipped in its on-hand inventories; the system required users to confirm the order shipments before the inventories were recorded as shipped—but didn't automatically take them to the confirmation screen.

So why have companies implemented ERP systems? The driving force behind most implementations was not to simplify the users' jobs but instead to improve the quality of the data, simplify system maintenance, and/or beat the Y2K problem. Ease of use wasn't a consideration, and what makes ERP systems so hard to use is that in the attempt to make them one-size-fits-all, developers had to include many little-used data items and processes. Instead of having a small custom system collecting only the data needed by the company itself (which could be condensed to fit one or two screens), companies now find themselves using a system designed to collect all the data items that any company could possibly use—data items that now require 6 to 12 screens.

Source: "ERP User Interfaces Drive Workers Nuts," *Computer World,* November 2, 1998. Vol. 32, No 4, pp. 1, 24.

QUESTION

Suppose you were a systems analyst at one of the leading ERP vendors (e.g., SAP, PeopleSoft, Baan). How could you apply the interface design principles and techniques in this chapter and Chapter 11 to improve the ease of use of your system?

items can be used. Both check boxes (for multiple selections) and radio buttons (for single selections) require that all list items be displayed at all times, thus requiring more screen space, but since they display all choices, they are often simpler for novice users.

Input Validation

All data entered into the system needs to be validated to ensure its accuracy. Input *validation* (also called *edit checks*) can take many forms. Ideally, computer systems

12-3 CAREER SERVICES

Consider a Web form that a student would use to input student information and resume information into a Career Services application at your university. First, sketch out how this form would look and iden- tify the fields that the form would include. What types of validity checks would you use to make sure that the correct information is entered into the system?

Type of Box	When to Use	Notes
Check box Presents a complete list of choices, each with a square box in front	When several items can be selected from a list of items	Check boxes are not mutually exclusive. Do not use negatives for box labels. Check box labels should be placed in some logical order, such as that defined by the business process, or failing that, alphabetically or most commonly used first. Use no more than 10 check boxes for any particular set of options. If you need more boxes, group them into subcategories.
Radio button Presents a complete list of mutually exclusive choices, each with a circle in front	When only one item can be selected from a set of mutually exclusive items	Use no more than six radio buttons in any one list; if you need more, use a drop-down list box. If there are only two options, one check box is usually preferred to two radio buttons, unless the options are not clear. Avoid placing radio buttons close to check boxes to prevent confusion between different selection lists.
On-screen list box Presents a list of choices in a box	Seldom or never—only if there is insufficient room for check boxes or radio buttons	This type of box can permit only one item to be selected (in which case it is an ugly version of radio buttons). This type of box can also permit many items to be selected (in which case it is an ugly version of check boxes), but users often fail to realize they can choose multiple items. This type of box permits the list of items to be scrolled, thus reducing the amount of screen space needed.
Drop-down list box Displays selected item in one-line box that opens to reveal list of choices	When there is insufficient room to display all choices	This type of box acts like radio buttons but is more compact. This type of box hides choices from users until it is opened, which can decrease ease of use; conversely, because it shelters novice users from seldom-used choices, it can improve ease of use. This type of box simplifies design if the number of choices is unclear, because it takes only one line when closed.
Combo box A special type of drop-down list box that permits user to type as well as scroll the list	Shortcut for experienced users	This type of box acts like drop-down list but is faster for experienced users when the list of items is long.
Slider Graphic scale with a sliding pointer to select a number	Entering an approximate numeric value from a large continuous scale	The slider makes it difficult for the user to select a precise number. Some sliders also include a number box to enable the user to enter a specific number.

FIGURE 12-9
Types of Selection Boxes

should not accept data that fails any important validation check to prevent invalid information from entering the system. However, this check can be very difficult, and invalid data often slips by data-entry operators and the users providing the information. It is up to the system to identify invalid data and either make changes or notify someone who can resolve the information problem.

There are six different types of validation checks: *completeness check, format check, range check, check digit check, consistency check,* and *database check* (see Figure 12-10). Every system should use at least one validation check on all entered data and ideally will perform all appropriate checks where possible.

Type of Validation	When to Use	Notes
Completeness check Ensures all required data have been entered	When several fields must be entered before the form can be processed	If required information is missing, the form is returned to the user unprocessed.
Format check Ensures data are of the right type (e.g., numeric) and in the right format (e.g., month, day, year)	When fields are numeric or contain coded data	Ideally, numeric fields should not permit users to type text data, but if this is not possible, the entered data must be checked to ensure it is numeric. Some fields use special codes or formats (e.g., license plates with three letters and three numbers) that must be checked.
Range check Ensures numeric data are within correct minimum and maximum values	With all numeric data, if possible	A range check permits only numbers between correct values. Such a system can also be used to screen data for "reasonableness"—e.g., rejecting birthdates prior to 1880 because people do not live to be a great deal over 100 years old (most likely, *1980* was intended).
Check digit check Check digits are added to numeric codes	When numeric codes are used	Check digits are numbers added to a code as a way of enabling the system to quickly validate correctness. For example, U.S. Social Security numbers and Canadian Social Insurance numbers assign only eight of the nine digits in the number. The ninth number—the check digit—is calculated using a mathematical formula from the first eight numbers. When the identification number is typed into a computer system, the system uses the formula and compares the result with the check digit. If the numbers don't match, then an error has occurred.
Consistency checks Ensure combinations of data are valid	When data are related	Data fields are often related. For example, someone's birth year should precede the year in which he or she was married. Although it is impossible for the system to know which data are incorrect, it can report the error to the user for correction.
Database checks Compare data against a database (or file) to ensure they are correct	When data are available to be checked	Data are compared against information in a database (or file) to ensure they are correct. For example, before an identification number is accepted, the database is queried to ensure that the number is valid. Because database checks are more "expensive" than the other types of checks (they require the system to do more work), most systems perform the other checks first and perform database checks only after the data have passed the previous checks.

FIGURE 12-10
Types of Input Validation

OUTPUT DESIGN

Outputs are the reports that the system produces, whether on the screen, on paper, or in other media, such as the Web. Outputs are perhaps the most visible part of any system because a primary reason for using an information system is to access the information that it produces.

Basic Principles

The goal of the output mechanism is to present information to users so that they can accurately understand it with the least effort. The fundamental principles for output design reflect how the outputs are used and ways to make it simpler for users to understand them.

Understand Report Usage   The first principle in designing reports is to understand how they are used. Reports can be used for many different purposes. In some cases—but not very often—reports are read cover to cover because all information is needed. In most cases, reports are used to identify specific items, or as references to find information, so the order in which items are sorted on the report or grouped within categories is critical. This is particularly important for the design of electronic or Web-based reports. Web reports that are intended to be read end-to-end should be presented in one long scrollable page, while reports that are used primarily to find specific information should be broken down into multiple pages, each with a separate link. Page numbers and the date on which the report was prepared also are important for reference reports.

The frequency of the report may also play an important role in its design and distribution. *Real-time reports* provide data that are accurate to the second or minute at which they were produced (e.g., stock market quotes). *Batch reports* are those that report historical information that may be months, days, or hours old, and they often provide additional information beyond the reported information (e.g., totals, summaries, historical averages).

There are no inherent advantages to real-time reports over batch reports. The only advantages lie in the time value of the information. If the information in a report is time critical (e.g., stock prices, air traffic control information), then real-time reports have value. This is particularly important because real-time reports often are expensive to produce, so unless they offer some clear business value, they may not be worth the extra cost.

Manage Information Load   Most managers get too much information, not too little. The goal of a well-designed report is to provide all of the information *needed* to support the task for which it was designed. This does not mean that the report needs to provide all the information *available* on the subject—just what the users decide they need in order to perform their jobs. In some cases, this may result in the production of several different reports on the same topics for the same users because they are used in different ways. This is a not bad design.

For users in westernized countries, the most important information should always be presented first in the top left corner of the screen or paper report. Information should be provided in a format that is usable without modification. The user should not need to re-sort the report's information, highlight critical information to find it more easily amid a mass of data, or perform additional mathematical calculations.

Minimize Bias   No analyst sets out to design a biased report. The problem with *bias* is that it can be very subtle; analysts can introduce it unintentionally. For example, bias can be introduced by the way lists of data are sorted, because entries that appear first in a list may receive more attention than those later in the list. Data often are sorted in alphabetical order, making those entries starting with the letter A more prominent. Data can be sorted in chronological order (or reverse chronological order), placing more emphasis on older (or most recent) entries. Data may be sorted by numeric value, placing more emphasis on higher or lower values. For example, consider a monthly sales report by state. Should the report be listed in alphabetical order by state name, in descending order by the amount sold, or in some other order (e.g., geographic region)? There are no easy answers to this question except to say that the order of presentation should match the way in which the information is used.

Graphical displays and reports can present particularly challenging design issues.[6] The scale on the axes in graphs is particularly subject to bias. For most types of graphs, the scale should always begin at zero; otherwise comparisons among values can be misleading. For example, does Figure 12-11 show that sales have increased by very much since 1993? The numbers in both charts are the same, but the visual images the two present are quite different. A glance at Figure 12-11*a* would suggest only minor changes, while a glance at Figure 12-11*b* might indicate that there have been some significant increases. In fact, sales have increased by a

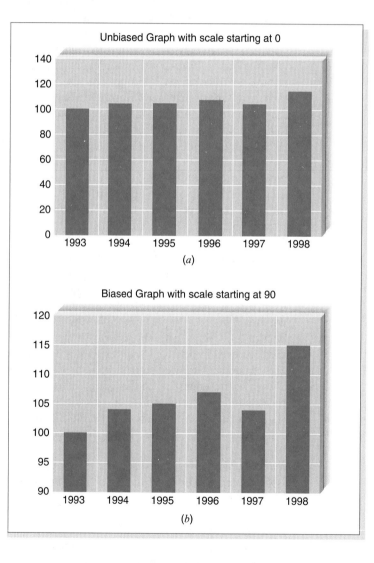

FIGURE 12-11
Bias in Graphs

[6] Some of the best books on the design of charts and graphical displays are by Edward R. Tufte, published by Graphics Press, Cheshire, Connecticut: *The Visual Display of Quantitative Information, Envisioning Information,* and *Visual Explanations.* Another good book is by William Cleveland, *Visualizing Data,* (Summit, NJ: Hobart Press, 1993).

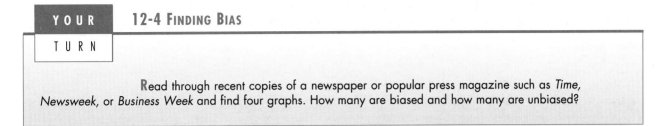

12-4 FINDING BIAS

Read through recent copies of a newspaper or popular press magazine such as *Time, Newsweek,* or *Business Week* and find four graphs. How many are biased and how many are unbiased?

total of 15% over five years, or 3% per year. Figure 12-11*a* presents the most accurate picture; Figure 12-11*b* is biased because the scale starts very close to the lowest value in the graph and misleads the eye into inferring that there have been major changes (i.e., more than doubling from "two lines" in 1993 to "five lines" in 1998). Figure 12-11*b* is the default graph produced by Microsoft Excel.

Types of Outputs

There are many different types of reports, including *detail reports, summary reports, turnaround documents* and *graphs.* (see Figure 12-12). Classifying reports is challenging because many reports have the characteristics of several different types. For example, some detail reports also produce summary totals, making them summary reports.

Media

Many different types of media are used to produce reports. The two dominant media in use today are paper and electronic with paper being the more traditional. For almost as long as there have been human organizations, there have been reports

12-B SELECTING THE WRONG STUDENTS

I helped a university department develop a small decision support system to analyze and rank students who applied to a specialized program. Some of the information was numeric and could easily be processed directly by the system (e.g., grade point average, standardized test scores). Other information required the faculty to make subjective judgments among the students (e.g., extracurricular activities, work experience). The users entered their evaluations of the subjective information via several data analysis screens in which the students were listed in alphabetical order.

To make the system "easier to use," it was designed so that the reports listing the results of the analysis were also presented in alphabetical order by student name rather than in order from the highest ranked student to the lowest ranked student. In a series of tests prior to installation, the users selected the wrong students to admit in 20% of the cases. They assumed, wrongly, that the students listed first were the highest ranked students and simply selected the first students on the list for admission. Neither the title on the report nor the fact that all the students' names were in alphabetical order made users realize that they had read the report incorrectly.   *Alan Dennis*

QUESTION

This system was biased because users assumed that the list of students implied ranking. Pretend that you are an analyst charged with minimizing bias in this application. Where else may you find bias in the application? How would you eliminate it?

Type of Reports	When to Use	Notes
Detail report Lists detailed information about all the items requested	When user needs full information about the items	This report is usually produced only in response to a query about items matching some criteria. This report is usually read cover to cover to aid understanding of one or more items in depth.
Summary report Lists summary information about all items	When user needs brief information on many items	This report is usually produced only in response to a query about items matching some criteria, but it can be a complete database. This report is usually read for the purpose of comparing several items to each other. The order in which items are sorted is important.
Turnaround document Outputs that "turn around" and become inputs	When a user (often a customer) needs to return an output to be processed	Turnaround documents are a special type of report that are both outputs and inputs. For example, most bills sent to consumers (e.g., credit card bills) provide information about the total amount owed and also contain a form that consumers fill in and return with payment.
Graphs Charts used in addition to and instead of tables of numbers	When users need to compare data among several items	Well-done graphs help users compare two or more items or understand how one has changed over time. Graphs are poor at helping users recognize precise numeric values and should be replaced by or combined with tables when precision is important. Bar charts tend to be better than tables of numbers or other types of charts when it comes to comparing values between items (but avoid three-dimensional charts that make comparisons difficult). Line charts make it easier to compare values over time, whereas scatter charts make it easier to find clusters or unusual data. Pie charts show proportions or the relative shares of a whole.

FIGURE 12-12
Types of Reports

on paper or similar media (e.g., papyrus, stone). Paper is permanent, easy to use, and accessible in most situations. It also is highly portable, at least for short reports.

Paper also has several rather significant drawbacks. It is inflexible. Once the report is printed, it cannot be sorted or reformatted to present a different view of the information. Likewise, if the information on the report changes, the entire report must be reprinted. Paper reports are expensive, hard to duplicate, and require considerable supplies (paper, ink) and storage space. Paper reports also are hard to move quickly through long distances (e.g., from a head office in Toronto to a regional office in Bermuda).

Many organizations are therefore shifting to electronic production of reports, whereby reports are "printed" but stored in electronic format on file servers or Web servers so that users can easily access them. Often the reports are available in more predesigned formats than their paper-based counterparts because the cost of producing and storing different formats is minimal. Electronic reports also can be produced on-demand as needed, and they enable the user to more easily search for certain words. Some users still print the electronic report on their own printers, but the reduced cost of electronic delivery over distances and the ease of enabling more users to access the reports than when they were only in paper form usually offsets the cost of local printing.

12-C CUTTING PAPER TO SAVE MONEY

One of the Fortune 500 firms with which I have worked had an 18-story office building for its world headquarters. It devoted two full floors of this building to nothing more than storing "current" paper reports (a separate warehouse was maintained outside the city for "archived" reports such as tax documents). Imagine the annual cost of office space in the headquarters building tied up in these paper reports. Now imagine how a staff member would gain access to the reports, and you can quickly understand the driving force behind electronic reports, even if most users end up printing them. Within one year of switching to electronic reports (for as many reports as practical), the paper report storage area was reduced to one small storage room.   *Alan Dennis*

QUESTION

What types of reports are most suited to electronic format? What types of reports are less suited to electronic reports?

APPLYING THE CONCEPTS AT CD SELECTIONS

In this section we continue with the interface design example started in Chapter 11. In Chapter 11, Alec developed use scenarios, the window navigation diagram, and interface standards (except for the interface template) for the Web-based processes. It is now time to complete the interface standards by designing the interface template, which will lead to interface prototyping and evaluation.

Interface Template Design

For the interface template, Alec decided on a simple, clean design that had a modern background pattern, with the CD Selections logo in the upper left corner. The template had two navigation areas: one menu across the top for navigation within the entire Web site (e.g., overall Web site home page, store locations) and one menu down the left edge for navigation within the Internet Sales System. The left edge menu contained the links to the top-level operations (see WND in Figure 11-11) as well as the "quick search" option. The center area of the screen is used for displaying forms and reports when the appropriate operation is chosen (see Figure 12-13).

At this point, Alec decided to seek some quick feedback on the interface structure and standards before investing time in prototyping the interface designs. Therefore, he met with Margaret Mooney, the project sponsor, and Chris Campbell, the consultant, to discuss the emerging design. Making changes at this point would be much simpler than after doing the prototype. Margaret and Chris had a few suggestions, so after the meeting Alec made the changes and moved into the design prototyping step.

Design Prototyping

Alec decided to develop a hypertext mark-up language (HTML) prototype of the system. The Internet Sales System was new territory for CD Selections and a strategic investment in a new business model, so it was important to make sure that no key issues were overlooked. The HTML prototype would provide the most detailed information and enable interactive evaluation of the interface.

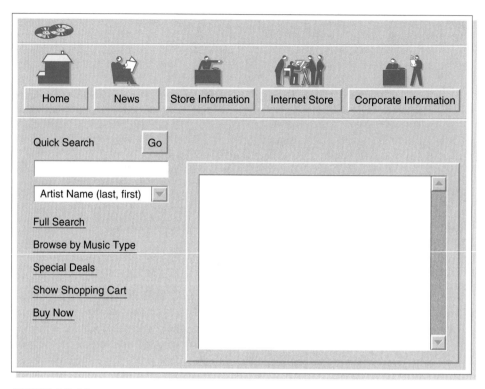

FIGURE 12-13
CD-Selections Interface Template for the Web portion of the Internet Sales system

In designing the prototype, Alec started with the home screen and gradually worked his way through all the screens. The process was very iterative, and he made many changes to the screens as he worked. Once he had an initial prototype designed, he posted it on CD Selections intranet and solicited comments from several friends with lots of Web experience. He revised it based on the comments he received. Figure 12-14 presents some screens from the prototype.

Interface Evaluation

The next step was interface evaluation. Alec decided on a two-phase evaluation. The first evaluation was to be an interactive evaluation (see Chapter 11) conducted by Margaret, her marketing managers, selected staff members, selected store managers, and Chris. They worked hands-on with the prototype and identified several ways to improve it. Alec modified the HTML prototype to reflect the changes suggested by the group and asked Margaret and Chris to review it again.

The second evaluation was another interactive evaluation, this time by a series of two focus groups of potential customers, one with little Internet experience and the other with extensive Internet experience. Once again, several minor changes were identified. Alec again modified the HTML prototype and asked Margaret and Chris to review it again. Once they were satisfied, the interface design was complete.

FIGURE 12-14
Sample Interfaces from the CD-Selections Design Prototype

Use-Case Name: *Place Order*	ID: *15*	Importance Level: *High*

Primary Actor: *The Browsing Customer*	Use-Case Type: *Detail, Real*

Stakeholders and Interests: *Customer – Wants to search web site to purchase CD*
EM Manager – Wants to maximize Customer satisfaction.

Brief Description: *This use case describes how customers can search the web site and place orders.*

Trigger: *Customer visits web site*

Type: *External*

Relationships:
Association: *Customer*
Include: *Checkout, Maintain Order*
Extend:
Generalization:

Normal Flow of Events:
1. *The Customer visits the Web site.*
2. *The System displays the Home Page*
   If the Customer wants to do a Full Search, execute S-1: Full Search
   If the Customer wants to Browse by Music Type, execute S-2: Browse by Music Type
   If the Customer wants to see any Special Deals, execute S-3: Special Deals
   If the Customer wants to see the contents of the Shopping Cart, execute S-4: Shopping Cart
   If the Customer wants to Buy Now, execute S-5: Buy Now
3. *The Customer leaves the site.*

Subflows:
S-1: Full Search
   1. *The Customer clicks the Full Search hyperlink*
   2. *The System displays the search type pop-up menu*
      If the Customer chooses an Artist search, execute S-1a: Artist List
      If the Customer chooses an Title search, execute S-1a: Title List
      If the Customer chooses an Composer search, execute S-1a: Composer List
S-1a: Artist List
   1. *The System displays the Artist List window in the Center Area of the Home Page.*
   2. *The Customer enters the Artist Name into the Search Item text box.*
   3. *The Customer presses the "Submit" button.*
   4. *The System executes S-2a: CD List.*
S-2a: CD List
   1. *The System displays the CD List hyperlink report.*
   2. *The Customer chooses a CD to review by clicking the CD link.*
   3. *The System executes S-2b: Display Basic Report*
   4. *Iterate over steps 2 and 3.*

Alternate/Exceptional Flows:

FIGURE 12-15
The Browsing Customer Real Use Case (partial listing only)

Navigation Design Documentation

The last step that Alec completed was to document the navigation design through the use of real use cases. To accomplish this, Alec gathered together the essential use case (see Figure 6-11), the use scenarios (see Figure 11-7), the window navigation diagram (see Figure 11-11), and the user interface prototype (see Figure 12-13 and 12-14). First, he copied the contents of the essential use case to the real use case. He changed the type from "detail, essential" to "detail, real" and the primary actor was specialized to "browsing customer" instead of simply "customer." Second, he wrote the specific set of steps and responses that described the interaction between the browsing customer and system. Figure 12-15 shows a partial listing of the steps in the Normal Flow of Events and Subflows sections of the real use case. Last, he repeated the steps for the hurry-up customer.

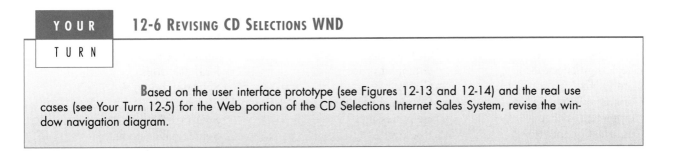

YOUR TURN

12-5 COMPLETING THE CD SELECTIONS REAL USE CASE

Based on the user interface prototype (see Figures 12-13 and 12-14), the use scenarios (see Figure 11-7), and the WND (see Figure 11-11) for the Web portion of the CD Selections Internet Sales System, complete the real use cases.

YOUR TURN

12-6 REVISING CD SELECTIONS WND

Based on the user interface prototype (see Figures 12-13 and 12-14) and the real use cases (see Your Turn 12-5) for the Web portion of the CD Selections Internet Sales System, revise the window navigation diagram.

SUMMARY

Navigation Design

The fundamental goal of the navigation design is to make the system as simple to use as possible by preventing the user from making mistakes, simplifying the recovery from mistakes, and using a consistent grammar order (usually object-action order). Command languages, natural languages, and direct manipulation are used in navigation, but the most common approach is menus (menu bar, drop-down menu, pop-up menu, tab menu, buttons and tool bars, and image maps). Error messages, confirmation messages, acknowledgment messages, delay messages, and help messages are common types of messages. Once the navigation design is agreed upon, it is documented in the form of window navigation diagrams and real use cases.

Input Design

The goal of the input mechanism is to simply and easily capture accurate information for the system, typically by using on-line or batch processing, capturing data at the source, and minimizing keystrokes. Input design includes both the design of input screens and all preprinted forms that are used to collect data before it is entered into the information system. There are many types of inputs, such as text fields, number fields, check boxes, radio buttons, on-screen list boxes, drop-down list boxes, and sliders. Most inputs are validated by using some combination of completeness checks, format checks, range checks, check digits, consistency checks, and database checks.

Output Design

The goal of the output mechanism is to present information to users so that they can accurately understand it with the least effort, usually by understanding how reports will be used and designing them to minimize information overload and bias. Output design means designing both screens and reports in other media, such as paper and the Web. There are many types of reports such as detail reports, summary reports, exception reports, turnaround documents, and graphs.

KEY TERMS

Acknowledgment message
Action-object order
Bar code reader
Batch processing
Batch report
Bias
Check box
Check digit check
Combo box
Command language
Completeness check
Confirmation message
Consistency check
Context-sensitive help
Database check
Data-entry operator
Default value
Delay message
Detail report
Direct manipulation
Drop-down list box
Drop-down menu

Edit check
Error message
Essential use cases
Exception report
Field
Field label
Format check
Grammar order
Graph
Graphical user interface (GUI)
Help message
Hot key
Image map
Input mechanism
Magnetic stripe readers
Menu
Natural language
Navigational mechanism
Number box
Object-action order
On-line processing
On-screen list box

Optical character recognition
Output mechanism
Pop-up menu
Radio button
Range check
Real-time information
Real-time report
Real use cases
Selection box
Slider
Smart card
Source data automation
Summary report
Tab menu
Text box
Transaction processing
Turnaround document
Use scenarios
Validation
Window navigation diagrams
    (WND)

QUESTIONS

1. Describe three basic principles of navigation design.
2. How can you prevent mistakes?
3. Explain the differences between object-action order and action-object order.
4. Describe four types of navigation controls.
5. Why are menus the most commonly used navigation control?
6. Compare and contrast four types of menus.
7. Under what circumstances would you use a drop-down menu versus a tab menu?
8. Under what circumstances would you use an image map versus a simple list menu?
9. Describe five types of messages.
10. What are the key factors in designing an error message?

11. What is context-sensitive help? Does your word processor have context-sensitive help?
12. How do the essential use case and real use case differ?
13. What is the relationship between essential use cases and use scenarios?
14. What is the relationship between real use cases and use scenarios?
15. Explain three principles in the design of inputs.
16. Compare and contrast batch processing and on-line processing. Describe one application that would use batch processing and one that would use on-line processing.
17. Why is capturing data at the source important?
18. Describe four devices that can be used for source data automation.

19. Describe five types of inputs.
20. Compare and contrast check boxes and radio buttons. When would you use one versus the other?
21. Compare and contrast on-screen list boxes and drop-down list boxes. When would you use one versus the other?
22. Why is input validation important?
23. Describe five types of input validation methods.
24. Explain three principles in the design of outputs.
25. Describe five types of outputs.

26. When would you use electronic reports rather than paper reports, and vice versa?
27. What three common mistakes do novice analysts make in navigation design?
28. What three common mistakes do novice analysts make in input design?
29. What three common mistakes do novice analysts make in output design?
30. How would you improve the form in Figure 11-5?

EXERCISES

A. Ask Jeeves (http://www.askjeeves.com) is an Internet search engine that uses natural language. Experiment with it and compare it to search engines that use key words.
B. Draw a window navigation diagram (WND) for the Your Turn Box 12-1 using the opposite grammar order from your original design (if you didn't do it do it, draw two WNDs, one in each grammar order). Which is "best"? Why?
C. In the Your Turn Box 12-1, you probably used menus. Design it again using a command language.
D. Design an interface template for Exercises D, E, F, and G in Chapter 11.
E. Develop an HTML prototype for Exercises D, E, F, and G in Chapter 11.
F. Develop a real use case based on Exercises D, E, F, and G in Chapter 11.
G. Design an interface template for Exercises H, I, J, and K in Chapter 11.

H. Develop an HTML prototype for Exercises H, I, J, and K in Chapter 11.
I. Develop a real use case based on Exercises H, I, J, and K in Chapter 11.
J. Design an interface template for Exercises L, M, N, and O in Chapter 11.
K. Develop an HTML prototype for Exercises L, M, N, and O in Chapter 11.
L. Develop a real use case based on Exercises L, M, N, and O Chapter 11.
M. Design an interface template for Exercises P, Q, R, S in Chapter 11.
N. Develop an HTML prototype for Exercises P, Q, R, S in Chapter 11.
O. Develop a real use case based on Exercises P, Q, R, S in Chapter 11.

MINICASES

1. The menu structure for Holiday Travel Vehicle's existing character-based system is shown below. Develop and prototype a new interface design for the system's functions using a graphical user interface. Also, develop a set of real use cases for your new interface. Assume the new system will need to include the same functions as those shown in the menus provided. Include any messages that will be produced as a user interacts with your interface (error, confirmation, status, etc.). Also, prepare a written summary that describes how your interface implements the principles of good interface design as presented in the textbook.

2 One aspect of the new system under development at Holiday Travel Vehicles will be the direct entry of the sales invoice into the computer system by the salesperson as the purchase transaction is being completed. In the current system, the salesperson fills out a paper form (shown on p. 367):

Design and prototype an input screen that will permit the salesperson to enter all the necessary information for the sales invoice. The following information

```
        Holiday Travel Vehicles                    Holiday Travel Vehicles

             Main Menu                               Sales Invoice Menu

         1 Sales Invoice                          1 Create Sales Invoice
         2 Vehicle Inventory                      2 Change Sales Invoice
         3 Reports                                3 Cancel Sales Invoice
         4 Sales Staff
                                              Type number of menu selection here:____
    Type number of menu selection here:____
```

```
        Holiday Travel Vehicles                    Holiday Travel Vehicles

          Vehicle Inventory Menu                       Reports Menu

     1 Create Vehicle Inventory Record           1 Commission Report
     2 Change Vehicle Inventory Record           2 RV Sales by Make Report
     3 Delete Vehicle Inventory Record           3 Trailer Sales by Make Report
                                                 4 Dealer Options Report
    Type number of menu selection here:____
                                              Type number of menu selection here:____
```

```
        Holiday Travel Vehicles

        Sales Staff Maintenance Menu

         1 Add Salesperson Record
         2 Change Salesperson Record
         3 Delete Salesperson Record

    Type number of menu selection here:____
```

may be helpful in your design process. Assume that Holiday Travel Vehicles sells recreational vehicles and trailers from four different manufacturers. Each manufacturer has a fixed number of names and models of RVs and trailers. For the purposes of your prototype, use this format:

Also, assume there are 10 different dealer options that could be installed on a vehicle at the customer's request. The company currently has 10 salespeople on staff.

Mfg-A	Name-1 Model-X		Mfg-C	Name-1 Model-X
Mfg-A	Name-1 Model-Y		Mfg-C	Name-1 Model-Y
Mfg-A	Name-1 Model-Z		Mfg-C	Name-1 Model-Z
Mfg-B	Name-1 Model-X		Mfg-C	Name-2 Model-X
Mfg-B	Name-1 Model-Y		Mfg-C	Name-3 Model-X
Mfg-B	Name-2 Model-X		Mfg-D	Name-1 Model-X
Mfg-B	Name-2 Model-Y		Mfg-D	Name-2 Model-X
Mfg-B	Name-2 Model-Z		Mfg-D	Name-2 Model-Y

Holiday Travel Vehicles
Sales Invoice

Invoice #: _____

Invoice Date: _____

Customer Name: _____

Address: _____

City: _____

State: _____

Zip: _____

Phone: _____

New RV/TRAILER
(circle one)

Name: _____

Model: _____

Serial #: _____ Year: _____

Manufacturer: _____

Trade-in RV/TRAILER
(circle one)

Name: _____

Model: _____

Year: _____

Manufacturer: _____

Options:	<u>Code</u>	<u>Description</u>	<u>Price</u>
	_____	_____	_____
	_____	_____	_____
	_____	_____	_____
	_____	_____	_____

Vehicle Base Cost: _____

Trade-in Allowance: _____

Total Options: _____

Tax: _____

License Fee: _____

Final Cost: _____

_____
(Salesperson Name)

_____
(Customer Signature)

PLANNING

ANALYSIS

DESIGN

- ☑ **Evolve Analysis Models**
- ☑ **Develop the Design Strategy**
- ☑ **Develop the Design**
- ☑ **Develop Infrastructure Design**
- ☑ **Develop Network Model**
- ☑ **Develop Hardware/Software Specification**
- ☑ **Develop Security Plan**
- ☑ **Develop Use Scenarios**
- ☑ **Design Interface Structure**
- ☑ **Design Interface Standards**
- ☑ **Design User Interface Template**
- ☑ **Design User Interface**
- ☑ **Evaluate User Interface**
- ☐ Select Object Persistence Format
- ☐ Map Objects to Object Persistence
- ☐ Optimize Object Persistence
- ☐ Size Object Persistence
- ☐ Specify Constraints
- ☐ Develop Contracts
- ☐ Develop Method Specifications

TASK CHECKLIST

PLANNING → ANALYSIS → DESIGN

CHAPTER 13

OBJECT PERSISTENCE DESIGN

A project team designs the object persistence component of the system using a two-step approach: selecting the format of the storage and then optimizing it to perform efficiently. This chapter first describes the different ways in which objects can be stored and several important characteristics that should be considered when choosing among object persistence formats. Because one of the most popular storage formats today is the relational database, the rest of this chapter focuses on the optimization of relational databases from both storage and access perspectives.

OBJECTIVES

- Become familiar with several object persistence formats.
- Be able to map problem domain objects to different object persistence formats.
- Be able to apply the steps of normalization to a relational database.
- Be able to optimize a relational database for object storage and access.
- Become familiar with indexes for relational databases.
- Be able to estimate the size of a relational database.

CHAPTER OUTLINE

IMPLEMENTATION

INTRODUCTION

As explained in Chapter 10, the work done by any application program can be divided into four general functions: data storage, data access logic, application logic, and presentation logic. Recall that the data storage and access logic reside on the data management layer, the application logic is on the problem domain layer, and the presentation logic is located on the human–computer interaction layer. The data storage component of the data management layer manages how data is stored and handled by the programs that run the system. This chapter describes how a project team designs the storage for objects (object persistence), on the data management layer, using a two-step approach: selecting the format of the storage and then optimizing it to perform efficiently.

Applications are of little use without the data that they support. How useful is a multimedia application that can't support images or sound? Why would someone log into a system to find information if it would take less time to locate the information manually? The design phase of the systems development life cycle (SDLC) includes two steps to object persistence design that decreases the chances of ending up with inefficient systems, long system response times, and users who cannot get to the information they need in the way they need it—all of which can affect the success of the project.

The first part of this chapter describes a variety of storage formats and explains how to select the appropriate one for your application. From a practical perspective, four basic types of formats can be used to store objects for application systems: files (sequential and random), object-oriented databases, object-relational databases, or relational databases.[1] Each type has certain characteristics that make it more appropriate for some types of systems over others.

Once the object persistence format is selected to support the system, it needs to be designed to optimize its processing efficiency, which is the focus of the second part of the chapter. One of the leading complaints by end users is that the final system is *too slow*. To avoid such complaints, project team members must allow time during the design phase to carefully make sure that the file or database performs as fast as possible. At the same time, the team must keep hardware costs down by minimizing the storage space that the application will require. The goals to maximize access to the objects and minimize the amount of space taken to store objects can conflict, and designing object persistence efficiency usually requires tradeoffs.

OBJECT PERSISTENCE FORMATS

There are four main types of object persistence formats: files (sequential and random access), object-oriented databases, object-relational databases, and relational databases. *Files* are electronic lists of data that have been optimized to perform a particular transaction. For example, Figure 13-1 shows a customer order file with information about customers' orders, in the form in which it is used, so that the information can be accessed and processed quickly by the system.

---

[1] There are other types of files, such as relative, indexed sequential, and multi-indexed sequential, and databases, such as hierarchical, network, and multidimensional. However, these formats typically are not used for object persistence.

Order Number	Date	Cust ID	Last Name	First Name	Amount	Tax	Total	Prior Customer	Payment Type
234	11/23/00	2242	DeBerry	Ann	$ 90.00	$ 5.85	$ 95.85	Y	MC
235	11/23/00	9500	Chin	April	$ 12.00	$ 0.60	$ 12.60	Y	VISA
236	11/23/00	1556	Fracken	Chris	$ 50.00	$ 2.50	$ 52.50	N	VISA
237	11/23/00	2242	DeBerry	Ann	$ 75.00	$ 4.88	$ 79.88	Y	AMEX
238	11/23/00	2242	DeBerry	Ann	$ 60.00	$ 3.90	$ 63.90	Y	MC
239	11/23/00	1035	Black	John	$ 90.00	$ 4.50	$ 94.50	Y	AMEX
240	11/23/00	9501	Kaplan	Bruce	$ 50.00	$ 2.50	$ 52.50	N	VISA
241	11/23/00	1123	Williams	Mary	$120.00	$ 9.60	$129.60	N	MC
242	11/24/00	9500	Chin	April	$ 60.00	$ 3.00	$ 63.00	Y	VISA
243	11/24/00	4254	Bailey	Ryan	$ 90.00	$ 4.50	$ 94.50	Y	VISA
244	11/24/00	9500	Chin	April	$ 24.00	$ 1.20	$ 25.20	Y	VISA
245	11/24/00	2242	DeBerry	Ann	$ 12.00	$ 0.78	$ 12.78	Y	AMEX
246	11/24/00	4254	Bailey	Ryan	$ 20.00	$ 1.00	$ 21.00	Y	MC
247	11/24/00	2241	Jones	Chris	$ 50.00	$ 2.50	$ 52.50	N	VISA
248	11/24/00	4254	Bailey	Ryan	$ 12.00	$ 0.60	$ 12.60	Y	AMEX
249	11/24/00	5927	Lee	Diane	$ 50.00	$ 2.50	$ 52.50	N	AMEX
250	11/24/00	2242	DeBerry	Ann	$ 12.00	$ 0.78	$ 12.78	Y	MC
251	11/24/00	9500	Chin	April	$ 15.00	$ 0.75	$ 15.75	Y	MC
252	11/24/00	2242	DeBerry	Ann	$132.00	$ 8.58	$140.58	Y	MC
253	11/24/00	2242	DeBerry	Ann	$ 72.00	$ 4.68	$ 76.68	Y	AMEX

FIGURE 13-1
Customer Order File

A *database* is a collection of groupings of information that are related to each other in some way (e.g., through common fields). Logical groupings of information could include such categories as customer data, information about an order, and product information. *A database management system* (DBMS) is software that creates and manipulates these databases (see Figure 13-2 for a relational database example). Such *end-user DBMSs* as Microsoft Access support small-scale databases that are used to enhance personal productivity, while *enterprise DBMSs,* such as DB2, Jasmine, and Oracle, can manage huge volumes of data and support applications that run an entire company. An end-user DBMS is significantly less expensive and easier for novice users to use than its enterprise counterpart, but it does not have the features or capabilities that are necessary to support mission critical or large-scale systems.

The next set of sections describe sequential and random access files, relational databases, object-relational databases, and object-oriented databases that can be used to handle a system's object persistence requirements. Finally, we describe a set of characteristics on which the different formats can be compared.

Sequential and Random Access Files

From a practical perspective, most object-oriented programming languages support sequential and random access files as part of the language.[2] In this section, we describe what sequential access and random access files are.[3] We also describe how sequential access and random access files are used to support an application. For

[2] For example, see the FileInputStream, FileOutputStream, and RandomAccessFile classes in the java.io package.

[3] For a more complete coverage of issues related to the design of files, see Owen Hanson, *Design of Computer Data Files* (Rockville, MD: Computer Science Press, 1982).

Customer

Cust ID	Last Name	First Name	Prior Customer
2242	DeBerry	Ann	Y
9500	Chin	April	Y
1556	Fracken	Chris	N
1035	Black	John	Y
9501	Kaplan	Bruce	N
1123	Williams	Mary	N
4254	Bailey	Ryan	Y
2241	Jones	Chris	N
5927	Lee	Diane	N

Tables related through Payment Type

Payment Type

Payment Type	Payment Desc
MC	Mastercard
VIS	VISA
AMEX	American Express

Order

Order Number	Date	Cust ID	Amount	Tax	Total	Payment Type
234	11/23/00	2242	$ 90.00	$ 5.85	$ 95.85	MC
235	11/23/00	9500	$ 12.00	$ 0.60	$ 12.60	VISA
236	11/23/00	1556	$ 50.00	$ 2.50	$ 52.50	VISA
237	11/23/00	2242	$ 75.00	$ 4.88	$ 79.88	AMEX
238	11/23/00	2242	$ 60.00	$ 3.90	$ 63.90	MC
239	11/23/00	1035	$ 90.00	$ 4.50	$ 94.50	AMEX
240	11/23/00	9501	$ 50.00	$ 2.50	$ 52.50	VISA
241	11/23/00	1123	$120.00	$ 9.60	$129.60	MC
242	11/24/00	9500	$ 60.00	$ 3.00	$ 63.00	VISA
243	11/24/00	4254	$ 90.00	$ 4.50	$ 94.50	VISA
244	11/24/00	9500	$ 24.00	$ 1.20	$ 25.20	VISA
245	11/24/00	2242	$ 12.00	$ 0.78	$ 12.78	AMEX
246	11/24/00	4254	$ 20.00	$ 1.00	$ 21.00	MC
247	11/24/00	2241	$ 50.00	$ 2.50	$ 52.50	VISA
248	11/24/00	4254	$ 12.00	$ 0.60	$ 12.60	AMEX
249	11/24/00	5927	$ 50.00	$ 2.50	$ 52.50	AMEX
250	11/24/00	2242	$ 12.00	$ 0.78	$ 12.78	MC
251	11/24/00	9500	$ 15.00	$ 0.75	$ 15.75	MC
252	11/24/00	2242	$132.00	$ 8.58	$ 140.58	MC
253	11/24/00	2242	$ 72.00	$ 4.68	$ 76.68	AMEX

Tables related through Cust ID

FIGURE 13-2
Customer Order Database

example, they can be used to support master files, look-up files, transaction files, audit files, and history files.

Sequential access files allow only sequential file operations to be performed, for example, read, write, and search. Sequential access files are very efficient for sequential operations, such as report writing. However, for random operations, such as finding or updating a specific object, they are very inefficient. On the average, 50% of the contents of a sequential access file will have to be searched through before finding the specific object of interest in the file. They come in two flavours: ordered and unordered.

An *unordered sequential access file* basically is an electronic list of information that is stored on disk. Unordered files are organized serially; that is, the order of the file is the order in which the objects are written to the file. Typically, new objects simply are added to the file's end.

Ordered sequential access files are placed into a specific sorted order, for example, in ascending order by customer number. However, there is overhead associated with keeping files in a particular sorted order. The file designer can keep the file in sorted order by always creating a new file each time a delete or addition occurs or he or she can keep track of the sorted order via the use of *pointers,* which are information about the location of the related record. A pointer is placed at the end of each record, and it "points" to the next record in a series or set. The underlying data/file structure in this case is a *linked list.*[4]

Random access files only allow random or direct file operations to be performed. This type of file is optimized for random operations, such as finding and updating a specific object. Random access files typically give a faster response time to find and update operations than any other type of file. However, since they do not support sequential processing, applications such as report writing are very inefficient. The various methods to implement random access files are beyond the scope of this book.[4]

At times it is necessary to be able to process files in both a sequential and random manner. One simple way to do this is to use a sequential file that contains a list of the keys (the field in which the file is to be kept in sorted order) and a random access file for the actual objects. This will minimize the cost of additions and deletions to a sequential file, while allowing the random file to be processed sequentially by simply passing the key to the random file to retrieve each object in sequential order. It also allows for fast random processing to occur by using only the random access file, thus optimizing the overall cost of file processing. However, if a file of objects needs to be processed in both a random and sequential manner, the developer should look toward using a database (relational, object-relational, or object-oriented) instead. There are many different application types of files, for example, master files, look-up files, transaction files, audit files, and history files. We discuss each application type next.

Master files store core information that is important to the business and, more specifically, to the application, such as order information or customer mailing information. They usually are kept for long periods of time, and new records are appended to the end of the file as new orders or new customers are captured by the system. If changes need to be made to existing records, programs must be written to update the old information.

Look-up files contain static values, such as a list of valid zip codes or the names of the U.S. states. Typically, the list is used for validation. For example, if a customer's mailing address is entered into a master file, the state name is validated against a look-up file that contains U.S. states to make sure that the operator entered the value correctly.

A *transaction file* holds information that can be used to update a master file. The transaction file can be destroyed after changes are added, or the file may be saved in case the transactions need to be accessed again in the future. Customer address changes, for one, are stored in a transaction file until a program is run that updates the customer address master file with the new information.

---

[4] For a more detailed look at the underlying data and file structures of the different types of files, see Mary E.S. Loomis, *Data Management and File Structures,* 2nd ed., (Englewood Cliffs, NJ: Prentice Hall, 1989) and Michael J. Folk and Bill Zoellick, *File Structures: A Conceptual Toolkit* (Reading, MA: Addison-Wesley, 1987).

For control purposes, a company might need to store information about how data changes over time. For example, as human resource clerks change employee salaries in a human resource system, the system should record the person who made the changes to the salary amount, the date, and the actual change that was made. An *audit file* records "before" and "after" images of data as it is altered so that an audit can be performed if the integrity of the data is questioned.

Sometimes files become so large that they are unwieldy, and much of the information in the file is no longer used. The *history file* (or archive file) stores past transactions (e.g., old customers, past orders) that are no longer needed by system users. Typically, the file is stored off-line, yet it can be accessed on an as-needed basis. Other files, such as master files, can then be streamlined to include only active or very recent information.

Relational Databases

The *relational database* is the most popular kind of database for application development today. It is based on collections of tables, with each table having a *primary key*—a field(s) whose value is unique for every row of the table. The tables are related to each other by placing the primary key from one table into the related table as a *foreign key* (see Figure 13-3).

Most *relational database management systems* (RDBMS) support *referential integrity,* or the idea of ensuring that values linking the tables together through the primary and foreign keys are valid and correctly synchronized. For example, if an order entry clerk using the tables in Figure 13-3 attempted to add order 254 for customer number 1111, he or she would have made a mistake because no customer with that number exists in the Customer table. If the RDBMS supported referential integrity, it would check the customer numbers in the Customer table; discover that the number 1111 was invalid; and return an error to the entry clerk. The clerk would then go back to the original order form and recheck the customer information. Can you imagine the problems that would occur if the RDBMS let the entry clerk add the order with the wrong information? There would be no way to track down the name of the customer for order 254.

Tables have a set number of columns and a variable number of rows that contain occurrences of data. *Structured query language* (SQL) is the standard language for accessing the data in the tables, and it operates on complete tables, as opposed to the individual rows in the tables. Thus, a query written in SQL is applied to all of the rows in a table all at once, which is different from a lot of

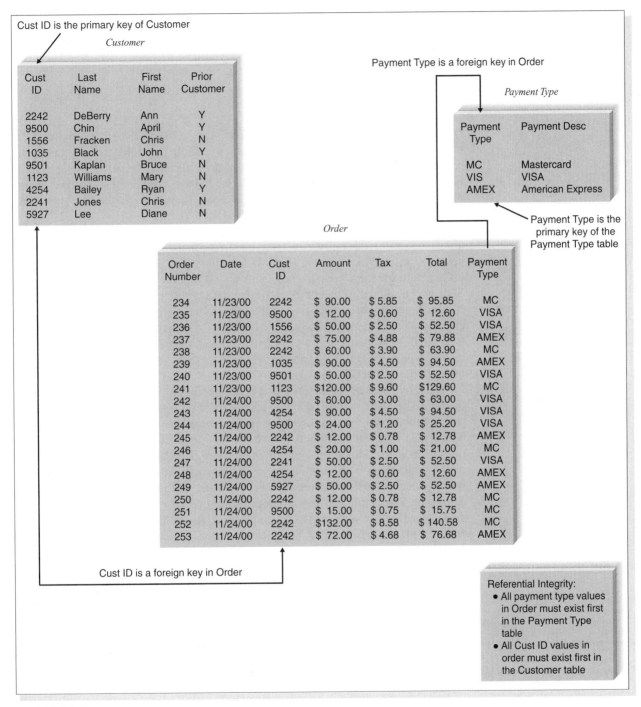

Cust ID is the primary key of Customer

Payment Type is a foreign key in Order

Payment Type is the primary key of the Payment Type table

Cust ID is a foreign key in Order

Referential Integrity:
- All payment type values in Order must exist first in the Payment Type table
- All Cust ID values in order must exist first in the Customer table

FIGURE 13-3
Relational Database

programming languages that manipulate data row by row. When queries must include information from more than one table, the tables first are *joined* together based on their primary key and foreign key relationships and treated as if they were one large table. Examples of RDBMS software are Microsoft Access, Oracle, DB2, Sybase, Informix, and Microsoft SQL Server.

To use a relational database management system to store objects, objects must be converted so that they can be stored in a table. From a design perspective, this entails mapping a UML class diagram to a relational database schema. We describe the mapping necessary later in this chapter.

Object-Relational Databases

Object-relational database management systems (ORDBMS) are relational database management systems with extensions to handle the storage of objects in the relational table structure. This is typically done through the use of user-defined types. For example, an attribute in a table could have a data type of image, a complex data type. In pure RDBMSs, attributes are limited to simple or atomic data types such as integers, floats, or characters.

ORDBMS, since they are simply extensions of their RDBMS counterpart, also have very good support for the typical data management operations that business has come to expect from RDBMSs, including an easy-to-use query language (SQL), authorization, concurrency-control, and recovery facilities. However, since SQL was designed to handle only simple data types, it too has been extended to handle complex object data. Currently, vendors have dealt with this issue in different ways. For example, some are moving toward supporting the next standard for SQL (SQL3), while others have simply added ad hoc extensions to the current version of SQL. DB2, Informix, and Oracle all have extensions that provide some level of support of objects.

Currently, many of the ORDBMSs on the market still do not support many of the object-oriented features that can appear in an object-oriented design, for example, multivalued attributes, repeating groups, or inheritance. One of the problems concerning inheritance support is that it is language dependent. For example, the manner in which Smalltalk supports inheritance is different from C++'s approach, which is different from Java's approach. As such, vendors currently must support many different versions of inheritance, one for each object-oriented language, or decide on a specific version and force developers to map their object-oriented design (and implementation) to their approach. As such, like RDBMSs, a mapping from a UML class diagram to an object-relational database schema is required. We describe the mapping necessary later in this chapter.

Object-Oriented Databases

The final type of database management system that we describe is the *object-oriented database management system (OODBMS)*. There have been two primary approaches to supporting object persistence within the OODBMS community: adding persistence extensions to an object-oriented programming language and creating an entirely separate database management system. In the case of OODBMS, the Object Data Management Group (ODMG) has begun the standardization process for defining (Object Definition Language—ODL), manipulating (Object Manipulating Language—OML), and querying (Object Query Language—OQL) objects in an OODBMS.[5]

With an OODBMS, collections of objects are associated with an extent. An *extent* is simply the set of instances associated with a particular class; that is, it is

[5] See www.odmg.org for more information.

the equivalent of a table in a RDBMS. Technically speaking, each instance of a class has a unique identifier assigned to it by the OODBMS: the *Object ID*. However, from a practical point of view, it is still a good idea to have a semantically meaningful primary key (even though from an OODBMS perspective this is unnecessary). Referential integrity is still very important. In an OODBMS, from the user's perspective, it looks as if the object is contained within the other object. However, the OODBMS actually keeps track of these relationships through use of the Object ID; as such, foreign keys are not necessary.[6]

OODBMSs provide support for some form of inheritance. However, as discussed above, inheritance tends to be language dependent. Currently, most OODBMSs are tied closely to either a particular *object-oriented programming language* (OOPL) or a set of OOPLs. Most originally supported either Smalltalk or C++. Today, many of the commercially available OODBMSs provide support for C++, Java, and Smalltalk.

OODBMSs also support the idea of *repeating groups* or *multivalued attributes*. These are supported through the use of *attribute sets* and *relationships sets*. RDBMSs explicitly do not allow for multivalued attributes or repeating groups. This is considered to be a violation of the first normal form (discussed later in this chapter) for relational databases. Some ORDBMSs do support repeating groups and multivalued attributes.

Until recently, OODBMSs have been used mainly to support multimedia applications or systems that involve complex data (e.g., graphics, video, and sound). Application areas such as computer-aided design and manufacturing (CAD/CAM), financial services, geographic information systems, health care, telecommunications, and transportation have been the most receptive to OODBMSs. They also are becoming popular technology for supporting electronic commerce, on-line catalogs, and large Web multimedia applications. Examples of pure OODBMSs include Gemstone, Jasmine, O2, Objectivity, ObjectStore, POET, and Versant.

Although pure OODBMS exist, most organizations currently invest in ORDBMS technology. The market for OODBMS is expected to grow, but its ORDBMS and RDBMS counterparts dwarf it. One reason for this situation is that there are many more experienced developers and tools in the RDBMS arena. Also, relational users find that OODBMS comes with a fairly steep learning curve.

Selecting an Object Persistence Format

Each of the file and database storage formats that have been presented has its strengths and weaknesses, and no one format is inherently better than the others are. In fact, sometimes a project team will choose multiple formats (e.g., a relational database for one, a file for another, and an object-oriented database for a third). Thus, it is important to understand the strengths and weaknesses of each format and when to use each one. Figure 13-4 presents a summary of the characteristics of each and the characteristics that can help identify when each type of database is more appropriate.

---

[6] Depending on the storage and updating requirements, it usually is a good idea to use a foreign key in addition to the Object ID. The Object ID has no semantic meaning. As such, in the case of needing to rebuild relationships between objects, Object IDs are difficult to validate. Foreign keys, on the other hand, should have some meaning outside of the DBMS.

	Sequential and Random Access Files	Relational DBMS	Object Relational DBMS	Object-Oriented DBMS
Major Strengths	Usually part of an object-oriented programming language Files can be designed for fast performance Good for short-term data storage	Leader in the database market Can handle diverse data needs	Based on established, proven technology (e.g., SQL) Able to handle complex data	Able to handle complex data Direct support for object-orientation
Major Weaknesses	Redundant data Data must be updated using programs (i.e., no manipulation or query language) No access control	Cannot handle complex data No support for object-orientation Impedance mismatch between tables and objects	Limited support for object-orientation Impedance mismatch between tables and objects	Technology is still maturing Skills are hard to find
Data Types Supported	Simple and Complex	Simple	Simple and Complex	Simple and Complex
Types of Application Systems Supported	Transaction processing	Transaction processing and decision making	Transaction processing and decision making	Transaction processing and decision making
Existing Storage Formats	Organization dependent	Organization dependent	Organization dependent	Organization dependent
Future Needs	Poor future prospects	Good future prospects	Good future prospects	Good future prospects

FIGURE 13-4

Comparison of Object Persistence Formats

Major Strengths and Weaknesses   The major strengths of files are as follows. Some support for sequential and random access files are normally part of an OOPL, files can be designed to be very efficient, and they are a good alternative for temporary or short-term storage. However, all file manipulation must be done through the OOPL. Files do not have any form of access control beyond that of the underlying operating system. Finally, in most cases, if files are used for permanent storage, redundant data most likely will result. This can cause many update anomalies.

RDBMs bring with them proven commercial technology. They are the leaders in the DBMS market. Furthermore, they can handle very diverse data needs. However, they cannot handle complex data types such as images. Therefore, all objects must be converted to a form that can be stored in tables composed of atomic or simple data. They provide no support for object-orientation. This lack of support causes an *impedance mismatch* between the objects contained in the OOPL and the data stored in the tables. An impedance mismatch refers to the amount of work, done by both the developer and DBMS, and the potential information loss that can occur when converting objects to a form that can be stored in tables.

Since ORDBMSs are typically object-oriented extensions to RDBMSs, they inherit the strengths of RDBMSs. They are based on established technologies, such as SQL, and unlike their predecessors, they can handle complex data types. However, they only provide limited support for object-orientation. The level of support varies among the vendors. Therefore, ORDBMSs also suffer from the impedance mismatch problem.

OODBMSs support complex data types and have the advantage of directly supporting object-orientation. Therefore, they do not suffer from the impedance mismatch that the previous DBMSs do. Even though the ODMG has released version 3.0 of their set of standards, the OODBMS community is still maturing. As such, this technology may still be too risky for some firms. The other major problem with OODBMS is the lack of skilled labor and the perceived steep learning curve of the RDBMS community.

Data Types Supported   The first issue is the type of data that will need to be stored in the system. Most applications need to store simple data types such as text, dates, and numbers, and all files and DBMSs are equipped to handle this kind of data. The best choice for simple data storage, however, is usually the RDBMS since the technology has matured over time and has continuously improved to handle simple data very effectively.

Increasingly, applications are incorporating complex data, such as video, images, or audio. ORDBMSs or OODBMSs are best able to handle data of this type. Complex data stored as objects can be manipulated much faster than with other storage formats.

Type of Application System   Many different kinds of application systems can be developed. *Transaction processing systems* are designed to accept and process many simultaneous requests (e.g., order entry, distribution, payroll). In transaction processing systems, the data continuously are updated by a large number of users, and the queries that are asked of the systems typically are predefined or targeted at a small subset of records (e.g., "List the orders that were backordered today." Or "What products did customer #1234 order on May 12, 2001?").

Another set of application systems are those designed to support decision making, such as *decision support systems* (DSS), *management information system* (MIS), *executive information systems* (EIS), and *expert systems* (ES). These decision-making support systems are built to support users who need to examine large amounts of read-only historical data. The questions that they ask are often ad hoc, and they include hundreds or thousands of records at a time (e.g., "List all customers in the West region who purchased a product costing more than $500 at least three times." or "What products had increased sales in the summer months that have not been classified as summer merchandise?").

Transaction processing and decision support systems thus have very different data storage needs. Transaction processing systems need data storage formats that are tuned for a lot of data updates and fast retrieval of predefined, specific questions. Files, relational databases, object-relational databases, and object-oriented databases can all support these kinds of requirements. By contrast, systems to support decision making are usually only reading data (not updating it), often in ad hoc ways. The best choices for these systems usually are DBMSs since these formats can be configured specially for needs that may be unclear and less apt to change the data.

Existing Storage Formats   The storage format should be selected primarily on the basis of the kind of data and application system being developed. However, project teams should consider the existing storage formats in the organization when making design decisions. In this way, they can better understand the technical skills that already exist and how steep the learning curve will be when the storage format is adopted. For example, a company that is familiar with RDBMS will have little

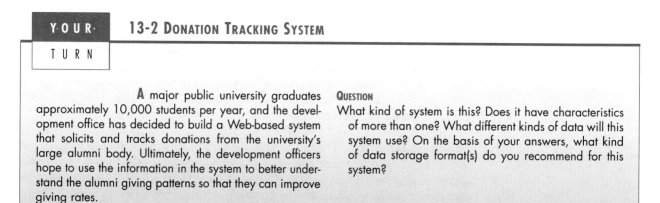

YOUR· **TURN**

13-2 DONATION TRACKING SYSTEM

A major public university graduates approximately 10,000 students per year, and the development office has decided to build a Web-based system that solicits and tracks donations from the university's large alumni body. Ultimately, the development officers hope to use the information in the system to better understand the alumni giving patterns so that they can improve giving rates.

QUESTION
What kind of system is this? Does it have characteristics of more than one? What different kinds of data will this system use? On the basis of your answers, what kind of data storage format(s) do you recommend for this system?

problem adopting a relational database for the project, whereas an OODBMS may require substantial developer training. In the latter situation, the project team may have to plan for more time to integrate the object-oriented database with the company's relational systems, or possibly consider moving toward an ORDBMS solution.

Future Needs   Not only should a project team consider the storage technology within the company, but they should also be aware of current trends and technologies that are being used by other organizations. A large number of installations of a specific type of storage format suggest that skills and products are available to support the format, and, therefore, the selection of that format is "safe." For example, it would probably be easier and less expensive to find RDBMS expertise when implementing a system than to find help with an OODBMS.

Other Miscellaneous Criteria   Other criteria that should be considered include cost, licensing issues, concurrency control, ease of use, security and access controls, version management, storage management, lock management, query management, language bindings, and APIs. We also should consider performance issues such as cache management, insertion, deletion, retrieval, and updating of complex objects. Finally, the level of support for object-orientation (such as objects, single inheritance, multiple inheritance, polymorphism, encapsulation and information hiding, methods, multivalued attributes, and repeating groups) is critical.

MAPPING PROBLEM DOMAIN OBJECTS TO OBJECT PERSISTENCE FORMATS

As described in the previous section, there are many different formats from which to choose to support object persistence. Each of the different formats can have some conversion requirements. Regardless of the object persistence format chosen, we suggest supporting primary keys and foreign keys by adding them to the problem domain at this point in time. However, this does imply that some additional processing is required. The developer will now have to set the value for the foreign key when adding the relationship to an object. In some cases, this overhead may be too costly. In those cases, this suggestion should be ignored. In the remainder of this section, we describe how to map the problem domain to the different object persistence formats. From a practical perspective, file formats are used only for temporary storage. As such, we do not consider them further.

We also recommend that the data management functionality specifics, such as retrieval and updating of data from the object storage, be included only in classes contained in the Data Management layer. This will ensure that the Data Management classes are dependent on the Problem Domain classes and not the other way around. Furthermore, this allows the design of Problem Domain classes to be independent of any specific implementation environment, thus increasing their portability and their potential for reuse. Like our previous recommendation, this one also implies additional processing. However, the increased portability and potential for reuse realized should more than compensate for the additional processing required.

Mapping Problem Domain Objects to an OODBMS Format

If we support object persistence with an OODBMS, the mappings between the problem domain objects and the data management objects tend to be a straightforward one-to-one mapping, that is, for each problem domain class, there will be a data management class. The data management class will contain all of the functionality required to manage the interaction between the OODBMS and the problem domain layer. For example, using the appointment system example from the previous chapters, the Patient class is associated with a DM-Patient class (see Figure 13-5). The Patient class essentially will be unchanged from the analysis phase. The DM-Patient class, however, will be a new class that is dependent on both the OODBMS used to store the instances of the Patient class and the Patient class itself. The DM-Patient class must be able to read from and write to the OODBMS. Otherwise, it will not be able to store and retrieve instances of the Patient class. Even though this does add overhead to the installation of the system, it allows the Problem Domain class to be

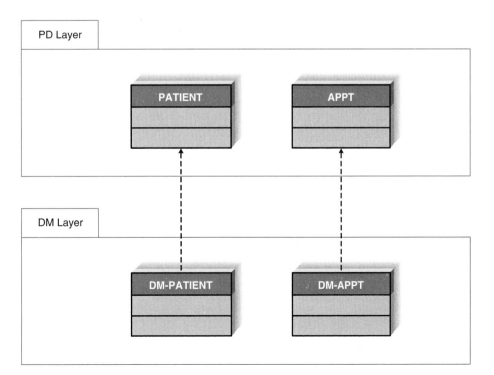

FIGURE 13-5

Appointment System Problem Domain and Data Management Layers independent of the OODBMS being used. As such, if at a latter point in time, another OODBMS or object persistence format is adopted, only the data management layer classes will have to be modified. This approach increases both the portability and the potential for reuse of the Problem Domain classes.

Even though we are implementing the data management layer in an OODBMS, depending on the level of support of inheritance in the OODBMS and the level of inheritance used in the problem domain classes, a mapping from the problem domain layer to the data management layer may be required. If multiple inheritance is used in the problem domain, but not supported by the OODBMS, then the multiple inheritance must be factored out of the data management classes. For each case of multiple inheritance, that is, more than one superclass, the following rules can be used to factor out the multiple inheritance effect.[7]

> RULE 1a: Create an instance of the additional data management superclass(es) and add an attribute that will contain the Object ID of the instance to each of the data management subclasses of the data management superclass(es). This is similar in concept to a foreign key in an RDBMS.
>
> or
>
> RULE 1b: Flatten the inheritance hierarchy of the data management layer by copying the attributes and methods of the additional data management super-class(es) down to all of the data management subclasses.

Figure 13-6 demonstrates the application of the above rules. Part A of Figure 13-6 portrays a simple example of multiple inheritance where Class 1 inherits from both SuperClass1 and SuperClass2. If we apply Rule 1a to Part A, we end up with the diagram in Part B where we have added the attribute SuperClass2 to the definition of Class1. This attribute will contain an instance of SuperClass2. If we apply Rule 1b to Part A, we end up with the diagram in Part C where all of the attributes of Super-Class2 have been copied down into Class1. In this later case, you may have to deal with the effects of inheritance conflicts. An inheritance conflict occurs when the name of an attribute (or method) in the subclass is the same as the name of one of the attributes (or methods) in one of the superclasses or the superclasses themselves could have a conflict among themselves. We describe this issue in more detail in Chapter 14.

In either case, additional processing will be required. In the first case, cascading of deletes will work, not only from the individual object to all of its elements, but also from the "superclass" instances to all of the "subclass" instances. Most OODBMSs do not support this type of deletion. However, to enforce the referential integrity of the system, it must be done. In the second case, there will be a lot of copying and pasting of the structure of the superclass to the subclasses. When a modification of the structure of the superclass is required, then the modification must be cascaded to all of the subclasses. Again, most OODBMSs do not support this type of modification cascading. Therefore, the developer must address it. However, multiple inheritance is rare in most business problems. As such, in most situations the above rule will never be necessary.

---

[7] The rules presented in this section are based on material in Ali Bahrami, *Object Oriented Systems Development Using the Unified Modeling Language* (New York: McGraw-Hill, 1999); Michael Blaha and William Premerlani, *Object-Oriented Modeling and Design for Database Applications* (Englewood Cliffs, NJ: Prentice Hall, 1998); Akmal B. Chaudri and Roberto Zicari, *Succeeding with Object Databases: A Practical Look at Today's Implementations with Java and XML* (New York: John Wiley and Sons, 2001); and Peter Coad and Edward Yourdon, *Object-Oriented Design* (Englewood Cliffs, NJ: Yourdon Press, 1991).

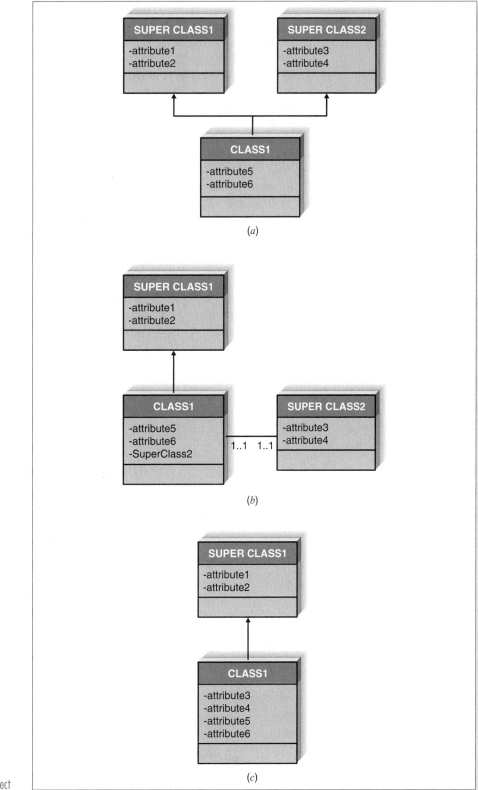

FIGURE 13-6
Factoring Out Multiple Inheritance Effect

When creating problem domain objects from data management objects, additional processing also will be required. The additional processing will be in the retrieval of the data management objects and taking their elements to create a problem domain object. Also, when storing the problem domain object, the conversion to a set of data management objects is required. Basically speaking, anytime that an interaction takes place between the database and the system, if multiple inheritance is involved and the OODBMS only supports single inheritance, a conversion between the two formats will be required.

Mapping Problem Domain Objects to an ORDBMS Format

If we support object persistence with an ORDBMS, then the mapping from the problem domain objects to the data management objects is much more involved. Depending on the level of support for object-orientation, different mapping rules are necessary. For our purposes, we assume that the ORDBMS supports Object IDs, multivalued attributes, and stored procedures. However, we assume that the ORDBMS does not provide any support for inheritance: Based on these assumptions, Figure 13-7 lists a set of rules that can be used to design the mapping from the problem domain objects to the ORDBMS-based data management layer tables.

First, all problem domain classes, abstract, concrete, and associative, must be mapped to data management classes and to the tables in the ORDBMS. For example in Figure 13-8, The Patient class has been mapped to both the DM-Patient class and Patient Table.

Rule 1: Map all problem domain classes (abstract, concrete, and associative) to data management classes and ORDBMS tables.

Rule 2: Map single valued attributes to columns of the ORDBMS tables and create a set of access methods in the data management classes.

Rule 3: Map methods and derived attributes to stored procedures or to program modules.

Rule 4: Map single-valued aggregation and association relationships to a column that can store an Object ID. Do this for both sides of the relationship.

Rule 5: Map multivalued attributes to a column that can contain a set of values.

Rule 6: Map repeating groups of attributes to a new table and create a one-to-many association from the original table to the new one.

Rule 7: Map multivalued aggregation and association relationships to a column that can store a set of Object IDs. Do this for both sides of the relationship.

Rule 8: For aggregation and association relationships of mixed type (one-to-many or many-to-one), on the single-valued side (1..1 or 0..1) of the relationship, add a column that can store a set of Object IDs. The values contained in this new column will be the Object IDs from the instances of the class on the multivalued side. On the multivalued side (1..* or 0..*), add a column that can store a single Object ID that will contain the value of the instance of the class on the single-valued side.

For generalization/inheritance relationships

Rule 9a: Add a column(s) to the table(s) that represents the subclass(es) that will contain an Object ID of the instance stored in the table that represents the superclass. Do this for each superclass.

OR

Rule 9b: Flatten the inheritance hierarchy by copying the superclass attributes and methods down to all of the subclasses.

FIGURE 13-7

Mapping Problem Domain Objects to ORDBMS Schema

Problem Domain Classes

Data Management Classes

ORDBMS Tables

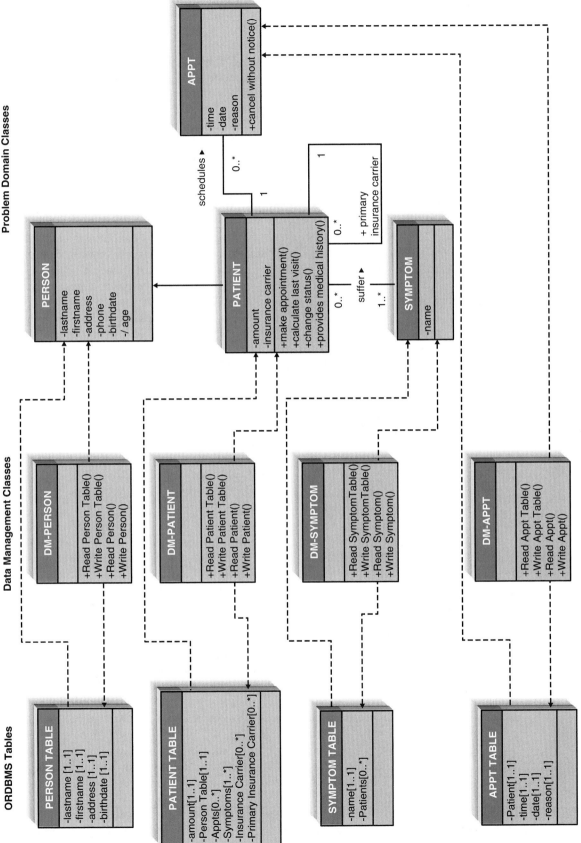

FIGURE 13-8
Mapping Problem Domain Objects to ORDBMS Schema Example

Second, single-valued attributes should be mapped to columns in the ORDBMS tables and to a set of access methods in the data management classes. Again, referring to Figure 13-8, we see that the amount attribute of the Patient class has been included in the Patient Table class and that a set of methods have been included in the DM-Patient class to handle the conversion from the problem domain class (Patient) to the ORDBMS table (PatientTable)—PDReadAmount, PDWriteAmount, TableReadAmount, and TableWriteAmount.

Third, depending on the level of support of stored procedures, the methods and derived attributes should be mapped either to stored procedures or program modules.

Fourth, single-valued (one-to-one) aggregation and association relationships should be mapped to a column that can store an Object ID. This should be done for both sides of the relationship.

Fifth, multivalued attributes should be mapped to columns that can contain a set of values. For example in Figure 13-8, since a patient may have more than one insurance carrier, the insurance carrier attribute in the Patient class may contain multiple values. As such, in the Patient Table, a multiplicity has been added to the insurance carrier attribute to portray this fact.

The sixth mapping rule addresses repeating groups of attributes in a problem domain object. In this case, the repeating group of attributes should be used to create a new table in the data management layer. Furthermore, it may imply a missing class in the problem domain layer. Normally, when a set of attributes repeat together as a group, it implies a new class. Finally, you should create a one-to-many association from the original table to the new one.

The seventh rule supports mapping multivalued (many-to-many) aggregation and association relationships to columns than can store a set of Object IDs. Basically, this is a combination of the fourth and fifth rules. Like the fourth rule, this should be done for both sides of the relationships. For example in Figure 13-8, the Symptom Table has a multivalued attribute that can contain multiple Object IDs to Patient Table objects and Patient Table has a multivalued attribute that can contain multiple Object IDs to Symptom Table objects.

The eighth rule combines the intentions of rules 4 and 7. In this case, the rule maps one-to-many and many-to-one relationships. On the single-valued side (1..1 or 0..1) of the relationship, a column that can store a set of Object IDs from the table on the multivalued side (1..* or 0..*) of the relationship should be added. On the multivalued side, a column should be added to the table that can store an Object ID from an instance stored in the table on the single-valued side of the relationship. For example in Figure 13-8, the Patient Table has a multivalued attribute that can contain multiple Object IDs to Appt Table objects while the Appt Table has a single-valued attribute that can contain an Object ID to Patient Table objects.

The ninth, and final, rule deals with the lack of support for generalization and inheritance. In this case, there are two different approaches. These approaches virtually are identical to the rules described with OODBMS object persistence formats above. In one case, a column is added to each table that represents a subclass for each of the subclass's superclasses. The columns will be used to store the Object IDs of the instances stored in the tables that represent each of the subclass's superclasses. For example in Figure 13-8, the Patient Table contains an attribute that can contain an Object ID for a Person Table object. In the other case, the inheritance hierarchy is flattened.

Of course, additional processing will be required whenever an interaction takes place between the database and the system. Every time an object must be created, retrieved from the database, updated, or deleted, the mapping between the problem domain and data management layers must be used to convert between the two different formats. The only other choice is to modify the problem domain objects. However, this can cause problems between the problem domain layer and the system architecture and human–computer interface layers. Generally speaking, the cost of conversion between the data management layer and the problem domain layer will be more than offset by the savings in development time associated with (1) the interaction between the problem domain and system architecture and human–computer interaction layers, and (2) the ease of maintenance of a semantically clean problem domain layer. In the long run, due to the necessary conversions, the development and production cost of using an OODBMS may be less than the development and production cost of having the data management layer implemented in an ORDBMS.

Mapping Problem Domain Objects to an RDBMS Format

If we support object persistence with an RDBMS, then the mapping from the problem domain objects to the data management objects is similar to the mapping to an ORDBMS. However, the assumptions made for an ORDBMS are no longer valid. Figure 13-9 lists a set of rules that can be used to design the mapping from the problem domain objects to the RDBMS-based data management layer tables.

The first four rules are basically the same set of rules that are used to map problem domain objects to ORDBMS-based data management objects. First, all problem domain classes, abstract, concrete, and associative, must be mapped to data management classes that represent tables in the RDBMS. Second, single-valued attributes should be mapped to columns in the RDBMS table and a set of access methods should be created in the data management classes. Third, methods and derived attributes should be mapped either to stored procedures or program modules, depending on the complexity of the method. Fourth, single-valued (one-to-one) aggregation and association relationships are mapped to columns that can store the foreign keys of the related tables. This should be done for both sides of the relationship.

The fifth rule addresses multivalued attributes and repeating groups of attributes in a problem domain object. In these cases, the attributes should be used to create new tables in the data management layer. Furthermore, as in the ORDBMS mappings, repeating groups of attributes may imply missing classes in the problem domain layer. As such, a new problem domain class may be required. Finally, you should create a one-to-many or zero-to-many association from the original table to the new one.

The sixth rule supports mapping multivalued (many-to-many) aggregation and association relationships to a new table that relates the two original tables together. In this case, the new table should contain foreign keys back to the original tables.

The seventh rule addresses one-to-many and many-to one relationships. With these types of relationships, the multivalued side (0..* or 1..*) should be mapped to a column in its table that can store a foreign key back to the single-valued side (0..1 or 1..1).

The eighth, and final, rule deals with the lack of support for generalization and inheritance. As in the case of an RDBMS, there are two different approaches. These approaches virtually are identical to the rules described with OODBMS and ORDBMS (Rule 9) object persistence formats. The first approach is to add a

> Rule 1: Map all problem domain classes (abstract, concrete, and associative) to data management classes and RDBMS tables.
>
> Rule 2: Map single-valued attributes to columns of the RDBMS tables and create a set of access methods in the data management classes.
>
> Rule 3: Map methods and derived attributes to stored procedures or to program modules.
>
> Rule 4: Map single-valued aggregation and association relationships to a column that can store the key of the related table; that is, add a foreign key to the table. Do this for both sides of the relationship.
>
> Rule 5: Map multivalued attributes and repeating groups to new tables and create a one-to-many association from the original table to the new ones.
>
> Rule 6: Map multivalued aggregation and association relationships to a new associative table that relates the two original tables together. Copy the primary key from both original tables to the new associative table, that is, add foreign keys to the table.
>
> Rule 7: For aggregation and association relationships of mixed type, copy the primary key from the single-valued side (1..1 or 0..1) of the relationship to a new column in the table on the multi-valued side (1..* or 0..*) of the relationship that can store the key of the related table; that is, add a foreign key to the table on the multivalued side of the relationship.
>
> For generalization/inheritance relationships
>
> Rule 8a: Add a column to the table(s) that represents the subclass(es) that will contain the primary key of the table that represents the superclass; that is, add a foreign key to the subclass tables. Do this for each superclass.
>
> or
>
> Rule 8b: Flatten the inheritance hierarchy by copying the superclass attributes and methods down to all of the subclasses.

FIGURE 13-9
Mapping Problem Domain Objects to RDBMS Schema

column to each table that represents a subclass for each of the subclass's super-classes. The columns will be used to store the foreign keys of the instances stored in the tables that represent each of the subclass's superclasses. Conversely, the inheritance hierarchy can be flattened.

As in the case of the ORDBMS approach, additional processing will be required anytime an interaction takes place between the database and the system. Every time an object must be created, retrieved from the database, updated, or deleted, the mapping between the problem domain and data management layers must be used to convert between the two different formats. In this case, a great deal of additional processing will be required. As such, it is recommended that OODBMS and ORDBMS approaches be used when possible. However, from a practical point of view, since RDBMSs are by far the most popular format in the marketplace, it is more likely that you will use a RDBMS for storage of objects than the other approaches. As such, we will focus on how to optimize the RDBMS format for object persistence in the remainder of this chapter.

OPTIMIZING RDBMS-BASED OBJECT STORAGE

Once the object persistence format is selected, the second step is to optimize the data management layer for processing efficiency. The methods of optimization will vary based on the format that you select; however, the basic concepts will remain the same. Once you understand how to optimize a particular type of object persistence, you will have some idea as to how to approach the optimization of other formats. This section focuses on the optimization of the most popular storage format: relational databases.

There are two primary dimensions in which to optimize a relational database: for storage efficiency and for speed of access. Unfortunately, these two goals often conflict because the best design for access speed may take up a great deal of storage space as compared to other less speedy designs. This section describes how to optimize the data management layer for storage efficiency using a process called normalization. The next section presents design techniques, such as denormalization and indexing, which can speed up the performance of the system. Ultimately, the project team will go through a series of tradeoffs until the ideal balance of the two optimization dimensions is reached. Finally, the project team must estimate the size of the data storage needed to ensure there is enough capacity on the server(s).

Optimizing Storage Efficiency

The most efficient tables in a relational database in terms of storage space have no redundant data and very few null values because the presence of these suggest that space is being wasted (and more data to store means higher data storage hardware costs). For example, notice that the table in Figure 13-10 repeats customer information, such as name and state, each time a customer places an order, and it contains many null values in the product-related columns. These nulls occur whenever a customer places an order for less than three items (the maximum number on an order).

In addition to wasting space, redundancy and null values also allow more room for error and increase the likelihood that problems will arise with the integrity of the data. What if customer 1035 moved from Maryland to Georgia? In the case of Figure 13-10, a program must be written to ensure that all instances of that customer are updated to show "GA" as the new state of residence. If some of the instances are overlooked, then the table will contain an *update anomaly* whereby some of the records contain the correctly updated value for state and other records contain the old information.

Nulls threaten data integrity because they are difficult to interpret. A blank value in the Order table's product fields could mean (1) the customer did not want more than one or two products on his or her order, (2) the operator forgot to enter in all three products on the order, (3) or the customer canceled part of the order and the products were deleted by the operator. It is impossible to be sure of the actual meaning of the nulls.

For both of these reasons—wasted storage space and data integrity threats—project teams should remove redundancy and nulls from the table. During the design phase, the class diagram is used to examine the design of the data management layer and optimize it for storage efficiency. If you follow the modeling instructions and guidelines that were presented in Chapter 7, you will have little trouble creating a design that is highly optimized in this way because a well-formed logical data model does not contain redundancy or many null values.

Sometimes, however, a project team needs to start with a model that was poorly constructed or with one that was created for files or a nonrelational type of format. In these cases, the project team should follow a series of steps that serve to check the model for storage efficiency. These steps make up a process called normalization.[8]

---

[8] Normalization also can be performed on the problem domain layer. However, the normalization process should only be used on the problem domain layer to uncover missing classes. Otherwise, optimizations that have nothing to do with the semantics of the problem domain can creep into this layer.

FIGURE 13-10
Optimizing Storage

ORDER

-Order Number : unsigned long
-Date : Date
-Cust ID : unsigned long
-Last Name : String
-First Name : String
-State : String
-Tax Rate : float
-Product 1 Number : unsigned long
-Product 1 Desc : String
-Product 1 Price : double
-Product 1 Qty : unsigned long
-Product 2 Number : unsigned long
-Product 2 Desc : String
-Product 2 Price : double
-Product 2 Qty : unsigned long
-Product 3 Number : unsigned long
-Product 3 Desc : String
-Product 3 Price : double
-Product 3 Qty : unsigned long

Sample Records:

Order Number	Date	Cust ID	Last Name	First Name	State	Tax Rate	Product 1 Number	Product 1 Desc	Product 1 Number	Product 1Qty	Product 2 Number	Product 2 Desc	Product 2 Price	Product 2 Qty	Product Numb
239	11/23/00	1035	Black	John	MD	0.05	555	Cheese Tray	$ 45.00	2					
260	11/24/00	1035	Black	John	MD	0.05	444	Wine Gift Pack	$ 60.00	1					
273	11/27/00	1035	Black	John	MD	0.05	222	Bottle Opener	$ 12.00	1					
241	11/23/00	1123	Williams	Mary	CA	0.08	444	Wine Gift Pack	$ 60.00	2					
262	11/24/00	1123	Williams	Mary	CA	0.08	222	Bottle Opener	$ 12.00	2					
287	11/27/00	1123	Williams	Mary	CA	0.08	222	Bottle Opener	$ 12.00	2					
290	11/30/00	1123	Williams	Mary	CA	0.08	555	Cheese Tray	$ 45.00	3					
234	11/23/00	2242	DeBerry	Ann	DC	0.065	555	Cheese Tray	$ 45.00	2	444	Wine Gift Pack	$ 60.00	1	
237	11/23/00	2242	DeBerry	Ann	DC	0.065	111	Wine Guide	$ 15.00	1					111
238	11/23/00	2242	DeBerry	Ann	DC	0.065	444	Wine Gift Pack	$ 60.00	1					
245	11/24/00	2242	DeBerry	Ann	DC	0.065	222	Bottle Opener	$ 12.00	1					
250	11/24/00	2242	DeBerry	Ann	DC	0.065	222	Bottle Opener	$ 12.00	1	444	Wine Gift Pack	$ 60.00	2	
252	11/24/00	2242	DeBerry	Ann	DC	0.065	222	Bottle Opener	$ 12.00	1	444	Wine Gift Pack	$ 60.00	1	
253	11/24/00	2242	DeBerry	Ann	DC	0.065	222	Bottle Opener	$ 12.00	1					
297	11/30/00	2242	DeBerry	Ann	DC	0.065	333	Jams &Jellies	$ 20.00	2					
243	11/24/00	4254	Bailey	Ryan	MD	0.05	555	Cheese Tray	$ 45.00	2					
246	11/24/00	4254	Bailey	Ryan	MD	0.05	333	Jams &Jellies	$ 20.00	3	333	Jams &Jellies	$ 20.00	2	
248	11/24/00	4254	Bailey	Ryan	MD	0.05	222	Bottle Opener	$ 12.00	1					
235	11/23/00	9500	Chin	April	KS	0.05	222	Bottle Opener	$ 12.00	1					
242	11/23/00	9500	Chin	April	KS	0.05	333	Jams &Jellies	$ 20.00	3					
244	11/24/00	9500	Chin	April	KS	0.05	222	Bottle Opener	$ 12.00	2					
251	11/24/00	9500	Chin	April	KS	0.05	111	Wine Guide	$ 15.00	2					

Redundant data

Null cells

Normalization is a process whereby a series of rules are applied to the data management layer to determine how well formed it is (see Figure 13-11). These rules help analysts identify tables that are not represented correctly. Here, we describe three normalization rules that are applied regularly in practice. Figure 13-10 shows a model in 0 Normal Form, which is an unnormalized model before the normalization rules have been applied.

A model is in *first normal form* (1NF) if it does not lead to *multivalued fields,* fields that allow a set of values to be stored, or *repeating fields,* fields that repeat within a table to capture multiple values. The rule for 1NF states that all tables must contain the same number of columns (i.e., fields) and that all of the columns must contain a single value. Notice that the model in Figure 13-10 violates 1NF since it causes product number, description, price, and quantity to repeat three times for each order in the table. The resulting table has many records that contain nulls in the product-related columns, and orders are limited to three products because there is no room to store information for more.

A much more efficient design (and one that conforms to 1NF) leads to a separate table to hold the repeating information; to do this, we create a separate table on the model to capture product order information. A zero-to-many relationship would then exist between the two tables. As shown in Figure 13-12, the new design eliminates nulls from the Order table and supports an unlimited number of products that can be associated with an order.

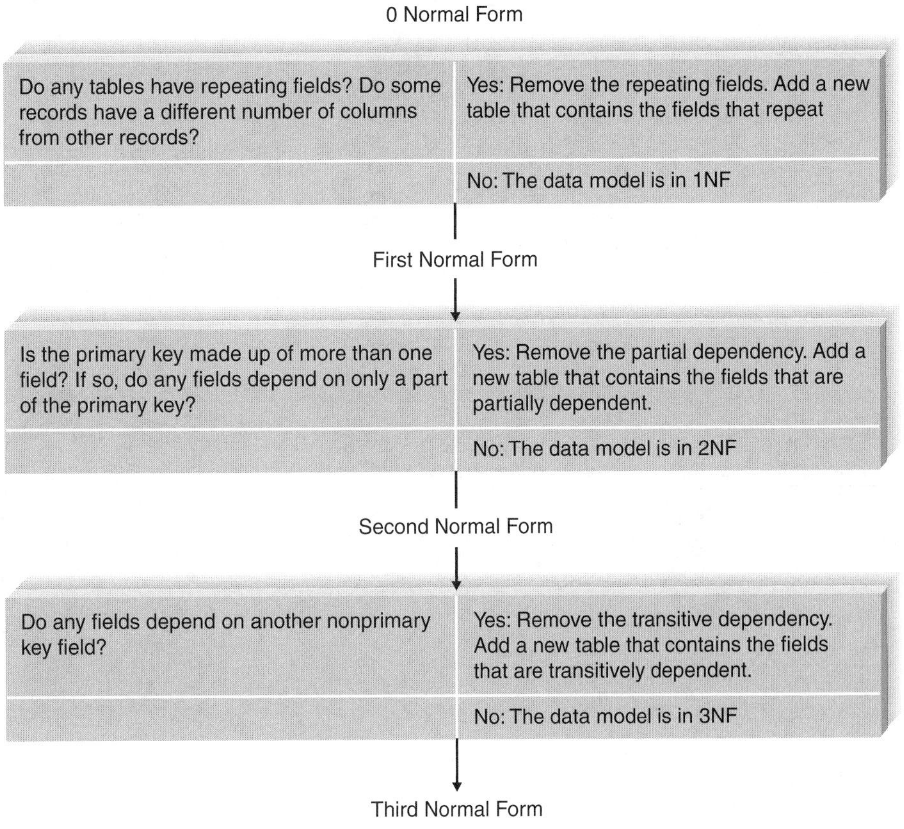

0 Normal Form

Do any tables have repeating fields? Do some records have a different number of columns from other records?	Yes: Remove the repeating fields. Add a new table that contains the fields that repeat
	No: The data model is in 1NF

First Normal Form

Is the primary key made up of more than one field? If so, do any fields depend on only a part of the primary key?	Yes: Remove the partial dependency. Add a new table that contains the fields that are partially dependent.
	No: The data model is in 2NF

Second Normal Form

Do any fields depend on another nonprimary key field?	Yes: Remove the transitive dependency. Add a new table that contains the fields that are transitively dependent.
	No: The data model is in 3NF

Third Normal Form

FIGURE 13-11
The Steps of Normalization

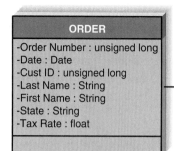

Note: Order Number will serve as part of the primary key of Order

Note: Cust ID also will serve as part of the primary key of Order

Note: Order Number will serve as part of the primary key of Product Order

Note: Order Number will serve as part of the primary key of Product Order

Note: Order Number also will serve as a foreign key in Product Order

Sample Records:

Order Table

Order 237 has 2 products

Product Order Table

Order Number	Date	Cust ID	Last Name	First Name	State	Tax Rate
239	11/23/00	1035	Black	John	MD	0.05
260	11/24/00	1035	Black	John	MD	0.05
273	11/27/00	1035	Black	John	MD	0.05
241	11/23/00	1123	Williams	Mary	CA	0.08
262	11/24/00	1123	Williams	Mary	CA	0.08
287	11/27/00	1123	Williams	Mary	CA	0.08
290	11/30/00	1123	Williams	Mary	CA	0.08
234	11/23/00	2242	DeBerry	Ann	DC	0.065
237	11/23/00	2242	DeBerry	Ann	DC	0.065
238	11/23/00	2242	DeBerry	Ann	DC	0.065
245	11/24/00	2242	DeBerry	Ann	DC	0.065
250	11/24/00	2242	DeBerry	Ann	DC	0.065
252	11/24/00	2242	DeBerry	Ann	DC	0.065
253	11/24/00	2242	DeBerry	Ann	DC	0.065
297	11/30/00	2242	DeBerry	Ann	DC	0.065
243	11/24/00	4254	Bailey	Ryan	MD	0.05
246	11/24/00	4254	Bailey	Ryan	MD	0.05
248	11/24/00	4254	Bailey	Ryan	MD	0.05
235	11/23/00	9500	Chin	April	KS	0.05
242	11/23/00	9500	Chin	April	KS	0.05
244	11/24/00	9500	Chin	April	KS	0.05
251	11/24/00	9500	Chin	April	KS	0.05

Order Number	Product Number	Product Desc	Product Price	Product Qty
239	555	Cheese Tray	$ 45.00	2
260	444	Wine Gift Pack	$ 60.00	1
273	222	Bottle Opener	$ 12.00	1
241	444	Wine Gift Pack	$ 60.00	2
262	222	Bottle Opener	$ 12.00	2
287	222	Bottle Opener	$ 12.00	2
290	555	Cheese Tray	$ 45.00	3
234	555	Cheese Tray	$ 45.00	2
237	111	Wine Guide	$ 15.00	1
237	444	Wine Gift Pack	$ 60.00	1
238	444	Wine Gift Pack	$ 60.00	1
245	222	Bottle Opener	$ 12.00	1
250	222	Bottle Opener	$ 12.00	1
252	222	Bottle Opener	$ 12.00	1
252	444	Wine Gift Pack	$ 60.00	2
253	222	Bottle Opener	$ 12.00	1
253	444	Wine Gift Pack	$ 60.00	1
297	333	Jams &Jellies	$ 20.00	2
243	555	Cheese Tray	$ 45.00	2
246	333	Jams &Jellies	$ 20.00	3
248	222	Bottle Opener	$ 12.00	1
248	333	Jams &Jellies	$ 20.00	2
248	111	Wine Guide	$ 15.00	1
235	222	Bottle Opener	$ 12.00	1
242	333	Jams &Jellie	$ 20.00	3
244	222	Bottle Opener	$ 12.00	2
251	111	Wine Guide	$ 15.00	2

Order 248 has 3 products

FIGURE 13-12

1NF: Remove Repeating Fields.

Second normal form (2NF) requires first that the data model is in 1NF and second that the data model leads to tables containing fields that are dependent on a *whole* primary key. This means that the primary key value for each record can determine the value for all of the other fields in the record. Sometimes fields are only dependent on part of the primary key (i.e., *partial dependency*), and these fields belong in another table.

For example, in the new Product Order table that was created in Figure 13-12, the primary key is a combination of the order number and product number, but the product description and price attributes are dependent only on product number. In other words, by knowing product number, you can identify the product description and price. However, knowledge of the order number and product number is required to identify the quantity. To rectify this violation of 2NF, a table is created to store product information and the description and price attributes are moved into the new table. Now, product description is stored only once for each instance of a product number as opposed to many times (every time a product is placed on an order).

A second violation of 2NF occurs in the Order table: customer first name and last name are dependent only on the customer ID, not the whole key (Cust ID and Order number). As a result, every time the customer ID appears in the order table, the names also appear. A much more economical way of storing the data is to create a customer table with the customer ID as the primary key and the other customer-related fields (i.e., last name and first name) listed only once within the appropriate record. Figure 13-13 illustrates how the model would look when placed in 2NF.

Third normal form (3NF) occurs when a model is in both 1NF and 2NF and in the resulting tables none of the fields are dependent on nonprimary key fields (i.e., *transitive dependency*). Figure 13-13 contains a violation of 3NF: the tax rate on the order depends on the state to which the order is being sent. The solution involves creating another table that contains state abbreviations serving as the primary key and the tax rate as a regular field. Figure 13-14 presents the end results of applying the steps of normalization to the original model from Figure 13-10.

YOUR TURN

13-3 NORMALIZING A STUDENT ACTIVITY FILE

Pretend that you have been asked to build a system that tracks student involvement in activities around campus. You have been given a file with information that needs to be imported into the system, and the file contains the following fields:

- Student Social Security number
- Student last name
- Student first name
- Student advisor name
- Student advisor phone
- Activity 1 code

- Activity 1 description
- Activity 1 start date
- Activity 2 code
- Activity 2 description
- Activity 2 start date
- Activity 3 code
- Activity 3 description
- Activity 3 start date

Normalize the file. Show how the logical data model would change at each step.

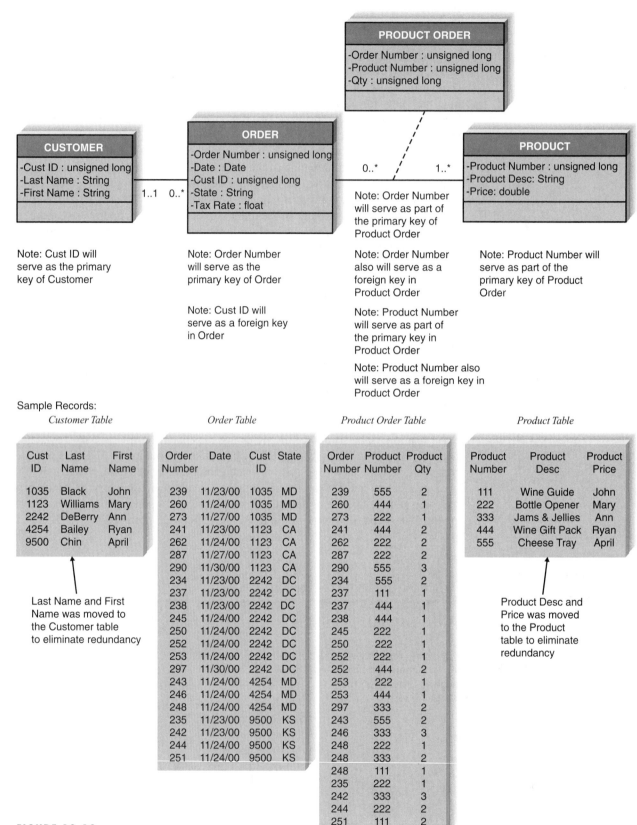

Sample Records:

Customer Table

Cust ID	Last Name	First Name
1035	Black	John
1123	Williams	Mary
2242	DeBerry	Ann
4254	Bailey	Ryan
9500	Chin	April

Last Name and First Name was moved to the Customer table to eliminate redundancy

Order Table

Order Number	Date	Cust ID	State
239	11/23/00	1035	MD
260	11/24/00	1035	MD
273	11/27/00	1035	MD
241	11/23/00	1123	CA
262	11/24/00	1123	CA
287	11/27/00	1123	CA
290	11/30/00	1123	CA
234	11/23/00	2242	DC
237	11/23/00	2242	DC
238	11/23/00	2242	DC
245	11/24/00	2242	DC
250	11/24/00	2242	DC
252	11/24/00	2242	DC
253	11/24/00	2242	DC
297	11/30/00	2242	DC
243	11/24/00	4254	MD
246	11/24/00	4254	MD
248	11/24/00	4254	MD
235	11/23/00	9500	KS
242	11/23/00	9500	KS
244	11/24/00	9500	KS
251	11/24/00	9500	KS

Product Order Table

Order Number	Product Number	Product Qty
239	555	2
260	444	1
273	222	1
241	444	2
262	222	2
287	222	2
290	555	3
234	555	2
237	111	1
237	444	1
238	444	1
245	222	1
250	222	1
252	222	1
252	444	2
253	222	1
253	444	1
297	333	2
243	555	2
246	333	3
248	222	1
248	333	2
248	111	1
235	222	1
242	333	3
244	222	2
251	111	2

Product Table

Product Number	Product Desc	Product Price
111	Wine Guide	John
222	Bottle Opener	Mary
333	Jams & Jellies	Ann
444	Wine Gift Pack	Ryan
555	Cheese Tray	April

Product Desc and Price was moved to the Product table to eliminate redundancy

FIGURE 13-13

2NF: Partial Dependencies Removed

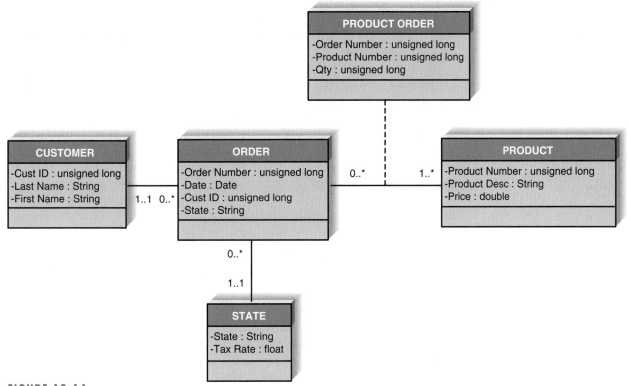

FIGURE 13-14
3NF Normalized Model

Optimizing Data Access Speed

After you have optimized the design of the data management layer for storage efficiency, the end result is that data are spread out across a number of tables. When data from multiple tables need to be accessed or queried, the tables must be first joined together. For example, before a user can print out a list of the customer names associated with orders, first the Customer and Order tables need to be joined together based on the customer number field (see Figure 13-14). Only then can both the order and customer information be included in the query's output. Joins can take a lot of time, especially if the tables are large or if many tables are involved.

Consider a system that stores information about 10,000 different products, 25,000 customers, and 100,000 orders, each averaging three products per order. If an analyst wanted to investigate whether there were regional differences in music preferences, he or she would need to combine all of the tables to be able to look at products that have been ordered while knowing the state of the customers placing the orders. A query of this information would result in a huge table with 300,000 rows (i.e., the number of products that have been ordered) and 11 columns (the total number of columns from all of the tables combined).

The project team can use several techniques to try to speed up access to the data: denormalization, clustering, indexing, and estimating the size of the data for hardware planning purposes.

Denormalization   After the data management layer is optimized in terms of storage, the project team may decide to denormalize, or add redundancy back into the design. *Denormalization* reduces the number of joins that need to be performed in a query, thus speeding up access. Figure 13-15 shows a denormalized model for customer orders. The customer last name was added back into the order table since the project team learned during the analysis phase that queries about orders usually require the customer last name field. Instead of joining the Order table repeatedly to the Customer table, the system now needs to access only the Order table because it contains all of the relevant information.

Denormalization should be applied sparingly for the reasons described in the previous section, but it is ideal in situations in which information is queried frequently yet updated rarely. There are three cases in which you may rely on denormalization to reduce joins and improve performance. First, denormalization can be applied in the case of look-up tables, which are tables that contain descriptions of values (e.g., a table of product descriptions or a table of payment types). Because descriptions of codes rarely change, it may be more efficient to include the description along with its respective code in the main table to eliminate the need to join the look-up table each time a query is performed (see Figure 13-16a).

Second, one-to-one relationships are good candidates for denormalization. Although logically two tables should be separated, from a practical standpoint the information from both tables may regularly be accessed together. Think about an order and its shipping information. Logically, it may make sense to separate the attributes related to shipping into a separate table, but as a result the queries regarding shipping likely will always need a join to the Order table. If the project team finds that certain shipping information like state and shipping method are needed when orders are accessed, they may decide to combine the tables or include some shipping attributes in the Order table (see Figure 13-16b).

Third, at times it will be more efficient to include a parent entity's attributes in its child entity on the physical data model. For example, consider the Customer and Order tables in Figure 13-15, which share a one-to-many relationship, with Customer as the parent and Order as the child. If queries regarding orders continuously require customer information, the most popular customer fields can be placed in Order to reduce the required joins to the Customer table as was done with customer last name.

Clustering Speed of access also is influenced by the way the data is retrieved. Think about going shopping in a grocery store. If you have a list of items to buy, but you are unfamiliar with the store's layout, then you need to walk down every

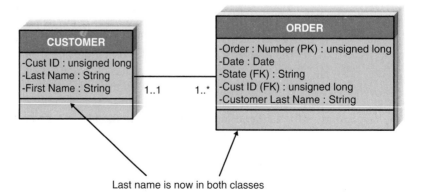

FIGURE 13-15
Denormalized Physical Data Model
(FK = foreign key; PK = primary key)

aisle to make sure that you don't miss anything from your list. Likewise, if records are arranged on a hard disk in no particular order (or in an order that is irrelevant to your data needs), then any query of the records results in a *table scan* in which the DBMS has to access every row in the table before retrieving the result set. The table scan is the most inefficient data retrieval method.

One way to improve access speed is to reduce the number of times that the storage medium needs to be accessed during a transaction. One method is to *cluster* records together physically so that similar records are stored close together. With *intra-file clustering,* like records in the table are stored together in some way, such

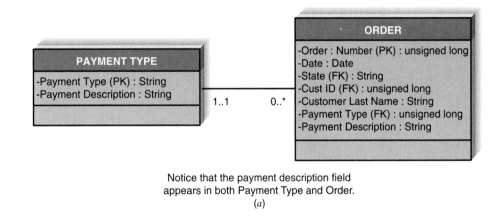

Notice that the payment description field
appears in both Payment Type and Order.
(a)

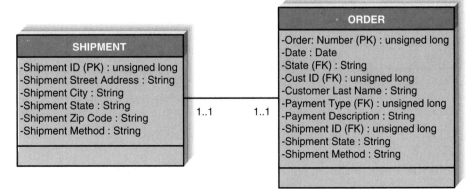

FIGURE 13-16

Denormalization Situations (FK = foreign key; PK = primary key); (a) Look-up table; (b) One-to-one relationship

Notice that the shipment state and shipment method
are included in both Shipment and Order.
(b)

as in order by primary key or, in the case of a grocery store, by item type. Thus, whenever a query looks for records, it can go directly to the right spot on the hard disk (or other storage medium) because it knows in what order the records are stored, just as we can walk directly to the bread aisle to pick up a loaf of bread. *Inter-file clustering* combines records from more than one table that typically are retrieved together. For example, if customer information is usually accessed with the related order information, then the records from the two tables may be physically stored in a way that preserves the customer order relationship. Returning to the grocery store scenario, an inter-file cluster would be similar to storing peanut butter, jelly, and bread next to each other in the same aisle because they are usually purchased together, not because they are similar types of items. Of course, each table can have only one clustering strategy because the records can be arranged physically in only one way.

Indexing   A time saver that you are familiar with is an index located in the back of a textbook, which points you directly to the page or pages that contain your topic of interest. Think of how long it would take you to find all of the times that "relational database" appears in this textbook if you didn't have the index to rely on! An *index* in data storage is like an index in the back of a textbook; it is a mini table that contains values from one or more columns in a table and the location of the values within the table. Instead of paging through the entire textbook, you can move directly to the right pages and get the information you need. Indexes are one of the most important ways to improve database performance. Whenever you have performance problems, the first place to look is an index.

A query can use an index to find the locations of only those records that are included in the query answer, and a table can have an unlimited number of indexes. Figure 13-17 shows an index that orders records by payment type. A query that searches for all of the customers who used American Express can use this index to find the locations of the records that contain American Express as the payment type without having to scan the entire Order table.

Project teams can make indexes perform even faster by placing them into the main memory of the data storage hardware. Retrieving information from memory is much faster than from another storage medium like a hard disk—think about how

---

CONCEPTS

IN ACTION

13-A MAIL ORDER INDEX

A Virginia-based mail order company sends out approximately 25 million catalogs each year using a Customer table with 10 million names. Although the primary key of the Customer table is customer identification number, the table also contains an index of customer last names. Most people who call to place orders know their last name, not their customer identification number, so this index is used frequently.

An employee of the company explained that indexes are critical to reasonable response times. A fairly complicated query was written to locate customers by the state in which they lived, and it took over three weeks to return an answer. A customer state index was created, and that same query provided a response in 20 minutes: that's 1,512 times faster!

QUESTION

As an analyst, how can you make sure that the proper indexes have been put in place so that users are not waiting for weeks to receive the answers to their questions?

Payment Type Index

Order

Payment Type	Pointer
AMEX	*
AMEX	*
AMEX	*
AMEX	*
AMEX	*
AMEX	*
MC	*
MC	*
MC	*
MC	*
MC	*
MC	*
VISA	*
VISA	*
VISA	*
VISA	*
VISA	*
VISA	*
VISA	*
VISA	*

Order Number	Date	Cust ID	Amount	Tax	Total	Payment Type
234	11/23/00	2242	$ 90.00	$ 5.85	$ 95.85	MC
235	11/23/00	9500	$ 12.00	$ 0.60	$ 12.60	VISA
236	11/23/00	1556	$ 50.00	$ 2.50	$ 52.50	VISA
237	11/23/00	2242	$ 75.00	$ 4.88	$ 79.88	AMEX
238	11/23/00	2242	$ 60.00	$ 3.90	$ 63.90	MC
239	11/23/00	1035	$ 90.00	$ 4.50	$ 94.50	AMEX
240	11/23/00	9501	$ 50.00	$ 2.50	$ 52.50	VISA
241	11/23/00	1123	$120.00	$ 9.60	$129.60	MC
242	11/24/00	9500	$ 60.00	$ 3.00	$ 63.00	VISA
243	11/24/00	4254	$ 90.00	$ 4.50	$ 94.50	VISA
244	11/24/00	9500	$ 24.00	$ 1.20	$ 25.20	VISA
245	11/24/00	2242	$ 12.00	$ 0.78	$ 12.78	AMEX
246	11/24/00	4254	$ 20.00	$ 1.00	$ 21.00	MC
247	11/24/00	2241	$ 50.00	$ 2.50	$ 52.50	VISA
248	11/24/00	4254	$ 12.00	$ 0.60	$ 12.60	AMEX
249	11/24/00	5927	$ 50.00	$ 2.50	$ 52.50	AMEX
250	11/24/00	2242	$ 12.00	$ 0.78	$ 12.78	MC
251	11/24/00	9500	$ 15.00	$ 0.75	$ 15.75	MC
252	11/24/00	2242	$132.00	$ 8.58	$ 140.58	MC
253	11/24/00	2242	$ 72.00	$ 4.68	$ 76.68	AMEX

FIGURE 13-17
Payment Type Index

much faster it is to retrieve a phone number that you have memorized versus one that you need to look up in a phone book. Similarly, when a database has an index in memory, it can locate records very, very quickly.

Of course, indexes require overhead in that they take up space on the storage medium. Also, they need to be updated as records in tables are inserted, deleted, or changed. Thus, although queries lead to faster access to the data, they slow down the update process. In general, you should create indexes sparingly for transaction systems or systems that require a lot of updates, but apply indexes generously when designing systems for decision support (see Figure 13-18).

Estimating Data Storage Size

Even if you have denormalized your physical data model, clustered records, and created indexes appropriately, the system will perform poorly if the database server

---

Use indexes sparingly for transaction systems.

Use many indexes to increase response times in decision support systems.

For each table, create a unique index that is based on the primary key.

For each table, create an index that is based on the foreign key to improve the performance of joins.

Create an index for fields that are used frequently for grouping, sorting, or criteria.

FIGURE 13-18
Guidelines for Creating Indexes

cannot handle its volume of data. Therefore, one last way to plan for good performance is to apply *volumetrics,* which means estimating the amount of data that the hardware will need to support. You can incorporate your estimates into the database server hardware specification to make sure that the database hardware is sufficient for the project's needs. The size of the database is based on the amount of *raw data* in the tables and the *overhead* requirements of the DBMS. To estimate size, you will need to have a good understanding of the initial size of your data as well as its expected growth rate over time.

Raw data refers to all of the data that are stored within the tables of the database, and it is calculated based on a bottom-up approach. First, write down the estimated average width for each column (field) in the table and sum the values for a total record size (see Figure 13-19). For example, if a variable width LAST NAME column is assigned a width of 20 characters, you can enter 13 as the average character width of the column. In Figure 13-19, the estimated record size is 49.

Next, calculate the overhead for the table as a percentage of each record. *Overhead* includes the room needed by the DBMS to support such functions as administrative actions and indexes. It should be assigned based on past experience, recommendations from technology vendors, or parameters that are built into software that was written to calculate volumetrics. For example, your DBMS vendor may recommend that you allocate 30% of the records' raw data size for overhead storage space, creating a total record size of 63.7 in the Figure 13-19 example.

Finally, record the number of initial records that will be loaded into the table, as well as the expected growth per month. This information should have been collected during the analysis phase. According to Figure 13-19, the initial space required by the first table is 3,185,000, and future sizes can be projected based on the growth figure. These steps are repeated for each table to get a total size for the entire database.

Many CASE tools will provide you with database size information based on how you set up the data management layer, and they will calculate volumetrics estimates automatically. Ultimately, the size of the database needs to be shared with the design team so that the proper technology can be put in place to support the sys-

Field	Average Size (Characters)
Order number	8
Date	7
Cust ID	4
Last name	13
First name	9
State	2
Amount	4
Tax rate	2
Record size	49
Overhead	30%
Total record size	63.7
Initial table size	50,000
Initial table volume	3,185,000
Growth rate/month	1,000
Table volume @ 3 years	5,478,200

FIGURE 13-19
Calculating Volumetrics

tem's data and potential performance problems can be addressed long before they affect the success of the system.

APPLYING THE CONCEPTS AT CD SELECTIONS

The CD Selections Internet Sales System needs to effectively present CD information to users and capture order data. Alec Adams, senior systems analyst and project manager for the Internet Sales System, recognized that these goals were dependent on a good design of the data management layer for the new application. He approached the design of the data management layer in two steps: first by selecting the object persistence format and later by optimizing it for processing efficiency.

Object Persistence Format Selection

The project team met to discuss two issues that would drive the object persistence format selection: what kind of objects would be in the system and how they would be used. Using a whiteboard, they listed the ideas that are presented in Figure 13-20. The project team agreed that the bulk of the data in the system would be the text and numbers that are exchanged with Web users regarding customers and orders. A relational database would be able to handle the data effectively, and the technology would be well received at CD Selections because relational technology is already in place throughout the organization.

They realized, however, that relational technology was not optimized to handle complex data, such as the images, sound, and video that the marketing facet of the system ultimately will require. Alec asked Brian, one of the staff members on

Data	Type	Use	Suggested Format
Customer information	Simple (mostly text)	Transactions	Relational
Order information	Simple (text and numbers)	Transactions	Relational
Marketing information	Both simple and complex (eventually the system will contain audio clips, video, etc.)	Transactions	Object add-on?
Information that will be exchanged with the distribution system	Simple text, formatted specifically for importing into the distribution system	Transactions	Transaction file
Temporary information	The Web component will likely need to hold information for temporary periods of time (e.g., the shopping cart will store order information before the order is actually placed)	Transactions	Transaction file

FIGURE 13-20
Types of Data in Internet Sales System

the Internet Sales System project, to look into relational databases that offered object add-on products, that is, an RDBMS that could become an ORDBMS. It might be possible for the team to invest in a RDBMS foundation and then upgrade to an ORDBMS version of the same product.

The team noted that it must also design two transaction files to handle the interface with the distribution system and the Web shopping cart program. The Internet Sales System will regularly download order information to the distribution system using a transaction file containing all the required information for that system. Also, the team must design the file that stored temporary order information on the Web server as customers shopped through the Web site. The file would contain the fields that ultimately would be transferred to an order object.

Alec realized that other data needs may arise over time, but he felt confident that the major data issues were identified (e.g., like the capability to handle complex data) and that the design of the data management layer would be based on the proper storage technologies.

Data Storage Optimization

After the meeting, Alec asked Brian to stay behind to discuss the data management layer model. Now that the team had a good idea of the type of object persistence formats that would be used, they were ready for the second step: optimizing the design for performance efficiency. Brian was the analyst in charge of the data management layer, and Alec wanted to be sure that the model was optimized for storage efficiency before the team discussed access speed issues.

Brian assured Alec that the current model was in third normal form. He was confident of this because the project team followed the modeling guidelines that lead to a well-formed model. Of course, a week or so ago, he did apply the three rules of normalization to the model as a check to make sure that no design errors were overlooked.

Brian then asked about the file formats for the two transaction files identified in the earlier meeting. Alec suggested that he normalize the files to better understand the various tables that would be involved in the import procedure. Figure 13-21 shows the initial file layout for the Distribution System import file as well as the steps that were taken as Brian applied each normalization rule.

The last step for the design of the data management layer was to optimize the design for access speed. Alec met with the analysts on the data management layer design team and talked about the techniques that were available to speed up access. Together the team listed all of the data that will be supported by the Internet Sales System and discussed how all of the data would be used. They developed the strategy that is laid out in Figure 13-22 to identify the specific techniques to put in place.

Data Sizing

Ultimately, clustering strategies, indexes, and denormalization decisions were applied to the physical data model, and a volumetrics report was run from the CASE tool to estimate the initial and projected size of the database. The report suggested that an initial storage capacity of about 450 megabytes would be needed for the expected one-year life of the first version of the system. Additional storage capacity would be needed for the second version that would include sound files for samples of the songs, but for the moment not much storage would be needed.

File Layout Required for Distribution System

Order no.	Cust Fname	Cust Lname	Cust Home address	Cust Ship address	Cust Pay	Item No*	Item Desc*	Item Price*	Item Qty*
A (7)	A (20)	A (20)	A (150)	A (150)	9999,99	A (7)	A (20)	9999,99	9999

* Item no, Item Desc, Item Price, and Item Qty repeats four (4) times.

Remove repeating groups of items and place them in a separate Item entity

1NF:

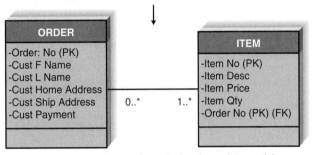

There is one partial dependency in the above data model since Item qty is dependent on the whole primary key while Item Desc and Item Price are dependent only on Item No.

2NF:

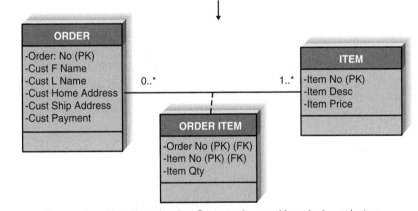

Remove transitive dependencies. Customer home address is dependent on Cust F Name and Cust L Name (and these do not serve as the identifier for the Order entity); therefore, a Customer entity is added to contain customer information. Order then contains only order information and necessary foreign keys.

3NF:

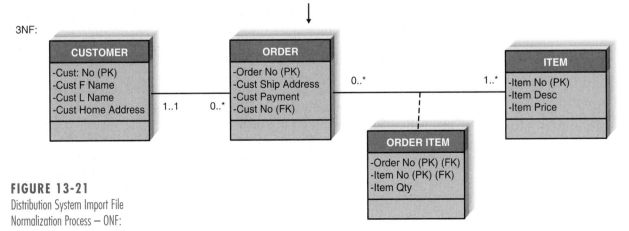

FIGURE 13-21
Distribution System Import File
Normalization Process — 0NF:

Target	Comments	Suggestions to Improve Data Access Speed
All tables	Basic table manipulation	• Investigate if records should be clustered physically by primary key. • Create indexes for primary keys. • Create indexes for foreign key fields.
All tables	Sorts and grouping	• Create indexes for fields that are frequently sorted or grouped.
CD information	Users will need to search CD information by title, artist, and category.	• Create indexes for CD title, artist, and category.
Order information	Operators should be able to locate information about a particular customer's order.	• Create an index in the Order table for orders by customer name.
Entire physical model	Investigate denormalization opportunities for all fields that are not updated very often.	• Investigate one-to-one relationships. • Investigate look-up tables. • Investigate one-to-many relationships.

FIGURE 13-22

Internet Sales System Performance

Alec gave the estimates to the analyst in charge of managing the server hardware acquisition so that the person could make sure that the technology could handle the expected volume of data for the Internet Sales System. The estimates would also go to the DBMS vendor during the implementation of the software so that the DBMS could be configured properly.

SUMMARY

Object Persistence Formats

There are four basic types of object persistence formats: files (sequential and random access), object-oriented databases, object-relational databases, and relational databases. Files are electronic lists of data that have been optimized to perform a particular transaction. There are two different access methods (sequential and random), and there are five different application types: master, look-up, transaction, audit, and history. Master files typically are kept for long periods of time because they store important business information, such as order information or customer mailing information. Look-up files contain static values that are used to validate fields in the master files, and transaction files temporarily hold information that will be used for a master file update. An audit file records before and after images of data as they are altered so that an audit can be performed if the integrity of the data is questioned. Finally, the history file stores past transactions (e.g., old customers, past orders) that are no longer needed by the system.

A database is a collection of groupings of information that is related to each other in some way, and a DBMS (database management system) is software that creates and manipulates these databases. Three types of databases are likely to be encountered during a project: relational, object-relational, and object-oriented.

The relational database is the most popular kind of database for application development today, and it is based on collections of tables that are related to each other through common fields, known as foreign keys. Object-relational databases are relational databases that have extensions that provide limited support for object-orientation. The extensions typically include some support for the storage of objects in the relational table structure. Object-oriented databases come in two flavors: full-blown DBMS products and extensions to an object-oriented programming language. Both approaches typically fully support object-orientation.

Selecting an Object Persistence Format
The application's data should drive the storage format decision. Relational databases support simple data types very effectively, whereas object databases are best for complex data. The type of system should also be considered when choosing among data storage formats; for example, relational databases have matured to support transactional systems. Although less critical to the format selection decision, the project team needs to consider what technology exists within the organization and the kind of technology likely to be used in the future.

Optimizing RDBMS-Based Object Storage
There are two primary dimensions in which to optimize a relational database: for storage efficiency and for speed of access. The most efficient relational database tables in terms of data storage are those that have no redundant data and very few null values. Normalization is the process whereby a series of rules are applied to the data management layer to determine how well formed it is. A model is in first normal form (1NF) if it does not lead to repeating fields, which are fields that repeat within a table to capture multiple values. Second normal form (2NF) requires that all tables be in 1NF and lead to fields whose values are dependent on the *whole* primary key. Third normal form (3NF) occurs when a model is in both 1NF and 2NF and none of the resulting fields in the tables is dependent on nonprimary key fields (i.e., transitive dependency). With each violation, additional tables should be created to remove the repeating fields or improper dependencies from the existing tables.

Once you have optimized the design of the data management layer for storage efficiency, the data may be spread out across a number of tables. To improve speed, the project team may decide to denormalize—add redundancy back into—the design. Denormalization reduces the number of joins that need to be performed in a query, thus speeding up data access. Denormalization is best in situations where data are accessed frequently and updated rarely.

Three modeling situations are good candidates for denormalization: look-up tables, entities that share one-to-one relationships, and entities that share one-to-many relationships. In all three cases, attributes from one entity are moved or repeated in another entity to reduce the joins that must occur while accessing the database.

Clustering occurs when similar records are stored close together on the storage medium to speed up retrieval. In intra-file clustering, similar records in the table are stored together in some way, such as in sequence. Inter-file clustering combines records from more than one table that typically are retrieved together. Indexes also can be created to improve the access speed of a system. An index is a mini table that contains values from one or more columns in a table, and it is where the values can be found. Instead of performing a table scan, which is the most inefficient way to retrieve data from a table, an index points directly to the records that match the requirements of a query.

Finally, the speed of the system can be improved if the right hardware is purchased to support it. Analysts can use volumetrics to estimate the current and future size of the database and then share these numbers with the people who are responsible for buying and configuring the database hardware.

KEY TERMS

Attribute sets
Audit file
Clustering
Database
Database management system
  (DBMS)
Decision support systems (DSS)
Denormalization
End-user DBMS
Enterprise DBMS
Executive information systems
  (EIS)
Expert system (ES)
Extent
File
First normal form (1NF)
Foreign key
History file
Impedance mismatch
Index

Inter-file cluster
Intra-file cluster
Join
Linked list
Look-up file
Management information system
  (MIS)
Master file
Multivalued attributes (fields)
Normalization
Object ID
Object-oriented database management system (OODBMS)
Object-oriented programming language (OOPL)
Object-relational database management system (ORDBMS)
Ordered sequential access file
Overhead
Partial dependency

Pointer
Primary key
Random access files
Raw data
Referential integrity
Relational database management
  system (RDBMS)
Relationship sets
Repeating groups (fields)
Second normal form (2NF)
Sequential access files
Structured query language (SQL)
Table scan
Third normal form (3NF)
Transaction file
Transaction processing system
Transitive dependency
Unordered sequential access file
Update anomaly
Volumetrics

QUESTIONS

1. Describe the two steps to object persistence design.
2. How are a file and a database different from each other?
3. What is the difference between an end-user database and an enterprise database? Provide an example of each one.
4. What are the differences between sequential and random access files?
5. Name five types of files and describe the primary purpose of each type.

6. What is the most popular kind of database today? Provide three examples of products that are based on this database technology.
7. What is referential integrity and how is it implemented in a RDBMS?
8. List some of the differences between an ORDBMS and a RDBMS.
9. What are the advantages of using an ORDBMS over a RDBMS?

10. List some of the differences between an ORDBMS and an OODBMS.
11. What are the advantages of using an ORDBMS over an OODBMS?
12. What are the advantages of using an OODBMS over a RDBMS?
13. What are the advantages of using an OODBMS over an ORDBMS?
14. What are the factors in determining the type of object persistence format that should be adopted for a system? Why are these factors so important?
15. Why should you consider the storage formats that already exist in an organization when deciding on a storage format for a new system?
16. When implementing the data management layer in an ORDBMS, what types of issues must you address?
17. When implementing the data management layer in an RDBMS, what types of issues must you address?
18. Name three ways in which null values can be interpreted in a relational database. Why is this problematic?
19. What are the two dimensions in which to optimize a relational database?
20. What is the purpose of normalization?
21. How does a model meet the requirements of third normal form?
22. Describe three situations that can be good candidates for denormalization.
23. Describe several techniques that can improve the performance of a database.
24. What is the difference between inter-file and intra-file clustering? Why are they used?
25. What is an index and how can it improve the performance of a system?
26. Describe what should be considered when estimating the size of a database.
27. Why is it important to understand the initial and projected size of a database during the design phase?
28. What are the key issues in deciding between using perfectly normalized database and denormalized databases?

EXERCISES

A. Using the Web or other resources, identify a product that can be classified as an end-user database and a product that can be classified as an enterprise database. How are the products described and marketed? What kinds of applications and users do they support? In what kinds of situations would an organization choose to implement an end-user database over an enterprise database?

B. Visit a commercial web site (e.g., CD Now, Amazon.com). If files were being used to store the data supporting the application, what types of files would be needed? What access type would be required? What data would they contain?

C. Using the Web, review one of the products listed below. What are the main features and functions of the software? In what companies has the DBMS been implemented, and for what purposes? According to the information that you found, what are three strengths and weaknesses of the product?
   1. Relational DBMS
   2. Object-relational DBMS
   3. Object-oriented DBMS

D. You have been given a file that contains the following fields relating to CD information. Using the steps of normalization, create a model that represents this file in third normal form. The fields include:
Musical group name
Musicians in group
Date group was formed
Group's agent
CD title 1
CD title 2
CD title 3
CD 1 length
CD 2 length
CD 3 length

Assumptions:
• Musicians in group contain a list of the members of the people in the musical group.

- Musical groups can have more than one CD, so both group name and CD title are needed to uniquely identify a particular CD.

E. Jim Smith's dealership sells Fords, Hondas, and Toyotas. The dealership keeps information about each car manufacturer with whom they deal so that they can get in touch with them easily. The dealership also keeps information about the models of cars that they carry from each manufacturer. They keep information such as list price, the price the dealership paid to obtain the model, and the model name and series (e.g., Honda Civic LX). They also keep information about all sales that they have made (for instance, they will record the buyer's name, the car they bought, and the amount they paid for the car). In order to contact the buyers in the future, contact information is also kept (e.g., address, phone number). Create a class diagram for this situation. Apply the rules of normalization to the class diagram to "check" the diagram for processing effciency.

F. Describe how you would denormalize the model that you created in Question E. Draw the new class diagram based on your suggested changes. How would performance be affected by your suggestions?

G. Examine the model that you created in Question F. Develop a clustering and indexing strategy for this model. Describe how your strategy will improve the performance of the database.

H. Calculate the size of the database that you created in Question F. Provide size estimates for the initial size of the database as well as for the database in one year's time. Assume that the dealership sells 10 models of cars from each manufacture to approximately 20,000 customers a year. The system will be set up initially with one year's worth of data.

MINICASES

1. The system development team at the Wilcon Company is working on developing a new customer order entry system. In the process of designing the new system, the team has identified the following class and its attributes:

INVENTORY ORDER

Order Number (PK)
Order Date
Customer Name
Street Address
City
State
Zip
Customer Type
Initials
District Number
Region Number
1 to 22 occurrences of:
Item Name
Quantity Ordered
Item Unit
Quantity Shipped
Item Out
Quantity Received

a. State the rule that is applied to place a class in first normal form. Revise the above class diagram so that it is in first normal form.

b. State the rule that is applied to place a class into second normal form. Revise the class diagram (if necessary) to place it in second normal form.

c. State the rule that is applied to place a class into third normal form. Revise the class diagram to place it in third normal form.

d. When planning for the physical design of this database, can you identify any likely situations where the project team may chose to denormalize the class diagram? After going through the work of normalizing, why would this be considered?

2. In the new system under development for Holiday Travel Vehicles, seven tables will be implemented in the new relational database. These tables are: New Vehicle, Trade-in Vehicle, Sales Invoice, Customer, Salesperson, Installed Option, and Option. The expected average record size for these tables and the initial record count per table are given below.

Table Name	Average Record Size	Initial Table Size (records)
New Vehicle	65 characters	10,000
Trade-in Vehicle	48 characters	7,500
Sales Invoice	76 characters	16,000
Customer	61 characters	13,000
Salesperson	34 characters	100
Installed Option	16 characters	25,000
Option	28 characters	500

Perform a volumetrics analysis for the Holiday Travel Vehicle system. Assume that the DBMS that will be used to implement the system requires 35% overhead to be factored into the estimates. Also, assume a growth rate for the company of 10% per year. The systems development team wants to ensure that adequate hardware is obtained for the next three years.

PLANNING

ANALYSIS

DESIGN

- ✔ **Evolve Analysis Models**
- ✔ **Develop the Design Strategy**
- ✔ **Develop the Design**
- ✔ **Develop Infrastructure Design**
- ✔ **Develop Network Model**
- ✔ **Develop Hardware/Software Specification**
- ✔ **Develop Security Plan**
- ✔ **Develop Use Scenarios**
- ✔ **Design Interface Structure**
- ✔ **Design Interface Standards**
- ✔ **Design User Interface Template**
- ✔ **Design User Interface**
- ✔ **Evaluate User Interface**
- ✔ **Select Object Persistence Format**
- ✔ **Map Objects to Object Persistence**
- ✔ **Optimize Object Persistence**
- ✔ **Size Object Persistence**
- ☐ **Specify Constraints**
- ☐ **Develop Contracts**
- ☐ **Develop Method Specifications**

TASK CHECKLIST

PLANNING   ANALYSIS   DESIGN

CHAPTER 14

CLASS AND
METHOD DESIGN

THE most important step of the design phase is designing the individual classes and methods. Object-oriented systems can be quite complex, so analysts need to create instructions and guidelines for programmers that clearly describe what the system must do. This chapter presents a set of criteria, activities, and techniques used to design classes and methods. Together they are used to ensure that the object-oriented design communicates how the system needs to be coded.

OBJECTIVES

- Become familiar with coupling, cohesion, and connascence.
- Be able to specify, restructure, and optimize object designs.
- Be able to identify the reuse of predefined classes, libraries, frameworks, and components.
- Be able to specify constraints and contracts.
- Be able to create a method specification.

CHAPTER OUTLINE

IMPLEMENTATION

INTRODUCTION

Class and method design is where all of the work actually gets done during the design phase. No matter which layer you are focusing on, the classes, which will be used to create the systems objects, must be designed. Some people believe that with reusable class libraries and off-the-shelf components, this type of low-level, or detailed, design is a waste of time, and they recommend that we should jump immediately into the "real" work: coding the system. However, if the past shows us anything, it reveals that low-level or detailed design is critical despite the use of libraries and components. Detailed design is still very important for two reasons. First, even preexisting classes and components need to be understood, organized, and pieced together. Second, it is still common for the project team to have to write some code (if not all) and produce original classes that support the application logic of the system.

Jumping right into coding will guarantee results that can be disastrous. For example, even though the use of layers can simplify the individual classes, they can increase the complexity of the interactions between them. As such, if the classes are not designed carefully, the resulting system can be very inefficient. Or worse, the instances of the classes, that is, the objects, will not be capable of communicating with each other, which, of course, will cause the system not to work properly.

Furthermore, in an object-oriented system, changes can take place at different levels of abstraction. These levels include Variable, Method, Class/Object, Partition/Package,[1] Library, and/or System levels (see Figure 14–1). The changes that take place at one level can impact other levels. For example, changes to a class can affect the partition/package level, which can affect both the system level and the library level, which in turn can cause changes back down at the class level. Finally, changes can be occurring at different levels at the same time.

The good news is that the detailed design of the individual classes and methods is fairly straight-forward and the interactions among the objects on the problem domain layer have been designed, in some detail, in the analysis phase (see Chapters 6 through 8). As far as the other layers go (system architecture, human–computer interaction, and data management), they will be highly dependent on the problem domain layer. Therefore, if we design the problem domain classes correctly, the design of the classes on the other layers will fall into place, relatively speaking.

Having stated the above, it has been our experience that many project teams are much too quick to jump into writing code for the classes without first designing them. Some of this has been caused by the fact that object-oriented systems analysis and design has evolved from object-oriented programming. Until recently, there has been a general lack of accepted guidelines on how to design and develop effective object-oriented systems. However, with the acceptance of UML as a standard object notation, standardized approaches based on the work of many object methodologists have begun to emerge.[2]

In this chapter, we begin by revisiting the principles of object-orientation. Next, we present a set of useful design criteria and activities that are applicable

---

[1] A package or partition is a group of collaborating objects. Other names for a partition/package include cluster, pattern, subject, and subsystem.

[2] For example, OPEN (I. Graham, B. Henderson-Seller, and H. Yanoussi, *The Open Process Specification,* Reading, MA: Addison-Wesley, 1997,) and RUP (P. Kruchten, *The Rational Unified Process: An Introduction,* 2nd ed., Reading, MA: Addison-Wesley, 2000).

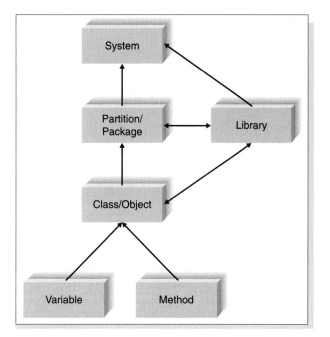

FIGURE 14-1

Levels of Abstraction in Object-Oriented Systems   *Source:* Adapted from David P. Tegarden, Steven D. Sheetz, and David E. Monarchi, "A Software Complexity Model of Object-Oriented Systems," *Decision Support Systems* 13 (March 1995): 241–262.

across any layer for class and method design. Finally, we present a set of techniques that are useful to design methods: contracts and method specifications.

REVISITING THE BASIC CHARACTERISTICS OF OBJECT-ORIENTATION

As pointed out in Chapter 1, object-oriented systems can be traced back to the Simula and Smalltalk programming languages. However, until the increase in processor power and the decrease in processor cost that occurred in the 1980s, object-oriented approaches were not practical. In this section, we review the basic characteristics of object-orientation.

Classes, Objects, Methods, and Messages

The basic building block of the system is the *object.* Objects are *instances* of *classes* that we use as templates to define objects. A class defines both the data and processes that each object contains. Each object has *attributes* that describe data about the object. Objects have *state,* which is defined by the value of its attributes and its relationships with other objects at a particular point in time. And each object has *methods* that specify what processes the object can perform. In order to get an object to perform a method (e.g., to delete itself), a *message* is sent to the object. A message is essentially a function or procedure call from one object to another object.

Encapsulation and Information Hiding

Encapsulation is the mechanism that combines the processes and data into a single object. *Information hiding* suggests that only the information required to use an object should be available outside the object. Exactly how the object stores data or performs methods is not relevant, as long as the object functions correctly. All that is required to use an object is the set of methods and the messages needed to be sent to trigger them. The only communication between objects should be through an object's methods. The fact that we can use an object by sending a message that calls methods is the key to reusability because it shields the internal workings of the object from changes in the outside system and it keeps the system from being affected when changes are made to an object.

Polymorphism

Polymorphism means having the ability to take several forms. By supporting polymorphism, object-oriented systems can send the same message to a set of objects, which can be interpreted differently by different classes of objects. And based on encapsulation and information hiding, an object does not have to be concerned with *how* something is done when using other objects. It simply sends a message to an object, and that object determines how to interpret the message. This is accomplished through the use of dynamic binding.

Dynamic binding refers to the ability of object-oriented systems to defer the data typing of objects to run time. For example, imagine that you have an array that contains instances of hourly employees and salaried employees. Both of these types of employees implement a "compute pay" method. An object can send the message to each instance contained in the array to compute the pay for that individual instance. Depending on whether the instance is an hourly employee or a salaried employee, a different method would be executed. The specific method is chosen at run time. With this ability, individual classes are easier to understand. However, the specific level of support for polymorphism and dynamic binding is language specific. Most object-oriented programming languages support dynamic binding of methods, and some support dynamic binding of attributes. As such, it is important to know what object-oriented programming language is going to be used.

Polymorphism can be a double-edged sword, however. Through the use of dynamic binding, there is no way to know before run time which specific object will be asked to execute its method. In effect, the system makes a decision that is not coded anywhere.[3] Furthermore, since all of these decisions are made at run time, it is possible to send a message to an object that it does not understand; that is, the object does not have a corresponding method. This can cause a run time error that, if not programmed to handle correctly, can cause the system to abort.

Finally, if the methods are not semantically consistent, the developer cannot assume that all methods with the same name will perform the same generic operation. For example, if the "create order" method in one type of object is implemented to print the object's descriptive information or some other function, that is, not to create an order, then semantic consistency across the interfaces of the objects no longer exists. As such, the semantics of each method must be determined individu-

---

[3] From a practical perspective, there is an implied case statement. The system chooses the method based on the type of object being asked to execute it.

ally. This substantially increases the difficulty of understanding individual objects. The key to controlling the difficulty of understanding object-oriented systems when using polymorphism is to ensure that all methods with the same name implement that same generic operation; that is, they are semantically consistent.

Inheritance

Inheritance allows developers to define classes incrementally by reusing classes previously defined as the basis for new classes. Although we could define each class separately, it might be simpler to define one general superclass that contains the data and methods needed by the subclasses, and then have these classes inherit the properties of the superclass. Subclasses inherit the appropriate attributes and methods from the superclasses "above" them. Inheritance makes it simpler to define classes.

Many different types of inheritance mechanisms have been associated with OO systems.[4] The most common inheritance mechanisms include different forms of single and multiple inheritance. *Single inheritance* allows a subclass to have only a single parent class. Currently, all object-oriented methodologies, databases, and programming languages permit extending the definition of the superclass through single inheritance.

Some object-oriented methodologies, databases, and programming languages allow a subclass to redefine some or all of the attributes and/or methods of its superclass. With *redefinition* capabilities, it is possible to introduce an *inheritance conflict,* that is, an attribute (or method) of a subclass with the same name as an attribute (or method) of a superclass. For example, in Figure 14-2, Class0 is a subclass of Class1. Both have attributes named attribute1 and methods named method1( ). This causes an inheritance conflict. Furthermore, when the definition of a superclass is modified, all of its subclasses are affected. This may introduce additional inheritance conflicts

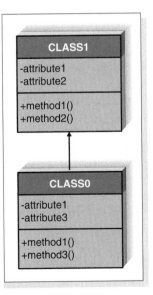

FIGURE 14-2
Single Inheritance Conflicts

---

[4] See, for example, M. Lenzerini, D. Nardi, and M. Simi, *Inheritance Hierarchies in Knowledge Representation and Programming Languages* (New York: John Wiley and Sons, 1991).

in one (or more) of the superclass's subclasses. For example, in Figure 14-2, Class1 could be modified to include an additional method, method3( ). This would add another inheritance conflict between Class1 and Class0. Therefore, developers must be aware of the effects of the modification not only in the superclass, but also in each subclass that inherits the modification.

Finally, through redefinition capabilities, it is possible for a programmer to arbitrarily cancel the inheritance of methods by placing stubs[5] in the subclass that will override the definition of the inherited method. If the cancellation of methods is necessary for the correct definition of the subclass, then it is likely that the subclass has been misclassified; that is, it is inheriting from the wrong superclass.

As you can see, from a design perspective, inheritance conflicts and redefinition can cause all kinds of problems with interpreting the final design and implementation.[6] However, most inheritance conflicts are due to poor classification of the subclass in the inheritance hierarchy; that is, the generalization (a-kind-of) semantics are violated, or the actual inheritance mechanism violates the encapsulation principle, so that the subclasses are capable of directly addressing the attributes or methods of a superclass. To address these issues, Jim Rumbaugh and his colleagues suggested the following guidelines.[7]

- Query operations should not be redefined.
- Methods that redefine inherited ones should only restrict the semantics of the inherited ones.
- The underlying semantics of the inherited method should never be changed.
- The signature (argument list) of the inherited method should never be changed.

However, many existing object-oriented programming languages violate these guidelines. When it comes to implementing the design, different object-oriented programming languages address inheritance conflicts differently. Therefore, it is important at this point in the development of the system to know what the programming language that you are going to use supports.

When considering the interaction of inheritance with polymorphism and dynamic binding, object-oriented systems provide the developer with a very powerful, but dangerous, set of tools.

Depending on the object-oriented programming language used, this interaction can allow the same object to be associated with different classes at different times. For example, an hourly employee can be treated as a generic employee, or any of its direct and indirect superclasses. Therefore, depending on whether static or dynamic binding is supported, the same object may execute different implementations of the same method at different times, or if the method only is defined with the hourly employee class and it is currently treated as a generic employee, the instance may cause a run time error to occur.[8] As stated previously, it is important to know what

---

[5] In this case, a stub is simply the minimal definition of a method to prevent syntax errors from occurring.

[6] For more inflormation, see Ronald J. Brachman, "I Lied about the Trees Or, Defaults and Definitions in Knowledge Representation," *AI Magazine* 5, no. 3 (Fall 1985): 80–93.

[7] J. Rumbaugh, M. Blaha, W. Premerlani, F. Eddy, and W. Lorensen, *Object-Oriented Modeling and Design* (Englewood Cliffs, NJ: Prentice Hall, 1991).

[8] This happens with novices quite regularly when using C++.

object-oriented programming language is going to be used so that these kinds of issues can be solved with the design, instead of the implementation, of the class.

With *multiple inheritance,* a subclass may inherit from more than one superclass. In this situation, the types of inheritance conflicts are multiplied. In addition to the possibility of having an inheritance conflict between the subclass and one (or more) of its superclasses, it is now possible to have conflicts between two (or more) superclasses. In this later case, three different types of additional inheritance conflicts can occur. These include:

- Two inherited attributes (or methods) have the same name and semantics.
- Two inherited attributes (or methods) have different names but with identical semantics; that is, they are *synonyms.*
- Two inherited attributes (or methods) have the same name but with different semantics; that is, they are *homonyms.* This also violates the proper use of polymorphism.

For example, in Figure 14-3, Class0 is a subclass of both Class1 and Class2. In this case, Class 1 and Class2 conflict with attribute1 and method1( ). Which one should Class0 inherit? Since they are the same, semantically speaking, does it really matter? It is also possible that Class1 and Class2 could have a semantic conflict on attribute2 (method2) and attribute3 (method3) if attribute2 (method2) and attribute3 (method3) are synonyms. Practically speaking, the only way to prevent this situation is for the developer to catch it during the design of the subclass. Finally, what if attribute4 (method4) of Class1 and attribute4 (method4) of Class2 are homonyms? Shouldn't Class0 inherit both of them? With the potential for these additional types of conflicts, there is a risk of decreasing the understandability in an object-oriented system, instead of increasing it, through the use of multiple inheritance. As such, great care should be taken when using multiple inheritance.

CONCEPTS   **14-A INHERITANCE ABUSES**

IN ACTION

Meilir Page-Jones, through his consulting company, identified a set of abuses of inheritance. In some cases, these abuses led to lengthy and bloody disputes, to gruesome implementations, and in one case, to the destruction of the development team. In all cases, the error was in not enforcing a generalization (a-kind-of) semantics. In one case, the inheritance hierarchy was inverted: BoardMember was a superclass of Manager, which was a superclass of Employee. However, in this case, an Employee is NOT A-Kind-Of Manager which is NOT A-Kind-Of BoardMember. In fact, the opposite was true. However, if you think of an Organization Chart, a BoardMember is "Super" to a Manager, which is "Super" to an Employee. In another example, the client's firm attempted to use inheritance to model a membership idea; for example, Student is a member of a club. However, the club should have had an attribute that contained the student members. In the other examples, inheritance was used to implement an association relationship and an aggregation relationship.

Source: Meilir Page-Jones, *Fundamentals of Object-Oriented Design in UML* (Reading, MA: Addison-Wesley, 2000).

QUESTION
As an analyst, how can you attempt to avoid these types of inheritance abuses?

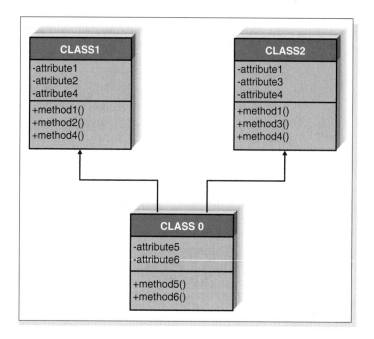

FIGURE 14-3
Additional Inheritance Conflicts with Multiple Inheritance

DESIGN CRITERIA

When considering the design of an object-oriented system, there is a set of criteria that can be used to determine whether the design is a good one or a bad one. According to Coad and Yourdon, "A good design is one that balances trade-offs to minimize the total cost of the system over its entire lifetime."[9] These criteria include coupling, cohesion, and connascence.

Coupling

Coupling refers to the level of interdependency or interrelationship among the modules (classes, objects, and methods) in a system. The higher the interdependency is, the more likely changes in part of a design can cause changes to be required in other parts of the design. For object-oriented systems, Coad and Yourdon[10] identified two types of coupling: interaction and inheritance.

Interaction coupling deals with the coupling among methods and objects through message passing. Lieberherr and Holland proposed the *Law of Demeter* as a guideline to minimize this type of coupling.[11] Essentially, the law minimizes the number of objects that can receive messages from a given object. The law states that an object should only send messages to itself, an object that is contained in an attribute of the object (or one of its superclasses), an object that is passed as a parameter to the method, an object that is created by the method, or an object that is

---

[9] Peter Coad and Edward Yourdon, *Object-Oriented Design,* (Englewood Cliffs, NJ: Yourdon Press, 1991), p. 128.

[10] Ibid.

[11] Karl J. Lieberherr and Ian M. Holland, "Assuring Good Style for Object-Oriented Programs," *IEEE Software* 6, no. 5 (September 1989): 38–48 and Karl J. Lieberherr, *Adaptive Object-Oriented Software: The Demeter Method with Propagation Patterns* (Boston, MA: PWS Publishing, 1996).

stored in a global variable. In each case, interaction coupling is increased. For example, the coupling increases between the objects if the calling method passes attributes to the called method or if the calling method depends on the value being returned by the called method.

There are six types of interaction coupling, each of which falls within different parts of a good to bad continuum. They range from no direct coupling to content coupling. Figure 14-4 presents the different types of interaction coupling. In general, interaction coupling should be minimized. The one possible exception is that non-problem domain classes must be coupled to their corresponding problem domain classes. For example, a report object (on the human–computer interaction layer) that displays the contents of an employee object (on the problem domain layer) will be dependent on the employee object. In this case, for optimization purposes, the report class may be even content or pathologically coupled to the employee class. However, problem domain classes should never be coupled to non-problem domain classes.

Inheritance coupling, as its name implies, deals with how tightly coupled the classes are in an inheritance hierarchy. Most authors tend to say simply that this type of coupling is desirable. However, depending on the issues raised previously with inheritance—inheritance conflicts, redefinition capabilities, and dynamic binding—a high level of inheritance coupling may not be a good thing. For example, should a method defined in a subclass be allowed to call a method defined in one of its superclasses? Or should a method defined in a subclass refer to an attribute defined in one of its superclasses? Or, even more confusing, can a method defined in an abstract superclass depend on its subclasses to define a method or attribute on which it is dependent? Depending on the object-oriented programming language being used, all of these options are possible.

Level	Type	Description
Good	No Direct Coupling	The methods do not relate to one another; that is, they do not call one another.
	Data	The calling method passes a variable to the called method. If the variable is composite, (i.e., an object), the entire object is used by the called method to perform its function.
	Stamp	The calling method passes a composite variable (i.e., an object) to the called method, but the called method only uses a portion of the object to perform its function.
	Control	The calling method passes a control variable whose value will control the execution of the called method.
	Common or Global	The methods refer to a "global data area" that is outside the individual objects.
Bad	Content or Pathological	A method of one object refers to the inside (hidden parts) of another object. This violates the principles of encapsulation and information hiding. However, C++ allows this to take place through the use of "friends."

Source: These types were adapted from Page-Jones, *The Practical Guide to Structured Systems Design*, 2nd ed, Englewood Cliffs, NJ: Yardon Press, 1988; and Glenford Myers, *Composite/Structured Design*. New York: Van Nostrand Reinhold, 1978.

FIGURE 14-4
Types of Interaction Coupling

As Alan Snyder has pointed out, most problems with inheritance revolve around the ability within the object–oriented programming languages to violate the encapsulation and information hiding principles.[12] Therefore, again, knowledge of which object-oriented programming language is to be used is crucial. From a design perspective, the developer will need to optimize the tradeoffs of violating the encapsulation and information hiding principles and increasing the desirable coupling between subclasses and its superclasses. The best way to solve this conundrum is to ensure that inheritance is used only to support generalization/specialization (a-kind-of) semantics. All other uses should be avoided.

Cohesion

Cohesion refers to how single-minded a module (class, object, or method) is within a system. A class or object should only represent one thing, and a method should only solve a single task. Coad and Yourdon have identified three general types of cohesion—method, class, and generalization/specialization—for object-oriented systems.[13]

Method cohesion addresses the cohesion within an individual method, that is, how single-minded a method is. Methods should do one and only one thing. A method that actually performs multiple functions is more difficult to understand than one that only performs a single function. Seven types of method cohesion have been identified (see Figure 14-5). They range from functional cohesion (good) to coincidental cohesion (bad). In general, method cohesion should be maximized.

Class cohesion is the level of cohesion among the attributes and methods of a class, that is, how single-minded a class is. A class should only represent one thing—an employee, a department, or an order, for example. All attributes and methods contained in a class should be required for the class to represent the thing. For example, an employee class should have attributes that deal with a social security number, last name, first names, middle initial, addresses, and benefits, but it should not have attributes such as door, engine, or hood. Furthermore, there should be no attributes or methods that are never used. In other words, a class should have only the attributes and methods necessary to fully define instances for the problem at hand. In this case, we have *ideal class cohesion.* Glenford Meyers suggested that a cohesive class[14] should

- contain multiple methods that are visible outside of the class and each visible method should only perform a single function (i.e., it has functional cohesion),
- have methods that only refer to attributes or other methods defined with the class or its superclass(es), and
- not have any control-flow couplings between its methods.

---

[12] Alan Snyder, "Encapsulation and Inheritance in Object-Oriented Programming Languages," in N. Meyrowitz, ed., *OOPSLA '86 Conference Proceedings, ACM SigPlan Notices,* 21, no. 11 (November 1986) and Alan Snyder, "Inheritance and the Development of Encapsulated Software Components," in B. Shriver and P. Wegner, eds., *Research Directions in Object-Oriented Programming* (Cambridge, MA: MIT Press, 1987).

[13] Coad and Yourdon, *Object-Oriented Design.*

[14] We have adapted his informational-strength module criteria from structured design to object-oriented design. (See Glenford J. Myers, *Composite/Structured Design.*

Level	Type	Description
Good	Functional	A method performs a single problem-related task (e.g., Calculate current GPA).
	Sequential	The method combines two functions in which the output from the first one is used as the input to the second one (e.g., format and validate current GPA).
	Communicational	The method combines two functions that use the same attributes to execute (e.g., calculate current and cumulative GPA).
	Procedural	The method supports multiple weakly related functions. For example, the method could calculate student GPA, print student record, calculate cumulative GPA, and print cumulative GPA.
	Temporal or Classical	The method supports multiple related functions in time (e.g., initialize all attributes).
	Logical	The method supports multiple related functions, but the choice of the specific function is chosen based on a control variable that is passed into the method. For example, the called method could open a checking account, open a savings account, or calculate a loan, depending on the message that is send by its calling method.
Bad	Coincidental	The purpose of the method cannot be defined or it performs multiple functions that are unrelated to one another. For example, the method could update customer records, calculate loan payments, print exception reports, and analyze competitor pricing structure.

Source: These types were adapted from Page-Jones, *The Practical Guide to Structured Systems* and Myers, *Composite/Structured Design.*

FIGURE 14-5
Types of Method Cohesion

Page-Jones[15] has identified three less than desirable types of class cohesion: mixed-instance, mixed-domain, and mixed-role (see Figure 14-6). An individual class can have a mixture of any of the three types.

Generalization/specialization cohesion addresses the sensibility of the inheritance hierarchy. How are the classes in the inheritance hierarchy related? Are the classes related through a generalization/specialization (a-kind-of) semantics? Or are they related via some association, aggregation, or membership type of relationship that was created for simple reuse purposes? Recall, all of the issues raised previously on the use of inheritance. Highly cohesive inheritance hierarchies should only support the semantics of generalization and specialization (a-kind-of).

Connascence

Connascence[16] generalizes the ideas of cohesion and coupling, and combines them with the arguments for encapsulation. To accomplish this, three levels of

[15] See Page-Jones, *Fundamentals of Object-Oriented Design in UML,* Reading, MA: Addison-Wesley, 2000.
[16] See Meilir Page-Jones, "Comparing Techniques by Means of Encapsulation and Connascence," *Communications of the ACM* 35, no. 9 (September 1992): 147–151.

Level	Type	Description
Good	Ideal	The class has none of the mixed cohesions.
	Mixed-Role	The class has one or more attributes that relate objects of the class to other objects on the same layer (e.g., the problem domain layer), but the attribute(s) have nothing to do with the underlying semantics of the class.
	Mixed-Domain	The class has one or more attributes that relate objects of the class to other objects on a different layer. As such, they have nothing to do with the underlying semantics of the thing that the class represents. In these cases, the offending attribute(s) belongs in another class located on one of the other layers. For example, a port attribute located in a problem domain class should be in a system architecture class that is related to the problem domain class.
Worse	Mixed-Instance	The class represents two different types of objects. The class should be decomposed into two separate classes. Typically, different instances only use a portion of the full definition of the class.

Source: Page-Jones, *Fundamentals of Object-Oriented Design in UML.*

FIGURE 14-6
Types of Class Cohesion

encapsulation have been identified. Level-0 encapsulation refers to the amount of encapsulation realized in an individual line of code; level-1 encapsulation is the level of encapsulation attained by combining lines of code into a method; and level-2 encapsulation is achieved by creating classes that contain both methods and attributes. Method cohesion and interaction coupling primarily address level-1 encapsulation. Class cohesion, generalization/specialization cohesion, and inheritance coupling only address level-2 encapsulation. Connascence, as a generalization of cohesion and coupling, addresses both level-1 and level-2 encapsulation.

But what exactly is connascence? Connascence literally means to be born together. From an object-oriented design perspective, it really means that two modules (classes or methods) are so intertwined that if you make a change in one, it is likely that a change will be requied in the other. On the surface, this is very similar to coupling and as such should be minimized. However, when you combine coupling with the encapsulation levels, it is not quite as simple as that. In this case, you want to (1) minimize overall connascence by eliminating any unnecessary connascence throughout the system, (2) minimize connascence across any encapsulation boundaries, such as method boundaries and class boundaries, and (3) maximize connascence within any encapsulation boundary.

Based on these guidelines, a subclass should never directly access any hidden attribute or method of a superclass; that is, a subclass should not have any special rights to the properties of its superclass(es). If direct access to the nonvisible attributes and methods of a superclass by its subclass is allowed, which is permitted in most object-oriented programming languages, and a modification to the superclass is made, then because of the connascence between the subclass and its superclass, it is likely that a modification to the subclass also will be required. In other words, the subclass has access to something across an encapsulation boundary (the class boundary between the subclass and the superclass). Practically speaking, you

Type	Description
Name	If a method refers to an attribute, it is tied to the name of the attribute. If the attribute's name changes, the content of the method will have to change.
Type or Class	If a class has an attribute of type A, it is tied to the type of the attribute. If the type of the attribute changes, the attribute declaration will have to change.
Convention	A class has an attribute in which a range of values has a semantic meaning (e.g., account numbers whose values range from 1000 to 1999 are assets). If the range would change, then every method that used the attribute would have to be modified.
Algorithm	Two different methods of a class are dependent on the same algorithm to execute correctly (e.g., insert an element into an array and find an element in the same array). If the underlying algorithm would change, then the insert and find methods would also have to change.
Position	The order of the code in a method or the order of the arguments to a method is critical for the method to execute correctly. If either is wrong, then the method will, at least, not function correctly.

Source: Page-Jones, *"Comparing Techniques by Means of Encapsulation and Connascence"* and Page-Jones, *Fundamentals of Object-Oriented Design in UML.*

FIGURE 14-7
Types of Connascence

should maximize the cohesion (connascence) within an encapsulation boundary and minimize the coupling (connascence) between the encapsulation boundaries. There are many possible types of connascence. Figure 14-7 describes five of them.

OBJECT DESIGN ACTIVITIES

The design activities for classes and methods are an extension of the analysis and evolution activities presented previously (see Chapters 6 through 9). In this case, we expand the descriptions of the partitions, layers, and classes. Practically speaking, the expanded descriptions are created through the activities that take place during the detailed design of the classes and methods. As such, we limit this discussion to the detailed design of classes and methods. The activities used to design classes and methods include additional specification of the current model, identifying opportunities for reuse, restructuring the design, and, finally, optimizing the design. Of course, any changes made to a class on one layer can cause the classes on the other layers that are coupled to it to be modified also.

Additional Specification

At this point in the development of the system, it is crucial to review the current set of structural and behavioral models.[17] First, we should ensure that the classes on the problem domain layer are both necessary and sufficient to solve the underlying problem. To do this, we need to be sure that there are no missing attributes or methods and no extra or unused attributes or methods in each class. Furthermore, are

[17] This has been mentioned before, but one cannot overemphasize the importance of constantly reviewing the evolving design.

there any missing or extra classes? If we have done our job well during the analysis phase, few, if any, attributes or methods will have to be added to the models. However, it is always better to be safe than sorry.

Second, we need to finalize the *visibility* (hidden or visible) of the attributes and methods in each class. Depending on the object-oriented programming language used, this could be predetermined. For example, in Smalltalk attributes are hidden and methods are visible.[18]

Third, we need to decide on the signature of every method in every class. The *signature* of a method consists of three parts: the name of the method, the parameters or arguments that must be passed to the method, and the type of value that the method will return to the calling method.

Fourth, we need to define any constraints that must be preserved by the objects, for example, an attribute of an object that can only have values in a certain range. The three different types of constraints—pre-conditions, post-conditions, and invariants—are captured in the form of contracts (described later in this chapter) and assertions added to CRC cards and class diagrams. We must also decide how to handle a violation of a constraint. Should the system simply abort? Should the system automatically undo the change that caused the violation? Should the system let the end user determine the approach to correct the violation? In other words, the designer must design the errors that the system is expected to handle. It is best to not leave these types of problems for the programmer to solve. Violations of a constraint are known as *exceptions* in languages such as C++ and Java.

The additional specification activities also are applicable to the other layers: system architecture (Chapter 10), human–computer interaction (Chapters 11 and 12), and data management (Chapter 13).

Identifying Opportunities for Reuse

Previously, we looked at possibly employing reuse in our models in the analysis phase through the use of *patterns* (see Chapter 7). In the design phase, additional opportunities arise for reuse through the use of frameworks, libraries, and components. The opportunities will vary depending on which layer is being reviewed. For example, a class library will probably not be of much help on the problem domain layer, but an appropriate class library could be of great help on the foundation layer. In this section, we describe the use of frameworks, libraries, and components.

A *framework* consists of a set of implemented classes that can be used as a basis for implementing an application. For example, frameworks available for CORBA and DCOM enable you to base the implementation of part of the system architecture layer. Most frameworks allow you to create subclasses to inherit from classes in the framework. There are object persistence frameworks that can be purchased and used to add persistence to the problem domain classes, which would be helpful on the data management layer. Of course, when inheriting from classes in a framework, you are creating a dependency, that is, increasing the inheritance coupling from the subclass to the superclass. Therefore, if you use a framework and the vendor makes changes to the framework, you will have to at least recompile the system when you upgrade to the new version of the framework.

A *class library* is similar to a framework in that you typically have a set of implemented classes that were designed for reuse. However, frameworks tend to

---

[18] By default, most object-oriented analysis and design approaches assume Smalltalk's approach.

be more domain specific. In fact, frameworks may be built using a class library. A typical class library could be purchased to support numerical or statistical processing, file management (data management layer), or user interface development (the human–computer interaction layer). In some cases, you will create instances of classes contained in the class library, and in other cases you will extend classes in the class library by creating subclasses based on them. Like frameworks, if you use inheritance to reuse the classes in the class library, you will run into all of the issues dealing with inheritance coupling and connascence. If you directly instantiate classes in the class library, you will create a dependency between your object and the library object based on the signatures of the methods in the library object. This will increase the interaction coupling between the class library object and your object.

A *component* is a self-contained, encapsulated piece of software that can be "plugged" into a system to provide a specific set of required functionalities. Today, many components are available for purchase that have been implemented using *ActiveX* or *JavaBean* technologies. A component has a well-defined *Application Program Interface* (API), which is essentially a set of method interfaces to the objects contained in the component. The internal workings of the component are hidden behind the API. Components can be implemented using class libraries and frameworks. However, components can also be used to implement frameworks. Unless the API changes between versions of the component, upgrading to a new version normally will require only linking the component back into the application. As such, recompilation typically is not required.

Which of these approaches should you use? It depends on what you are trying to build. In general, frameworks are used mostly to aid in developing objects on the system architecture, human–computer interaction, or data management layers; components are used primarily to simplify the development of objects on the problem domain and human–computer interaction layers; and class libraries are used to develop frameworks and components and to support the foundation layer.

Restructuring the Design

Once the individual classes and methods have been specified and class libraries, frameworks, and components have been incorporated into the evolving design, you should use factoring to restructure the design. *Factoring,* if you recall from Chapter 9, is the process of separating out aspects of a method or class into a new method or class to simplify the overall design. For example, when reviewing a set of classes on a particular layer, you might discover that a subset of them shares a similar definition. In that case, it may be useful to factor out the similarities and create a new class. Based on the issues related to cohesion, coupling, and connascence, the new class may be related to the old classes via inheritance (generalization) or through an aggregation or association relationship.

Up until this point in the development of the system, it has been assumed that the classes and methods in the models would be implemented directly in an object-oriented programming language. However, at this point in time, it is important to map the current design to the capabilities of the programming language used. For example, if you have used multiple inheritance in your design, but you are implementing in a language that only supports single inheritance, then the multiple inheritance must be factored out of the design. The easiest way to do this is to use the following rule:

RULE 1a: Convert the additional inheritance relationships to association relationships. Add appropriate methods to ensure that all information is still available to the original class.

RULE 1b: Flatten the inheritance hierarchy by copying the attributes and methods of the additional superclass(es) down to all of the subclasses.

This rule is similar to the one described in Chapter 13 which handled multiple inheritance mapping to single inheritance in regards to OODBMSs. The same type of problems associated with creating, reading, updating, and deleting instances discussed in Chapter 13 apply here.

If the implementation is to be done in an object-based language, one that does not support inheritance,[19] or a nonobject-based language, such as C or Pascal, mappings similar to those described in Chapter 13 for ORDBMS and RDBMS must be done. From a practical perspective, you are much better off implementing the design in an object-oriented programming language such as Smalltalk, C++, or Java.

Another useful process for restructuring the evolving design is *normalization*. Normalization was described in Chapter 13 in relation to relational databases. However, normalization can sometimes be useful to identify potential classes that are missing from the design. Also, related to normalization is the requirement to implement the actual association and aggregation relationships as attributes. Virtually no object-oriented programming language differentiates between attributes and association and aggregation relationships. Therefore, all association and aggregation relationships must be converted to attributes in the classes.

Finally, all inheritance relationships should be challenged to ensure that they only support a generalization/specialization (a-kind-of) semantics. Otherwise, all of the problems mentioned previously in connection with inheritance coupling, class cohesion, and generalization/specialization cohesion will come to pass.

Optimizing the Design[20]

Up until now, we have focused our energy on developing an understandable design. With all of the classes, patterns, collaborations, partitions, and layers designed and the class libraries, frameworks, and components included in the design, understandability has rightly been our primary focus. However, increasing the understandability of a design typically creates an inefficient design. Conversely, focusing on efficiency issues will deliver a design that is more difficult to understand. A good practical design will manage the inevitable tradeoffs that must occur to create an acceptable system. In addition to the optimization material covered in Chapter 13, in this section, we describe a set of simple optimizations that can be used to create a more efficient design.

The first optimization to consider is to review the access paths between objects. In some cases, a message from one object to another may have a long path

---

[19] In this case, we are talking about implementation inheritance, not the so-called interface inheritance. Interface inheritance supported by Visual Basic and Java only supports inheriting the requirements to implement certain methods, not any implementation. Java and Visual Basic.Net also support single inheritance as described in this text, whereas Visual Basic 6 only supports interface inheritance.

[20] The material contained in this section is based on James Rumbaugh, Michael Blaha, William Premerlani, Frederick Eddy, and William Lorensen, *Object-Oriented Modeling and Design* (Englewood Cliffs, NJ: Prentice-Hall, 1991) and Bernd Brugge and Allen H. Dutoit, *Object-Oriented Software Engineering: Conquering Complex and Changing Systems,* (Englewood Cliffs, NJ: Prentice-Hall, 2000).

to traverse; that is, it goes through many objects. If the path is long and the message is sent frequently, a redundant path should be considered. Adding an attribute to the calling object that will store a direct connection to the object at the end of the path can accomplish this.

A second optimization is to review each attribute of each class. Which methods use the attributes and which objects use the methods should be determined. If the only methods that use an attribute are read and update methods and only instances of a single class send messages to read and update the attribute, the attribute may belong with the calling class instead of the called class. Moving the attribute to the calling class will substantially speed up the system.

A third optimization is to review the direct and indirect fan-out of each method. *Fan-out* refers to the number of messages sent by a method. The direct fan-out is the number of messages sent by the method itself, whereas the indirect fan-out also includes the number of messages sent by the methods called by the other methods in a message tree. If the fan-out of a method is high relative to that of other methods in the system, the method should be optimized.

A fourth optimization is to look at the execution order of the statements in often-used methods. In some cases, it may be possible to rearrange some of the statement to be more efficient. For example, if it is known, based on the objects in the system, that a search routine can be narrowed by searching on one attribute before another one, then the search algorithm should be optimized by forcing it to always search in a predefined order.

A fifth optimization is to avoid recomputation by creating a *derived attribute* (or *active value*), for example, a total, that will store the value of the computation. This is also known as caching computational results. It can be accomplished by adding a *trigger* to the attributes contained in the computation, that is, those attributes in which the derived attribute is dependent. This would require a recomputation to take place only when one of the attributes that go into the computation is changed. Another approach would be to mark the derived attribute and delay the recomputation until the next time the derived attribute is accessed. This last approach delays the recomputation as long as possible. In this manner, a computation does not occur unless it has to occur. Otherwise, every time a derived attribute needs to be accessed, a computation will be required.

CONSTRAINTS AND CONTRACTS

Contracts were introduced in Chapter 7 in association with collaborations. As you will recall, a *contract* formalizes the interactions between the client and server objects, where a *client (consumer)* object is an instance of a class that sends a message to a *server (supplier)* object that executes one of its methods in response to the request. Contracts are modeled on the legal notion of a contract where both parties, client and server objects, have obligations and rights. Practically speaking, a contract is a set of constraints and guarantees. If the constraints are met, then the server object will guarantee certain behavior.[21] Constraints can be written in either a natural language (such as English), Structured English, pseudocode, or a formal

---

[21] The idea of using contracts in design evolved from the "Design by Contract" technique developed by Bertrand Meyer. See Bertrand Meyer, *Object-Oriented Software Construction* (Englewood Cliffs, NJ: Prentice Hall, 1988).

language (such as, UML's Object Constraint Language.)[22] Three different types of *constraints* are typically captured in object-oriented design: pre-conditions, post-conditions, and invariants.

Contracts are used primarily to establish the pre-conditions and post-conditions for a method to be able to execute properly. A *pre-condition* is a constraint that must be met for a method to execute. For example, the parameters passed to a method must be valid for the method to execute; otherwise, an exception should be raised. A *post-condition* is a constraint that must be met after the method executes or the effect of the method execution must be undone. For example, the method cannot make any of the attributes of the object take on an invalid value. In this case, an exception should be raised, and the effect of the method's execution should be undone.

Where pre-conditions and post-conditions model the constraints on an individual method, *invariants* model constraints that must always be true for all instances of a class. Examples of invariants include domains or types of attributes, multiplicity of attributes, and the valid values of attributes. This includes the attributes that model association and aggregation relationships. For example, if an association relationship is required, an invariant should be created that will force it to have a valid value for the instance to exist. Invariants are normally attached to the class. As such, we can attach invariants to the CRC cards or class diagram by adding a set of assertions to them.

In Figure 14–8, the back of the CRC card constrains the attributes of an Order to specific types. For example, Order Number must be an unsigned long, and Customer must be an instance of the Customer class. Furthermore, additional invariants were added to four of the attributes. For example, Cust ID must not only be an unsigned long, but it must also have one and only one value, that is, a multiplicity of (1..1), and it must have the same value as the result of the GetCustID( ) message sent to the instance of Customer stored in the Customer attribute. This is an example of having a foreign key as part of an object (see Chapter 13). Also shown is the constraint for an instance to exist, an instance of the Customer class, an instance of the State class, and at least one instance of the Product class must be associated with the Order object (see Relationships section of the CRC card). Figure 14-9 portrays the

[22] See Jos Warmer and Anneke Kleppe, *The Object Constraint Language: Precise Modeling with UML* (Reading, MA: Addison-Wesley, 1999).

Front:

Class Name: *Order*		ID: 2	Type: *Concrete, Domain*
Description: *an individual that needs to receive or has received medical attention*			Associated Use Cases: 3

Responsibilities	Collaborators
Calculate Subtotal	
Calculate tax	
Calculate shipping	
Calculate total	

Back:

Attributes:

Order Number	(1..1)	(unsigned long)	
Date	(1..1)	(Date)	
Sub Total	(0..1)	(double)	{Sub Total = sum (Product Order. GetExtension())}
Tax	(0..1)	(double)	{Tax = State. Get Tax Rate()* Sub Total}
Shipping	(0..1)	(double)	
Total	(0..1)	(double)	
Customer	(1..1)	(Customer)	
Cust ID	(1..1)	(unsigned long)	{Cust ID = Customer. Get Cust ID()}
State	(1..1)	(State)	
StateName	(1..1)	(String)	{State Name = State. Get State()}

Relationships:

    Generalization (a-kind-of):

    Aggregation (has-parts):

    Other Associations: Product {1..*} Customer {1..1} State {1..1}

FIGURE 14-8

Invariants on a CRC Card

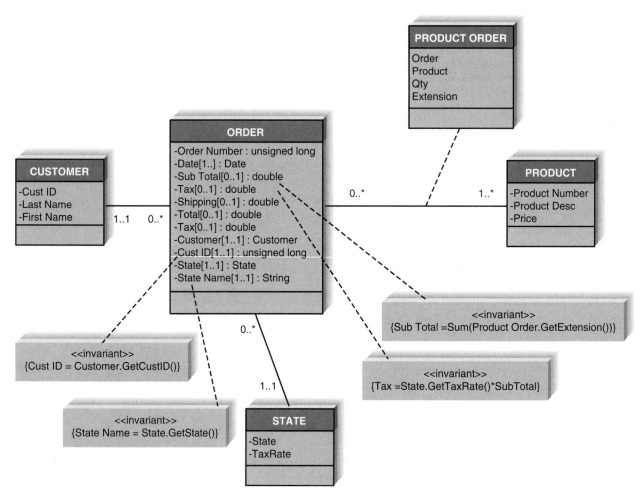

FIGURE 14-9
Invariants on a Class Diagram

same set of constraints on a class diagram. However, if all invariants are placed on a class diagram, the diagram will become very difficult to understand. As such, we recommend extending the CRC card to document the invariants instead of attaching them all to the class diagram.

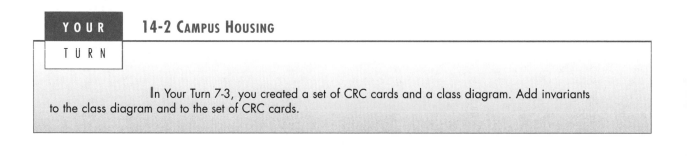

YOUR TURN

14-2 CAMPUS HOUSING

In Your Turn 7-3, you created a set of CRC cards and a class diagram. Add invariants to the class diagram and to the set of CRC cards.

Elements of a Contract

Contracts document the message passing that takes place between objects. Technically speaking, a contract should be created for each message sent and received by each object, one for each interaction. However, there would be quite a bit of duplication if this were done. In practice, a contract is created for each method that can receive messages from other objects, that is, one for each visible method.

A contract should contain the information necessary for a programmer to understand what a method is to do; that is, they are declarative in nature. This information includes the method name, class name, ID number, client objects, associated use cases, description, arguments received, type of data returned, and the pre- and post-conditions.[23] Contracts do not have a detailed algorithmic description of how the method is to work; they are not procedural in nature. Detailed algorithmic descriptions typically are documented in a method specification (this is described later in this chapter). In other words, a contract is composed of the information required for the developer of a client object to know what messages can be sent to the server objects and what the client can expect in return. Figure 14-10 shows a sample format for a contract.

Since each contract is associated with a specific method and a specific class, the contract must document them. The ID number of the contract is used to provide a unique identifier for every contract. The Clients (Consumers) element of a contract is a list of classes and methods that send a message to this specific method. This list is determined by reviewing the sequence diagrams associated with the server class. The Associated Use Cases element is a list of use cases in which this method is used to realize the implementation of the use case. The use cases listed here can be found by reviewing the server class's CRC card and the associated sequence diagrams.

The Description of Responsibilities provides an informal description of *what* the method is to perform, not *how* it is to do it. The arguments received are the data types of the parameters passed to the method, and the value returned is the data type of the value that the method will return to its clients. Together with the method name, they form the signature of the method.

The pre-condition and post-condition element is where the pre- and post-conditions for the method are recorded. Recall that pre- and post-conditions can be

---

YOUR   **14-3 CONTRACT**

TURN

   Using the CRC card in Figure 14-8, the class diagram in Figure 14-9, and the sample contract format in Figure 14-10 as guides, create contracts for the Calculate Subtotal, Calculate tax, Calculate shipping, and Calculate total methods.

---

[23] Currently, there is no standard format for a contract. The contract in Figure 14-10 is based on material contained in Ian Graham, *Migrating to Object Technology* (Workingham, England: Addison-Wesley, 1995); Craig Larman, *Applying UML and Patterns: An Introduction to Object-Oriented Analysis and Design* (Englewood Cliffs, NG: Prentice-Hall, 1998); Meyer, *Object-Oriented Software Construction;* and R. Wirfs-Brock, B. Wilkerson, and L. Wiener, *Designing Object-Oriented Software* (Englewood Cliffs, NJ: Prentice Hall, 1990).

Method Name:	Class Name:		ID:
Clients (Consumers):			
Associated Use Cases:			
Description of Responsibilities:			
Arguments Received:			
Type of Value Returned:			
Pre-Conditions:			
Post-Conditions:			

FIGURE 14-10
Sample Contract Format

written in natural language, Structured English, pseudocode, or a formal language. It does not matter which one you use. However, the more precisely they are written, the least likely that the programmer will misunderstand them. As such, we recommend that you use either pseudocode or a formal language such as UML's Object Constraint Language.[24]

METHOD SPECIFICATION

Once the analyst has communicated the big picture of how the system needs to be put together, he or she needs to describe the individual classes and methods in enough detail so that programmers can take over and begin writing code. Methods on the CRC cards, class diagram, and contracts are described using *method specifications.* Method specifications are written documents that include explicit instructions on how to write the code to implement the method. Typically, project team members write a specification for each method and then pass them along to programmers, who write the code during the implementation phase of the project. Specifications need to be very clear and easy to understand, or programmers will be slowed down trying to decipher vague or incomplete instructions.

Syntax

There is no formal syntax for a method specification, so every organization uses its own format, often using a form like the one in Figure 14-11. Typical method specification forms contain four components that convey the information that pro-

[24] See Warmer and Kleppe, *The Object Constraint Language.*

Method Name:	Class Name:	ID:
Contract ID:	Programmer:	Date Due:

Programming Language:
❏ Visual Basic   ❏ Smalltalk   ❏ C++   ❏ Java

Triggers/Events:

Arguments Received:
        Data Type: Notes:

Messages Sent & Arguments Passed:
    ClassName.MethodName: Data Type: Notes:

Argument Returned:
        Data Type: Notes:

Algorithm Specification:

Misc.Notes:

FIGURE 14-11
Method Specification Form

grammers will need to write the appropriate code: general information, events, message passing, and algorithm specification.

General Information   The top of the form in Figure 14-11 contains general information, such as the name of the method, the name of the class in which this implementation of the method will reside, the ID number, the Contract ID that identifies the contract associated with this method implementation, the programmer assigned, the date due, and the target programming language. This information is used to help manage the programming effort.

Events   The second section of the form is used to list the events that trigger the method. An *event* is an action that happens or takes place. Clicking the mouse generates a mouse event, and pressing a key generates a keystroke event; in fact, almost everything the user does generates an event.

In the past, programmers used procedural programming languages (e.g., COBOL, C) that contained instructions implemented in a predefined order, as determined by the computer system, and users were not allowed to deviate from the order. Many programs today are *event-driven* (e.g., programs written in languages such as Visual Basic, Smalltalk, C++, or Java), and event-driven programs include methods that are executed in response to an event initiated by the user, system, or another method. After initialization, the system waits for some kind of event to happen, and when it does, a method is fired which carries out the appropriate task, and then the system waits once again.

We have found that many programmers still use method specifications when programming in event-driven languages, and they include the event section on the form to capture when the method will be invoked. Other programmers have switched to other design tools that capture event-driven programming instructions, such as the statechart diagram described in Chapter 8.

Message Passing   The next set of sections of the method specification describes the message passing to and from the method, which are identified on the sequence and collaboration diagrams. Programmers will need to understand what arguments are being passed into, from, and returned by the method since the arguments ultimately will translate into attributes and data structures within the actual method.

Algorithm Specification   Algorithm specifications can be written in Structured English, some form of pseudocode, or some type of formal language.[25] *Structured English* is simply a formal way of writing instructions that describe the steps of a process. Because it is the first step toward implementation of the method, it looks much like a simple programming language. Structured English uses short sentences that clearly describe exactly what work is performed on what data. There are many versions of Structured English because there are no formal standards; each organization has its own type of Structured English.

Figure 14-12 shows some examples of commonly used Structured English statements. *Action statements* are simple statements that perform some action. *If statements* control actions that are performed under different conditions, a *For statement* (or *a While statement*) performs some actions until some condition is reached. Finally, a *Case statement* is an advanced form of an If statement that has several mutually exclusive branches.

Pseudocode is a language that contains logical structures, including sequential statements, conditional statements, and iteration. It differs from Structured English in that pseudocode contains details that are programming-specific, like initialization instructions or linking, and it also is more extensive so that a programmer can write the module by mirroring the pseudocode instructions. In general, pseudocode is much more like real code, and its audience is the programmer as opposed to the analyst. Its format is not as important as the information it conveys. Figure 14-13 shows a short example of pseudocode for a module that is responsible for getting CD information.

Writing good pseudocode can be difficult. Imagine creating instructions that someone else can follow without having to ask for clarification or make a wrong

---

[25] For our purposes, Structured English or pseudocode will suffice. However, there has been some work with the Catalysis, Fusion, and Syntropy methodologies to include formal languages, such as VDM and Z, into specifying object-oriented systems.

Common Statements	Example
Action Statement	Profits = Revenues − Expenses Generate Inventory-Report
If Statement	IF Customer Not in the Customer Object Store THEN Add Customer record to Customer Object Store ELSE Add Current-Sale to Customer's Total-Sales 　Update Customer record in Customer Object Store
For Statement	FOR all Customers in Customer Object Store DO 　Generate a new line in the Customer-Report 　Add Customer's Total-Sales to Report-Total
Case Statement	CASE 　IF Income < 10,000: Marginal-tax-rate = 10 percent 　IF Income < 20,000: Marginal-tax-rate = 20 percent 　IF Income < 30,000: Marginal-tax-rate = 31 percent 　IF Income < 40,000: Marginal-tax-rate = 35 percent 　ELSE Marginal-Tax-Rate = 38 percent ENDCASE

FIGURE 14-12
Structured English

assumption. For example, have you ever given a friend directions to your house, and they ended up getting lost? To you, the directions may have been very clear, but that is because of your personal assumptions. To you, the instruction "take the first left turn" may really mean, "take a left turn at the first stoplight." Someone else may have made the interpretation of "take a left turn at the first road, with or without a light." Therefore, when writing pseudocode, pay special attention to detail and readability.

The last section of the method specification provides space for other information that needs to be communicated to the programmer, such as calculations, special business rules, calls to subroutines or libraries, and other relevant issues. This can also point out changes or improvements that will be made to any of the other design documentation based on problems that the analyst detected during the specification process.[26]

Y O U R   T U R N

14-4 METHOD SPECIFICATION

Using the CRC card in Figure 14-8, the class diagram in Figure 14-9, and the contract created in Your Turn 14-3 as guides, create method specifications for the Calculate subtotal, Calculate tax, Calculate shipping, and Calculate total methods.

[26] Remember that the development process is very incremental and iterative in nature. Therefore, changes could be cascaded back to any point in the development process, for example, to essential use cases, use-case diagrams, CRC cards, class diagrams, object diagrams, sequence diagrams, collaboration diagrams, state-chart diagrams, package diagrams, window navigation diagrams, use scenarios, or real use cases.

```
(Get_CD_Info module)
    Accept (CD.Title) {Required}
    Accept (CD.Artist) {Required}
    Accept (CD.Category) {Required}
    Accept (CD.Length)
Return
```

FIGURE 14-13
Pseudocode

APPLYING THE CONCEPTS AT CD SELECTIONS

Alec Adams, the senior systems analyst and project manager for the Internet Sales System, and his team began the detailed object design process by reviewing the class and package diagram for the problem domain layer (see Figures 7-12 and 9-10). Their review revealed that there were quite a few many-to-many (*..*) association relationships on the class diagram. Alec questioned whether this was a correct representation of the actual situation. Brian, one of the staff members on the Internet Sales System project, admitted that when they put together the class diagram, they had decided to model most of the associations as a many-to-many multiplicity, figuring that this could be easily fixed at a later point in time when they had more precise information. Based on Brian's performance on normalizing the transaction data that the Internet Sales System sent to the distribution system (see Chapter 13), Alec assigned him to evaluate the multiplicity of each association in the model and to restructure and optimize the model.

Figure 14-14 shows the updated version of the class diagram. Once he had corrected the multiplicity of the associations, Brian decided to keep the Mkt Info and Review, Artist Info, and Sample Clip classes separate to prevent nulls from appearing in their database. Furthermore, even though there is a one-to-one relationship between the CD class and the Mkt Info class, he decided to keep them separate to keep the retrieval of CD information as fast as possible. He reasoned this by assuming that not all customers would want to see all of the Mkt Info associated with every CD. He based this assumption on the window navigation diagram in Figure 11-11.

Upon reviewing the new revised class diagram, Alec assigned different members of his team to the different packages identified (see Figure 9-10). Since Brian had already spent quite a bit of time on the classes in the CD package, Alec assigned it to him. The classes in the CD package were CD, Vendor, Mkt Info, Review, Artist Info, and Sample Clip.

As Brian delved deeper into the design of the CD package, he realized that since they had decided to implement the system on top of a RDBMS, he should further normalize the package. In doing so, he uncovered an additional class: Tracks. He decided that this type of addition to the problem domain layer would not add that much complexity to the model, but it would allow the problem domain and data management layer objects to be in closer agreement.

Next, Brian added invariants, pre-conditions, and post-conditions to the classes and their methods. For example, Figure 14-15 portrays the back of the CRC card for the CD class. He decided to add only the invariant information to the CRC cards and not the class diagram, thereby keeping the class diagram as simple and as easy to understand as possible. Notice the additional set of multiplicity, domain, and referential integrity invariants added to the attributes and relationships. Furthermore, he

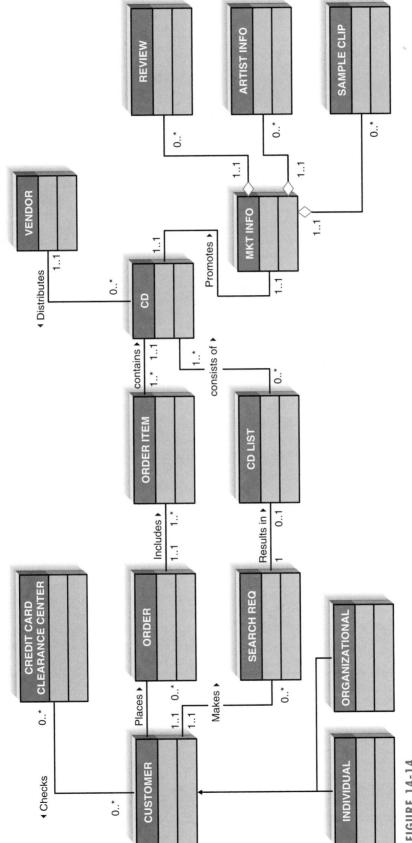

FIGURE 14-14

Revised CD Selections Internet Sales System Class Diagram (Place Order Use-Case View)

Back:

Attributes:			
CD Number	(1..1)	(unsigned long)	
CD Name	(1..1)	(String)	
Pub Date	(1..1)	(Date)	
Artist Name	(1..1)	(String)	
Artist Number	(1..1)	(unsigned long)	
Vendor	(1..1)	(Vendor)	
Vendor ID	(1..1)	(unsigned long)	{Vendor ID = Vendor. Get VendorID()}

Relationships:

Generalization (a-kind-of): _____

_____

Aggregation (has-parts): _____

_____

Other Associations: <u>Tracks {1..*} Vendor {1..1} Mkt {1..1}</u>

_____

FIGURE 14-15
Back of CD CRC Card

needed to add contracts for each method. Figure 14-16 portrays the contract for the GetReview( ) method associated with the Mkt Info class. Notice that there is a precondition for this method to succeed — Review attribute not Null.

Upon completing the CRC cards and contracts, Brian moved on to specify the detailed design for each method. For example, the method specification for the

Method Name: GetReview()	Class Name:	ID:

Clients (Consumers): CD Detailed Report

Associated Use Cases: Places Order

Description of Responsibilities: Return review objects for the Detailed Report Screen to display.

Arguments Received:

Type of Value Returned: List of Review objects

Pre-Conditions: Review attribute not Null

Post-Conditions:

FIGURE 14-16
Get Review Method Contract

Method Name: Get Review()	Class Name: MKT Info	ID: 453
Contract ID: 89	Programmer: John Smith	Date Due: 7/7/02

Programming Language:
❏ Visual Basic   ❏ Smalltalk   ❏ C++   ■ Java

Triggers/Events:
     Detail Button on Basic Report is pressed

Arguments Received:	
Data Type:	Notes:

Messages Sent & Arguments Passed:		
ClassName.MethodName:	Data Type:	Notes:

Argument Returned:	
Data Type:	Notes:
List	List of Review objects

Algorithm Specification:
     IF Review Not Null
           Return Review
     Else
           Throw Null Exception

Misc.Notes:

FIGURE 14-17
Create Review Method Specification

GetReview( ) method is given in Figure 14-17. Brian developed this specification by reviewing the Place Order use case (see Figure 6-11), the sequence diagram (see Figure 8–5), the window navigation diagram (see Figure 11-11), the real use case (see Figure 12-15), and the contract (see Figure 14-16). Notice that Brian is enforcing the pre-condition on the contract by testing to see whether or not the Review attribute that contains the list of reviews contains a value. Since the method is to be implemented in Java, he has specified that an exception is to be "thrown" if there are no reviews.

Finally, based on all of the modifications Brian made to the models, he developed a new class diagram for the CD package (see Figure 14-18). As you can see, the Tracks class is now part of the CD package, and it is related to the CD class.

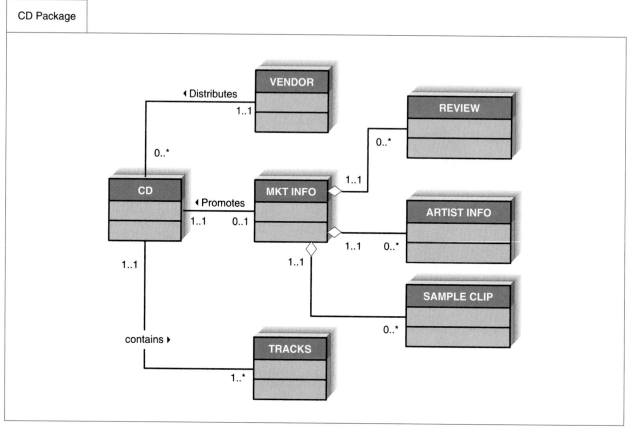

FIGURE 14-18
Revised Package Diagram for the CD Package on the PD Layer of CD Selections Internet Sales System

SUMMARY

Revisiting the Basic Characteristics of Object-Oriented Systems
A class is a template on which objects can be instantiated. An object is a person, place, or thing about which we want to capture information. Each object has attributes and methods. The methods are executed by objects sending messages that trigger them. Encapsulation and information hiding allow an object to conceal its inner processes and data from the other objects. Polymorphism and dynamic binding allow a message to be interpreted differently by different kinds of objects. However, if polymorphism is not used in a semantically consistent manner, it can make an object design incomprehensible. Classes can be arranged in a hierarchical fashion in which subclasses inherit attributes and methods from superclasses to reduce the redundancy in development. However, through redefinition capabilities or multiple inheritance, inheritance conflicts can be introduced into the design.

Design Criteria
Coupling, cohesion, and connascence have been proposed as a set of criteria on which a design of an object-oriented system can be evaluated. Two types of coupling—interaction and inheritance—and three types of cohesion—method, class,

and generalization/specialization—were described. Interaction coupling deals with the communication that takes place among the objects, and inheritance coupling deals with the innate dependencies in the use of inheritance in object-oriented systems. Method cohesion addresses how single-minded a method is. The fewer things a method does, the more cohesive it is. Class cohesion does the same for classes. A class should be a representation of one and only one thing. Generalization/specialization cohesion deals with how good an inheritance hierarchy is. Good inheritance hierarchies only support generalization and specialization (a-kind-of) semantics. Connascence generalizes coupling and cohesion and then combines them with the different levels of encapsulation. The general rule of thumb is to maximize the cohesion (connascence) within an encapsulation boundary and minimize the coupling (connascence) between the encapsulation boundaries.

Object Design Activities

There are four basic object design activities. First, additional specification is possible by carefully reviewing the models, deciding on the proper visibility of the attributes and methods, setting the signature for each method, and identifying any constraints associated with the classes or the classes' methods. Second, look for opportunities for reuse by reviewing the model and looking at possible patterns, class libraries, frameworks, and components that could be used to enhance the system. Third, restructure the model through the use of factoring and normalization.

Be sure to take the programming language into consideration. It may be necessary to map the current design into the restricted capabilities of the languages; for example, the language only supports single inheritance. Also, be certain that the inheritance in your model only supports generalization/specialization (a-kind-of) semantics. Fourth, optimize the design. However, be careful in the process of optimizing the design. Optimizations typically decrease the understandability of the model.

Constraints and Contracts

Three types of constraints are associated with object-oriented design: invariants, preconditions, and post-conditions. Invariants capture constraints that must always be true for all instances of a class, for example, domains and values of attributes or multiplicity of relationships. Typically, invariants are attached to class diagrams and CRC cards. However, for clarity purposes, we suggest only placing them on the CRC cards.

Contracts formalize the interaction between objects, that is, the message passing. As such, they include the pre- and post-conditions that must be enforced for a method to execute properly. Contracts provide an approach to model the rights and obligations when client and server objects interact. From a practical perspective, every interaction that can take place between all possible client and server objects is not modeled on a separate contract. Instead, a single contract is drawn up for using each visible method of a server object.

Method Specification

A method specification is a written document that provides clear and explicit instructions on how a method is to behave. Without clear and unambiguous method specifications, critical design decisions will have to be made by programmers instead of designers. Even though there is no standard format for a method specification, typically four types of information are captured. First, there is general information such as the name of the method, name of the class, Contract ID, programmer assigned, date due, and target programming language. Second, due to the rise in popularity of GUI-based and event-driven systems, events also are captured. Third, the information that deals with the signature of the method, data received, data passed on to other methods, and data returned by the method is recorded. Finally, an unambiguous specification of the algorithm is given. The algorithm typically is modeled using Structured English, pseudocode, or a formal language.

KEY TERMS

Action statement
Active value
ActiveX
Application Program Interface
Attribute
Case statement
Class
Class cohesion
Class library
Client
Cohesion
Component
Connascence
Constraint
Consumer
Contract
Coupling
Derived attribute
Dynamic binding
Encapsulation
Event
Event-driven

Exceptions
Factoring
Fan-out
For statement
Framework
Generalization/specialization
 cohesion
Homonyms
Ideal class cohesion
If statement
Information hiding
Inheritance
Inheritance conflict
Inheritance coupling
Instance
Interaction coupling
Invariant
JavaBean
Law of Demeter
Message
Method
Method cohesion

Method specification
Multiple inheritance
Normalization
Object
Patterns
Polymorphism
Post-condition
Pre-condition
Pseudocode
Redefinition
Server
Signature
Single inheritance
State
Structured English
Supplier
Synonyms
Trigger
Visibility
While statement

QUESTIONS

1. What are the basic characteristics of object-oriented systems?
2. What is dynamic binding?
3. Define polymorphism. Give one example of a "good" use of polymorphism and one example of a "bad" use of polymorphism.
4. What is an inheritance conflict? How does an inheritance conflict affect the design?
5. Why is cancellation of methods a bad thing?
6. Give the guidelines to avoid problems with inheritance conflicts.
7. How important is it to know which object-oriented programming language is going to be used to implement the system?
8. What additional types of inheritance conflicts are there when using multiple inheritance?

9. What is the Law of Demeter?
10. What are the six types of interaction coupling? Give one example of "good" interaction coupling and one example of "bad" interaction coupling.
11. What are the seven types of method cohesion? Give one example of "good" method cohesion and one example of "bad" method cohesion.
12. What are the four types of class cohesion? Give one example of each type.
13. What are the five types of connascence described in your text? Give one example of each type.
14. When designing a specific class, what types of additional specification for a class could be necessary?
15. What are constraints? What are the three different types of constraints?
16. What are exceptions?

17. What are patterns, frameworks, class libraries, and components? How are they used to enhance the evolving design of the system?

18. How are factoring and normalization used in designing an object system?

19. What are the different ways to optimize a object system?

20. What is the typical downside of system optimization?

21. What is the purpose of a contract? How is it used?

22. What is an invariant? How are invariants modeled in a design of a class? Give an example of an invariant for an hourly employee class.

23. Create a contract for a "compute-pay" method associated with an hourly employee class.

24. How do you specify a method's algorithm? Give an example of an algorithm specification for a "compute-pay" method associated with an hourly employee class.

25. How are methods specified? Give an example of a method specification for a "compute-pay" method associated with an hourly employee class.

EXERCISES

A. For the Health Club exercises in Chapter 6 (N,O), Chapter 7 (O), Chapter 8 (C,D), Chapter 9 (D), Chapter 11 (D,E,F,G), and Chapter 12 (D,E,F), choose one of the classes and create a set of invariants for attributes and relationships and add them to the CRC card for the class.

B. Choose one of the methods in the class that you chose for Exercise A and create a contract and a method specification for it. Use Structured English for the algorithm specification.

C. For the Picnics R Us exercises in Chapter 6 (P,Q), Chapter 7 (P), Chapter 8 (H), Chapter 9 (F), Chapter 11 (H,I,J,K), and Chapter 12 (G,H,I), choose one of the classes and create a set of invariants for attributes and relationships and add them to the CRC card for the class.

D. Choose one of the methods in the class that you chose for Exercise C and create a contract and a method specification for it. Use Structured English for the algorithm specification.

E. For the Of-The-Month Club exercises in Chapter 6 (R,S), Chapter 7 (Q), Chapter 8 (I), Chapter 9 (G), Chapter 11 (L,M,N,O), and Chapter 12 (J,K,L), choose one of the classes and create a set of invariants for attributes and relationships and add them to the CRC card for the class.

F. Choose one of the methods in the class that you chose for Exercise E and create a contract and a method specification for it. Use Structured English for the algorithm specification.

G. For the Library exercises in Chapter 6 (T,U), Chapter 7 (R), Chapter 8 (J), Chapter 9 (E), Chapter 11 (P,Q,R,S), and Chapter 12 (M,N,O), choose one of the classes and create a set of invariants for attrib-

utes and relationships and add them to the CRC card for the class.

H. Choose one of the methods in the class that you chose for Exercise G and create a contract and a method specification for it. Use Structured English for the algorithm specification.

I. Describe the difference in meaning between the following two class diagrams. Which is a "better" model? Why?

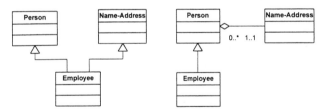

J. From a cohesion, coupling, and connascence perspective, is the following class diagram a "good" model? Why or why not?

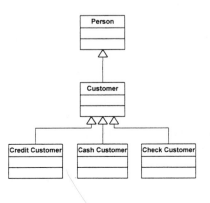

K. From a cohesion, coupling, and connascence perspective, are the following class diagrams "good" models? Why or why not?

L. Create a set of inheritance conflicts for the two inheritance structures in the class diagrams of Exercise K.

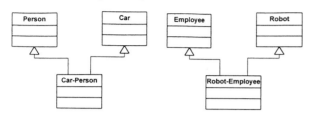

MINICASES

1. Your boss has been in the software development field for 30 years. He has always prided himself on his ability to adapt his skills from one approach to developing software to the next approach. For example, he had no problem learning structured analysis and design in the early 1980s and information engineering in the early 1990s. He even understands the advantage of rapid application development. But the other day, when you and he were talking about the advantages of object-oriented approaches, he became totally confused. He thought that characteristics such as polymorphism and inheritance were an advantage for object-oriented systems. However, when you explained the problems with inheritance conflicts, redefinition capabilities, and the need for semantic consistency across different implementations of methods, he was ready to simply give up. To make matters worse, you then went on to explain the importance of contracts in controlling the development of the system. At this point in the conservation, he basically threw in the towel. As he walked off, you heard him say something like "I guess it's true, it's too hard to teach an old dog new tricks."

   Being a loyal employee and friend, you decided to write a short tutorial to give your boss on object-oriented systems development. As a first step, create a detailed outline for the tutorial. As a subtle example, use good design criteria, such as coupling and cohesion, in the design of your tutorial outline.

2. You have been working with the Holiday Travel Vehicle problem for quite a while. You should go back and refresh your memory about the problem before attempting to solve this situation. Refer back to your solutions for Minicase 2 in Chapter 7, Minicase 2 in Chapter 8, Minicases 1 and 2 in Chapter 11, and Minicase 2 in Chapter 13.

   In the new system for Holiday Travel Vehicles, the system users follow a two-stage process to record complete information on all of the vehicles sold. When an RV or trailer first arrives at the company from the manufacturer, a clerk from the inventory department creates a new vehicle record for it in the computer system. The data entered at this time include basic descriptive information on the vehicle such as manufacturer, name, model, year, base cost, and freight charges. When the vehicle is sold, the new vehicle record is updated to reflect the final sales terms and the dealer-installed options added to the vehicle. This information is entered into the system at the time of sale when the salesperson completes the sales invoice.

   When it is time for the clerk to finalize the new vehicle record, the clerk will select a menu option from the system, which is called Finalize New Vehicle Record. The tasks involved in this process are described below.

   When the user selects the "Finalize New Vehicle Record" from the system menu, the user is immediately prompted for the serial number of the new vehicle. This serial number is used to retrieve the new vehicle record for the vehicle from system storage. If a record cannot be found, the serial number is probably invalid. The vehicle serial number is then used to retrieve the option records that describe the dealer-installed options that were added to the vehicle at the customer's request. There may be zero or more options. The cost of the option specified on the option record(s) is totaled. Then, the dealer cost is calculated using the vehicle's base cost, freight charge, and total option cost. The completed new vehicle record is passed back to the calling module.

   a. Update the structural model (CRC cards and class diagram) with this additional information.

   b. For each class in the structural model create a set of invariants for attributes and relationships and add them to the CRC cards for the classes.

   c. Choose one of the classes in the structural model. Create a contract for each method in that class. Be as complete as possible.

   d. Create a method specification for each method in the class you chose for question c. Use Structured English for the algorithm specification.

Programs

Test Plan

Tested
System

User
Documentation

Conversion
Plan

Change
Management
Plan

Support
Plan

Delivered
System

Project
Assessment

PART FOUR
IMPLEMENTATION PHASE

PROJECT BINDER

CHAPTER 15

Construction

CHAPTER 16

Installation

The final phase in the SDLC is the Implementation Phase, during which the system is actually built (or purchased, in the case of a packaged software design). The first step in implementation is system construction, during which the system is built and tested to ensure it performs as designed. Deliverables during this step include Programs, the Test Plan, User Documentation, and the Tested System. Conversion is the process by which the old system is turned off and the new one turned on, during which the project team creates a Conversion Plan, a Change Management Plan, a Support Plan, and a Project Assessment. At the end of Implementation, the final system is delivered to the project sponsor and approval committee.

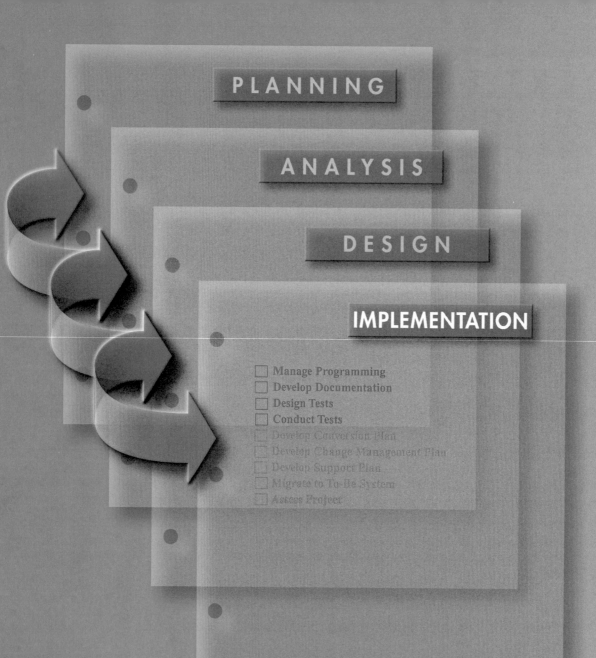

PLANNING

ANALYSIS

DESIGN

IMPLEMENTATION

- Manage Programming
- Develop Documentation
- Design Tests
- Conduct Tests
- Develop Conversion Plan
- Develop Change Management Plan
- Develop Support Plan
- Migrate to To-Be System
- Assess Project

TASK CHECKLIST

PLANNING   ANALYSIS   DESIGN

CHAPTER 15

CONSTRUCTION

THIS chapter discusses the activities needed to successfully build the information system: programming, testing, and documenting the system. Programming is time-consuming and costly, but except in unusual circumstances, it is the simplest for the systems analyst because it is well understood. For this reason, the system analyst focuses on testing (proving that the system works as designed) and developing documentation during this part of the SDLC.

OBJECTIVES

- Be familiar with the system construction process.
- Understand different types of tests and when to use them.
- Understand how to develop documentation.

CHAPTER OUTLINE

IMPLEMENTATION

INTRODUCTION

When people first learn about developing information systems, usually they immediately think about writing programs. Programming can be the largest single component of any systems development project in terms of both time and cost. However, it also can be the best understood component and therefore—except in rare circumstances—offers the least problems of all aspects of the SDLC. When projects fail, it is usually not because the programmers were unable to write the programs but because the analysis, design, installation, and/or project management were done poorly. In this chapter, we focus on the construction and testing of the software and the documentation.

Construction is the development of all parts of the system, including the software itself, documentation, and new operating procedures. Programming is often seen as the focal point of system development. After all, system development *is* writing programs. It is the reason why we do all the analysis and design. And it's fun. Many beginning programmers see testing and documentation as bothersome afterthoughts. Testing and documentation aren't fun, so they often receive less attention than the creative activity of writing programs.

Programming and testing, however, are very similar to writing and editing. No professional writer (or student writing an important term paper) would stop after writing the first draft. Rereading, editing, and revising the initial draft into a good paper is the hallmark of good writing. Likewise, thorough testing is the hallmark of professional software developers. Most professional organizations devote more time and money to testing (and the subsequent revision and retesting) than to writing the programs in the first place.

The reasons are simple economics: downtime and failures caused by software bugs[1] are extremely expensive. Many large organizations estimate the costs of downtime of critical applications at $50,000 to $200,000 *per hour.*[2] One serious bug that causes an hour of downtime can cost more than one year's salary of a programmer—and how often are bugs found and fixed in one hour? Testing is therefore a form of insurance. Organizations are willing to spend a lot of time and money to prevent the possibility of major failures after the system is installed.

Therefore, programs are not usually considered finished until the test for that program is passed. For this reason, programming and testing are tightly coupled, and since programming is the primary job of the programmer (not the analyst), testing (not programming) often becomes the focus of the construction stage for the systems analysis team.

In this chapter, we discuss three parts of the construction step: programming, testing, and writing the documentation. Because programming is primarily the job of programmers, not systems analysts, and because this is not a programming book, we devote less time to programming than to testing and documentation.

---

[1] When I was an undergraduate, I had the opportunity to hear Admiral Grace Hopper tell how the term *bug* was introduced. She was working on one of the early Navy computers when suddenly it failed. The computer would not restart properly so she began to search for failed vacuum tubes. She found a moth inside one tube and recorded in the log book that a bug had caused the computer to crash. From then on, every computer crash was jokingly blamed on a bug (as opposed to programmer error), and eventually the term *bug* entered the general language of computing.

[2] See Billie Shea, "Quality Patrol: Eye on the Enterprise," *Application Development Trends,* (November 5, 1998): 31–38.

CONCEPTS   15-A THE COST OF A BUG

IN ACTION

My first programming job in 1977 was to convert a set of application systems from one version of COBOL to another version of COBOL for the government of Prince Edward Island. The testing approach was to first run a set of test data through the old system and then run it through the new system to ensure that the results from the two matched. If they matched, then the last three months of production data were run through both to ensure that they too matched.

Things went well until I began to convert the Gas Tax system that kept records on everyone authorized to purchase gasoline without paying tax. The test data ran fine, but the results using the production data were peculiar. The old and new systems matched, but rather than listing several thousand records, the report listed only 50. I checked the production data file and found it listed only 50 records, not the thousands that were supposed to be there.

The system worked by copying the existing gas tax records file into a new file and making changes in the new file. The old file was then copied to tape backup. There was a bug in the program such that if there were no changes to the file, a new file was created, but no records were copied into it.

I checked the tape backups and found one with the full set of data that was scheduled to be overwritten three days after I discovered the problem. The government was only three days away from losing all gas tax records. *Alan Dennis*

QUESTION

What would have been the cost of this bug if it hadn't been caught?

MANAGING PROGRAMMING

In general, systems analysts do not write programs; programmers write programs. Therefore, the primary task of the systems analysts during programming is … waiting. However, the project manager is usually very busy *managing* the programming effort by assigning the programmers, coordinating the activities, and managing the programming schedule.[3]

Assigning Programmers

The first step in programming is assigning modules to the programmers. As discussed in Chapter 14, each module (class, object, or method) should be as separate and distinct as possible from the other modules. The project manager first groups together classes that are related so that each programmer is working on related classes. These groups of classes are then assigned to programmers. A good place to start is to look at the package diagrams.

One rule of system development is that the more programmers who are involved in a project, the longer the system will take to build. This is because as the size of the programming team increases, the need for coordination increases exponentially, and the more coordination that is required, the less time programmers can spend actually writing systems. The best size is the smallest possible programming team. When projects are so complex that they require a large team, the

---

[3] One of the best books on managing programming (even though it was first written 25 years ago) is that by Frederick P. Brooks, Jr., *The Mythical Man-Month*, 20th Anniversary edition (Reading, MA: Addison-Wesley, 1995).

best strategy is to try to break the project into a series of smaller parts that can function as independently as possible.

Coordinating Activities

Coordination can be done through both high-tech and low-tech means. The simplest approach is to have a weekly project meeting to discuss any changes to the system that have arisen during the past week—or just any issues that have come up. Regular meetings, even if they are brief, encourage the widespread communication and discussion of issues before they become problems.

Another important way to improve coordination is to create and follow standards that can range from formal rules for naming files to forms that must be completed when goals are reached to programming guidelines (see Chapter 3). When a team forms standards and then follows them, the project can be completed faster because task coordination is less complex.

The analysts also must put mechanisms in place to keep the programming effort well-organized. Many project teams set up three "areas" in which programmers can work: a development area, a testing area, and a production area. These areas can be different directories on a server hard disk, different servers, or different physical locations, but the point is that files, data, and programs are separated based on their status of completion. At first, programmers access and build files within the development area and then copy them to the testing area when the programmers are "finished." If a program does not pass a test, it is sent back to development. Once all programs and the like are tested and ready to support the new system, they are copied into the production area—the location where the final system will reside.

Keeping files and programs in different places based on completion status helps manage *change control,* the action of coordinating a system as it changes through construction. Another change control technique is keeping track of which programmer changes which classes and packages by using a *program log.* The log is merely a form on which programmers "sign-out" classes and packages to write and "sign-in" when they are completed. Both the programming areas and program log help the analysts understand exactly who has worked on what and to determine the system's current status. Without these techniques, files can be put into production without the proper testing; two programmers, for example, could start working on the same class or package at the same time.

If a CASE tool is used during the construction step, it can be very helpful for change control because many CASE tools are set up to track the status of programs and help manage programmers as they work. In most cases, maintaining coordination is not conceptually complex. It just requires a lot of attention and discipline to track small details.

Managing the Schedule

The time estimates that were produced during the initial planning phase and refined during the analysis and design phases will almost always need to be refined as the project progresses during construction because it is virtually impossible to develop an exact assessment of the project's schedule. As we discussed in Chapter 3, a well-done set of time estimates will usually have a 10% margin of error by the time you reach the construction step. It is critical that the time estimates be revised as the construction step proceeds. If a program module takes longer to develop than expected, then the prudent response is to move the expected completion date later by the same amount.

One of the most common causes of schedule problems is scope creep. Scope creep occurs when new requirements are added to the project after the system design was finalized. Scope creep can be very expensive because changes made late in the SDLC can require much of the completed system design (and even programs already written) to be redone. Any proposed change during the construction phase must require the approval of the project manager and should only be done after a quick cost-benefit analysis has been done.

Another common cause is the unnoticed day-by-day slippage in the schedule. One package is a day later here, another one a day late there. Pretty soon these minor delays add up, and the project is noticeably behind schedule. Once again, the key to managing the programming effort is to watch these minor slippages carefully and update the schedule accordingly.

Typically, a project manager will create a risk assessment that tracks potential risks along with an evaluation of their likelihood and potential impact. As the construction step moves to a close, the list of risks will change as some items are removed and others surface. The best project managers, however, work hard to keep risks from having an impact on the schedule and costs associated with the project.

PRACTICAL
TIP

15-1 AVOIDING CLASSIC IMPLEMENTATION MISTAKES

In previous chapters, we discussed classic mistakes and how to avoid them. Here, we summarize four classic mistakes in the implementation phase.[4]

1. **Research-oriented Development:** Using state-of-the-art technology requires research-oriented development that explores the new technology because "bleeding edge" tools and techniques are not well understood, are not well documented, and do not function exactly as promised.
   Solution: If you use state-of-the-art technology, you need to significantly increase the project's time and cost estimates even if (some experts would say *especially if*) such technologies claim to reduce time and effort.

2. **Using Low-Cost Personnel:** You get what you pay for. The lowest cost consultant or the staff member is significantly less productive than the best staff. Several studies have shown that the best programmers produce software six to eight times faster than the least productive (yet cost only 50 to 100% more).

   Solution: If cost is a critical issue, assign the best, most expensive personnel; never assign entry-level personnel in an attempt to save costs.

3. **Lack of Code Control:** On large projects, programmers need to coordinate changes to the program source code (so that two programmers don't try to change the same program at the same time and overwrite the other's changes). Although manual procedures appear to work (e.g., sending e-mail notes to others when you work on a program to tell them not to), mistakes are inevitable.
   Solution: Use a source code library that requires programmers to "check out" programs and prohibits others from working on them at the same time.

4. **Inadequate Testing:** The number one reason for project failure during implementation is ad hoc testing—where programmers and analysts test the system without formal test plans.
   Solution: Always allocate sufficient time in the project plan for formal testing.

---

[4] Adapted from Steve McConnell, *Rapid Development,* (Redmond, WA: Microsoft Press, 1996).

DESIGNING TESTS[5]

In object-oriented systems, the temptation is to minimize testing. After all, through the use of patterns, frameworks, class libraries, and components, much of the system has been tested previously. Therefore, we should not have to test as much. Right? Wrong!! Testing is more critical to object-oriented systems than to systems developed in the past. Based on encapsulation (and information hiding), polymorphism (and dynamic binding), inheritance, and reuse, thorough testing is much more difficult and critical. Testing must be done systematically and the results documented so that the project team knows what has and has not been tested.

The purpose of testing is not to demonstrate that the system is free of errors. Indeed, it is not possible to prove that a system is error free, especially object-oriented systems. This is similar to theory testing. You cannot prove a theory. If a test fails to find problems with a theory, your confidence in the theory is increased. However, if a test succeeds in finding a problem, then the theory has been falsified. Software testing is similar in that it can only show the existence of errors. As such, the purpose of testing is to uncover as many errors as feasible.[6]

There are four general stages of tests: unit tests, integration tests, system tests, and acceptance tests. Although each application system is different, most errors are found during integration and system testing (see Figure 15-1).

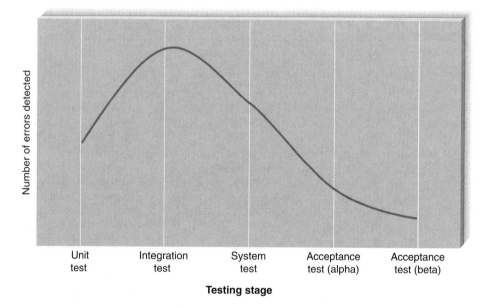

FIGURE 15-1

Error discovery rates for different stages of tests

[5] This section is based on material contained in Imran Bashir and Amrit L. Goel, *Testing Object-Oriented Software: Life Cycle Solutions* (New York, NY: Springer Verlag, 1999); Bernd Bruegge and Allen H. Dutoit, *Object-Oriented Software Engineering: Conquering Complex and Changing Systems,* (Englewood Cliffs, NJ: Prentice Hall, 2000); Philippe Kruchten, *The Rational Unified Process: An Introduction,* 2nd ed., (Reading, MA: Addison-Wesley, 2000); and John D. McGregor and David A. Sykes, *A Practical Guide to Testing Object-Oriented Software* (Reading, MA: Addison-Wesley, 2001). For more information on testing object-oriented software, see Robert V. Binder, *Testing Object-Oriented Systems: Models, Patterns, and Tools* (Reading, MA: Addison-Wesley, 1999); and Shel Sieget, *Object-Oriented Software Testing: A Hierarchical Approach* (New York: John Wiley & Sons, 1996).

[6] It is not cost-effective to get each and every error out of the software. Except in simple examples, it is in fact impossible. There are simply too many combinations to check.

In each of the following sections, we describe the four stages. However, before doing this, we describe the effect that the object-oriented characteristics have on testing and the necessary planning and management activities that must take place to have a successful testing program.

Testing and Object-Orientation

Most testing techniques have been developed to support nonobject-oriented development. Object-oriented development is still relatively new. As such, most of the testing approaches have had to be adapted to object-oriented systems. The characteristics of object-oriented systems that affect testing the most are encapsulation (and information hiding), polymorphism (and dynamic binding), inheritance, and the use of patterns, class libraries, frameworks, and components. Also, the sheer volume of products that come out of a typical object-oriented development process has increased the importance of testing in object-oriented development.

Encapsulation and Information Hiding   Encapsulation and information hiding allow processes and data to be combined to create holistic entities (i.e., objects). Furthermore, they support hiding everything behind a visible interface. Although this allows the system to be modified and maintained in an effective and efficient manner, it makes testing the system problematic. What do you need to test to build confidence in the system's ability to meet the user's need? Answer, you need to test the business process that is represented in the use cases. However, the business process is distributed over a set of collaborating classes and contained in the methods of those classes. The only way to know the effect that a business process has on a system is to look at the state changes that take place in the system. But in object-oriented systems, the instances of the classes hide the data behind a class boundary. How is it possible then to see the impact of a business process?

A second issue raised by encapsulation and information hiding is the definition of a "unit" for unit testing. What is the unit to be tested? Is it the package, class, or method? In traditional approaches, the answer would be the process that is contained in a function. However, the process in object-oriented systems is distributed over a set of classes. Therefore, testing individual methods makes no sense. The answer is the class. This dramatically changes the manner in which unit testing is done.

A third issue raised is the impact on integration testing. In this case, objects can be aggregated to form aggregate objects; for example, a car has many parts, or they can be grouped together to form collaborations. Furthermore, they can be used in class libraries, frameworks, and components. Based on all of these different ways in which classes can be grouped together, how does one effectively do integration testing?

Polymorphism and Dynamic Binding   Polymorphism and dynamic binding dramatically impact both unit and integration testing. Since an individual business process is implemented through a set of methods distributed over a set of objects, as shown above, the unit test makes no sense to be at the method level. However, with polymorphism and dynamic binding, the same method (a small part of the overall business process) can be implemented in many different objects. Therefore, testing individual implementations of methods makes no sense. Again, the unit that makes sense to test is the class. Furthermore, except for trivial cases, dynamic binding makes it impossible to know which implementation is going to be executed until the system does it. Therefore, integration testing becomes very challenging.

Inheritance   When considering the issues raised about inheritance (see Chapter 14), it should be no surprise that inheritance impacts the testing of object-oriented systems. Through the use of inheritance, bugs can be propagated from a superclass to all of its direct and indirect subclasses instantaneously. However, the tests that are applicable to a superclass are also applicable to all of its subclasses. As usual, inheritance is a double-edged sword. Finally, even though we have stated this many times before, inheritance should only support a generalization and specialization type of semantics. Remember, when using inheritance the principle of substitutability is critical (see Chapter 7). All of these issues impact unit and integration testing.

Reuse   On the surface, reuse should decrease the amount of testing required. However, each time a class is used in a different context, the class must be tested again. Therefore, anytime a class library, framework, or component is used, unit testing and integration testing are important. In the case of a component, the unit to test is the component itself. Remember that a component has a well-defined Application Program Interface (API) that hides the details of its implementation.

Object-Oriented Development Process and Products   In virtually all textbooks, including this one, testing is always covered near the end of the SDLC. This seems to imply that testing takes place only after the programming has ended. However, each and every product[7] that comes out of the object-oriented development process must be tested. For example, it is a lot easier to ensure that the requirements are captured and modeled correctly through testing the use cases. Furthermore, it is a lot cheaper to catch this type of error back in the analysis phase than it is in the implementation phase. Obviously, this is true for testing collaborations as well. By the time we have implemented a collaboration as a set of layers and partitions, we could have expended a great deal of time, and time is money, on implementing the wrong thing. So, testing collaborations by role-playing the CRC cards back in the analysis phase will actually save the team lots of time and money.

Testing must take place throughout the SDLC, not simply at the end. However, the type of testing that can take place on nonexecutable representations, such as use cases and CRC cards, is different from those that take place on code written in an object-oriented programming language. The primary approach to testing nonexecutable representations is some form of an inspection or walk-through of the representation.[8] Role-playing the CRC cards based on use cases is an example of one type of walk-through.

Test Planning

Testing starts with the development of a *test plan* that defines a series of *tests* that will be conducted. Since testing takes place through the development of an object-oriented system, a test plan should be developed at the very beginning of the SDLC and continuously updated as the system evolves. The test plan should address all products that are created during the development of the system. For example, tests

---

[7] For example, use-case descriptions, use-case diagrams, CRC cards, class diagrams, object diagrams, sequence diagrams, collaboration diagrams, statechart diagrams, package diagrams, use scenarios, window navigation diagrams, real use cases, contracts, method specifications, and source code.

[8] See Michael Fagan, "Design and Code Inspections to Reduce Errors in Program Development," *IBM Systems Journal* 15 no. 3, (1976): 105–211.

should be created that can be used to test the completeness of a CRC card. Figure 15-2 shows a typical unit test plan form for a class. For example, Figure 15-3 shows a partial list of the invariant test specifications for a class. Each individual test has a specific objective and describes a set of very specific *test cases* to examine. In the case of invariant-based tests, a description of the invariant, the original values of the attribute, the event that will cause the attribute value to change, the actual results observed, the expected results, and whether it passed or failed are shown. *Test specifications* are created for each type of constraint that must be met by the class. Also, similar types of specifications are done for integration, system, and acceptance tests.

Not all classes are likely to be finished at the same time, so the programmer usually writes *stubs* for the unfinished classes to enable the classes around them to be tested. A stub is a placeholder for a class that usually displays a simple test message on the screen or returns some *hardcoded* value[9] when it is selected. For example, consider an application system that provides the five standard functions discussed in Chapter 6 for some object such as CDs, patients, or employees: creating, changing, deleting, finding, and printing (whether on the screen or on a printer). Depending on the final design, these different functions could end up in different

---

Class Test Plan		Page _____ of _____
Class Name: _____	Version Number: _____	CRC Card ID: _____
Tester: _____	Date Designed: _____	Date Conducted: _____
Class Objective:		
Associated Contract IDs: _____		
Associated Use Case IDs: _____		
Associated Superclass(es): _____		
Testing Objectives:		
Walk-through Test Requirements:		
Invariant-Based Test Requirements:		
State-Based Test Requirements:		
Contract-Based Test Requirements:		

FIGURE 15-2
Class Test Plan

---

[9] The word "hardcoded" means written into the program. For example, suppose you were writing a unit to calculate the net present value of a loan. The stub might be written so that it would always display (or return to the calling module) a value of 100 regardless of the input values.

Class Invariant Test Specification

Page _____ of _____

| Class Name: _____ | Version Number: _____ | CRC Card ID: _____ |
| Tester: _____ | Date Designed: _____ | Date Conducted: _____ |

Testing Objectives:

Test Cases

Invariant Description	Original Attribute Value	Event	New Attribute Value	Expected Result	Result P/F
Attribute Name:					
1) _____	_____	_____	_____	_____	_____
2) _____	_____	_____	_____	_____	_____
3) _____	_____	_____	_____	_____	_____
Attribute Name:					
1) _____	_____	_____	_____	_____	_____
2) _____	_____	_____	_____	_____	_____
3) _____	_____	_____	_____	_____	_____
Attribute Name:					
1) _____	_____	_____	_____	_____	_____
2) _____	_____	_____	_____	_____	_____
3) _____	_____	_____	_____	_____	_____

FIGURE 15-3

Class Invariant Test Specification

objects on different layers. Therefore, to test the functionality associated with the classes on the problem domain layer, a stub would be written for each of the classes on the other layers that interact with the problem domain classes. These stubs would contain the minimal interface necessary to be able to test the problem domain classes. For example, they would have methods that could receive the messages being sent by the problem domain layer objects and methods that could send messages to the problem domain layer objects. Typically, the methods would display a message on the screen notifying the tester that the method had been successfully reached (e.g., "Delete item from Database method reached"). In this way, the problem domain classes could pass class testing before the classes on the other layers were completed.

Unit Tests

Unit tests focus on a single unit—the class. There are two approaches to unit testing: black-box and white-box (see Figure 15-4). *Black-box testing* is the most commonly used since each class represents an encapsulated object. Black-box testing is driven by the CRC cards, statechart diagrams, and contracts associated with a class, not by the programmers' interpretation. In this case, the test plan is developed directly from the specification of the class: each item in the specification becomes

Stage	Types of Tests	Test Plan Source	When to Use	Notes
Unit Testing	**Black-Box Testing** Treats program as "black-box"	CRC Cards Class Diagrams Contracts	For normal unit testing	• Tester focuses on whether the class meets the requirements stated in the specifications.
	White-Box Testing Looks inside the program to test its major elements.	Method Specifications	When complexity is high	• By looking inside the unit to review the code itself the tester may discover errors or assumptions not immediately obvious to someone treating the class as a black box.
Integration Testing	**User Interface Testing** The tester tests each interface function.	Interface Design	For normal integration testing	• Testing is done by moving through each and every menu item in the interface either in a top-down or bottom-up manner.
	Use-Case Testing The tester tests each use case.	Use Cases	When the user interface is important	• Testing is done by moving through each use case to ensure they work correctly • Usually combined with user interface testing because it does not test all interfaces.
	Interaction Testing Tests each process in a step-by-step fashion.	Class Diagrams Sequence Diagrams Collaboration Diagrams	When the system performs data processing	• The entire system begins as a set of stubs. Each class is added in turn and the results of the class compared to the correct result from the test data; when a class passes, the next class is added and the test rerun. This is done for each package. Once each package has passed all tests, then the process repeats integrating the packages.
	System Interface Testing Tests the exchange of data with other systems	Use-Case Diagram	When the system exchanges data	• Because data transfers between systems are often automated and not monitored directly by the users it is critical to design tests to ensure they are being done correctly.
System Testing	**Requirements Testing** Tests to whether original business requirements are met.	System Design, Unit Tests, and Integration Tests	For normal system testing	• Ensures that changes made as a result of integration testing did not create new errors. • Testers often pretend to be uninformed users and perform improper actions to ensure the system is immune to invalid actions (e.g., adding blank records).
	Usability Testing Tests how convenient the system is to use	Interface Design and Use Cases	When user interface is important	• Often done by analyst with experience in how users think and in good interface design. • Sometimes uses formal usability testing procedures discussed in Chapter 11.
	Security Testing Tests disaster recovery & unauthorized access	Infrastructure Design	When the system is important	• Security testing is a complex task, usually done by an infrastructure analyst assigned to the project. • In extreme cases, a professional firm may be hired
	Performance Testing Examines the ability to perform under high loads	System Proposal Infrastructure Design	When the system is important	• High volumes of transactions are generated and given to the system • Often done by using special purpose testing software
	Documentation Testing Tests the accuracy of the documentation	Help System, Procedures, Tutorials	For normal system testing	• Analysts spot check or check every item on every page in all documentation to ensure the documentation items and examples work properly.
Acceptance Testing	**Alpha Testing** Conducted by users to ensure they accept the system.	System Tests	For normal acceptance testing	• Often repeats previous tests but are conducted by users themselves to ensure they accept the system
	Beta Testing Uses real data, not test data	No plan	When the system is important	• Users closely monitor system for errors or useful improvements

FIGURE 15-4
Types of Tests

a test, and several test cases are developed for it. *White-box testing* is based on the method specifications associated with each class. However, white-box testing has had limited impact in object-oriented development. This is due to the rather small size of the individual methods in a class. As such, most approaches to testing classes use black-box testing to assure their correctness.

Class tests should be based on the invariants on the CRC cards, the statechart diagrams associated with each class, and the pre- and post-conditions contained on each method's contract. Assuming all of the constraints have been captured on the CRC cards and contracts, individual test cases can be developed fairly easily. For example, suppose the CRC card for an order class gave an invariant that the order quantity must be between 10 and 100 cases. The tester develops a series of test cases to ensure that the quantity is validated before the system accepts it. It is impossible to test every possible combination of input and situation; there are simply too many possible combinations. In this example, the test requires a minimum of three test cases: one with a valid value (e.g., 15), one with an invalid value too low (e.g., 7), and one with invalid value too high (e.g., 110). Most tests would also include a test case with a nonnumeric value to ensure the data types were checked (e.g., ABCD). A really good test would include a test case with nonsensical but potentially valid data (e.g., 21.4).

Using a statechart diagram is a useful way to identify tests for a class. Any class that has a statechart diagram associated with it will have a potentially complex life cycle. As such, it is possible to create a series of tests to guarantee that each state can be reached.

Tests also can be developed for each contract associated with the class. In the case of a contract, a set of tests for each pre- and post-condition is required. Furthermore, if the class is a subclass of another class, then all of the tests associated with the superclass must be executed again. Also, the interactions among the constraints, invariants, and pre- and post-conditions in the subclass and the superclass(es) must be addressed.

Finally, owing to good object-oriented design, in order to fully test a class, special testing methods may have to be added to the class being tested. For example, how can invariants be tested? The only way to really test them is through methods that are visible to the outside of the class that can be used to manipulate the values of the class's attributes. However, adding these types of methods to a class does two things. First, they add to the testing requirements since they themselves will have to be tested. Second, if they are not removed from the deployed version of the system, the system will be less efficient and the advantage of information hiding effectively will be lost. As is readily apparent, testing classes is complex. Therefore, great care must be taken when designing tests for classes.

Integration Tests

Integration tests assess whether a set of classes that must work together do so without error. They ensure that the interfaces and linkages between different parts of the system work properly. At this point, the classes have passed their individual unit tests, so the focus now is on the flow of control among the classes and on the data exchanged among them. Integration testing follows the same general procedures as unit testing: the tester develops a test plan that has a series of tests that in turn have tests. Integration testing is often done by a set of programmers and/or systems analysts.

From an object-oriented systems perspective, integration testing can be difficult. A single class can be in many different aggregations, because of the way in which objects can be combined to form new objects, class libraries, frameworks, components, and packages. Where is the best place to start the integration? Typically, the answer is to begin with the set of classes, or collaboration, used to support the highest priority use case (see Chapter 6). Also, dynamic binding makes it crucial to design the integration tests carefully to ensure that the combinations of methods are tested.

There are four approaches to integration testing: *user interface testing, use-case testing, interaction testing,* and *system interface testing* (see Figure 15-4). Most projects use all four approaches.

System Tests

System tests are usually conducted by the systems analysts to ensure that all classes work together without error. System testing is similar to integration testing but is much broader in scope. Whereas integration testing focuses on whether the classes work together without error, system tests examine how well the system meets business requirements and its usability, security, and performance under a heavy load (see Figure 15-4). It also tests the system's documentation.

Acceptance Tests

Acceptance testing is done primarily by the users with support from the project team. The goal is to confirm that the system is complete, meets the business needs that prompted the system to be developed, and is acceptable to the users. Acceptance testing is done in two stages: *alpha testing,* in which users test the system using made-up data, and *beta testing,* in which users begin to use the system with real data but are carefully monitored for errors (see Figure 15-4).

DEVELOPING DOCUMENTATION

Like testing, developing documentation of the system must be done throughout the SDLC. There are two fundamentally different types of documentation: system documentation and user documentation. *System documentation* is intended to help programmers and systems analysts understand the application software and enable them to build it or maintain it after the system is installed. System documentation is largely a byproduct of the systems analysis and design process and is created as the project unfolds. Each step and phase produces documents that are essential in understanding how the system is constructed or is to be built, and these documents are stored in the Project Binder(s). In many object-oriented development environments, it is possible to somewhat automate the creation of detailed documentation for classes and methods. For example, in Java, if the programmers use *javadoc*-style comments, it is possible to create HTML pages that document a class and its methods automatically by using the *javadoc* utility.[10] Since most programmers look on documentation with much distaste, anything that can make documentation easier to create is useful.

User documentation (such as user's manuals, training manuals, and on-line help systems) is designed to help the user operate the system. Although most project teams expect users to have received training and to have read the user's manuals before operating the system, unfortunately this is not always the case. It is more common today—especially in the case of commercial software packages for microcomputers—for users to begin using the software without training or reading the user's manuals. In this section, we focus on user documentation.[11]

User documentation is often left until the end of the project, which is a dangerous strategy. Developing good documentation takes longer than many people expect because it requires much more than simply writing a few pages. Producing documentation requires designing the documents (whether on paper or on-line), writing the text, editing them, and testing them. For good-quality documentation, this process usually takes about three hours per page (single-spaced) for paper-based documentation or two hours per screen for on-line documentation. Thus, a "simple" set of documentation such as a 10-page user's manual and a set of 20 help screens takes 70 hours. Of course, lower quality documentation can be produced faster.

The time required to develop and test user documentation should be built into the project plan. Most organizations plan for documentation development to start once the interface design and program specifications are complete. The initial draft

---

[10] For those who have used Java, *javadoc* is how the JDK documentation from Sun is created.

[11] For more information on developing documentation, see Thomas T. Barker, *Writing Software Documentation,* (Boston, MA: Allyn and Bacon, 1998).

of documentation is usually scheduled for completion immediately after the unit tests are complete. This reduces the chance that the documentation will need to be changed due to software changes (but doesn't eliminate it) and still leaves enough time for the documentation to be tested and revised before the acceptance tests are started.

Although paper-based manuals are still important, on-line documentation is becoming more important. Paper-based documentation is simpler to use because it is more familiar to users, especially novices who have less computer experience; on-line documentation requires the users to learn one more set of commands. Paper-based documentation is also easier to flip through and gain a general understanding of its organization and topics and can be used far away from the computer itself.

There are four key strengths of on-line documentation that all but guarantee it will be the dominant form for the next century. First, searching for information is often simpler (provided the help search index is well designed) because the user can type in a variety of keywords to view information almost instantaneously rather than having to search through the index or table of contents in a paper document. Second, the same information can be presented several times in many different formats, so that the user can find and read the information in the most informative way (such redundancy is possible in paper documentation, but the cost and intimidating size of the resulting manual make it impractical). Third, on-line documentation enables many new ways for the user to interact with the documentation that are not possible in static paper documentation. For example, it is possible to use links or "tool tips" (i.e., pop-up text; see Chapter 12) to explain unfamiliar terms, and one can write "show-me" routines that demonstrate on the screen exactly what buttons to click and text to type. Finally, on-line documentation is significantly less expensive to distribute than paper documentation.

Types of Documentation

There are three fundamentally different types of user documentation: reference documents, procedures manuals, and tutorials. *Reference documents* (also called the help system) are designed to be used when the user needs to learn how to perform a specific function (e.g., updating a field, adding a new record). Often people read reference information when they have tried and failed to perform the function; writing reference documents requires special care because often the user is impatient or frustrated when he or she begins to read them.

Procedures manuals describe how to perform business tasks (e.g., printing a monthly report, taking a customer order). Each item in the procedures manual typically guides the user through a task that requires several functions or steps in the system. Therefore, each entry is typically much longer than an entry in a reference document.

Tutorials—obviously—teach people how to use major components of the system (e.g., an introduction to the basic operations of the system). Each entry in the tutorial is typically longer still than the entries in procedures manuals and is usually designed to be read in sequence (whereas entries in reference documents and procedures manuals are designed to be read individually).

Regardless of the type of user documentation, the overall process for developing it is similar to the process of developing interfaces (see Chapters 11 and 12). The developer first designs the general structure for the documentation and then develops the individual components within it.

Designing Documentation Structure

In this section, we focus on the development of on-line documentation because we believe it will become the most common form of user documentation. The general structure used in most on-line documentaion, whether reference documents, procedures manuals, or tutorials is to develop a set of *documentation navigation controls* that lead the user to *documentation topics.* The documentation topics are the material the user wants to read, while the navigation controls are the way in which the user locates and accesses a specific topic.

Designing the structure of the documentation begins by identifying the different types of topics and navigation controls that need to be included. Figure 15-5 shows a commonly used structure for on-line reference documents (i.e., the help system). The documentation topics generally come from three sources. The first and most obvious source of topics is the set of commands and menus in the user interface. This set of topics is very useful if the user wants to understand how a particular command or menu is used.

The users often don't know what commands to look for or where they are in the system's menu structure. Instead, users have tasks they want to perform, and rather than thinking in terms of commands, they think in terms of their business tasks. Therefore, the second and often more useful set of topics focuses on how to perform certain tasks—usually those in the use scenarios, window navigation diagram, and the real use cases from the user interface design (see Chapters 11 and 12). These topics walk the user through the set of steps (often involving several keystrokes or mouse clicks) needed to perform some task.

The third set of topics is definitions of important terms. These terms are usually the use cases and classes in the system, but sometimes they also include commands.

There are five general types of navigation controls for topics, but not all systems use all five types (see Figure 15-5). The first is the table of contents that organizes the information in a logical form, as though the users were to read the reference documentation from start to finish.

The index provides access to the topics based on important keywords, in the same way that the index at the back of a book helps you find topics. Text search provides the ability to search through the topics either for any text the user types or for words that match a developer-specified set of words that is much larger than the words in the index. Unlike the index, text search typically provides no organization to the words (other than alphabetical). Some systems provide the ability to use an intelligent agent to help in the search (e.g., the Microsoft Office Assistant—also known as the paper clip guy). The fifth and final navigation control to topics are the Web-like links between topics that enable the user to click and move among topics.

Procedures manuals and tutorials are similar but often simpler in structure. Topics for procedures manuals usually come from the use scenarios, window navigation diagram, and the real use cases developed during interface design and from other basic tasks the users must perform. Topics for tutorials are usually organized around major sections of the system and the level of experience of the user. Most tutorials start with the basic, most commonly used commands and then move into more complex and less frequently used commands.

Writing Documentation Topics

The general format for topics is fairly similar across application systems and operating systems (see Figure 15-6). Topics typically start with very clear titles, fol-

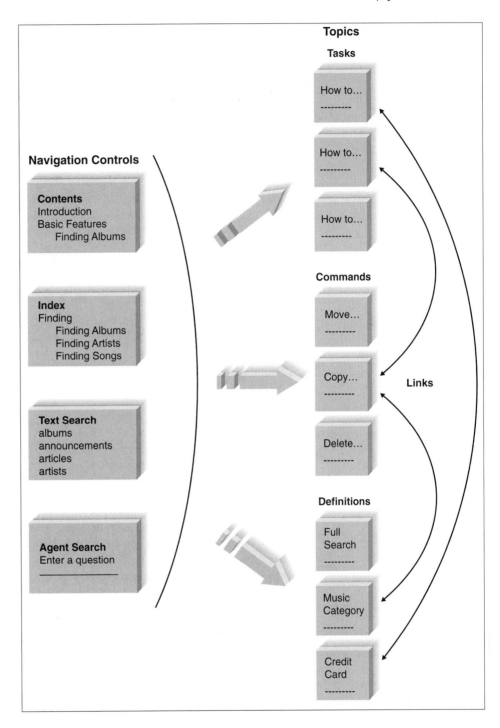

FIGURE 15-5
Organizing On-line Reference
Documents

lowed by some introductory text that defines the topic, and then provide detailed, step-by-step instructions on how to perform what is being described (where appropriate). Many topics include screen images to help the user find items on the screen; some also have "show-me" examples in which the series of keystrokes and/or mouse movements and clicks needed to perform the function are demonstrated to the user. Most also include navigation controls to enable movement among topics,

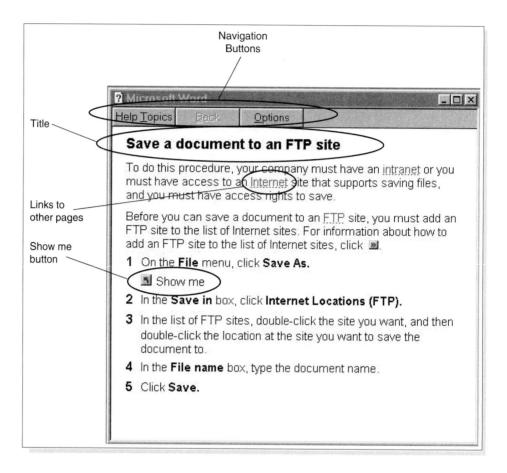

FIGURE 15-6
A Help Topic in Microsoft Word

usually at the top of the window, plus links to other topics. Some also include links to "related topics" that include options or other commands and tasks the user may want to perform in concert with the topic being read.

Writing the topic content can be challenging. It requires a good understanding of the user (or more accurately the range of users) and a knowledge of what skills the user(s) currently have and can be expected to import from other systems and tools they are using or have used (including the system the new system is replacing). Topics should always be written from the viewpoint of the user and describe what the user wants to accomplish, not what the system can do. Figure 15-7 provides some general guidelines to improve the quality of documentation text.[12]

Identifying Navigation Terms

As you write the documentation topics you also begin to identify the terms that will be used to help users find topics. The table of contents is usually the most straightforward because it is developed from the logical structure of the documentation topics, whether reference topics, procedure topics, or tutorial topics. The items for the index and search engine require more care because they are developed from the

---

[12] One of the best books to explain the art of writing is William Strunk and E.B. White, *Elements of Style,* 4th ed. (Boston: Allyn and Bacon, 2000).

Guideline	Before the Guideline	After the Guideline
Use the active voice: The active voice creates more active and readable text by putting the subject at the start of the sentence, the verb in the middle, and the object at the end.	Finding albums is done using the album title, the artist's name, or a song title.	You can find an album by using the album title, the artist's name, or a song title.
Use e-prime style: E-prime style creates more active writing by omitting all forms of the verb *to be*.	The text you want to copy must be selected before you click on the copy button.	Select the text you want to copy before you click on the copy button.
Use consistent terms: Always use the same term to refer to the same items, rather than switching among synonyms (e.g., change, modify, update).	Select the text you want to copy. Pressing the *copy* button will copy the marked text to the new location.	Select the text you want to copy. Pressing the *copy* button will copy the selected text to the new location.
Use simple language: Always use the simplest language possible to accurately convey the meaning. This does not mean you should "dumb down" the text but that you should avoid artificially inflating its complexity. Avoid separating subjects and verbs and try to use the fewest words possible. (When you encounter a complex piece of text, try eliminating words; you may be surprised at how few words are really needed to convey meaning.)	The Georgia Statewide Academic and Medical System (GSAMS) is a cooperative and collaborative distance learning network in the state of Georgia. The organization in Atlanta that administers and manages the technical and overall operations of the currently more than 300 interactive audio and video teleconferencing classrooms throughout Georgia system is the Department of Administrative Service (DOAS). (56 words)	The Department of Administrative Service (DOAS) in Atlanta manages the Georgia Statewide Academic and Medical System (GSAMS), a distance learning network with more than 300 teleconferencing classrooms throughout Georgia. (29 words)
Use friendly language: Too often, documentation is cold and sterile because it is written in a very formal manner. Remember, you are writing for a person, not a computer.	Blank disks have been provided to you by Operations. It is suggested that you ensure your data is not lost by making backup copies of all essential data.	You should make a backup copy of all data that is important to you. If you need more diskettes, contact Operations.
Use parallel grammatical structures: Parallel grammatical structures indicate the similarity among items in list and help the reader understand content.	Opening files Saving a document How to delete files	Opening a file Saving a file Deleting a file
Use steps correctly: Novices often intersperse action and the results of action when describing a step-by-step process. Steps are always actions.	1. Press the *customer* button. 2. The customer dialogue box will appear. 3. Type the customer ID and press the *submit* button and the customer record will appear.	1. Press the *customer* button. 2. Type the customer ID in the customer dialogue box when it appears. 3. Press the *submit* button to view the customer record for this customer.
Use short paragraphs: Readers of documentation usually quickly scan text to find the information they need, so the text in the middle of long paragraphs is often overlooked. Use separate paragraphs to help readers find information more quickly.		

Source: Adapted from T. T. Barker, *Writing Software Documentation*, Boston: Allyn & Bacon, 1998.

FIGURE 15-7
Guidelines for Crafting Documentation Topics

Y O U R	15-2 DOCUMENTATION FOR AN **ATM**
T U R N	

Pretend you are a project manager for a bank developing software for ATMs. Develop an on-line help system.

major parts of the system and the users' business functions. Every time you write a topic, you must also list the terms that will be used to find the topic. Terms for the index and search engine can come from four distinct sources.

The first source is the set of the commands in the user interface, such as open file, modify customer, and print open orders. All commands contain two parts (action and object). It is important to develop the index for both parts because users could search for information using either part. A user looking for more information about saving files, for example, might search by using the term "save" or the term "files."

The second source is the set of major concepts in the system, which are often use cases and classes. In the case of CD Selections, for example, this might include music category, album, and shipping costs.

A third source is the set of business tasks the user performs, such as ordering replacement unit or making an appointment. Often these will be contained in the command set, but sometimes they require several commands and use terms that always appear in the system. Good sources for these terms are the use scenarios and real use cases developed during interface design (see Chapters 11 and 12).

A fourth, often controversial, source is the set of synonyms for the three sets of items above. Users sometimes don't think in terms of the nicely defined terms used by the system. They may try to find information on how to "stop" or "quit" rather than "exit," or "erase" rather than "delete." Including synonyms in the index increases the complexity and size of the documentation system but can greatly improve the usefulness of the system to the users.

APPLYING THE CONCEPTS AT CD SELECTIONS

Managing Programming

Three programmers were assigned by CD Selections to develop the three major parts of the Internet Sales System. The first was the Web interface, both the client side (browser) and the server side. The second was the client-server-based management system (managing the CD information and marketing materials databases). The third was the interfaces between the Internet Sales System and CD Selections' existing distribution system and the credit card clearance center. Programming went smoothly and despite a few minor problems proceeded according to plan.

Testing

While the programmers were working, Alec began developing the test plans and user documentation. The test plans for the three components were similar but slightly more intensive for the Web interface component (see Figure 15-8). Unit

Test Stage	Web Interface	System Management	System Interfaces
Unit tests	Black-box tests	Black-box tests	Black-box tests
Integration tests	User interface tests; use scenario tests	User interface tests; use scenario tests	System interface tests
System tests	Requirements tests; security tests; performance tests; usability tests	Requirements tests; security tests	Requirements tests; security tests; performance tests
Acceptance tests	Alpha test; beta test	Alpha test; beta test	Alpha test; beta test

FIGURE 15-8
CD Selections' Test Plan

testing would use black-box testing based on the CRC cards, class diagrams, and contracts for all components. Figure 15-9 shows part of the class-invariant test specification for the order class in the Web interface component.

Integration testing for the Web interface and system management component would be subjected to all user interface and use-case tests to ensure that the interface would work properly. The system interface component would undergo system interface tests to ensure that the system performed calculations properly and was capable of exchanging data with the CD Selections' other systems and the credit card clearance center.

Class Invariant Test Specification

Page __5__ of __15__

Class Name:	Order	Version Number:	3	CRC Card ID:	15
Tester:	Susan Doe	Date Designed:	9/9	Date Conducted:	

Testing Objectives:

Ensure that the information entered by the customer on the place order form is valid.

Test Cases

Invariant Description	Original Attribute Value	Event	New Attribute Value	Expected Result	Result P/F
Attribute Name: CD Number					
1) (1..1)	Null	CreateOrder	Null	F	
2) (unsigned long)	Null	CreateOrder	ABC	F	
3) (unsigned long)	Null	CreateOrder	123	p	
Attribute Name: CD Name					
1) (1..1)	Null	CreateOrder	Null	F	
2) (String)	Null	CreateOrder	ABC	P	
3) (String)	Null	CreateOrder	123	P	
Attribute Name: Artist Name					
1) (1..1)	Null	CreateOrder	Null	F	
2) (String)	Null	CreateOrder	ABC	P	
3) (String)	Null	CreateOrder	123	P	

FIGURE 15-9
Partial Class Invariant Test Specification for the Order Class

Systems tests are by definition tests of the entire system—all components together. However, not all parts of the system would receive the same level of testing. Requirements tests would be conducted on all parts of the system to ensure that all requirements were met. Because security was a critical issue, the security of all aspects of the system would be tested. Security tests would be developed by CD Selections' infrastructure team, and once the system passed those tests, an external security consulting firm would be hired to attempt to break in to the system.

Performance was an important issue for the parts of the system used by the customer (the Web interface and the system interfaces to the credit card clearance and inventory systems) but not as important for the management component that would be used by staff, not customers. The customer-facing components would undergo rigorous performance testing to see how many transactions (whether searching or purchasing) they could handle before they were unable to provide a response time of two seconds or less. Alec also developed an upgrade plan so that as demand on the system increased, there was a clear plan for when and how to increase the processing capability of the system.

Finally, formal usability tests would be conducted on the Web interface portion of the system with six potential users (both novice and expert Internet users).

Acceptance tests would be conducted in two stages, alpha and beta. Alpha tests would be done during the training of CD Selections' staff. The Internet Sales manager would work together with Alec to develop a series of tests and training exercises to train the Internet Sales group staff in how to use the system. They would then load the real CD data into the system and begin adding marketing materials. These same staff and other CD Selections staff members would also pretend to be customers and test the Web interface.

Beta testing would be done by "going live" with the Web site but its existence would only be announcied to CD Selections' employees. As an incentive to try the Web site (rather than buying from the store in which they worked), employees would be offered triple their normal employee discount for all products ordered from the Web site. The site would also have a prominent button on every screen that would enable employees to e-mail comments to the project team, and the announcement would encourage employees to report problems, suggestions, and compliments to the project team. After one month, assuming all went well, the beta test would be completed, and the Internet Sales site linked to the main Web site and advertised to the general public.

Developing User Documentation

Three types of documentation (reference documents, procedures manuals, and tutorials) could be produced for the Web interface and the management component. Since the number of CD Selections' staff using the system management component would be small, Alec decided to produce only the reference documentation (an online help system). He felt that an intensive training program and a one-month beta test period would be sufficient without tutorials and formal procedures manuals. Similarly, he felt that the process of ordering CDs and the Web interface itself was simple enough to not require a tutorial on the Web—a help system would be sufficient (and a procedures manual didn't make sense).

Alec decided that the reference documents for both the Web interface and system management components would contain help topics for user tasks, commands, and definitions. He also decided that the documentation component would contain four types of navigation controls: a table of contents, an index, a find, and links to

Tasks	Commands	Terms
Find an album	Find	Album
Add an album to my shopping cart	Browse	Artist
Placing an order	Quick search	Music type
How to buy	Full search	Special deals
What's in my shopping cart?		Cart
		Shopping cart

FIGURE 15-10
Sample Help Topics for CD Selections

definitions. He did not feel that the system was complex enough to benefit from a search agent.

After these decisions were made, Alec delegated the development of the reference documents to a technical writer assigned to the project team. Figure 15-10 shows examples of a few of topics that the writer developed. The tasks and commands were taken directly from the interface design. The list of definitions was developed once the tasks and commands were developed based on the writer's experience in understanding what terms might be confusing to the user.

Once the topic list was developed, the technical writer then began writing the topics themselves as well as the navigation controls to access. Figure 15-11 shows an example of one topic taken from the task list: how to place an order. This topic presents a brief description of what it is and then leads the user through the step-by-step process needed to complete the task. The topic also lists the navigation controls that will be used to find the topic in terms of the table of contents entries, the index entries, and search entries. It also lists what words in the topic itself will have links to other topics (e.g., shopping cart).

Help Topic	Navigation Controls
How to Place an Order	**Table of Contents** list:
	How to Place an Order
When you are ready to pay for the merchandise you have selected (the items in your shopping cart) you can place your order. There are four steps.	**Index** list:
	Credit Card
1. Move to the Place order Page	Order
	Pay
Click on the ⬤Place order button to move to the place order page.	Place order
2. Make sure you are ordering what you want	**Search** find by:
	Credit Card
The place order screen displays all the items in your <u>shopping cart</u>. Read through the list to make sure these are what you want because once you submit your credit card information you cannot change the order.	Delete Items
	Order
	Pay
	Place order
	Shopping Cart
	Verify Order
you can delete an item by	**Links**:
	Shopping Cart

FIGURE 15-11
Example Documentation Topic for CD Selections

SUMMARY

Managing Programming

Programming is done by programmers, so systems analysts have few responsibilities during this stage. The project manager, however, is usually very busy. The first task is to assign the programmers, ideally the fewest possible to complete the project because coordination problems increase as the size of the programming team increases. Coordination can be improved by having regular meetings, ensuring that standards are followed, implementing change control, and using CASE tools effectively. One of the key functions of the project manager is to manage the schedule and adjust it for delays. Two common causes of delays are scope creep and minor slippages that go unnoticed.

Test Planning

Tests must be carefully planned because the cost of fixing one major bug after the system is installed can easily exceed the annual salary of a programmer. A test plan contains several tests that examine different aspects of the system. A test, in turn, specifies several test cases that will be examined by the testers. A unit test examines a class within the system; test cases come from the class specifications or the class code itself. An integration test examines how well several classes work together; test cases come from the interface design, use cases, the use case diagram, sequence diagrams, and collaboration diagrams. A system test examines the system as a whole and is broader than the unit and integration tests; test cases come from the system design, the infrastruture design, and the unit and integration tests. Acceptance testing is done by the users to determine whether the system is acceptable to them; it draws on the system test plans (alpha testing) and on the real work the users perform (beta testing).

Documentation

Documentation, both user documentation and system documentation, is moving away from paper-based documents to on-line documentation. There are three types of user documentation: reference documents are designed to be used when the user needs to learn how to perform a specific function (e.g., an on-line help system), procedures manuals describe how to perform business tasks, while tutorials teach people how to use the system. Documentation navigation controls (e.g., a table of contents, index, find, intelligent agents, or links between pages) enable users to find documentation topics (e.g., how to perform a function, how to use an interface command, an explanation of a term).

KEY TERMS

Acceptance test
Alpha test
Beta test
Black-box testing
Change control
Construction
Documentation navigation
   control
Documentation topic
Hardcoded
Integration test

Interaction testing
Procedures manual
Program log
Reference document
Requirements testing
Security testing
Stub
System documentation
System interface testing
System test
Test

Test case
Test plan
Test specification
Tutorial
Unit test
Usability testing
Use-case testing
User documentation
User interface testing
White-box testing

QUESTIONS

1. Why is testing important?
2. What is the primary role of systems analysts during the programming stage?
3. In *The Mythical Man-Month,* Frederick Brooks argues that adding more programmers to a late project makes it later. Why?
4. What is the purpose of testing?
5. Describe how object-orientation impacts testing.
6. Compare and contrast the terms *test, test plan,* and *test case.*
7. What is a stub, and why is it used in testing?
8. What is the primary goal of unit testing?
9. Compare and contrast black-box testing and white-box testing.
10. How are the test cases developed for unit tests?
11. What are the different types of class tests?
12. What is the primary goal of integration testing?
13. How are the test cases developed for integration tests?
14. What is the primary goal of system testing?
15. How are the test cases developed for system tests?
16. What is the primary goal of acceptance testing?
17. How are the test cases developed for acceptance tests?
18. Compare and contrast alpha testing and beta testing.
19. Compare and contrast user documentation and system documentation.
20. Why is on-line documentation becoming more important?
21. What are the primary disadvantages of on-line documentation?
22. Compare and contrast reference documents, procedures manuals, and tutorials.
23. What are five types of documentation navigation controls?
24. What are the commonly used sources of documentation topics? Which is the most important? Why?
25. What are the commonly used sources of documentation navigation controls? Which is the most important? Why?

EXERCISE

A. Create an invariant test specification for the class you chose for the Health Club in Exercise A in Chapter 14.

B. Create an invariant test specification for the class you chose for the Picnics R Us in Exercise C in Chapter 14.

C. Create an invariant test specification for the class you chose for the Of-The-Month Club (OTMC) in Exercise E in Chapter 14.

D. Create an invariant test specification for the class you chose for the Library in Exercise G in Chapter 14.

E. If the registration system at your university does not have a good on-line help system, develop one for one screen of the user interface.

F. Examine and prepare a report on the on-line help system for the calculator program in Windows (or a similar on the Mac or Unix). (You will likely be surprised at the amount of help for such a simple program.)

G. Compare and contrast the on-line help at two different Web sites that enable you to perform some function (e.g., make travel reservations, order books).

MINICASES

1. Pete is a project manager on a new systems development project. This project is Pete's first experience as a project manager, and he has led his team successfully to the programming phase of the project. The project has not always gone smoothly, and Pete has made a few mistakes, but he is generally pleased with the progress of his team and the quality of the system being developed. Now that programming has begun, Pete has been hoping for a little break in the hectic pace of his workday.

   Prior to beginning programming, Pete recognized that the time estimates made earlier in the project were too optimistic. However, he was firmly committed to meeting the project deadline because of his desire for his first project as project manager to be a success. In anticipation of this time pressure problem, Pete arranged with the Human Resources department to bring in two new college graduates and two college interns to beef up the programming staff. Pete would have liked to find some staff with more experience, but the budget was too tight, and he was committed to keeping the project budget under control.

   Pete made his programming assignments, and work on the programs began about two weeks ago. Now, Pete has started to hear some rumbles from the programming team leaders that may signal trouble. It seems that the programmers have reported several instances where they wrote programs, only to be unable to find them when they went to test them. Also, several programmers have opened programs that they had written, only to find that someone had changed portions of their programs without their knowledge.

   a. Is the programming phase of a project a time for the project manager to relax? Why or why not?

   b. What problems can you identify in this situation? What advice do you have for the project manager? How likely does it seem that Pete will achieve his desired goals of being on time and within budget if nothing is done?

2. The systems analysts are developing the test plan for the user interface for the Holiday Travel Vehicles system. As the salespeople are entering a sales invoice into the system, they will be able to either enter an option code into a text box, or select an option code from a drop-down list. A combo box was used to implement this, since it was felt that the salespeople would quickly become familiar with the most common option codes and would prefer entering them directly to speed up the entry process.

It is now time to develop the test for validating the option code field during data entry. If the customer did not request any dealer-installed options for the vehicle, the salesperson should enter "none"; the field should not be blank. The valid option codes are four-character alphabetic codes and should be matched against a list of valid codes.

Prepare a test plan for the test of the option code field during data entry.

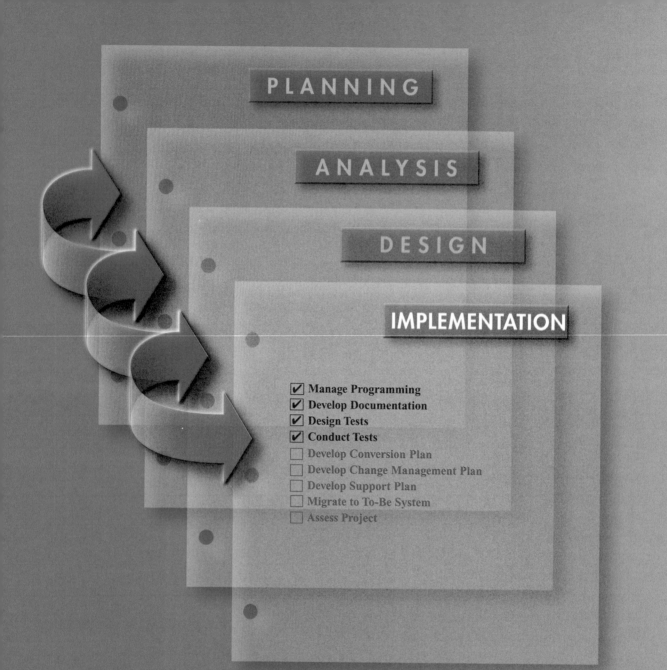

PLANNING

ANALYSIS

DESIGN

IMPLEMENTATION

- ☑ **Manage Programming**
- ☑ **Develop Documentation**
- ☑ **Design Tests**
- ☑ **Conduct Tests**
- ☐ Develop Conversion Plan
- ☐ Develop Change Management Plan
- ☐ Develop Support Plan
- ☐ Migrate to To-Be System
- ☐ Assess Project

T A S K C H E C K L I S T

PLANNING   ANALYSIS   DESIGN

CHAPTER 16

INSTALLATION

THIS chapter examines the activities needed to install the information system and successfully convert the organization to using it. It also discusses post-implementation activities, such as system support, system maintenance, and project assessment. Installing the system and making it available for use from a technical perspective is relatively straightforward. However, the training and organizational issues surrounding the installation are more complex and challenging because they focus on people, not computers.

OBJECTIVES

- Be familiar with the system installation process.
- Understand different types of conversion strategies and when to use them.
- Understand several techniques for managing change.
- Be familiar with postinstallation processes.

CHAPTER OUTLINE

IMPLEMENTATION

INTRODUCTION

It must be remembered that there is nothing more difficult to plan, more doubtful of success, nor more dangerous to manage than the creation of a new system. For the initiator has the animosity of all who would profit by the preservation of the old institution and merely lukewarm defenders in those who would gain by the new.

—Machiavelli, *The Prince,* 1513

Although written almost 500 years ago, Machiavelli's comments are still true today. Managing the change to a new system—whether it is computerized or not—is one of the most difficult tasks in any organization. Because of the challenges involved, most organizations begin developing their conversion and change management plans while the programmers are still developing the software. Leaving conversion and change management planning to the last minute is a recipe for failure.

In many ways, using a computer system or set of work processes is much like driving on a dirt road. Over time with repeated use, the road begins to develop ruts in the most commonly used parts of the road. Although these ruts show where to drive, they make change difficult. As people use a computer system or set of work processes, those systems/work processes begin to become habits or norms; people learn them and become comfortable with them. These systems or work processes then begin to limit people's activities and make it difficult for them to change because they begin to see their jobs in terms of these processes rather than of the final business goal of serving customers.

One of the earliest models for managing organizational change was developed by Kurt Lewin.[1] Lewin argued that change is a three-step process: unfreeze, move, refreeze (Figure 16-1). First, the project team must *unfreeze* the existing habits and norms (the As-Is system) so that change is possible. Most of the systems development life cycle (SDLC) to this point has laid the groundwork for unfreezing. Users are aware of the new system being developed, some have participated in an analysis of the current system (and so are aware of its problems), and some have helped design the new system (and so have some sense of the potential benefits of the new system). These activities have helped to unfreeze the current habits and norms.

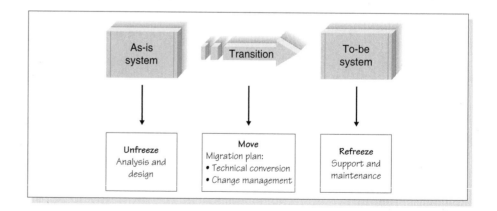

FIGURE 16-1
Implementing Change

[1] Kurt Lewin, "Frontiers in Group Dynamics," *Human Relations,* 1, no. 5 (1947:5–41), and Kurt Lewin, "Group Decision and Social Change" in E.E. Maccoby, T.M. Newcomb, and E.L. Hartley (eds.), *Readings in Social Psychology,* (New York: Holt, Rinehart & Winston, 1958), pp. 197–211.

The second step is to help the organization move to the new system via a *migration plan.* The migration plan has two major elements. One is technical, which includes how the new system will be installed and how data in the As-Is system will be moved into the To-Be system; this is discussed in the *Conversion* section of this chapter. The second component is organizational, which includes helping users understand the change and motivating them to adopt it; this is discussed in the *Change Management* section of this chapter.

The third step is to *refreeze* the new system as the habitual way of performing the work processes—ensuring that the new system successfully becomes the standard way of performing the business function it supports. This refreezing process is a key goal of the *Post-implementation Activities* discussed in the final section of this chapter. By providing ongoing support for the new system and immediately beginning to identify improvements for the next version of the system, the organization helps solidify the new system as the new habitual way of doing business. Post-implementation activities include *system support,* which means providing help desk and telephone support for users with problems; *system maintenance,* which means fixing bugs and improving the system after it has been installed; and *project assessment,* which means the process of evaluating the project to identify what went well and what could be improved for the next system development project.

Change management is the most challenging of the three components because it focuses on people, not technology, and because it is the one aspect of the project that is the least controllable by the project team. Change management means winning the hearts and minds of potential users and convincing them that the new system actually provides value.

Maintenance is the most costly aspect of installation process, because the cost of maintaining systems usually greatly exceeds the initial development costs. It is not unusual for organizations to spend 60 to 80% of their total IS development budget on maintenance. Although this may sound surprising initially, think about the software you use. How many software packages do you use that are the very first version? Most commercial software packages become truly useful and enter widespread use only in their second or third version. Maintenance and continual improvement of software is ongoing, whether it is a commercially available package or software developed in-house. Would you buy software if you knew that no new versions were going to be produced? Of course, commercial software is somewhat different from custom in-house software used by only one company, but the fundamental issues remain.

Project assessment is probably the least commonly performed part of the SDLC but is perhaps the one that has the most long-term value to the IS department. Project assessment enables project team members to step back and consider what they did right and what they could have done better. It is an important component in the individual growth and development of each member of the team, because it encourages team members to learn from their successes and failures. It also enables new ideas or new approaches to system development to be recognized, examined, and shared with other project teams to improve their performance.

CONVERSION

Conversion is the technical process by which the new system replaces the old system. Users are moved from using the As-Is business processes and computer programs to the To-Be business processes and programs. The *migration plan* specifies what activities will be performed when and by whom and includes both technical aspects (such as installing hardware and software and converting data from the As-Is system to the To-Be system) and organizational aspects (such as training and motivating the users to embrace the new system). Conversion refers to the technical aspects of the migration plan. Organizational aspects are discussed in the *Change Management* section.

There are three major steps to the conversion plan before commencement of operations: install hardware, install software, and convert data (Figure 16-2). Although it may be possible to do some of these steps in parallel, usually they must be done sequentially at any one location.

The first step in the conversion plan is to buy and install any needed hardware. In many cases, no new hardware is needed, but sometimes the project requires such new hardware as servers, client computers, printers, and networking equipment. It is critical to work closely with vendors who are supplying needed hardware and software to ensure that the deliveries are coordinated with the conversion schedule so that the equipment is available when it is needed. Nothing can stop a conversion plan in its tracks as easily as the failure of a vendor to deliver needed equipment.

Once the hardware is installed, tested, and certified as being operational, the second step is to install the software. This includes the To-Be system under development, and sometimes additional software that must be installed to make the system operational. For example, the CD Selections Internet Sales System needs Web server software. At this point, the system is usually tested again to ensure that it operates as planned.

The third step is to convert the data from the As-Is system to the To-Be system. Data conversion is usually the most technically complicated step in the migration plan. Often, separate programs must be written to convert the data from the As-Is system to the new formats required in the To-Be system and store it in the To-Be system files and databases. This process is often complicated by the fact that the files and databases in the To-Be system do not exactly match the files and databases in the As-Is system (e.g., the To-Be system may use several tables in a database to store customer data that was contained in one file in the As-Is system). Formal test plans are always required for data conversion efforts (see Chapter 15).

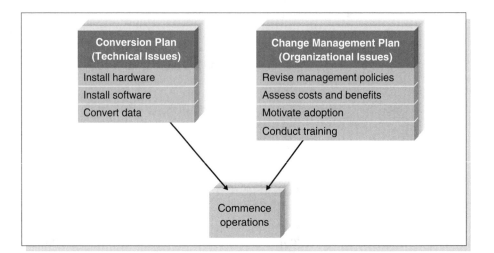

FIGURE 16-2

Elements of a Migration Plan

Conversion can be thought of along three dimensions: the style by which the conversion is done, what location or work groups are converted at what time, and what modules of the system are converted at what time (Figure 16-3).

Conversion Style

The *conversion style* is the way in which users are switched between the old and new systems. There are two fundamentally different approaches to the style of conversion: direct conversion and parallel conversion.

Direct Conversion   With *direct conversion* (sometimes called cold turkey, big bang, or abrupt cutover), the new system instantly replaces the old system. The new system is turned on and the old system is immediately turned off. This is the approach that you probably use when you upgrade commercial software (e.g., Microsoft Word) from one version to another; you simply begin using the new version and stop using the old version.

Direct conversion is the simplest and most straightforward. However, it is the most risky, because any problems with the new system that have escaped detection during testing may seriously disrupt the organization.

Parallel Conversion   With *parallel conversion,* the new system is operated side by side with the old system; both systems are used simultaneously. For example, if a new accounting system is installed, the organization enters data into both the old system and the new system and then carefully compares the output from both systems to ensure that the new system is performing correctly. After some time period (often one to two months) of parallel operation and intense comparison between the two systems, the old system is turned off and the organization continues using the new system.

This approach is more likely to catch any major bugs in the new system and prevent the organization from suffering major problems. If problems are discovered in the new system, the system is simply turned off and fixed and then the conversion process starts again. The problem with this approach is the added expense of operating two systems that perform the same function.

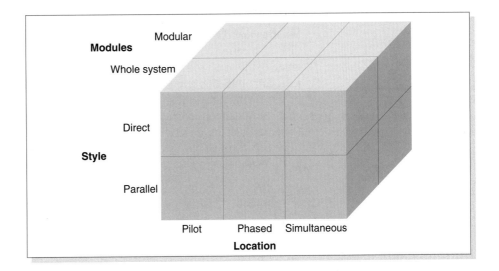

FIGURE 16-3
Conversion Strategies

Conversion Location

Conversion location refers to those parts of the organization that are converted at what points in time. Often, parts of the organization are physically located in different offices (e.g., Toronto, Atlanta, Los Angeles). In other cases, location refers to different organizational units located in different parts of the same office complex (e.g., order entry, shipping, purchasing). There are at least three fundamentally different approaches to selecting the way in which different organizational locations are converted: pilot conversion, phased conversion, and simultaneous conversion.

Pilot Conversion   With a *pilot conversion,* one or more locations or units/work groups within a location are selected to be converted first as part of a pilot test. The locations participating in the pilot test are converted (using either direct or parallel conversion). If the system passes the pilot test, then the system is installed at the remaining locations (again using either direct or parallel conversion).

Pilot conversion has the advantage of providing an additional level of testing before the system is widely deployed throughout the organization, so that any problems with the system affect only the pilot locations. However, this type of conversion obviously requires more time before the system is installed at all organizational locations. Also, it means that different organizational units are using different versions of the system and business processes, which may make it difficult for them to exchange data.

Phased Conversion   With *phased conversion,* the system is installed sequentially at different locations. A first set of locations are converted, then a second set, then a third set, and so on, until all locations are converted. Sometimes there is a deliberate delay between the different sets (at least between the first and the second), so that any problems with the system are detected before too much of the organization is affected. In other cases, the sets are converted back-to-back so that as soon as those converting one location have finished, the project team moves to the next and continues the conversion.

Phased conversion has the same advantages and disadvantages of pilot conversion. In addition, it means that a smaller set of people are required to perform the actual conversion (and any associated user training) than if all locations were converted at once.

Simultaneous Conversion   *Simultaneous conversion,* as the name suggests, means that all locations are converted at the same time. The new system is installed and made ready at all locations, and at a preset time, all users begin using the new system. Simultaneous conversion is often used with direct conversion, but it can also be used with parallel conversion.

Simultaneous conversion eliminates problems with having different organizational units using different systems and processes. However, it also means that the organization must have sufficient staff to perform the conversion and train the users at all locations simultaneously.

Conversion Modules

Although it is natural to assume that systems are usually installed in their entirety, this is not always the case.

CONCEPTS

IN ACTION

16-A CONVERTING TO THE EURO (PART 1)

When the European Union decided to introduce the euro, the European Central Bank had to develop a new computer system (called Target) to provide a currency settlement system for use by investment banks and brokerages. Prior to the introduction of the euro, settlement was performed between the central banks of the countries involved. After the introduction, the Target system, which consists of 15 national banking systems, would settle trades and perform currency conversions for cross-border payments for stocks and bonds.

Source: Thomas Hoffman, "Debut of Euro Nearly Flawless," *Computerworld,* 33, no. 2 (January 11, 1999): 16.

QUESTION:
Implementing Target was a major undertaking for a number of reasons. If you were an analyst on the project, what kinds of issues would you have to address to make sure the conversion happened successfully?

Whole System Conversion   *Whole-system conversions,* in which the entire system is installed at one time, is the most common. It is simple and the easiest to understand. However, if the system is large and/or extremely complex (e.g., an enterprise resource planning system such as SAP or PeopleSoft), the whole system may prove too difficult for users to learn in one conversion step.

Modular Conversion   When the *modules*[2] within a system are separate and distinct, organizations sometimes choose to convert to the new system one module at a time—that is, *modular conversion.* Modular conversion requires special care in developing the system (and usually adds extra cost). Each module either must be written to work with both the old and new systems or object wrappers (see Chapter 9) must be used to encapsulate the old system from the new. When modules are tightly integrated, this is very challenging and therefore seldom done. However, when there is only a loose association between modules, this becomes easier. For example, consider a conversion from an old version of Microsoft Office to a new version. It is relatively simple to convert from the old version of Word to the new version without simultaneously having to change from the old to the new version of Microsoft Excel.

Modular conversion reduces the amount of training required to begin using the new system. Users need training in only the new module being implemented. However, modular conversion does take longer and has more steps than does the whole-system process.

Selecting the Appropriate Conversion Strategy

Each of the three dimensions in Figure 16-3 is independent, so that a conversion strategy can be developed to fit in any one of the boxes in this figure. Different boxes can also be mixed and matched into one *conversion strategy.* For example, one commonly used approach is to begin with a pilot conversion of the whole system using parallel conversion in a handful of test locations. Once the system has passed the pilot test at these locations, it is then installed in the remaining locations using phased conversion

[2] In this case, a module is typically a component or a package, that is, a set of collaborating classes.

	Conversion Style		Conversion Location			Conversion Modules	
Characteristic	**Direct Conversion**	**Parallel Conversion**	**Pilot Conversion**	**Phased Conversion**	**Simultaneous Conversion**	**Whole-System Conversion**	**Modular Conversion**
Risk	High	Low	Low	Medium	High	High	Medium
Cost	Low	High	Medium	Medium	High	Medium	High
Time	Short	Long	Medium	Long	Short	Short	Long

FIGURE 16-4
Characteristics of Conversion Strategies

with direct cutover. There are three important factors to consider in selecting a conversion strategy: risk, cost, and the time required (Figure 16-4).

Risk   After the system has passed a rigorous battery of unit, system, integration, and acceptance testing, it should be bug free … maybe. Because humans make mistakes, nothing built by people is ever perfect. Even after all these tests, there may still be a few undiscovered bugs. The conversion process provides one last step in which to catch these bugs before the system goes live and the bugs have the chance to cause problems.

Parallel conversion is less risky than is direct conversion because it has a greater chance of detecting bugs that have gone undiscovered in testing. Likewise, pilot conversion is less risky than is phased conversion or simultaneous conversion because if bugs do occur, they occur in pilot test locations whose staff are aware that they may encounter bugs. Since potential bugs affect fewer users, there is less risk. Likewise, converting a few modules at a time lowers the probability of a bug because there is more likely to be a bug in the whole system than in any given module.

How important the risk is depends on the system being implemented—the combination of the probability that bugs remain undetected in the system and the potential cost of those undetected bugs. If the system has indeed been subjected to extensive methodical testing, including alpha and beta testing, then the probability of undetected bugs is lower than if the testing was less rigorous. However, there still might have been mistakes made in the analysis process, so that although there might be no software bugs, the software might fail to properly address the business needs.

Assessing the cost of a bug is challenging, but most analysts and senior managers can make a reasonable guess at the relative cost of a bug. For example, the cost of a bug in an automated stock market trading program or a heart-lung machine keeping someone alive is likely to be much greater than a bug in a computer game or word processing program. Therefore, risk is likely to be a very important factor in the conversion process if the system has not been as thoroughly tested as it might have been and/or if the cost of bugs is high. If the system has been thoroughly tested and/or the cost of bugs is not that high, then risk becomes less important to the conversion decision.

Cost   As might be expected, different conversion strategies have different costs. These costs can include things such as salaries for people who work with the system (e.g., users, trainers, system administrators, external consultants), travel expenses, operation expenses, communication costs, and hardware leases. Parallel conversion is more expensive than direct cutover because it requires that two sys-

YOUR

TURN

16-1 DEVELOPING A CONVERSION PLAN

Suppose you are leading the conversion from one word processor to another at your university. Develop a conversion plan (i.e., technical issues only). You have also been asked to develop a conversion plan for the university's new Web-based course registration system. How would the second conversion plan be similar to and different from the one you developed for the word processor?

tems (the old and the new) be operated at the same time. Employees must now perform twice the usual work because they have to enter the same data into both the old and the new systems. Parallel conversion also requires the results of the two systems to be completely cross-checked to make sure there are no differences between the two, which entails additional time and cost.

Pilot conversion and phased conversion have somewhat similar costs. Simultaneous conversion has higher costs because more staff are required to support all the locations as they simultaneously switch from the old to the new system. Modular conversion is more expensive than whole system conversion because it requires more programming. The old system must be updated to work with selected modules in the new system, and modules in the new system must be programmed to work with selected modules in both the old and new systems.

Time The final factor is the amount of time required to convert between the old and the new system. Direct conversion is the fastest because it is immediate. Parallel

CONCEPTS

IN ACTION

16-B U.S. ARMY INSTALLATION SUPPORT

Throughout the 1960s, 1970s, and 1980s, the U.S. Army automated its installations (army bases, in civilian terms). Automation was usually a local effort at each of the more than 100 bases. Although some bases had developed software together (or borrowed software developed at other bases), each base often had software that performed different functions or performed the same function in different ways. In 1989, the army decided to standardize the software so that the same software would be used everywhere. This would greatly reduce software maintenance and also reduce training when soldiers were transferred between bases.

The software took four years to develop. The system was quite complex and the project manager was concerned that there was a high risk that not all requirements of all installations had been properly captured. Cost and time were less important, since the project had already run four years and cost $100 million.

Therefore, the project manager chose a modular pilot conversion using parallel conversion. The manager selected seven installations, each representing a different type of army installation (e.g., training base, arsenal, depot) and began the conversion. All went well, but several new features were identified that had been overlooked during the analysis, design, and construction. These were added and the pilot testing resumed. Finally, the system was installed in the rest of the army installations using a phased direct conversion of the whole system. *Alan Dennis*

QUESTION:
1. Do you think the conversion strategy was appropriate?
2. Regardless of whether you agree, what other conversion strategy could have been used?

conversion takes longer because the full advantages of the new system do not become available until the old system is turned off. Simultaneous conversion is fastest because all locations are converted at the same time. Phased conversion usually takes longer than pilot conversion because usually (but not always) once the pilot test is complete all remaining locations are simultaneously converted. Phased conversion proceeds in waves, often requiring several months before all locations are converted. Likewise, modular conversion takes longer than whole system conversion because the models are introduced one after another.

CHANGE MANAGEMENT

In the context of a systems development project, *change management* is the process of helping people to adopt and adapt to the To-Be system and its accompanying work processes without undue stress.[3] There are three key roles in any major organizational change. The first is the *sponsor* of the change—the person who wants the change. This person is the business sponsor who first initiated the request for the new system (see Chapter 2). Usually, the sponsor is a senior manager of the part of the organization that must adopt and use the new system. It is critical that the sponsor be active in the change management process because a change that is clearly being driven by the sponsor, not by the project team or the IS organization, has greater legitimacy. The sponsor has direct management authority over those who adopt the system.

The second role is that of the *change agent*—the person(s) leading the change effort. The change agent, charged with actually planning and implementing the change, is usually someone outside of the business unit adopting the system and therefore has no direct management authority over the potential adopters. Because the change agent is an outsider from a different organizational culture, he or she has less credibility than do the sponsor and other members of the business unit. After all, once the system has been installed, the change agent usually leaves and thus has no ongoing impact.

The third role is that of *potential adopter* or target of the change—the people who actually must change. These are the people for whom the new system is designed and who will ultimately choose to use or not use the system.

In the early days of computing, many project teams simply assumed that their job ended when the old system was converted to the new system at a technical level. The philosophy was "build it and they will come." Unfortunately, that happens only in the movies. Resistance to change is common in most organizations. Therefore, the change management plan is an important part of the overall installation plan that glues together the key steps in the change management process. Successful change requires that people want to adopt the change and are able to adopt the change. The change management plan has four basic steps: revising management policies, assessing the cost and benefit models of potential adopters, motivating adoption, and enabling people to adopt through training (see Figure 16-2). How-

---

[3] Many books have been written on change management. Some of our favorites are the following: Patrick Connor and Linda Lake, *Managing Organizational Change,* 2nd ed. (Westport, CT: Praeger, 1994); Douglas Smith, *Taking Charge of Change* (Reading, MA: Addison-Wesley, 1996); and Daryl Conner, *Managing at the Speed of Change* (New York: Villard Books, 1992).

ever, before we can discuss the change management plan, we must first understand why people resist change.

Understanding Resistance to Change

People resist change—even change for the better—for very rational reasons.[4] What is good for the organization is not necessarily good for the people who work there. For example, consider an order-processing clerk who used to receive orders to be shipped on paper shipping documents but now uses a computer to receive the same information. Rather than typing shipping labels with a typewriter, the clerk now clicks on the print button on the computer and the label is produced automatically. The clerk can now ship many more orders each day, which is a clear benefit to the organization. The clerk, however, probably doesn't really care how many packages are shipped. His or her pay doesn't change; it's just a question of which the clerk prefers to use, a computer or typewriter. Learning to use the new system and work processes—even if the change is minor—requires more effort than continuing to use the existing well-understood, system and work processes.

So why do people accept change? Simply put, every change has a set of costs and benefits associated with it. If the benefits of accepting the change outweigh the costs of the change, then people change. And sometimes the benefit of change is avoidance of the pain that you would experience if you did not adopt the change (e.g., if you don't change, you are fired, so one of the benefits of adopting the change is that you still have a job).

In general, when people are presented with an opportunity for change, they perform a cost–benefit analysis (sometime consciously, sometimes subconsciously) and decide the extent to which they will embrace and adopt the change. They identify the costs of and benefits from the system and decide whether the change is worthwhile. However, it is not that simple, because most costs and benefits are not certain. There is some uncertainty as to whether a certain benefit or cost will actually occur; so both the costs of and benefits from the new system will need to be weighted by the degree of certainty associated with them (Figure 16-5). Unfortu-

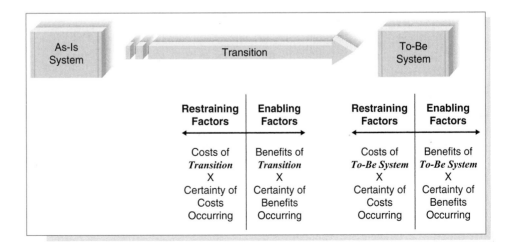

FIGURE 16-5
The Costs and Benefits of Change

---

[4] This section benefited from conversations with Dr. Robert Briggs, research scientist at the Center for the Management of Information at the University of Arizona.

nately, most humans tend to overestimate the probability of costs and underestimate the probability of benefits.

There are also costs and sometimes benefits associated with the actual transition process itself. For example, suppose you found a nicer house or apartment than your current one. Even if you liked it better, you might decide not to move simply because the cost of moving outweighed the benefits from the new house or apartment itself. Likewise, adopting a new computer system might require you to learn new skills, which could be seen as a cost to some people, or as a benefit to others, if they perceived that those skills would somehow provide other benefits beyond the use of the system itself. Once again, any costs and benefits from the transition process must be weighted by the certainty with which they will occur (see Figure 16-5).

Taken together, these two sets of costs and benefits (and their relative certainties) affect the acceptance of change or resistance to change that project teams encounter when installing new systems in organizations. The first step in change management is to understand the factors that inhibit change—the factors that affect the *perception* of costs and benefits and certainty that they will be generated by the new system. It is critical to understand that the "real" costs and benefits are far less important than the perceived costs and benefits. People act on what they *believe* to be true, not on what *is* true. Thus, any understanding of how to motivate change must be developed from the viewpoint of the people expected to change, not from the viewpoint of those leading the change.

Revising Management Policies

The first major step in the change management plan is to change the management policies that were designed for the As-Is system to new management policies designed to support the To-Be system. *Management policies* provide goals, define how work processes should be performed, and determine how organizational members are rewarded. No computer system will be successfully adopted unless management policies support its adoption. Many new computer systems bring changes to business processes; they enable new ways of working. Unless the policies that provide the rules and rewards for those processes are revised to reflect the new opportunities that the system permits, potential adopters cannot easily use it.

Management has three basic tools for structuring work processes in organizations.[5] The first are the *standard operating procedures (SOPs)* that become the habitual routines for how work is performed. The SOPs are both formal and informal. Formal SOPs define proper behavior. Informal SOPs are the norms that have developed over time for how processes are actually performed. Management must ensure that the formal SOPs are revised to match the To-Be system. The informal SOPs will then evolve to refine and fill in details absent in the formal SOPs.

The second aspect of management policy is defining how people assign meaning to events. What does it mean to "be successful" or "do good work"? Policies help people understand meaning by defining *measurements* and *rewards*. Measurements explicitly define meaning because they provide clear and concrete evi-

---

[5] This section builds on the work of Anthony Giddons, *The Constitution of Society: Outline of the Theory of Structure* (Berkeley: University of California Press, 1984). A good summary of Giddons's theory that has been revised and adapted for use in understanding information systems is an article by Wanda Orlikowski and Dan Robey: "Information Technology and the Structuring of Organizations," *Information Systems Research* 2, no. 2 (1991): 143–169.

YOUR	**16-2 STANDARD OPERATING PROCEDURES**
TURN	

Identify and explain three standard operating procedures for the course in which you are using this book. Discuss whether they are formal or informal.

dence about what is important to the organization. Rewards reinforce measurements because "what gets measured gets done" (an overused but accurate saying). Measurements must be carefully designed to motivate desired behavior. The IBM credit example (Your Turn 4-3 in Chapter 4) illustrates the problem with flawed measurements' driving improper behavior (when the credit analysts became too busy to handle credit requests, they would find nonexistent errors so they could return them unprocessed).

A third aspect of management policy is *resource allocation*. Managers can have clear and immediate impacts on behavior by allocating resources. They can redirect funds and staff from one project to another, create an infrastructure that supports the new system, and invest in training programs. Each of these activities has both a direct and symbolic effect. The direct effect comes from the actual reallocation of resources. The symbolic effect shows that management is serious about its intentions. There is less uncertainty about management's long-term commitment to a new system when potential adopters see resources being committed to support it.

Assessing Costs and Benefits

The next step in developing a change management plan is to develop two clear and concise lists of costs and benefits provided by the new system (and the transition to it) compared with the As-Is system. The first list is developed from the perspective of the organization, which should flow easily from the business case developed during the feasibility study and refined over the life of the project (see Chapter 2). This set of organizational costs and benefits should be distributed widely so that everyone expected to adopt the new system should clearly understand why the new system is valuable to the organization.

The second list of costs and benefits is developed from the viewpoints of the different potential adopters expected to change, or stakeholders in the change. For example, one set of potential adopters may be the frontline employees, another may be the first-line supervisors, and yet another might be middle management. Each of these potential adopters/stakeholders may have a different set of costs and benefits associated with the change—costs and benefits that can differ widely from those of the organization. In some situations, unions may be key stakeholders that can make or break successful change.

Many systems analysts naturally assume that frontline employees are the ones whose set of costs and benefits are the most likely to diverge from those of the organization and thus are the ones who most resist change. However, they usually bear the brunt of problems with the current system. When problems occur, they often experience them firsthand. Middle managers and first-line supervisors are the most likely to have a divergent set of costs and benefits and therefore resist change because new computer systems often change how much power they have. For example, a new computer system may improve the organization's control over a work process (a benefit to the organization) but reduce the decision-making power of middle management (a clear cost to middle managers).

An analysis of the costs and benefits for each set of potential adopters/stakeholders will help pinpoint those who will likely support the change and those who may resist the change. The challenge at this point is to try to change the balance of the costs and benefits for those expected to resist the change so that they support it (or at least do not actively resist it).

This analysis may uncover some serious problems that have the potential to block the successful adoption of the system. It may be necessary to reexamine the management policies and make significant changes to ensure that the balance of costs and benefits is such that important potential adopters are motivated to adopt the system.

Figure 16-6 summarizes some of the factors that are important to successful change. The first and most important reason is a compelling personal reason to change. All change is made by individuals, not organizations. If there are compelling reasons for the key groups of individual stakeholders to want the change, then the change is more likely to be successful. Factors such as increased salary, reduced unpleasantness, and—depending on the individuals—opportunities for promotion and personal development can be important motivators. However, if the change makes current skills less valuable, individuals may resist the change because they have invested a lot of time and energy in acquiring those skills and anything that diminishes those skills may be perceived as diminishing the individual (because important skills bring respect and power).

There must also be a compelling reason for the organization to need the change; otherwise, individuals become skeptical that the change is important and are less certain it will in fact occur. Probably the hardest organization to change is an organization that has been successful because individuals come to believe that what worked in the past will continue to work. By contrast, in an organization that is on the brink of bankruptcy, it is easier to convince individuals that change is

---

CONCEPTS   16-C UNDERSTANDING RESISTANCE TO A DECISION SUPPORT SYSTEM

IN ACTION

One of the first commercial software packages I developed was a decision support system to help schedule orders in a paper mill (see Concepts in Action 4-A in Chapter 4). The system was designed to help the person who scheduled orders decide when to schedule particular orders to reduce waste in the mill. This was a very challenging problem—so challenging, in fact, that it usually took the scheduler a year or two to really learn how to do the job well.

The software was tested by a variety of paper mills over the years and always reduced the amount of waste, usually by about 25% but sometimes by 75% when a scheduler new to the job was doing the scheduling. Although we ended up selling the package to most paper

mills that tested it, we usually encountered significant resistance from the person doing the scheduling (except when the scheduler was new to the job and the package clearly saved a significant amount). At the time, I assumed that the resistance to the system was related to the amount of waste reduced: the less waste reduced, the more resistance because the payback analysis showed it took longer to pay for the software. *Alan Dennis*

QUESTION:
1. What is another possible explanation for the different levels of resistance encountered at different mills?
2. How might this be addressed?

	Factor	Examples	Effects	Actions to Take
Benefits of To-Be system	*Compelling* personal reason(s) for change	Increased pay, fewer unpleasant aspects, opportunity for promotion, most existing skills remain valuable	If the new system provides clear personal benefits to those who must adopt it, they are more likely to embrace the change.	Perform a cost–benefit analysis from the viewpoint of the stakeholders, make changes where needed, and actively promote the benefits.
Certainty of benefits	*Compelling* organizational reason(s) for change	Risk of bankruptcy, acquisition, government regulation	If adopters do not understand why the organization is implementing the change, they are less certain that the change will occur.	Perform a cost–benefit analysis from the viewpoint of the organization and launch a vigorous information campaign to explain the results to everyone.
	Demonstrated top management support	Active involvement, frequent mentions in speeches	If top management is not seen to actively support the change, there is less certainly that the change will occur.	Encourage top management to participate in the information campaign.
	Committed and involved business sponsor	Active involvement, frequent visits to users and project team, championing	If the business sponsor (the functional manager who initiated the project) is not seen to actively support the change, there is less certainty that the change will occur.	Encourage the business sponsor to participate in the information campaign and play an active role in the change management plan.
	Credible top management and business sponsor	Management and sponsor who do what they say instead of being members of the "management fad of the month" club	If the business sponsor and top management have credibility in the eyes of the adopters, the certainty of the claimed benefits is higher.	Ensure that the business sponsor and/or top management has credibility so that such involvement will help; if there is no credibility involvement will have little effect.
Costs of transition	Low personal costs of change	Few new skills needed	The cost of the change is not borne equally by all stakeholders; the costs are likely to be higher for some.	Perform a cost–benefit analysis from the viewpoint of the stakeholders, make changes where needed, and actively promote the low costs.
Certainty of costs	Clear plan for change	Clear dates and instructions for change, clear expectations	If there is a clear migration plan, it will likely lower the perceived costs of transition.	Publicize the migration plan.
	Credible change agent	Previous experience with change, does what he/she promises to do	If the change agent has credibility in the eyes of the adopters, the certainty of the claimed costs is higher.	If the change agent is not credible, then change will be difficult.
	Clear mandate for change agent from sponsor	Open support for change agent when disagreements occur	If the change agent has a clear mandate from the business sponsor, the certainty of the claimed costs is higher.	The business sponsor must actively demonstrate support for the change agent.

FIGURE 16-6

Major Factors in Successful Change

needed. Commitment and support from credible business sponsors and top management are also important in increasing the certainty that the change will occur.

The likelihood of successful change is increased when the cost of the transition to individuals who must change is low. The need for significantly different new skills or disruptions in operations and work habits may create resistance. A clear migration plan developed by a credible change agent who has support from the business sponsor are important factors in increasing the certainty about the costs of the transition process.

Motivating Adoption

The single most important factor in motivating a change is providing clear and convincing evidence of the need for change. Simply put, everyone who is expected to adopt the change must be convinced that the benefits from the To-Be system outweigh the costs of changing.

There are two basic strategies to motivating adoption: informational and political. Both strategies are often used simultaneously. With an *informational strategy,* the goal is to convince potential adopters that the change is for the better. This strategy works when the cost–benefit set of the target adopters has more benefits than costs. In other words, there really are clear reasons for the *potential adopters* to welcome the change.

Using this approach, the project team provides clear and convincing evidence of the costs and benefits of moving to the To-Be system. The project team writes memos and develops presentations that outline the costs and benefits of adopting the system from the perspective of the organization and from the perspective of the target group of potential adopters. This information is disseminated widely throughout the target group, much like an advertising or public relations campaign. It must emphasize the benefits as well as increase the certainty in the minds of potential adopters that these benefits will actually be achieved. In our experience, it is always easier to sell painkillers than vitamins; that is, it is easier to convince potential adopters that a new system will remove a major problem (or other source of pain) than that it will provide new benefits (e.g., increase sales). Therefore, informational campaigns are more likely to be successful if they stress the reduction or elimination of problems, rather than focusing on the provision of new opportunities.

The other strategy to motivate change is a *political strategy.* With a political strategy, organizational power, not information, is used to motivate change. This approach is often used when the cost–benefit set of the target adopters has more costs than benefits. In other words, although the change may benefit the organization, there are no reasons for the potential adopters to welcome the change.

The political strategy is usually beyond the control of the project team. It requires someone in the organization who holds legitimate power over the target group to influence the group to adopt the change. This may be done in a coercive manner (e.g., "adopt the system or you're fired") or in a negotiated manner, in which the target group gains benefits in other ways that are linked to the adoption of the system (e.g., linking system adoption to increased training opportunities). Management policies can play a key role in a political strategy by linking salary to certain behaviors desired with the new system.

In general, for any change that has true organizational benefits, about 20% to 30% of potential adopters will be *ready adopters.* They recognize the benefits, quickly adopt the system, and become proponents of the system. Another 20% to 30% are *resistant adopters.* They simply refuse to accept the change and they fight against it, either because the new system has more costs than benefits for them personally or because they place such a high cost on the transition process itself that no amount of benefits from the new system can outweigh the change costs. The remaining 40% to 60% are *reluctant adopters.* They tend to be apathetic and will

16-D OVERCOMING RESISTANCE TO A DECISION SUPPORT SYSTEM

How would you motivate adoption if you were the developer of the decision support system described in Concepts in Action 16-C earlier in this chapter?

go with the flow to either support or resist the system, depending on how the project evolves and how their coworkers react to the system. Figure 16-7 illustrates the actors who are involved in the change management process.

The goal of change management is to actively support and encourage the ready adopters and help them win over the reluctant adopters. There is usually little that can be done about the resistant adopters because their set of costs and benefits may be divergent from those of the organization. Unless there are simple steps that can be taken to rebalance their costs and benefits or the organization chooses to adopt a strongly political strategy, it is often best to ignore this small minority of resistant adopters and focus on the larger majority of ready and reluctant adopters.

Enabling Adoption: Training

Potential adopters may want to adopt the change, but unless they are capable of adopting it, they won't. Adoption is enabled by providing the skills needed to adopt the change through careful *training*. Training is probably the most self-evident part of any change management initiative. How can an organization expect its staff members to adopt a new system if they are not trained? However, we have found that training is one of the most commonly overlooked parts of the process. Many organizations and project managers simply expect potential adopters to find the system easy to learn. Since the system is presumed to be so simple, it is taken for granted that potential adopters should be able to learn with little effort. Unfortunately, this is usually an overly optimistic assumption.

Every new system requires new skills either because the basic work processes have changed (sometimes radically in the case of business process reengineering [BPR]; see Chapter 4) or because the computer system used to support the processes is different. The more radical the changes to the business processes, the more important it is to ensure the organization has the new skills required to operate the new business processes and supporting information sys-

FIGURE 16-7
Actors in the Change Management Process

Sponsor	Change Agent	Potential Adopters
The sponsor wants the change to occur.	The change agent leads the change effort.	Potential adopters are the people who must change.
		20–30% are ready adopters.
		20–30% are resistant adopters.
		40–60% are reluctant adopters.

tems. In general, there are three ways to get these new skills. One is to hire new employees who have the needed skills that the existing staff does not. Another is to outsource the processes to an organization that has the skills that the existing staff does not. Both of these approaches are controversial and are usually considered only in the case of BPR when the new skills needed are likely to be the most different from the set of skills of the current staff. In most cases, organizations choose the third alternative: training existing staff in the new business processes and the To-Be system. Every training plan must consider what to train and how to deliver the training.

What to Train   What training should you provide to the system users? It's obvious: how to use the system. The training should cover all the capabilities of the new system so users understand what each module does, right?

Wrong. Training for business systems should focus on helping the users to accomplish their jobs, not on how to use the system. The system is simply a means to an end, not the end in itself. This focus on performing the job (i.e., the business processes), not using the system, has two important implications. First, the training must focus on those activities around the system, as well as on the system itself. The training must help the users understand how the computer fits into the bigger picture of their jobs. The use of the system must be put in context of the manual business processes as well as of those that are computerized, and it must also cover the new management policies that were implemented along with the new computer system.

Second, the training should focus on what the user needs to do, not what the system can do. This is a subtle—but very important—distinction. Most systems will provide far more capabilities than the users will need to use (e.g., when was the last time you wrote a macro in Microsoft Word?). Rather than attempting to teach the users all the features of the system, training should instead focus on the much smaller set of activities that users perform on a regular basis and ensure that users are truly expert in those. When the focus is on the 20% of functions that the users will use 80% of the time (instead of attempting to cover all functions), users become confident about their ability to use the system. Training should mention the other little-used functions, but only so that users are aware of their existence and know how to learn about them when their use becomes necessary.

One source of guidance for designing training materials is the use cases. The use cases outline the common activities that users perform and thus can be helpful in understanding the business processes and system functions that are likely to be most important to the users.

How to Train   There are many ways to deliver training. The most commonly used approach is *classroom training* in which many users are trained at the same time by the same instructor. This has the advantage of training many users at one time with only one instructor and creates a shared experience among the users.

It is also possible to provide *one-on-one training* in which one trainer works closely with one user at a time. This is obviously more expensive, but the trainer can design the training program to meet the needs of individual users and can better ensure that the users really do understand the material. This approach is typically used only when the users are very important or when there are very few users.

Y O U R

T U R N

16-3 DEVELOPING A TRAINING PLAN

Suppose you are leading the conversion from one word processor to another in your organization. Develop an outline of topics that would be included in the training. Develop a plan for training delivery.

Another approach that is becoming more common is to use some form of *computer-based training* (CBT), in which the training program is delivered via computer, either on CD or over the Web. CBT programs can include text slides, audio, and even video and animation. CBT is typically more costly to develop but is cheaper to deliver because no instructor is needed to actually provide the training.

Figure 16-8 summarizes four important factors to consider in selecting a training method. CBT is typically more expensive to develop than one-on-one or classroom training, but it is less expensive to deliver. One-on-one training has the most impact on the user because it can be customized to the user's precise needs, knowledge, and abilities, whereas CBT has the least impact. However, CBT has the greatest reach—the ability to train the most users over the widest distance in the shortest time—because it is much simpler to distribute, than classroom and one-on-one training, because no instructors are needed.

Figure 16-8 suggests a clear pattern for most organizations. If there are only a few users to train, one-on-one training is the most effective. If there are many users to train, many organizations turn to CBT. We believe that the use of CBT will increase in the future. Quite often, large organizations use a combination of all three methods. Regardless of which approach is used, it is important to leave the users with a set of easily accessible materials that can be referred to long after the training has ended (usually a quick reference guide and a set of manuals, whether on paper or in electronic form).

POST-IMPLEMENTATION ACTIVITIES

The goal of post-implementation activities is to *institutionalize* the use of the new system—that is to make it the normal, accepted, routine way of performing the

FIGURE 16-8
Selecting a Training Method

	One-on-One Training	Classroom Training	Computer-Based Training
Cost to develop	Low–medium	Medium	High
Cost to deliver	High	Medium	Low
Impact	High	Medium–high	Low–medium
Reach	Low	Medium	High

business processes. The *post-implementation* activities attempt to refreeze the organization after the successful transition to the new system. Although the work of the project team naturally winds down after implementation, the business sponsor and sometimes the project manager are actively involved in refreezing. These two—and ideally many other stakeholders—actively promote the new system and monitor its adoption and usage. They usually provide a steady flow of information about the system and encourage users to contact them to discuss issues.

In this section, we examine three key post-implementation activities: support (providing assistance in the use of the system), maintenance (continuing to refine and improve the system), and project assessment (analyzing the project to understand what activities were done well—and should be repeated—and what activities need improvement in future projects).

System Support

Once the project team has installed the system and performed the change management activities, the system is officially turned over to the *operations group*. This group is responsible for the operation of the system, whereas the project team was responsible for development of the system. Members of the operations group usually are closely involved in the installation activities because they are the ones who must ensure that the system actually works. After the system is installed, the project team leaves but the operations group remains.

Providing system support means helping the users to use the system. Usually, this means providing answers to questions and helping users understand how to perform a certain function; this type of support can be thought of as *on-demand training*.

On-line support is the most common form of on-demand training. This includes the documentation and help screens built into the system, as well as separate Web sites that provide answers to *frequently asked questions (FAQs)* that enable users to find answers without contacting a person. Obviously, the goal of most systems is to provide sufficiently good on-line support so that the user doesn't need to contact a person, because providing on-line support is much less expensive than is providing a person to answer questions.

Most organizations provide a *help desk* that provides a place for a user to talk with a person who can answer questions (usually over the phone, but sometimes in person). The help desk supports all systems, not just one specific system, so it receives calls about a wide variety of software and hardware. The help desk is operated by *level 1 support* staff who have very broad computer skills and are able to respond to a wide range of requests, from network problems and hardware problems to problems with commercial software and problems with the business application software developed in-house.

The goal of most help desks is to have the level 1 support staff resolve 80% of the help requests they receive on the first call. If the issue cannot be resolved by level 1 support staff, a *problem report* (Figure 16-9) is completed (often using a special computer system designed to track problem reports) and passed to a *level 2 support* staff member.

The level 2 support staff members are people who know the application system well and can provide expert advice. For a new system, they are usually selected during the implementation phase and become familiar with the system as it is being tested. Sometimes, the level 2 support staff members participate in training during

- Time and date of the report
- Name, e-mail address, and telephone number of the support person taking the report
- Name, e-mail address, and telephone number of the person who reported the problem
- Software and/or hardware causing problem
- Location of the problem
- Description of the problem
- Action taken
- Disposition (problem fixed or forwarded to system maintenance)

FIGURE 16-9
Elements of a Problem Report

the change management process to become more knowledgeable with the system, the new business processes, and the users themselves.

The level 2 support staff works with users to resolve problems. Most problems are successfully resolved by the level 2 staff. However, sometimes, particularly in the first few months after the system is installed, the problem turns out to be a bug in the software that must be fixed. In this case, the problem report becomes a *change request* that is passed to the system maintenance group (see the next section).

System Maintenance

System maintenance is the process of refining the system to make sure it continues to meet business needs. Substantially more money and effort is devoted to system maintenance than to the initial development of the system, simply because a system continues to change and evolve as it is used. Most beginning systems analysts and programmers work first on maintenance projects; usually only after they have gained some experience are they assigned to new development projects.

CONCEPTS

IN ACTION

16-E CONVERTING TO THE EURO (PART 2)

When the European Union decided to introduce the euro, the European Central Bank had to develop a new computer system (called Target) to provide a currency settlement system for use by investment banks and brokerages. The euro opened at an exchange rate of U.S. $1.167. However, a rumor that the Target system malfunctioned sent the value of the euro plunging two days later.

That evening, it was determined that the malfunction was not due to system problems. Instead, operators at some German banks had misunderstood how to use the system and had entered incorrect data. Once the problems were identified and the operators quickly retrained, the Target system continued to operate and the euro quickly regained its lost value.

Source: Thomas Hoffman, "Debut of euro nearly flawless," *Computerworld,* 33(2), p. 16, January 11, 1999.

QUESTION:

Target could be considered a high risk system because of its effects on the European economy. What kinds of system support activities could be put in place to mitigate problems with Target?

Every system is "owned" by a project manager in the IS group (Figure 16-10). This individual is responsible for coordinating the systems maintenance effort for that system. Whenever a potential change to the system is identified, a change request is prepared and forwarded to the project manager. The change request is a "smaller" version of the *system request* discussed in Chapters 1 and 2. It describes the change requested and explains why the change is important.

Changes can be small or large. Change requests that are likely to require a significant effort are typically handled in the same manner as system requests: they follow the same process as the project described in this book, starting with project initiation in Chapter 2 and following through installation in this chapter. Minor changes typically follow a "smaller" version of this same process. There is an ini-

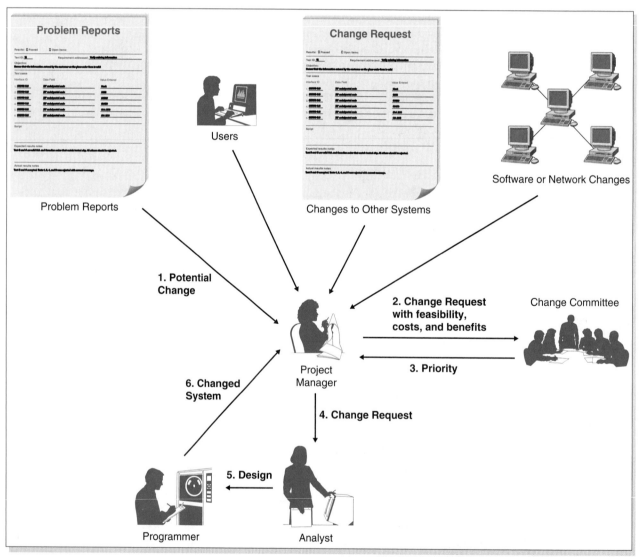

FIGURE 16-10

Processing a Change Request

CONCEPTS | **16-F SOFTWARE BUGS**

IN ACTION

The awful truth is that every operating system and application system is defective. System complexity, the competitive pressure to hurry applications to market, and simple incompetence contribute to the problem.

Will software ever be bug free? Not likely. Microsoft Windows Group General Manager Chris Jones believes that bigger programs breed more bugs. Each revision is usually bigger and more complex than its predecessor, which means new places for bugs to hide. Former Microsoft product manager Richard Freedman agrees that the potential for defects increases as software becomes more complex, but he believes users ultimately win more than they lose. "I'd say the features have gotten exponentially better, and the product quality has degraded a fractional amount."

Still, the majority of users who responded to our survey said they'd buy a software program with fewer features if it were bug free. This sentiment runs counter to what most software developers believe. "People buy features, plain and simple," explains Freedman. "There have been attempts to release stripped-down word processors and spreadsheets, and they don't sell." Freedman says a trend toward smaller, less-bug-prone software with fewer features will "never happen."

Eventually, the software ships and the bug reports start rolling in. What happens next is what separates the companies you want to patronize from the slackers. While almost every vendor provides bug fixes eventually, some companies do a better job of it than others. Some observers view Microsoft's market dominance as a roadblock to bug-free software. Todd Paglia, an attorney with the Washington, D.C.–based Consumer Project on Technology, says, "If actual competition for operating systems existed and we had greater competition for some of the software that runs on the Microsoft operating system, we would have higher quality than we have now."

Source: "Software Bugs Run Rampant," *PC World* 17, no. 1 (January 1999): 46.

QUESTION:

If commercial systems contain the amount of bugs that the above article suggests, what are the implications for systems developed in-house? Would in-house systems be more likely to have a lower or higher quality than commercial systems? Explain.

tial assessment of feasibility and of costs and benefits, and the change request is prioritized. Then a systems analyst (or a programmer/analyst) performs the analysis, which may include interviewing users, and prepares an initial design before programming begins. The new (or revised) program is then extensively tested before the system is converted from the old system to the revised one.

Change requests typically come from five sources. The most common source is problem reports from the operations group that identify bugs in the system that must be fixed. These are usually given immediate priority because a bug can cause significant problems. Even a minor bug can cause major problems by upsetting users and reducing their acceptance of and confidence in the system.

The second most common source of change requests is enhancement to the system from users. As users work with the system, they often identify minor changes in the design that can make the system easier to use or identify additional functions that are needed. Such enhancements are important in satisfying the users and are often key in ensuring that the system changes as the business requirements change. Enhancements are often given second priority after bug fixes.

A third source of change requests is other system development projects. For example, as part of CD Selections' Internet Sales System project, CD Selections likely had to make some minor changes to the distribution system to ensure that the two systems would work together. These changes required by the need to integrate two systems are generally rare but are becoming more common as system integration efforts become more common.

A fourth source of change requests are those that occur when underlying software or networks change. For example, new versions of Windows often require an application to change the way it interacts with Windows, or enables application systems to take advantage of new features that improve efficiency. While users may never see these changes (because most changes are inside the system and do not affect its user interface or functionality), these changes can be among the most challenging to implement because analysts and programmers must learn about the new system characteristics, understand how application systems use (or can use) those characteristics, and then make the needed programming changes.

The fifth source of change requests is senior management. These change requests are often driven by major changes in the organization's strategy (e.g., the CD Selections Internet Sales project) or operations. These significant change requests are typically treated as separate projects, but the project manager responsible for the initial system is often placed in charge of the new project.

Project Assessment

The goal of *project assessment* is to understand what was successful about the system and the project activities (and therefore should be continued in the next system or project) and what needs to be improved. Project assessment is not routine in most organizations, except for military organizations, which are accustomed to preparing after-action reports. Nonetheless, assessment can be an important component in organizational learning because it helps organizations and people understand how to improve their work. It is particularly important for junior staff members because it helps promote faster learning. There are two primary parts to project assessment—project team review and system review.

Project Team Review   *Project team review* focuses on the way in which the project team carried out its activities. Each project member prepares a short two- to three-page document that reports and analyzes his or her performance. The focus is on performance improvement, not penalties for mistakes made. By explicitly identifying mistakes and understanding their causes, project team members will, it is hoped, be better prepared for the next time they encounter a similar situation—and less likely to repeat the same mistakes. Likewise, by identifying excellent performance, team members will be able to understand why their actions worked well and how to repeat them in future projects.

The documents prepared by each team member are assessed by the project manager, who meets with the team members to help them understand how to improve their performance. The project manager then prepares a summary document that outlines the key learnings from the project. This summary identifies what actions should be taken in future projects to improve performance but is careful not to identify team members who made mistakes. The summary is widely circulated among all project managers to help them understand how to manage their projects better. Often, it is also circulated among regular staff members who did not work on the project so that they, too, can learn from other projects.

System Review   The focus of the *system review* is to understand the extent to which the proposed costs and benefits from the new system that were identified during project initiation were actually recognized from the implemented system. Project team review is usually conducted immediately after the system is installed, while key events are still fresh in team members' minds, but system review is often undertaken several months after the system is installed because it often takes a while before the system can be properly assessed.

System review starts with the system request and feasibility analysis prepared at the start of the project. The detailed analyses prepared for the expected business value (both tangible and intangible) as well as the economic feasibility analysis are reexamined and a new analysis is prepared after the system has been installed. The objective is to compare the anticipated business value against the actual realized business value from the system. This helps the organization assess whether the system actually provided the value it was planned to provide. Whether or not the system provides the expected value, future projects can benefit from an improved understanding of the true costs and benefits.

A formal system review also has important behavior implications for project initiation. Since everyone involved with the project knows that all statements about business value and the financial estimates prepared during project initiation will be evaluated at the end of the project, they have an incentive to be conservative in their assessments. No one wants to be the project sponsor or project manager for a project that goes radically over budget or fails to deliver promised benefits.

PRACTICAL 16-1 BEATING BUGGY SOFTWARE

TIP

How do you avoid bugs in the commercial software you buy? Here are six tips:

1. **Know your software:** Find out if the few programs you use day in and day out have known bugs and patches, and track the Web sites that offer the latest information on them.
2. **Back up your data:** This dictum should be tattooed on every monitor. Stop reading right now and copy the data you can't afford to lose onto a floppy disk, second hard disk or Web server. We'll wait.
3. **Don't upgrade—yet:** It's tempting to upgrade to the latest and greatest version of your favorite software, but why chance it? Wait a few months, check out other users' experiences with the upgrade on Usenet newsgroups or the vendor's own discussion forum, and then go for it. But only if you must.

4. **Upgrade slowly:** If you decide to upgrade, allow yourself at least a month to test the upgrade on a separate system before you install it on all the computers in your home or office.
5. **Forget the betas:** Installing beta software on your primary computer is a game of Russian roulette. If you really have to play with beta software, get a second computer.
6. **Complain:** The more you complain about bugs and demand remedies, the more costly it is for vendors to ship buggy products. It's like voting—the more people participate, the better the results.

Source: "Software Bugs Run Rampant," *PC World* 17, no. 1 (January 1999): 46.

APPLYING THE CONCEPTS AT CD SELECTIONS

Installation of the Internet Sales System at CD Selections was somewhat simpler than the installation of most systems because the system was entirely new; there was no As-Is system for the new system to replace. Also there was not a large number of staff members who needed to be trained on the operation of the new system.

Conversion

Conversion went smoothly. First, the new hardware was purchased and installed. Then the software was installed on the Web server and on the client computers to be used by the staff of the Internet sales group. There was no data conversion per se, although the system started receiving data downloaded from the distribution system every day as it would during normal operations.

Alec Adams, senior systems analyst and project manager for CD Selections' Internet Sales System, decided on a direct conversion (because there was no As-Is system) in the one location (because there was only one location) of all system modules. The conversion, if you could call it that, went smoothly through the alpha and beta tests described in Chapter 15, and the system was declared technically ready for operation.

Change Management

There were few change management issues because there were no existing staff members who had to change. All new staff were hired, most by internal transfer from other groups within CD Selections. The most likely stakeholders to be concerned by the change would be managers and employees in the traditional retail stores who might see the Internet Sales System as a threat to their stores. Alec therefore developed an information campaign (distributed through the employee newsletter and internal Web site) that discussed the reasons for the change and explained that the Internet Sales System was seen as a complement to the existing stores, not as a competitor. The system was instead targeted at Web-based competitors, such as Amazon.com and CDnow.

The new management policies were developed, along with a training plan that encompassed both the manual work procedures and computerized procedures. Alec decided to use classroom training for the Internet Sales System personnel because there was a small number of them and it was simpler and more cost effective to train them all together in one classroom session.

Post-Implementation Activities

Support of the system was turned over to the CD Selections operations group, who had hired four additional support staff members with expertise in networking and Web-based systems. System maintenance began almost immediately, with Alec designated as the project manager responsible for maintenance of this version of the system plus the development of the next version. Alec began the planning to develop the next version of the system.

Project team review uncovered several key learnings, mostly involving Web-based programming and the difficulties in linking to existing Structured Query Language (SQL) databases. The project was delivered on budget (see Figure 2-13 in Chapter 2), with the exception that more was spent on programming than was anticipated.

A preliminary system review was conducted after two months of operations. Sales were $40,000 for the first month and $60,000 for the second, showing a gradual increase (remember that the goal for the first year of operations was $1,000,000). Operating expenses averaged $60,000 per month, a bit higher than the projected average, owing to startup costs and the initial marketing campaign. Nonetheless, Margaret Mooney, vice president of marketing and the project sponsor, was quite pleased. She approved the feasibility study for the follow-on project to develop the second version of the Internet Sales System, and Alec began the SDLC all over again.

SUMMARY

Conversion

Conversion, the technical process by which the new system replaces the old system, has three major steps: install hardware, install software, and convert data. Conversion style, the way in which users are switched between the old and new systems, can be via either direct conversion (in which users stop using the old system and immediately begin using the new system) or parallel conversion (in which both systems are operated simultaneously to ensure the new system is operating correctly). Conversion location, what parts of the organization are converted when, can be via a pilot conversion in one location; via a phased conversion, in which locations are converted in stages over time; or via simultaneous conversion, in which all locations are converted at the same time. The system can be converted module by module or as a whole at one time. Parallel and pilot conversion are less risky because they have a greater chance of detecting bugs before the bugs have widespread effect, but parallel conversion can be expensive.

Change Management

Change management is the process of helping people to adopt and adapt to the To-Be system and its accompanying work processes. People resist change for very rational reasons, usually because they perceive the costs to themselves of the new system (and the transition to it) to outweigh the benefits. The first step in the change management plan is to change the management policies (such as standard operating procedures), devise measurements and rewards that support the new system, and allocate resources to support it. The second step is to develop a concise list of costs and benefits to the organization and all relevant stakeholders in the change, which will point out who is likely to support and who is likely to resist the change. The third step is to motivate adoption both by providing information and by using political strategies—using power to induce potential adopters to adopt the new system. Finally, training, whether classroom, one-on-one, or computer based, is essential to enable successful adoption. Training should focus on the primary functions the users will perform and look beyond the system itself to help users integrate the system into their routine work processes.

Post-Implementation Activities

System support is performed by the operations group, which provides on-line and help desk support to the users. System support has both a level 1 support staff, which answers the phone and handles most of the questions, and level 2 support staff, which follows up on challenging problems and sometimes generates change requests for bug fixes. System maintenance responds to change requests (from the system support staff, users, other development project teams, and senior management) to fix bugs and improve the business value of the system. The goal of project assessment is to understand what was successful about the system and the project activities (and therefore should be continued in the next system or project) and what needs to be improved. Project team review focuses on the way in which the project team carried out its activities and usually results in documentation of key lessons learned. System review focuses on understanding the extent to which the proposed costs and benefits from the new system that were identified during project initiation were actually recognized from the implemented system.

KEY TERMS

Change agent
Change management
Change request
Classroom training
Computer-based training (CBT)
Conversion
Conversion location
Conversion modules
Conversion strategy
Conversion style
Direct conversion
Frequently asked question (FAQ)
Help desk
Informational strategy
Installation
Institutionalization
Level 1 support
Level 2 support

Management policies
Measurements
Migration plan
Modular conversion
On-demand training
One-on-one training
On-line support
Operations group
Parallel conversion
Phased conversion
Pilot conversion
Political strategy
Post-implementation
Potential adopter
Problem report
Project assessment
Project team review
Ready adopters

Refreeze
Reluctant adopters
Resistant adopters
Resource allocation
Rewards
Simultaneous conversion
Sponsor
Standard operating procedure
 (SOP)
System maintenance
System request
System review
System support
Training
Transition
Unfreeze
Whole-system conversion

QUESTIONS

1. What are the three basic steps in managing organizational change?
2. What are the major components of a migration plan?
3. Compare and contrast direct conversion and parallel conversion.
4. Compare and contrast pilot conversion, phased conversion, and simultaneous conversion.
5. Compare and contrast modular conversion and whole-system conversion.

6. Explain the tradeoffs among selecting between the types of conversion in questions 3, 4, and 5.
7. What are the three key roles in any change management initiative?
8. Why do people resist change? Explain the basic model for understanding why people accept or resist change.
9. What are the three major elements of management policies that must be considered when implementing a new system?

10. Compare and contrast an information change management strategy with a political change management strategy. Is one better than the other?
11. Explain the three categories of adopters you are likely to encounter in any change management initiative.
12. How should you decide what items to include in your training plan?
13. Compare and contrast three basic approaches to training.
14. What is the role of the operations group in the systems development life cycle (SDLC)?
15. Compare and contrast two major ways of providing system support.
16. How is a problem report different from a change request?
17. What are the major sources of change requests?
18. Why is project assessment important?
19. How is project team review different from system review?
20. What do you think are three common mistakes that novice analysts make in migrating from the As-Is to the To-Be system?
21. Some experts argue that change management is more important than any other part of the SDLC. Do you agree or not? Explain.
22. In our experience, change management planning often receives less attention than conversion planning. Why do you think this happens?

EXERCISES

A. Suppose you are installing a new accounting package in your small business. What conversion strategy would you use? Develop a conversion plan (i.e., technical aspects only).
B. Suppose you are installing a new room reservation system for your university that tracks which courses are assigned to which rooms. Assume that all the rooms in each building are "owned" by one college or department and only one person in that college or department has permission to assign them. What conversion strategy would you use? Develop a conversion plan (i.e., technical aspects only).
C. Suppose you are installing a new payroll system in a very large multinational corporation. What conversion strategy would you use? Develop a conversion plan (i.e., technical aspects only).
D. Consider a major change you have experienced in your life (e.g., taking a new job, starting a new school). Prepare a cost–benefit analysis of the change in terms of both the change and the transition to the change.
E. Suppose you are the project manager for a new library system for your university. The system will improve the way in which students, faculty, and staff can search for books by enabling them to search over the Web, rather than using only the current text-based system available on the computer terminals in the library. Prepare a cost–benefit analysis of the change in terms of both the change and the transition to the change for the major stakeholders.

F. Prepare a plan to motivate the adoption of the system in Exercise E.
G. Prepare a training plan that includes both what you would train and how the training would be delivered for the system in Exercise E.
H. Suppose you are leading the installation of a new decision support system to help admissions officers manage the admissions process at your university. Develop a change management plan (i.e., organizational aspects only).
I. Suppose you are the project leader for the development of a new Web-based course registration system for your university that replaces an old system in which students had to go to the coliseum at certain times and stand in line to get permission slips for each course they wanted to take. Develop a migration plan (including both technical conversion and change management).
J. Suppose you are the project leader for the development of a new airline reservation system that will be used by the airline's in-house reservation agents. The system will replace the current command-driven system designed in the 1970s that uses terminals. The new system uses PCs with a Web-based interface. Develop a migration plan (including both conversion and change management) for your telephone operators.
K. Develop a migration plan (including both conversion and change management) for the independent travel agencies who use your system.

MINICASES

1. Nancy is the IS department head at MOTO Inc., a human resources management firm. The IS staff at MOTO Inc. completed work on a new client management software system about a month ago. Nancy was impressed with the performance of her staff on this project because the firm had not previously undertaken a project of this scale in-house. One of Nancy's weekly tasks is to evaluate and prioritize the change requests that have come in for the various applications used by the firm.

   Right now, Nancy has five change requests for the client system on her desk. One request is from a system user who would like some formatting changes made to a daily report produced by the system. Another request is from a user who would like the sequence of menu options changed on one of the system menus to more closely reflect the frequency of use for those options. A third request came in from the Billing Department. This department performs billing through the use of a billing software package. A major upgrade of this software is being planned, and the interface between the client system and the bill system will need to be changed to accommodate the new software's data structures. The fourth request seems to be a system bug that occurs whenever a client cancels a contract (a rare occurrence, fortunately). The last request came from Susan, the company president. This request confirms the rumor that MOTO Inc. is about to acquire another new business. The new business specializes in the temporary placement of skilled professional and scientific employees, and represents a new business area for MOTO Inc. The client management software system will need to be modified to incorporate the special client arrangements that are associated with the acquired firm.

   How do you recommend that Nancy prioritize these change requests for the client/management system?

2. Sky View Aerial Photography offers a wide range of aerial photographic, video, and infrared imaging services. The company has grown from its early days of snapping pictures of client houses to its current status as a full-service aerial image specialist. Sky View now maintains numerous contracts with various governmental agencies for aerial mapping and surveying work. Sky View has its offices at the airport where its fleet of specially-equipped aircraft are hangared. Sky View contracts with several free-lance pilots and photographers for some of its aerial work and also employs several full-time pilots and photographers.

   The owners of Sky View Aerial Photography recently contracted with a systems development consulting firm to develop a new information system for the business. As the number of contracts, aircraft, flights, pilots, and photographers increased, the company experienced difficulty keeping accurate records of its business activity and the utilization of its fleet of aircraft. The new system will require all pilots and photographers to swipe an ID badge through a reader at the beginning and conclusion of each photo flight, along with recording information about the aircraft used and the client served on that flight. These records would be reconciled against the actual aircraft utilization logs maintained and recorded by the hangar personnel.

   The office staff was eagerly awaiting the installation of the new system. Their general attitude was that the system would reduce the number of problems and errors that they encountered and would make their work easier. The pilots, photographers, and hangar staff were less enthusiastic, being unaccustomed to having their activities monitored in this way.

   a. Discuss the factors that may inhibit the acceptance of this new system by the pilots, photographers, and hangar staff.

   b. Discuss how an informational strategy could be used to motivate adoption of the new system at Sky View Aerial Photography.

   c. Discuss how a political strategy could be used to motivate adoption of the new system at Sky View Aerial Photography.

INDEX

Abstract classes, 189, 247
Abstraction, 247–248
Acceptance tests, 460
Access, unauthorized, 297
Acknowledgment messages, 343
Action-object order, 339
Actions design, 322
Action statements, 434
Active value, 427
ActiveX, 425
Actors
  in sequence diagramming, 218
  specialized, 161, 162
  in use-case description, 154
  in use-case modeling, 160
Adjusted project complexity score, 63
Aesthetics, in interface design, 313–315
Aggregation
  class diagrams and, 198
  relationships, 190–191
A-kind-of relationships, 190
Algorithm specification, 434–435
Alpha testing, 460
Alternative flows, in use-case modeling, 157
Analysis models, 6
  evolving into design models, 247–251
  factoring in, 247–248
  layers, 249–251

partitions and collaborations, 248
Analysis phase, in systems development life cycle, 5–6
A-part-of relationships, 191
API. *See* Application program interface
Application familiarity, feasibility analysis and, 39
Application logic, 279
Application program interface, 425
Approval committee, 5, 32, 34
Architecture centric approach, 14, 243
As-Is system, 5
  document analysis and, 135–136
  information gathering and, 120, 138–140
  interviewing and, 122, 124, 125
  JAD sessions and, 131
  understanding, 142
Assessment, project, 498–499
Associations. *See also* Relationships
  class, 197
  in collaboration diagrams, 224
  in use-case modeling, 155
Attributes
  analysis, 189
  class diagrams and, 195
  private, 195
  public, 195

Attribute sets, 377
Audit file, 374

Batch processing, 348
Batch reports, 355
Behavioral modeling
  collaboration diagrams, 223–228
  interaction diagrams, 217–222
  sequence diagrams, 217–222
  statechart diagrams, 229–233
  use cases and, 216
Benchmarking, 142
Benefits
  assigning value to, 41–42
  identifying, 39–41
Beta testing, 460
Bias
  in graphs, 356
  output design and, 355–357
Bottom-up questioning strategies, 124, 125
Build, in phased delivery research and design, 244
Business analyst, role of, 21
Business need, 34
  design strategy and, 264–265
Business process
  automation, 120
  reengineering, 120